HUMAN EVOLUTIONARY PSYCHOLOGY

Human Evolutionary Psychology

Louise Barrett
Robin Dunbar
John Lycett

Princeton University Press
Princeton and Oxford

Published in the United States and Canada by Princeton University Press,
41 William Street, Princeton, New Jersey 08540

Published in North America under license from Palgrave Publishers Ltd,
Houndmills, Basingstoke, Hants RG21 6XS, United Kingdom

Library of Congress Control Number 2001097709

ISBN-13: 978-0-691-09622-3

ISBN-10: 0-691-09622-8

Printed on acid-free paper. ∞

www.pup.princeton.edu

Printed in the United States of America

10   9   8   7   6   5   4   3   2   1

10   9   8   7   6
(Pbk.)

Contents

For

John Hurrell Crook

il maggiore fabbro

Preface

Necessity, they say, is the mother of invention and, quite simply, that is why we wrote this book. We had been teaching undergraduate courses in evolutionary psychology for nearly a decade and were frustrated by the fact that there was nothing we could use as a textbook that was sufficiently broad to cover all our interests. Consequently, we invariably ended up directing students to tackle the primary literature head-on, a prospect that often frightened the life out of them. Many had never taken a course with an evolutionary theme and, even among those that had, the majority were unfamiliar with a number of the techniques used to study human behaviour and psychology. We decided that a text that could provide a firm grounding in evolutionary theory as applied to humans, taking on board both anthropological and psychological approaches, was desperately needed.

Initially, we set out to provide a gentle introduction to the literature but, as we went along, we began to realise that we were being faced with a real opportunity to try to synthesize what had become, in a remarkably short space of time, a surprisingly fractionated field – not to say one that had become seriously misunderstood by both the popular press and academics in other fields (including evolutionary psychology's own parent disciplines, psychology and biology). We hope we have risen to that challenge by providing a synthesis with sufficient depth to appeal to a wider audience. Our aim has been to encourage a more sophisticated view of evolution and human behaviour than might otherwise have been the case. As a result of all this, the book has taken much longer to finish than we imagined and is considerably fatter than we originally envisaged.

We would like to thank Frances Arnold, our editor at Palgrave, for her exemplary tolerance and patience as we promised yet again that the finished manuscript was definitely on its way. We would also like to thank the publishers' reviewers for their detailed comments, which helped to improve the book enormously (although, of course, any errors that remain are entirely our own). We are also grateful to Lisa Hannah-Stewart for diligently checking through all the references, to Craig Roberts, Peter Henzi and Kevin Laland for reading several of the draft chapters, to Arran Dunbar for drawing Figure 11.1, to Linda Boothroyd for preparing the index and Pauline Snelson for her efficient and exemplary copy-editing. Finally, we'd like to thank all undergraduate and postgraduate students who have taken our courses in human evolution and behaviour over the past few years for all their comments: they helped to make the whole thing as reader-friendly as possible – which, after all, was the whole point.

LB, RD, JL
Liverpool, April 2001

Acknowledgements

The publisher and authors would like to thank the organisations listed below for permission to reproduce material from their publications. Full bibliographic information for all sources is given in the 'References' section at the end of the book.

Academic Press for Figure 2.2 from Hoogland, J. L., 1983, 'Nepotism and alarm calling in the black-tailed prairie dog (*Cynomys ludo-vicianus*)', *Animal Behaviour*, 31: 472–9.

Macmillan Magazines for Figure 2.7: Reprinted with permission from *Nature*, 299: 818–20. © 1982 Macmillan Magazines Limited.

Elsevier Science for Figure 3.5 from Wang, X., 1996, 'Evolutionary hypotheses of risk sensitive choice: age differences and perspective change'. *Ethology and Sociobiology*, 17: 1–15. © 1996, with permission from Elsevier Science.

Aldine de Gruyter for Figure 3.9. reprinted with permission from Martin Daly and Margo Wilson. *Homicide*. (New York: Aldine de Gruyter). Copyright © 1988 by Aldine de Gruyter.

Thomson Learning for Figure 4.2 from Hames, R., 1979, 'Relatedness and interaction among the Ye'kwana: A preliminary analysis'. In N. A. Chagnon and W. Irons (eds) *Evolutionary Biology and Human Social Behavior: An Anthropological Perspective*. North Scituate, Mass.: Duxberry Press. © 1979 by Wadsworth Inc., Belmont.

Westview Press for Figures 4.3 and 4.4 from Hames, R., 1990, 'Sharing among the Yanomamö: Part I, the effects of risk', in E. Cashdan, (ed.) *Risk and Uncertainty in Tribal and Peasant Economies*. © 1990 by Westview Press, Inc.

Elsevier Science for Figures 4.7 and 4.8 from Hawkes, K., 1991, 'Showing Off: Tests of another hypothesis about men's foraging goals'. *Evolution and Human Behavior* (formerly *Ethology and Sociobiology*), 11:29–54. © 1991, with permission from Elsevier Science.

Macmillan Magazines for Figures 5.4 and 5.6: Reprinted with permission from *Nature*, 293, 55–7: © 1981 Macmillan Magazines Limited, and *Nature*, 403, 156: © 2000 Macmillan Magazines, respectively.

Kluwer Academic Publishers for Figures 5.7 and 5.16 from Baker, R. R., and Bellis, M. A., 1995, *Human Sperm Competition: Copulation, Masturbation and Infidelity*. London: Chapman and Hall. © 1995 R. Robin Baker and Mark A. Bellis. With kind permission from Kluwer Academic Publishers.

Cambridge University Press for Figure 5.10 from Pawlowski, B., and Dunbar, R. I. M., 2001, in R. Noë, J.A.R.A.M. van Hooff, P. Hammerstein (eds) *Economics in Nature: Social dilemmas, Mate choice and Biological Markets*. © 2001 Cambridge University Press.

Elsevier Science for Figure 5.13 from Pawlowski, B., and Dunbar, R. I. M., 1999, 'Withholding age as putative deception in mate search tactics'. *Evolution and Human Behavior*, 20: 52–9. © 1999, with permission from Elsevier Science.

The University of Chicago Press for Figure 5.14 from Borgerhoff Mulder, M., 1988c, 'Reproductive success in three Kipsigis cohorts', in T. H. Clutton-Brock (ed.) *Reproductive Success*. © 1988 The University of Chicago.

Elsevier Science for Figures 6.3 and 6.4 from Blurton-Jones, N., 1986, 'Bushman birth spacing: a test for optimal interbirth intervals'. *Ethology and Sociobiology*, 7: 91–105. © 1986, with permission from Elsevier Science.

Elsevier Science for Figures 6.5 and 6.6 from Mace, R., 1996, 'When to have another baby'. *Ethology and Sociobiology*, 17: 263–73. © 1996, with permission from Elsevier Science.

Kluwer Academic Publishers for Figures 6.7 and 6.8 from Flinn, M. V., 1989, 'Household composition and female reproductive strategies in a Trinidadian village', in A. E. Rasa, C. Vogel and E. Voland (eds) *The Sociobiology of Sexual and Reproductive Strategies*, pp. 224, 225, Figures 12.11 and 12.12. © 1989 Anne E. Rasa, Christian Vogel and Eckart Voland. With kind permission from Kluwer Academic Publishers.

Elsevier Science for Figures 6.9, 6.10 and 6.11 from Rogers, A. R., 1990, 'Evolutionary economies of human reproduction'. *Ethology and Sociobiology*, 11: 479–95. © 1990, with permission from Elsevier Science.

The Royal Society for Figures 6.13a and 6.13b from Mace, R., 1998, 'The coevolution of human fertility and wealth inheritance strategies'. Philosophical Transactions of the Royal Society London, B353: 389–397.

American Anthropological Association for Figures 7.7 and 7.8 from Cronk, L., 1989, 'Low socioeconomic status and female-biased parental investment: the Mukogodo example'. *American Anthropologist*, 91: 414–29. With permission from the American Anthropological Association.

The University of Chicago Press for Figure 7.10 from Smith, E. A. and Smith, S. A., 1994, 'Inuit sex-ratio variation'. *Current Anthropology*, 35: 595–624. © 1994 by The Wenner-Gren Foundation for Anthropological Research.

Cambridge University Press for Figure 7.14 from Turke, P. W., 1988, 'Helpers at the nest: childcare networks on Ifaluk', in L. Betzig, M. Borgerhoff Mulder and P. W. Turke (eds) *Human Reproductive Behaviour: A Darwinian Perspective*. © 1988 Cambridge University Press.

Springer-Verlag for Figure 7.15 from Margulis, S. W., Altmann, J., and Ober, C., 1993, 'Sex-biased lactation in a human population and its reproductive costs'. *Behavioural Ecology and Sociobiology*, 32: p. 43, Figure 2. © 1993 Springer-Verlag.

The University of Chicago Press for Figure 7.16 from Voland, E., Dunbar, R. I. M., Engel, C., and Stephen, P., 1997, 'Population increase and sex-biased parental investment in humans: evidence from 18th- and 19th-century Germany'. *Current Anthropology*, 38: 129–35. © 1997 by The Wenner-Gren Foundation for Anthropological Research.

Cambridge University Press for Figure 8.1 from Hartung, J. 1985, 'Matrilineal inheritance. New theory and analysis'. *Behavioral and Brain Sciences*, 8: 661–88. Reprinted with permission of Cambridge University Press.

Springer-Verlag for Figures 8.2, 8.3 and 8.4 from Mace, R., 1996, 'Biased parental investment and reproductive success in Gabbra pastoralists'. *Behavioural Ecology and Sociobiology*, 38: pp. 77, 78 and 80, Figures 2, 3 and 5. © 1996 Springer-Verlag.

Aldine de Gruyter for Figures 8.5 and 8.6 reprinted with permission from Eckart Voland and Robin I. M. Dunbar. 'Resource competition and reproduction: the relationship between economic and parental strategies in the Krummhörn population'. *Human Nature*, 6: 33–49. Copyright © 1995 by Aldine de Gruyter.

Academic Press for Figures 9.1 and 9.2 from Belovsky, G. E., 1987, 'Hunter-gatherer foraging: a linear programming approach'. *Journal of Anthropological Archaeology*, 6: 29–76.

Aldine de Gruyter for Figure 9.4 reprinted with permission from Robin I. M. Dunbar and Matt Spoors. 'Social networks, support cliques, and kinship'. *Human Nature*, 6: 273–90. Copyright © 1995 by Aldine de Gruyter.

Academic Press for Figures 9.5 and 9.6 from Enquist, M., and Leimar, O., 1993, 'The evolution of cooperation in mobile organisms', *Animal Behaviour*, 45: 747–57.

Edinburgh University Press for Figures 9.7, and 9.8 from Dunbar, R. I. M., Knight, C., and Power, C. (eds), 1999, *The Evolution of Culture*. With permission from Edinburgh University Press.

The University of Chicago Press for Figure 9.9 from Nettle, D. and Dunbar, R. I. M., 1997, 'Social markers and the evolution of reciprocal exchange'. *Current Anthropology*, 38: 93–8. © 1997 by The Wenner-Gren Foundation for Anthropological Research.

Edinburgh University Press for Figure 10.4 from Dunbar, R. I. M., Knight, C., and Power, C. (eds), 1999, *The Evolution of Culture*. With permission from Edinburgh University Press.

The MIT Press for Figures 10.3 and 11.2 from Baron-Cohen, S., 1995, *Mindblindness: an Essay on Autism and Theory of Mind*. Cambridge (MA): MIT Press. © 1995 Massachusetts Institute of Technology.

Basil Blackwell for Figure 11.3 from Wellman, H. M., 1991, 'From desires to beliefs: Acquisition of a theory of mind', in A. Whiten (ed.) *Natural Theories of Mind*. © 1991 Basil Blackwell. By permission from Basil Blackwell.

The MIT Press for Figures 11.5 and 11.8 from Leslie, A. M., 2000, '"Theory of Mind" as a mechanism of selective attention', in M. Gazzaniga (ed.) *The New Cognitive Neurosciences*. Cambridge (MA): MIT Press. © 2000 Massachusetts Institute of Technology.

The British Psychological Society for Figure 11.6 from Kinderman, P., Dunbar, R. and Bentall, R., 'Theory-of-mind deficits and causal attributions'. *British Journal of Psychology*, Vol. 89, 2: 191–204. Reproduced with permission from the British Journal of Psychology © The British Psychological Society.

Basil Blackwell for Figure 11.7 from Leslie, A. M., 1991, 'The theory of mind impairment in autism: Evidence for a modular mechanism of development?' in A. Whiten (ed.) *Natural Theories of Mind*. © 1991 Basil Blackwell. By permission from Basil Blackwell.

The British Psychological Society for Figure 11.10 from Kinderman, P., Dunbar, R. I. M., and Bentall, R. P., in 1998, 'Theory-of-mind deficits and causal attributions'. *British Journal of Psychology*, 89: 191–204. © 1998 The British Psychological Society.

Aldine de Gruyter for Figure 12.2 reprinted with permission from Robin I. M. Dunbar, Neil Duncan and Daniel Nettle. 'Size and structure of freely forming conversational groups'. *Human Nature*, 6: 67–78. Copyright © 1995 by Aldine de Gruyter.

Plenum Press for Figure 12.3 from Dunbar, R. I. M., 1997, 'Groups, gossip and the evolution of language', in A. Schmitt, K. Atwanger, K. Grammer and K. Schafer (eds) *New Aspects of Human Ethology*. © 1997 Plenum Press, New York.

The MIT Press for Figure 12.6 from Wilson, D. S., Wilczynski, C., Wells, A., and Weiser, L., 2000, 'Gossip and other aspects of language as group-level adaptations', in C. Heyes and L. Huber (eds) *The Evolution of Cognition*. Cambridge (MA): MIT Press. © 2000 Massachusetts Institute of Technology.

Elsevier Science for Figure 13.1 from Boyd, R., and Richerson, P. J., 1994, 'Why does culture increase human adaptability?' *Ethnology and Sociobiology*, 16: 125–43. © 1994, with permission from Elsevier Science.

Oxford University Press for Figure 13.4 from Nettle, D., 1999, *Linguistic Diversity*. © Daniel Nettle 1999. By permission of Oxford University Press.

Academic Press for Figure 13.5 from Hinde, R. A. and Barden, L. A., 1985, 'The evolution of the teddy bear'. *Animal Behaviour*, 33: 1371–3.

Princeton University Press for Figure 13.6 from Cavalli-Sforza, L. L. and Feldman, M. W., 1981, *Cultural Transmission and Evolution: A Quantitative Approach*, Princeton, NJ; Princeton University Press © 1981 by Princeton University Press.

The University of Chicago Press for Figure 13.7 Voland, E., Dunbar, R. I. M., Engel, C., and Stephen, P., 1997, 'Population increase and sex-biased parental investment in humans: evidence from 18th- and 19th-century Germany'. *Current Anthropology*, 38: 129–35. © 1997 by The Wenner-Gren Foundation for Anthropological Research.

Every effort has been made to obtain necessary permission with reference to copyright material. The publisher and authors apologise if, inadvertently, any sources remain unacknowledged and will be glad to make the necessary arrangements at the earliest opportunity.

Note to the reader: Words in bold text are defined in the Glossary at the end of the book.

The evolutionary approach to human behaviour

CONTENTS

Why do some women of a certain age opt for plastic surgery in an attempt to preserve their youth? Why do the husbands of Dogon women in Mali insist that their wives spend five days a month living alone in a small dark hut? Why are step-children at greater risk of fatal abuse than a parent's natural offspring? And just what is it that makes a man with a fast car and a strong chin that much more attractive than your basic Mr Average?

At first glance, these would seem to be four entirely unrelated questions, each requiring a completely different explanation. But, as in most things, first impressions can be misleading. In fact, there is a theory that explains all of these phenomena, that reveals the natural connections that exist between them. This is the theory of evolution by natural selection. Our aim in this book is to demonstrate that by adopting an evolutionary perspective on human behaviour and psychology, we can provide a coherent unified explanation of human social evolution and adaptation.

In order to do this, we first have to recognise that humans are animals like any other, and that we can thus explain our behaviour using the same models used to explain the behaviour of lions or blackbirds or baboons. Inevitably, some people find this suggestion disturbing. They don't really like being lumped in with the rest of the animal kingdom. Even Alfred Russel Wallace, co-founder with Darwin of the theory of evolution by natural selection, couldn't accept that humans were actually animals. He preferred to think that, at the crucial point, God intervened to place humans on the side of the angels, so placing us a cut above the rest of creation. The same argument still persists today, although 'culture' now replaces God as the means by which we are able to rise above the beasts. Of course, in a very real sense, this is true the impact of culture on human behaviour is enormous and not to be underestimated. The very fact that you are sitting here reading this book is testament to that fact. As clever as our closest relatives, the chimpanzees, are, they do not write books, play musical instruments, undergo psychoanalysis, build skyscrapers or launch spaceshuttles. Only we do.

As humans, we have been able to transform the natural world to suit us and, by virtue of our capacity for language, we have also been able to create and live in 'virtual worlds' – worlds where intangible ideas and imaginary flights of fancy are as important and meaningful as solid objects. Ever since modern humans first evolved, we have been transforming both nature and, as a consequence, ourselves to the extent that we have become less dominated by nature, with culture playing a more prominent role. Consequently, understanding human nature is not a problem for biology alone. As Malik (2000) puts it: 'Culture is not a mere encrustation upon human nature, like dirt on a soiled shirt. It is an integral part of it because human nature can only be expressed through human culture' (see also Plotkin 1998).

On the other hand, human nature and culture both have biological roots. Unless you are a Creationist, you have to accept that humans have been subject to the same processes of evolutionary change as all other living things on earth. A full understanding of human nature therefore requires an understanding of biological as well as sociological processes. Indeed, it is actually impossible to separate the two. We are products of an interaction between biology and culture, or to put it in its more familiar guise, nature and nurture, genes and environment. To separate the two is a false dichotomy. Many would argue that human nature cannot be reduced to mere biological processes – and they would be right. But to infer from this, as many do, that biology is now completely irrelevant (see many of the papers in Rose and Rose 2000) is to commit an egregious logical error. In what follows, we shall try to show that those who espouse this view could not be more wrong.

The resistance to biological explanations of our behaviour is in part a reaction to an over-enthusiastic application of evolutionary theory to humans in a way that seems to leave no room for cultural influences (see, for example, Pinker 1997, Baker 2000, Dennett 1995). It smacks too heavily of genetic determinism for some people and therefore questions human morality and free will. Their view seems to be that we must resist acknowledging our biological roots because, if we accept them, this must mean that our biological inheritance is solely responsible for determining our behaviour: biology as destiny. This is to commit what has been dubbed the 'naturalistic fallacy' (that the way things are is the way they ought to be) so that criminals are 'born, not made', and men can't help philandering because 'it's in their genes'.

But to understand our evolutionary history and recognise its antecedents in the animal kingdom is not to deny what it is to be human. In fact, it can only add to our understanding of the human condition, and possibly even help us overcome human frailty. It can explain why we have to teach our children to share (since they won't do it naturally); it can shed light on why people prefer to gossip about their neighbours than solve problems in differential calculus; it can even help explain why our seas are over-fished despite our best efforts to regulate such practices.

In fact, an evolutionary perspective on human behaviour and psychology, far from promoting the view that we are automatons driven relentlessly by our genes, actually highlights our inherent flexibility or '**phenotypic plasticity**' – the ability to vary responses according to circumstances, to learn from experience, to recognise and exploit opportunities as they arise. Above all else, we shall show that phenotypic plasticity

PHENOTYPIC PLASTICITY

is the most important of the human evolutionary adaptations, and that any accusations of genetic determinism are simply misplaced. Before we can begin to look at human behaviour from an evolutionary perspective, however, we need to be clear about what we mean by the term 'evolutionary'.

NATURAL SELECTION

The first thing to establish is that evolution is not a theory, but a fact. The fossil record shows that species have changed through time; they have diverged and transmuted and become entirely new species. This is all (literally) hard evidence, and as such very difficult to question. The 'theory' bit of evolution comes in with respect to the process by which these changes occurred. This was Charles Darwin's (Darwin 1859) and Alfred Russel Wallace's great insight: the theory of natural selection. As theories go, this one is particularly straightforward and easy to grasp, being based on just three premises and their logical consequence (Dunbar 1982):

Premise 1: All individuals of a particular species show variation in their behavioural, morphological and/or physiological traits – their '**phenotype**'. (This is usually known as the Principle of Variation).

PHENOTYPE

Premise 2: A part of this variation between individuals is '**heritable**': some of that variation will be passed on from one generation to the next or, to put it even more simply, offspring will tend to resemble their parents more than they do other individuals in the population. (The Principle of Inheritance).

HERITABLE

Premise 3: There is competition among individuals for scarce resources such as food, mates and somewhere to live, and some of these variants allow their bearers to compete more effectively. This competition occurs because organisms have a great capacity to increase in numbers, and can produce far more offspring than ever give rise to breeding individuals – just think of frogspawn, for example. (The Principle of Adaptation).

Consequence: As a result of being more effective competitors, some individuals will leave more offspring than others because the particular traits they possess give them some sort of edge: they are more successful at finding food or mating, or avoiding predators. The offspring of such individuals will have inherited these successful traits from their parents, and 'natural selection' can be said to have taken place. Through this process, organisms become 'adapted' to their environment. The success with which a trait is propagated in future generations relative to other variants of that trait is called its **fitness**. Fitness is a measure of relative reproductive success – that is, relative to alternative variants of the same trait; strictly speaking, it is a property of traits. (This is sometimes known as the Principle of Evolution).

FITNESS

It is important to notice here that we have deliberately avoided mentioning the terms **DNA** (the genetic code) and **gene** or anything suggesting that the mechanism of inheritance in Premise (2) entails a particular biochemical process. This is because the theory of natural selection as originally conceived by Darwin and Mendel (who identified the mechanism of inheritance missing in Darwin's original formulation) makes no mention of genes as we know them today. As Dawkins (1983) has pointed out, Mendel's theory of inheritance is constructed entirely in terms of phenotypic characters and makes no assumptions about the process of heredity other than that

DNA, GENE

there is fidelity of copying between parents and offspring. Any mechanism that allows fidelity of copying ensures that natural selection will take place. In so far as the theory of natural selection is concerned, learning is as much a bona fide mechanism of evolutionary inheritance as the genetic code.

This perhaps surprising conclusion is important for much of what follows for two reasons. First, it allows us to avoid unnecessarily fruitless arguments about whether or not a particular behaviour is genetically determined. This frees us up to consider behavioural strategies as genuine Darwinian entities subject to the influence of natural selection – a device that evolutionary biologists like Maynard Smith (1982) have long exploited without any sense of discomfort (see Dunbar 1995a). Second, as a consequence, it allows us to consider culture (which is transmitted only by learning: see Chapter 13) as part and parcel of the Darwinian world, and hence a legitimate object for evolutionary analysis.

BOX 1.1

Speciation and the evolutionary processes

Not surprisingly perhaps, the theory of evolution has been dominated by what we might properly refer to as genetically determined characters. This is because biologists have been mainly concerned to explain the evolution of species, and these are defined by their phenotypic traits (that is, appearance). In this respect, genes are the proper mode of inheritance.

The phenotype is produced by an interaction between the individual's genetic makeup (or '**genotype**') and the environment. The source of variation is genetic mutation, whereby physical changes occur in DNA (the genetic code). These mutations result in changes in protein synthesis and ultimately to changes in the way that phenotypic traits are expressed in the organism. Selection acting on the phenotypic characters results in those genes that produce these characters being passed on to the next generation in greater numbers.

One of the consequences of natural selection is that, over time, individuals tend to track their environments and the ecological niches that become available to them. For example, among birds, a seed-eating niche requires a different beak shape (thick and robust for cracking seeds and nuts) to a nectar-feeding niche (a long thin pointed beak that can get into the nectaries of flowers). Exactly these kinds of changes in beak morphology are found among the finches of the Galapagos islands.

The finches were discovered by Darwin himself and helped him to formulate his theory of the origin of species. On the Galapagos islands today, there are 14 different species of finch that are all descended from a single ancestral species. Radiation into all the available niches and subsequent reproductive isolation between individuals of the original ancestral species gave rise to this diverse array and provide us with one of our best examples of evolution in action. Grant and Grant (1993) have shown that small changes in beak shape and size among these bird populations from one year to the next can be attributed directly to the effects that these have on birds' abilities to survive and reproduce as climatic and vegetation conditions change.

Although genetic mutation is the engine of natural selection, the processes of adaptive radiation and reproductive isolation are essential elements in the origin of new species. Reproductive isolation occurs when individuals within populations are prevented from breeding and, consequently, genes are not freely exchanged throughout the population. This can occur because of the formation of geographical barriers: for example, a new mountain range may arise and divide a species' population in two with the result that mating can only occur within each sub-population instead of throughout the entire population as before. As a consequence, the two populations diverge from each other genetically due to the action of mutation, natural selection (that is, adaptation to local conditions) and 'drift' (random changes in gene frequencies not driven by natural selection).

ASKING THE RIGHT QUESTIONS

While a good understanding of evolutionary theory is obviously essential if we are to understand why humans behave in the way that they do, it is also important to realise that just as there is more than one way to skin a cat so there are a number of different reasons for asking 'why?' in the first place. Understanding the reason why a particular question is being asked is all important since this determines the kind of the answer that one can expect. In a seminal paper, the ethologist Niko Tinbergen (1963) identified four ways in which to ask the seemingly simple question: why?

First, one might wish to know what motivates an animal or a person to behave in a particular way at a particular moment in time; that is, what is the immediate or *proximate* cause of the behaviour. An answer to this question might be couched in terms of the impact that stimuli have on the nervous system and the manner in which this triggers the appropriate response in the organism. It answers questions about mechanisms that produce behaviour, and hence is sometimes referred to as the *mechanistic* cause.

Alternatively, one might wish to know why an individual performed the behaviour in a particular way; what was it about their upbringing or development that led to them adopting a particular way of performing actions (the developmental or *ontogenetic* cause of the behaviour)? An answer to this type of question would require an investigation into the factors that shape development throughout the lifespan, including both their genetic inheritance and the impact of learning on the individual.

Another reason for asking why is to understand the evolutionary history of the behaviour; when did it arise in the first place and why did it follow the particular evolutionary path that it did? This is known as the *phylogenetic* or historical cause. To answer this kind of question, one needs to look back at the fossil record and identify the changes that occurred through evolutionary time.

Finally, one can ask why the behaviour increases an animal's ability to survive and reproduce. This is known as the functional or *ultimate* cause. This is the causal explanation most closely linked to natural selection. Since natural selection works by a process of differential reproduction across individuals, we need to understand why a particular behaviour promotes (or hinders) the production and survival of offspring in order to identify and assess the impact of particular selection pressures.

For example, take the question: why does a woman suckle her baby? This can be answered in terms of:

(i) *Proximate or mechanistic cause*: the baby was crying and/or the mother's breasts were full of milk.

(ii) *Developmental or ontogenetic cause*: the mother learned to care for babies while she was growing up by observing other females suckling their babies. In addition, she may have an innate (built-in) tendency to show positive caring behaviours toward infants that is triggered by the presence of a young baby.

(iii) *Phylogenetic or historical cause*: humans are mammals and like all members of this group, they produce milk with which to feed their offspring. This explanation would also include an account of how mammals evolved from their non-mammalian ancestors: what sequence of changes was involved in moving from a species that laid eggs (and perhaps reared its young in a nest) to one that could nurture and grow its offspring inside its body and then feed the young with milk once they were born?

(iv) *Functional or ultimate cause*: By suckling her offspring, a mother provides them with all the nutrients and energy they need to survive and grow, thus increasing their chances of surviving to maturity, thereby passing the mother's genes on to future generations.

As should be apparent, each of these explanations tackles a different 'level' of explanation that is logically quite independent of the others. An understanding at one level does not presuppose or necessitate an understanding at any of the other levels; nor, more importantly, does it commit us to any *particular* explanation at any of these other levels (the same function may be subserved by several different proximate mechanisms, and may arise either genetically or by learning). On the other hand, being able to provide answers at two or more levels at the same time can be helpful. If we can provide mutually consistent answers at all four levels, for example, then we can be fairly confident that we have achieved a full account of the phenomenon under study. For example, demonstrating the existence of a proximate mechanism that would produce the functional consequences we have inferred strengthens the case for both levels of explanation.

The thing to avoid at all costs, however, is confusing one level of explanation with another. To argue about whether mothers suckle their babies in order to stop them crying or in order to ensure that they survive (and so propagate their mother's genes) is pointless: both explanations are right, and there is no reason, other than personal bias, to think that any one level is 'more correct' or should take precedence over any of the others.

When studying non-human animals, it is usually quite easy to keep levels of explanation separate. However, with humans it is often much more difficult – especially with regard to proximate and ultimate levels of explanation. This may be partly a consequence of human consciousness and self-awareness. We are often aware that there is a functional explanation underlying our behaviour, even though we recognise that any particular instance is motivated by more proximal factors. To go back to our previous example, mothers may feed their infants because they are crying, but they may also be aware that feeding is essential to promote the growth and continued survival of the baby and this could therefore be regarded as a proximate cue prompting them to suckle their offspring. This ability to recognise and understand the long-term consequences of our actions may explain why we occasionally confuse different kinds of explanation.

The flip-side of this is that we often assume that because we are, for the most part, conscious of our motivations, then we must be conscious of all our decision-making processes. This leads people to question whether we would be able to work out the sometimes complicated calculations that seem necessary to explain behaviour: for example, calculating kinship relations (see Chapter 3) or maximising the rate of energy intake (Chapter 4). However, these are calculations for which evolution has worked out the answer, so we don't actually have to do them in our heads. People who question the abilities of humans to make these calculations often have no problem believing that desert ants find their way back to their nestholes using polarised light and trigonometry. With the ant, it is more obvious that natural selection has created animals with this ability programmed into them and that the ant's brain (such as it is) has very little to do with it. However, certain aspects of human behaviour may operate in exactly the same manner. Studies of humans therefore have to be very explicit about (a) the type of explanation they are attempting to provide and (b) whether evolution has selected for a cognitive ability or an unconscious pre-programmed 'rule of thumb'.

BOX 1.2

Reductionism vs holism

Opponents of the evolutionary approach to the study of human behaviour often argue that biology is a reductionist science (explanations for phenomena are given in terms of lower level phenomena, for example genes or, even more extreme, chemistry). In contrast, they argue, human behaviour is complex and can only be studied holistically in terms of cultural or sociological explanations. In part, this view derives from the work of the French sociologist Émile Durkheim who argued (at the end of the nineteenth century) that cultural phenomena cannot be studied biologically, but rather must be explained by reference to other cultural phenomena.

The claim that the evolutionary approach is necessarily reductionist rests on a misunderstanding of what evolutionary explanations entail, probably because of the significance attached to genes in all evolutionary explanations. However, reference to the (genetic) fitness of traits (or behaviour) does not necessarily imply that the trait (or behaviour) is genetically determined, but rather that it has genetic *consequences* in terms of the numbers of extra offspring it allows the bearer to produce. In effect, the reductionist argument confuses two different levels of explanation: ontogenetic arguments (genes as *developmental* determinants of behaviour) with functional arguments (gene replication as the measurable *consequence* of behaviour). In such cases, the genetic consequence of a well-chosen behavioural decision could simply be the propagation of the gene(s) for a brain complex enough to make smart decisions; it would thus have nothing at all to do with the particular behavioural outcomes.

In addition, it is important to note that evolutionary explanations are never couched solely in terms of lower level phenomena such as genes or other chemical processes. The genetic fitness of behaviour is the outcome of a decision (which requires some kind of cognitive machinery to support it) in the context of a whole range of ecological, demographic and social factors. The latter are crucial to a proper evolutionary understanding of behaviour because they determine the costs and benefits that the organism assesses when choosing between two or more courses of action. As a result, the behavioural strategies of most higher organisms (and, a fortiori, humans) are very flexible and are fine-tuned to the particular circumstances in which the individual finds itself.

Were this not so, such large-bodied long-lived species as birds and mammals would not be able to survive since they experience many variations in their environment over the course of their lives which require a flexible response. Indeed, behavioural flexibility of this kind may have been crucial to the evolutionary success of these species (a phenomenon known as the **Baldwin Effect**).

In effect, then, evolutionary explanations of behaviour (sometimes referred to as *behavioural ecology*) are necessarily holistic in that they inevitably refer not just to lower level disciplines (genetics, chemistry, cognitive psychology) but also to other 'higher' level disciplines (history, economics, cultural processes) as well as to variables at the same logical level (the behaviour of competitors and predators, the interests of offspring and allies). For further discussion, see Dunbar (1995a).

Once the nature of the question being asked has been established, the next thing to do is decide on an appropriate research strategy to determine the answer. This may sound straightforward, but in the field of evolutionary psychology it has led to some rather heated debates. Essentially, the arguments put forward hinge on how the term 'biological adaptation' is defined. It is to these that we now turn.

APPROACHES TO THE STUDY OF HUMAN BEHAVIOUR

The broad field of study that we characterise as human evolutionary psychology (that is, the evolutionary-oriented study of human behaviour and cognition) is currently divided into two quite distinct camps who disagree fundamentally on some key issues. In this section, we summarise their basic positions. In the following section, we will try to draw them together into a unified framework.

HUMAN BEHAVIOURAL ECOLOGY

On one side of the fence, there are those individuals who take a functional perspective and consider a trait to be biologically adapted if it increases the fitness (the number of genes passed to future generations) of those who bear the trait relative to those who do not. Individuals working in this field adopt an approach that is virtually identical to that taken by behavioural ecologists who study non-human animals (see, for example, Krebs and Davies 1997). That is, human behavioural ecology (HBE) focuses on measuring differences in reproductive success between individuals in relation to differences in the behavioural strategies that they follow (Smith et al. 2000). Because many of those who adopt this perspective were originally trained in anthropology, they are sometimes referred to as 'Darwinian anthropologists' (DA).

This kind of study usually involves observing and quantifying the study subject's natural behaviour and correlating this with measures of their reproductive output or some appropriate proxy of this. Alternatively, in the time-honoured tradition of comparative analyses established by Darwin himself, it can take the form of large cross-cultural studies based on survey or census data. Due to its focus on reproductive outcomes it has been dubbed the 'counting babies' approach (Crawford 1993).

The HBE approach makes extensive use of formal mathematical modelling to generate testable predictions. Since mathematics has a unique capacity to turn people off, we would enter a plea here for its importance. Mathematical modelling allows us to do two important things, namely (1) to evaluate the effect of our assumptions and (2) to explore the interactions between several contributing factors that influence a given behavioural outcome. The first is important because it forces us to specify exactly how we think a process works in a way that is difficult to fudge. The second is important because humans are notoriously bad at being able to visualise the consequences of more than one explanatory variable at a time: we have trouble thinking in anything more than two dimensions and this limits us to thinking about one cause and one effect. Evolutionary analyses are particularly susceptible to this problem because they invariably involve trade-offs between two driving variables (a cost and a benefit) as well as a pay-off (the fitness consequences) even when there is only one benefit for a particular action.

Because the use of mathematical modelling in this way is a feature of all mature sciences (Dunbar 1995a), we shall not shirk from presenting mathematical treatments where these are relevant. However, in deference to most people's tolerance of these matters, we will confine our treatments to verbal expositions wherever possible. Indeed, it is a maxim of good modelling that any mathematical model that cannot be explained in simple English is probably so poorly understood by the modeller as to be wrong.

Many of the studies conducted using the HBE approach have been conducted in environments that are thought to have remained stable for many thousands of years –

for example, the !Kung San tribe of Southern Africa (Lee 1979, Howell 1979) or the Yanomamö of Amazonia (Chagnon 1974, 1988) – or at least largely uninfluenced by western culture – for instance, the Kipsigis (Borgerhoff Mulder 1988a, b) and the Gabbra of Kenya (Mace 1996a, b). There is thus a temptation to assume that studying these tribal peoples is close to studying our ancestors, and that the traits possessed today can therefore be considered to have increased the fitness of individuals in the past as well as in the present day. In other words, that selection pressures operating today are identical to those that operated in past environments.

This may not always be true, of course, for any one of several reasons. First, there is much archaeological evidence to show that even hunter-gatherer lifestyles have changed considerably over the course of the last 10 000 years or so (Caro and Borgerhoff Mulder 1987, Shennan 2000). Second, there are examples where individual behaviour is found to be sub-optimal, suggesting that we might be witnessing time-lag effects as individuals adjust to new selection pressures (for example, Borgerhoff Mulder 1988a, 1995). Finally, behaviours that confer selective advantages today need not have been selected for that purpose in the past: they may have evolved for another reason entirely, and later become secondarily co-opted into their present role by a change in environmental circumstances – a process known as exaptation (Gould and Vrba 1983).

In some cases, of course, the inference that a behavioural trait that is adaptive today was also adaptive in the past may, in fact, be perfectly accurate. However, there is no way to prove this if the only evidence available is based on current reproductive differentials. To be able to state with any confidence that behavioural traits accorded the same reproductive benefits to individuals in the past as they do in the present, we must be able to demonstrate that past conditions were the same as those which promote the behaviour in contemporary populations, and that the trait has been transmitted faithfully across the generations. This is true not only for studies of humans, but also for non-human animals. Such evidence is obviously very hard to come by, especially for humans, due to our sketchy knowledge of palaeoenvironments and the ethical restrictions preventing human breeding experiments (something which is much less of a problem for studies of non-human animals).

It is thus very difficult to validate claims concerning traits as the product of selection, and any such claims should be treated rather cautiously. Essentially, all behavioural ecologists, whether they study human or non-human animals, play what Grafen (1984) has called 'the **phenotypic gambit**'. As Smith (2000) puts it: 'this means taking a calculated risk to ignore the (generally unknown) details of inheritance (genetic or cultural), cognitive mechanisms, and phylogenetic history that may pertain to a given decision rule and behavioural domain in the hope that they don't matter in the end result'. Evidence of current fit between behaviour and its functional consequences is then taken as sufficient evidence to justify the claim that this provides an explanation for the evolution of the behaviour or trait in question. Taking shortcuts of this kind is considered quite normal in science (Dunbar 1995a).

PHENOTYPIC
GAMBIT

These limitations aside, 'counting babies' and estimating current reproductive benefits of a particular trait provide valuable information about how natural selection is acting in the present and can provide insights into the evolutionary process overall. It is an approach that has proved enormously successful in the study of animal behaviour and shows no sign of being any less productive when applied to human behaviour. Furthermore, as Caro and Borgerhoff Mulder (1987) point out, concentrating on current fitness differentials is a cautious approach to adopt when asking evolutionary questions, since it relies only on features that can actually be measured. This is not to deny the impact of past selection pressures, for, as Betzig (1989) has warned, we shouldn't ignore the history in natural history. But equally, we shouldn't extrapolate beyond the bounds of our data. Our ability to demonstrate that a trait is a product of past selection is much more limited than our ability to identify the pressures that maintain a trait in the present. Conversely, of course, being able to demonstrate a selective advantage that is likely to maintain a trait in the present is no small achievement, since the evolutionary processes that operate in the present are those that guide a trait's evolution in the future.

In short, studies that look only at current adaptiveness do not ignore our evolutionary history; rather, they are simply not designed to confront the issue at that particular level of analysis. Their real interests lie in determining whether or not evolutionary considerations (such as maximising fitness) underpin individual organisms' behavioural decisions and in studying the dynamics of the evolutionary process.

EVOLUTIONARY PSYCHOLOGY

On the other side of the fence from the human behavioural ecologists – and facing in an entirely different direction – are those who consider themselves to be practising evolutionary (or Darwinian) psychology (EP). As might be expected, workers in this area study human adaptation from a largely psychological perspective and their parent discipline is not behavioural ecology but cognitive psychology. The aim of EP is to identify the selection pressures that have shaped the human psyche over the course of evolutionary time, and then test whether our psychological mechanisms actually show the features one would expect if they were designed to solve these particular adaptive problems (for example, choosing mates or detecting cheats). Accordingly, the human psyche is envisaged as being composed of a number of specialised 'domain-specific modules' or 'mental algorithms' rather than a small number of generalised mechanisms that can cope with the whole range of adaptive problems.

In contrast to the HBE approach, EP does not consider the demonstration of reproductive benefits necessary to determine whether or not a particular feature is an adaptation. Instead, they look for evidence of 'good design' that points to the operation of selection in the past. Evolutionary psychologists thus focus on identifying the design features of human psychological adaptations, and make no attempt to determine whether particular traits contribute to fitness differentials in the present. Consequently, most of their studies are conducted in the lab, using batteries of psychological tests or questionnaires. To date, relatively few studies in this field have looked at subjects' behaviour in a natural environment.

Clearly, this is quite a different approach to that adopted by the behavioural ecologists, and to some extent reflects the large differences between the parent disciplines from which the two arise (zoology and behavioural ecology in the case of HBE, cognitive psychology in the case of EP). While the 'phenotypic gambit' means that HBE can be agnostic about the actual psychological mechanisms that humans use to make their decisions and thus focus solely on outcomes, EP is committed to identifying these mechanisms in precise detail (while having much less interest in the outcomes). This difference in focus led to misunderstandings between the two camps when EP first emerged as a discipline and spawned a large amount of rather acrimonious debate (for example, Symons 1989, 1990; Tooby and Cosmides 1990; Turke 1990), some of which continues (albeit in more polite vein) to this day (Daly and Wilson 1999, 2000; Smith 2000; Smith et al. 2000; Sherman and Reeve 1997).

BOX 1.3

The problem of external validity

Most EP (and some HBE) studies rely on what are sometimes referred to as 'pencil-and-paper' methods rather than the direct observation of behaviour (as is perhaps more typical of most HBE studies). These typically involve giving subjects questionnaires or short vignettes of particular situations and asking them to say how they would respond if they found themselves in that particular situation. Two important criticisms can be made of this approach.

One is that what people say they will do in the benign conditions of an interview or questionnaire may be an entirely different thing to what they actually do when faced with the real circumstances. Similar problems arise when we use written sources (such as historical records or advertisements in personal columns) as sources of data on people's behaviour or intentions. As socio-cultural anthropologists have frequently learned to their cost, humans the world over are notorious for the fact that they will often tell you what they think you want to hear rather than what they actually believe to be the case. In some cases, this may be out of a sense of politeness to the interviewer, but in other cases it may be because the way they would actually behave when push came to shove would be considered morally reprehensible.

Although this will always be a problem with studies that rely on self-report by subjects, it is not an insurmountable problem providing we are aware of it. Questionnaires, for example, can be designed with questions that allow lying to be detected, while written sources can always be treated with a healthy dose of scepticism (if only because the victors in a contest invariably seek to rewrite the history of an event in their favour). However, lying is in itself an evolutionarily interesting phenomenon for two reasons: (a) it forms part and parcel of the cognitively complex social processes underpinned by Machiavellian intelligence (see Box 6.1) and (b) it reflects the role of culture in modulating the conflict between individual selfishness and the benefits of group co-operation (see Chapters 2, 4 and 9).

The second criticism of the pencil-and-paper approach is that such tests are invariably carried out on a very small sample of modern humans. In terms of sheer number of studies, the workhorse of these studies is the North American undergraduate, most of whom are white, middle class and (culturally) Euro-American. Evolutionary psychologists would be inclined to argue that, since what they study are the universal aspects of the way the human mind is designed, one group of subjects ought to be much the same as any other, just as livers from one group of humans are much the same as the livers from all other humans.

While this is a defendable position (and conveniently avoids accusations of racism), two notes of caution should be sounded. First, there are genetic differences between human races that have real behavioural and fitness consequences (Cavalli-Sforza et al. 1994). (Two examples that we discuss further in Chapter 13 are sickle cell anaemia – a defence against malaria – and lactose tolerance – the ability to drink milk as an adult.) It may be heuristically unwise not to check that all human minds really do have the same construction. Second, everything depends on whether we are measuring a fundamental *cognitive mechanism* (such as the way memories are coded and stored in the brain) or the *behavioural outcomes* of a cognitive mechanism. While the former may well be universal traits, the latter are strongly affected by current circumstances (in particular, the costs and benefits that drive behavioural decisions). Moreover, socio-cultural anthropologists would insist that the construction of the human mind itself is influenced by local cultural perspectives. While the latter position is certainly debatable (see Dunbar 1995a), it may be unwise to ignore this possibility altogether since, at least at a superficial level, it may turn out to have some validity.

ENVIRONMENT OF EVOLUTIONARY ADAPTEDNESS

One important source of disagreement between the two approaches centres around the concept of the **Environment of Evolutionary Adaptedness** (or **EEA**). This issue arises out of the fact that EP and HBE disagree about the extent to which we can expect humans to be adapted to current environments. HBE argues that humans are likely to be well adapted to current environments due to a capacity for rapid shifts in phenotype as a consequence of increases in brain size and a capacity for flexible 'off-line' planning of action (Smith 2000). EP on the other hand takes the view that the massive cultural changes that have taken place in the last 10 000 years have occurred at a pace that is simply too fast to allow human brains (and hence behaviour) to adapt. The psychological adaptations we possess today were selected for in our past environment of evolutionary adaptedness (EEA) (Bowlby 1969, 1973) and are not geared for the modern world. Consequently, EP argues that, a priori, there is no reason to expect any modern behaviour to be adaptive since present environments are so different from those in which the behaviour evolved (Cosmides and Tooby 1987).

As defined by Tooby and Cosmides (1990), the EEA is no particular point in time or space, but is 'a statistical composite of the adaptation-relevant properties of the ancestral environments encountered by members of ancestral populations'. Put more simply, the EEA is the conglomeration of selection pressures that have operated on humans over the course of evolution and, in this sense, it is really quite uncontroversial. However, most authors (including Tooby and Cosmides themselves: see Tooby and Cosmides 2000, p. 1170) have operationalised this definition by placing the EEA at

some time in the **Pleistocene** (roughly the last 2.5 million years) prior to the advent of agriculture 10 000 years ago and the cultural revolution of the last 40 000 or so years. By doing so, they imply that modern environments and modern selection pressure do not form part of the human EEA. Consequently, we are, as Eaton et al. (1988) would have it, 'Stone agers in the fast lane'. The world to which we are adapted no longer exists, but, due to evolutionary lags, we continue to behave as though it does; consequently, our psychological mechanisms inevitably result in behaviour that no longer produces reproductively successful outcomes.

All this is, of course, entirely possible, and examples can no doubt be quoted from studies of animals. However, it is equally plausible that (as HBE assumes) human behaviour does, in fact, produce adaptive outcomes in the modern world, and there is at least as much evidence from studies of animal behaviour (including, for example, optimal foraging theory) to support this claim. The a priori assumption that behaviour is currently maladaptive has no firm evidence to support it (other than the studies which assume that this is the case in the first place).

In addition, there are two further problems with the EEA concept. One is that human evolutionary history has in fact been a mosaic process in which different components arose at different stages under very different ecological environments (Foley 1995a; Strassman and Dunbar 1999). Identifying the EEA with a particular period in time overlooks the fact that some components of a phenomenon may predate others. Second, as Malik (2000) points out, 'we humans have not simply been dropped into an alien environment. We *created* that environment . . . If the brain is "wired up" to create modernity, why is it not wired up to cope with it?' In other words, there is no a priori reason to suppose that current behaviour shouldn't be adaptive. It is, instead, an empirical issue that can only be tested by measuring current fitness differentials.

BOX 1.4

Human evolution

The last common ancestor between the great apes and the line leading to modern humans (the hominids) lived around 5 to 7 million years ago. The earliest certain members of our lineage, the australopithecines (meaning 'southern apes'), are found as fossils dating from 4.5 to around 2 million years ago (MYA).

The australopithecines walked on two legs (bipedal) like modern humans, but had brains approximately the same size as modern great apes. Although capable of fully bipedal locomotion, they had long arms and rather long, curved finger and toe bones suggesting they were also at home in the trees. The australopithecines are often viewed as the first primitive step towards modern humanity, but they were, in fact, a highly adapted species in their own right. Their unique form of locomotion, for example, was not just a 'transitory' step between life in trees to life on the ground, but was a stable and successful adaptation in itself that remained essentially unchanged during the period that australopithecines were alive.

The australopithecines were a highly successful group of animals in their time, persisting for nearly three million years and diversifying into a number of different species. They can be divided into two main groups, the gracile forms and the robust forms. The gracile forms were small and slender, with a diet that included both plant and animal matter. The robust forms, by contrast, became highly specialised for a tough, vegetarian diet, evolving massive jaws and enormous cheek-teeth (molars). The robust forms survived the longest (until around 2 MYA according to the fossil record) and were contemporaneous with the earliest fossil specimens of our own genus, *Homo*.

The genus *Homo* is thought to have arisen from one of the gracile australopithecines. The earliest *Homo*, distinguished by an increase in brain size relative to the australopithecines, belongs to the species *Homo habilis,* from around 2.5 to 2 MYA. *Homo habilis* means 'handy man', and was so named because the first evidence of tool use was associated with these fossils (however, there is now some evidence to suggest that australopithecines may have used simple tools: Asfaw et al. 1999). When first discovered, *H. habilis* was considered to be a distinct species and one of our direct ancestors. As the number of fossil finds has increased, evidence is accumulating to suggest that the fossils making up *H. habilis* should in fact be regarded as at least two (if not more) species (*H. habilis* and *H. rudolfensis*) – see Tattersall (2000) for review – and some workers dispute whether they are, in fact, members of the genus *Homo* at all (Wood and Collard 1999). This increase in the number of species makes it much harder to know which (if any) was ancestral to the line that led to modern humans. The human family tree is very bushy, with lots of side branches representing an adaptive radiation of hominid species in the newly emerging savannah environments of the Pleistocene.

A more advanced form of hominid, *Homo erectus*, with a further increase in brain size and larger stature, arose around 2 MYA. Like *H. habilis*, *H. erectus* was initially thought to be a single species, but more recent finds suggest that there are at least two different species (*H. erectus*, and an earlier species *H. ergaster* that was found only in Africa). *H. erectus* is the first hominid to be found outside Africa (there are many sites in southeast Asia with *H. erectus* fossils) and it is the first to be associated with fire. They also made use of more advanced tools than *H. habilis* but the form of the tools remains remarkably stable through time; there are no advances or innovations made to improve their form. This stasis has fascinated archaeologists and palaeoanthropologists alike, and there has been some speculation about the psychological capacities and attributes of these individuals based on tool form (Mithen 1996).

Irons (1998) recommends adopting a different concept in which to situate human behavioural and psychological studies: the **adaptively relevant environment**. He argues that the adaptively relevant environment of an adaptation consists of those features of the environment that organisms must interact with to confer a reproductive advantage and that, as a rule, there are only a few key features that need to be present for an adaptation to confer its advantage on its possessors. Environmental novelty may therefore disrupt some adaptations but not others; it all depends on whether or not the key features with which the adaptation needs to interact have been affected. So there are undoubtedly some modern behaviours that are non-adaptive (deliberate lifelong use of contraceptives to avoid all chance of reproduction, addictive drug use and modern eating habits, to name but a few) due to a very recent historical origin (for example, only 40 years or so in the case of the contraceptive pill), but some traits (such as mate choice or parenting strategies) continue to be adaptive since the adaptively relevant environment still persists.

Another point to remember, and one very nicely pointed out by Sherman and Reeve (1997), is that a reliance on the concept of the EEA means that the EP approach is just as limited as that of HBE when it comes to identifying traits as the products of selection. As we have pointed out, HBE studies of current reproductive success and adaptive function may not enlighten us about past adaptation, since we cannot know for certain that the environment has remained constant, or that the behaviour actually evolved for the purpose it now serves. However, these are arguments that can also be levelled at the study of psychological mechanisms, since these are just as much an aspect of the phenotype as particular behavioural traits (that is, they are both products of a gene-environment interaction), and as such are just as likely to be confounded by significant changes in the environment (Sherman and Reeve 1997).

Identifying psychological mechanisms as adaptations thus requires a good knowledge of the EEA in which the traits evolved, and evidence that the traits have been transmitted faithfully through time for the same purpose that they serve today. If the EEA cannot be characterised satisfactorily, then it is, at best, a heuristic convenience and, at worst, an unsupported assumption. Consequently, EP will then be as limited as HBE in its ability to illuminate the process of evolutionary adaptation. By the same token, if we can be confident about the nature of past environments and the EEA, then both the EP and HBE approaches may be equally useful, if used cautiously.

Finally, a note of caution. It is important to remember that identifying selection pressures that create or maintain adaptations can often be extremely difficult. We should beware of concluding that a trait is maladaptive simply because we cannot see an obvious advantage to it. Sometimes, the selection processes involved can only be identified after a very careful detailed analysis. The so-called 'demographic transition' that has resulted in a dramatic reduction in birth rates among those who live in western industrialised countries may be a case in point (see Chapter 6). The claim that a trait has no function or is maladaptive may simply be a statement of ignorance. As a general rule of thumb, therefore, the conclusion that a trait is maladaptive should be an explanation of last resort after all other possible adaptive explanations have been excluded (Dunbar 1982).

TOWARDS A UNIFIED APPROACH

In what follows, it may appear as though we are defending the position of HBE against that of EP. This is only because workers in EP have tended to be more vocal and upfront in their criticism of the HBE approach. However, this should not be taken to imply that we find the HBE approach preferable to that of EP. As we have pointed out, different questions require different answers and both EP and HBE are needed to answer those concerned with human adaptation. Rather, our aim is to try to bridge the gap between the two approaches and develop a single coherent theoretical framework for the study of human behaviour and cognition. Indeed, on a broader front, what follows in the rest of the book might be seen as an attempt to provide a single overarching theoretical framework for the behavioural sciences, in particular psychology (a discipline that is notoriously fragmented into a number of sub-disciplines that spend most of their time trying to ignore each other).

One of the main sources of antagonism between the HBE and EP camps focuses on the issue of whether current behaviour and reproductive differentials are relevant to the study of adaptation. The view of the EP camp is that studies of current reproductive differentials tell us little about the process of adaptation, because (by definition) adaptations are the result of past selection pressures. The pressures that operate today on a particular trait to produce reproductive differentials tell us nothing about the pressures that led to the evolution of the trait in the first place. The human eyeball, for example, displays abundant evidence that it is adapted for visual perception, and an investigation of visual processes reveals more about the nature of that adaptation than a comparison of the reproductive success of sighted versus blind individuals. Thus, in order to discover the nature of human psychological adaptation, EP looks for similar evidence of 'good design' (Williams 1966). This approach is perfectly valid and quite uncontroversial: if one wants to look for evidence of good design, reproductive differentials are not particularly illuminating.

However, in the early days of EP, this assertion was often made in a way that implied that not only were reproductive differentials not useful to the EP approach, but they were not useful *at all* in the study of human behaviour and adaptation. Symons (1990), for example, asserted that 'studies of adaptiveness (that is, current fitness differentials) have no significance *in and of themselves*'. Some even went so far as to suggest that studies of current fitness were not sufficiently Darwinian. John Tooby was quoted by Symons (1990), for example, as stating that 'the study of adaptiveness merely draws metaphysical inspiration from Darwinism, whereas the study of adaptation is Darwinian'.

This was very unfortunate since in most cases, the EP contingent was not, in fact, arguing that measuring reproductive differentials was pointless per se, only that it was not a useful way to help *them* to achieve their particular goals (see also the recent debate between Daly and Wilson [2000] and Smith et al. [2000]). Indeed, in a paper that has been held up as a defining example of the EP view, Tooby and Cosmides (1990) insist that 'there is nothing wrong per se with documenting correspondences [between behaviour and reproductive success] and in fact such investigations can be very worthwhile' – a more conciliatory (if condescending) view than that with which they are usually credited. However, for the most part, this recognition of the value of HBE was not apparent in many early EP writings (Symons 1989, 1990; Barkow 1990) and, in some cases, still isn't today (see Buss 1995).

Not surprisingly, human behavioural ecologists were a little irritated by the implication that they didn't understand evolution properly. With their background in animal behavioural ecology and theoretical evolutionary biology, human behavioural ecologists can legitimately claim to understand evolution extremely well. If a particular study does suggest that their data on current function can necessarily explain past selection without offering supporting evidence for this assumption, then this is, properly

BOX 1.5

Modern human origins

The traditional view assumes that modern humans arose from the populations of *H. erectus* that emerged out of Africa around 1.0 to 1.5 MYA. After colonising Eurasia, these pre-human populations subsequently underwent independent evolution in different parts of the world to produce the various human races seen today. This is known as the **multi-regional hypothesis**. Although evolution occurred independently, proponents of the multi-regional hypothesis assume that there was sufficient gene flow between the populations to prevent reproductive isolation and speciation from occurring (although this would have required gene exchange across a geographical range that biologists would consider unusual, even for highly mobile species like birds).

During the late 1980s, an alternative hypothesis was proposed, based on evidence from molecular genetics. Known as the '**Out-of-Africa**' or 'African Eve' hypothesis (see Stringer and McKie 1996), this suggested a much more recent origin for modern humans. This hypothesis argued that all living humans share a recent common ancestor (or very small number of ancestors) that lived in Africa some time between 100 000 and 200 000 years ago. After occupying virtually the whole of sub-Saharan Africa, one population crossed the Levant landbridge around 70 000 years ago and, over the next 30 000 years, spread rapidly across Eurasia and into Australia, finally breaching the Bering Strait to cross into the New World by around 15 000 years ago.

The principal evidence on which the Out-of-Africa hypothesis was based came from molecular genetics, and particularly from an analysis of variations in the molecular composition of mitochondrial DNA (Cann et al. 1987, Stoneking and Cann 1989). Mitochondria are the tiny elements within living cells that are primarily responsible for providing the cell's energy; thought to have originated as bacteria that successfully invaded living cells at an early stage in the evolution of life on earth, they are not part of the DNA that makes up the chromosomes in the cell's nucleus, but instead are passed on only through the maternal line in the cellular matter (cytoplasm) that surrounds the nucleus in the egg.

Comparison of the number of differences in the base pair sequences of mitochondrial DNA from individuals of different living races suggested that all modern humans share a recent common ancestor (or very small group of ancestors). In addition, all non-African peoples (plus a small number of Africans) share a smaller number of mitochondrial DNA variants, suggesting an even more recent common ancestor (dated to around 70 000 years ago).

Because mitochondria are inherited only through the maternal line, this means their evolution is not affected by the complexities of inter-sexual selection. Hence, their evolution represents a relatively unblemished record of the descent history of particular lineages. In addition, because their function as the cell's powerhouses buffers them against the impact of natural selection due to environmental change, any changes that do occur in their genetic code are likely to be a consequence of random mutations rather than active selection. Since mutations normally occur at random, the number of differences between the mitochondrial DNA of two individuals can be used (with appropriate corrections for back mutations and other statistical effects) to estimate the length of time since they last shared a common ancestor (the so-called **molecular clock**).

The Out-of-Africa hypothesis has a number of important implications for how we interpret human ancestry. First, it suggests that the Neanderthals of Europe and western Asia could not have been direct ancestors of modern humans (Europeans or otherwise) – a fact confirmed by molecular evidence that their DNA is sufficiently different from that of all modern humans to indicate a much deeper common ancestor around 600 000 years ago (Krings et al. 1997). Second, it rules out the possibility that any modern human races evolved out of different populations of *Homo erectus*. Third, the speed with which early modern humans colonised Eurasia and Australia suggests that they were characterised by an

BOX 1.5 cont'd

Modern human origins

extraordinary level of behavioural flexibility in the face of environmental and geographical challenges for which their long evolutionary history in Africa could not have prepared them.

While these two views of modern human origins have remained locked in sometimes vitriolic dispute for the past decade (for example, Templeton 1993), the weight of genetic (including Y-chromosome sequence data) and fossil evidence has, over the intervening years, come down increasingly strongly in favour of the Out-of-Africa hypothesis (or something very close to it) with a consensus date for the origin of all modern humans somewhere around 150 000 years (Stoneking 1993, Aiello 1993; Cavalli-Sforza et al. 1994, Lahr and Foley 1994, Relethford 1995, Hammer and Zegura 1996, Ingman et al. 2000). More importantly, both mitochondrial and nuclear DNA (for example, the Y-chromosome, which is passed down only through the male line) suggest that the ancestral breeding population at the common origin was very small (about 5000 individuals of each sex). The latter represents the individuals whose DNA has contributed to all living humans, not necessarily the total number of individuals alive at the time.

speaking, a failing on the part of an individual researcher rather than an indictment of the HBE programme as a whole. However, we caution once again about drawing an absolute rift between past and current function (or between the selection forces responsible for a trait's original evolution and those currently responsible for its maintenance): in some cases (and this may be especially true of behaviour), current function does reflect past function.

However, we could take the more ambitious view that, so far from being irrelevant, studies of current fitness can help to provide a more complete explanation than the EP approach can achieve alone. If studies of current fitness and psychological mechanism coincide, then we can be much more confident in the evolutionary explanations we advance since, in essence, what we are doing is providing an explanation at both the proximate and ultimate levels of explanation. Because the two approaches are (and should be) complementary, cooperation between them should yield an outcome in terms of understanding that is more than just the sum of its parts. Indeed, when all is said and done, biologists actually use both fitness (the HBE approach) *and* evidence of design (the EP approach) as equally appropriate alternatives for identifying adaptation (Dunbar 1993a).

Another source of antagonism between EP and HBE was (and still is) related to the adoption of the phenotypic gambit (Grafen 1984) in behavioural ecological studies. Behavioural ecologists assume that individuals behave 'as if' they are attempting to maximize their fitness. As pointed out above, when they do this, behavioural ecologists make no assumptions about the underlying cognitive mechanisms that produce behaviour. Instead, they choose to ignore these processes altogether and focus on the outcomes of behaviour. However, EP researchers (for example, Symons 1990) have sometimes over-interpreted the nature of the 'as if' assumption. They take it to imply that the human mind operates using one general-purpose rule that states 'maximize the number of offspring raised to maturity'. Such an all-purpose rule would be unlikely to result in adaptive behaviour, since, as the EPs point out, efficient functioning in everyday life requires achieving a large number of proximate goals that are only distantly related to reproductive goals (Symons 1990). Again, this is a misunderstanding of what each approach is trying to achieve. The question as to whether or not the mind is made up of a number of 'domain-specific' modules is irrelevant to HBE studies in just the same way that studies of fitness differentials are irrelevant to the study of good phenotypic design. Behavioural ecologists have always made it clear that their use of the language of conscious decision-making is a convenient metaphor for evolutionary processes and does not imply anything about the underlying cognitive processes that might be involved.

The emphasis on behavioural outcomes in HBE studies has also been criticised on the grounds that 'Natural selection cannot select for behaviour *per se*; it can only select for the mechanisms that produce behaviour' (Tooby and Cosmides 1990). This has been interpreted as meaning that behaviour is not an appropriate focus for the study of human adaptation. As before, the message of the EPs was actually less condemnatory than is often reported. Tooby and Cosmides (1990), for example, state that 'Turke [1990] argues that behaviour can be an adaptation just as much as any other phenotypic property can be and, depending on exactly what is meant by the word behaviour, we agree with him'. If behaviour is taken to be the manifest phenotypic expression of an underlying cognitive trait, then, for Tooby and Cosmides (1990), it is appropriate to consider it an adaptation. However, it seems likely that they want to argue for a stronger interpretation than this, namely that behaviours cannot themselves be viewed as something whose design is honed by natural selection (in the sense that eyeballs represent a good design for vision); only psychological mechanisms can be viewed in this way. This is certainly a defensible position, and in fact, most studies of HBE implicitly make the same assumption when they play the phenotypic gambit.

However, one could equally take the position that expressed behaviour *is* 'visible' to natural selection in a way that brain processes are not. It is behaviour that operates 'out in the world' and, since natural selection acts in the world, it is behaviour that maximises fitness. Furthermore, the nature of neurobiological growth and learning mechanisms means that different patterns of neural activity can lead to the same behaviour and, by the same token, different behaviours can occur as the result of the same neurological processes. If this is the case, then what exactly is being selected – neurological structures or the behaviours they produce? While it makes sense to argue that the physical structure of the brain can be subject to evolutionary change, it is also clear that behaviour can be subject to the process of natural selection. That, after all, is exactly what learning is all about.

This particular aspect of the dispute between EP and HBE is especially puzzling since just those kinds of human behaviour that the behavioural ecologists concentrate on have long been a legitimate focus of interest within psychology where it traditionally falls under the rubric of social psychology. Within traditional psychology, social psychologists and cognitive psychologists generally get along fine (or, more accurately perhaps, simply ignore each other) and view their particular specialisations as complementary. They study different aspects of psychology using different methodologies but, as far as we know, neither has ever accused the other of being completely misguided or irrelevant to the issue of understanding human psychology. The different types of psychologist can adopt this neutral view of each other because they constitute quite separate fields within modern psychology, and do not have any overarching framework (such as evolutionary theory) in common.

However, if EP – which is in effect cognitive (plus developmental?) psychology with evolution added – has greater explanatory power as a consequence of taking a strong evolutionary stance, then HBE could be seen as social psychology with evolution added. Indeed, evolutionarily informed studies of human psychology provide the role model for how the various branches of psychology could be reunited under a single intellectual umbrella. HBE studies provide a perfect example of how social psychology could be improved enormously by conducting studies that lie within an immensely solid theoretical framework. In short, evolutionary psychology is not just the study of universal cognitive mechanisms, but rather the wrapping together in a single unified framework of all of psychology's rather disparate sub-disciplines.

Having said this, however, some human behavioural ecologists (for example, Smith et al. 2000) wish to remain distinct from the evolutionary psychologists, arguing that theirs is an older more established discipline that can and should stand alone. While this is a fair point, we feel that a truly evolutionary psychology should encompass HBE studies, both for the reason outlined above and also because human behaviour is more heavily influenced by culture than that of other animals. HBE studies will not be able to provide a comprehensive explanation of human behaviour if aspects of cognition are not accounted for. Interestingly enough, the need to abandon part of the phenotypic gambit and pay more attention to psychological mechanisms is something that is becoming increasingly common in studies of animal behavioural ecology (Krebs and Kacelnik 1991, Kacelnik and Krebs 1997, Guilford and Dawkins 1992).

One further plea should be entered at this point. Not a few of those who consider themselves to be evolutionary psychologists (in the broad sense rather than the narrower EP sense) have argued quite insistently that a true evolutionary psychology should include observational and experimental studies of other animals besides humans (for example, Byrne 1995). After all, if the past explains the present, this must surely mean our distant evolutionary past as well as our more recent purely human history. In a similar vein, Heyes (2000) argues that 'human nativist evolutionary psychology' is just one of four possible routes to study the evolution of cognitive mechanisms, emphasising that comparative, developmental and phylogenetic approaches should also be considered a part of evolutionary psychology. Indeed, it would be a shame if evolutionary psychology became associated solely with the study of universal psychological traits in humans. If we seem to play down this point in what follows, it is simply because constraints of space oblige us to do so.

Another important thing to remember when considering the value of studying psychological mechanisms as opposed to behaviour is that the former often requires the verbal reporting of things that occur inside people's heads (since there is no other way to get in there). Since verbal behaviour is a phenotypic feature like any other kind of behaviour, it will also have been subject to selection, and may not provide an entirely transparent 'window to the mind' (Sherman and Reeve 1997). Trivers (1985) has pointed out that self-deception may have evolved in order to improve our ability to deceive others; after all, what could make a lie more convincing than your belief that you are actually telling the truth? Consequently (and rather ironically), evolutionary psychologists should not be dismissing behaviour as a focus of study; instead, they should be attempting to incorporate behavioural data into their research programme as a check on whether self-reports match up with actual behaviour.

One final point is that neither EP nor HBE, even when rigorously applied, can provide a complete explanation of human behaviour, since not all human behaviour is either necessarily adaptive (fitness-promoting) or an adaptation. As we pointed out at the beginning of the chapter, cultural processes must be taken into account if we are to get a complete picture of human nature. In fact, there are those (for example, Boyd and Richerson 1985) who argue that the study of cultural processes (including their mechanisms of transmission) constitutes a 'Third Way' (which they refer to as **Dual Inheritance Theory**, or **DIT**). Unlike HBE and EP, DIT explicitly takes culture

DUAL INHERITANCE
THEORY

into account and then explores the effect that this has on genetic inheritance and transmission. According to DIT, culture and genes are independent but interacting systems of evolutionary change with cultural influences affecting traditional genetic selection in sometimes unexpected ways. This is partly because of differences in the way that cultural information is transmitted, how it varies between individuals and the kinds of fitness effects it produces. For example, cultural information can be passed from parent to offspring (vertical transmission), between peers (horizontal) or from teacher to pupil (oblique), whereas genetic information is only ever transmitted vertically. One of the consequences of this is that it speeds up the pace of cultural evolution compared to genetic evolution (see Chapter 13).

Although some theoretical advances have been made in this area (see Chapter 13), few studies in either HBE or EP have, as yet, acknowledged the importance of culture as an explanation for many facets of human behaviour. Indeed, even in this book, we are concerned mainly with exploring how far a biological explanation can take us, rather than trying to present a fully rounded explanation of human behaviour. Such a book will not appear for quite a few years yet, we imagine – for reasons that will perhaps become clear in Chapter 13. In the main, this is because something similar to the phenotypic gambit is being played in respect of culture by both EP and HBE: neither deny the importance of cultural processes, but they nonetheless choose to ignore them for the purposes of their studies and take the risk that the outcome won't be too badly affected by so doing. As long as this is recognised by all – biologists and non-biologists alike – there should be no problem. It is only when the 'biological' or 'scientific' gambit is misconstrued as a rejection of factors that operate at the societal or cultural level that trouble arises.

What should be apparent from this debate is that although both camps have made mistakes as to what particular research strategies can and cannot achieve, the two approaches are however entirely compatible, and could be combined to great effect. Studies that aim to look at current reproductive differentials provide extremely valuable information on the process of selection, and can highlight the plasticity of human behavioural strategies under variable environmental conditions. In addition, they provide an important empirical test of the evolutionary psychologists' assumption that current behaviour is unlikely to be adaptive and, in those instances where this assumption is upheld, they can be used to identify the component of fitness which acts as the 'sticking point' that leads behaviour off-track.

Hence, in our view, one of the most valuable contributions that evolutionary psychology can make may not be understanding the process of human adaptation as such, but the tying together of human behavioural ecology with psychological mechanisms. As we pointed out earlier, if we can answer a particular question at more than one level of explanation, our understanding will inevitably be more secure than if we can only provide an answer at just one level. Thus, if we can show that, say, our functional explanation of bridewealth payments among the Kipsigis (Borgerhoff Mulder 1988a, b, 1995) is underpinned by a proximate psychological mechanism that explains male mate selection preferences, then our functional explanation will be strengthened considerably.

*

We hope we have made clear our belief that the most powerful tests of evolutionary theory will come from applying both the HBE and the EP approaches together. There are a number of interesting questions regarding human social evolution that simply cannot be answered by using either the HBE or EP approach alone. All we can do is echo Blurton-Jones (1990) in saying 'there is plenty to do on evolution and human behaviour without bickering among ourselves' and fully endorse Turke's (1990) plea that we should get on and 'just do it'.

Chapter summary

- Though originally conceived in terms of the genetic inheritance of morphological traits, Darwin's theory of evolution by natural selection does not strictly speaking make any assumptions about the mechanism of inheritance; recognising that learning is a Darwinian process allows us to explore behavioural decisions and culture using an evolutionary perspective without having to assume genetic determinism.

- Human behaviour and psychology are the products of evolution and can be investigated profitably using an evolutionary framework, although any approach that ignores the fact that culture is an integral part of the biological process will, of necessity, be incomplete.

- When investigating human behaviour, it is imperative to be clear about the level of explanation at which research is focused.

- Human behavioural ecology investigates the manner in which variation in human phenotypic expression influences reproductive outcomes.

- Evolutionary psychology investigates the design of cognitive mechanisms.

- Debate about the appropriateness of the different approaches has been intense, but often reflects a misunderstanding between what the two approaches are trying to achieve rather than a serious divide between them. Combining the two in order to provide explanations at more than one level of analysis will be the key to providing more satisfactory accounts of human behaviour.

Further reading

Barkow, J., Cosmides, L., Tooby, J. (1992) *The Adapted Mind: Evolutionary Psychology and the Generation of Culture.* Oxford: Oxford University Press.

Cavalli-Sforza, L. L., Menozzi, P. and Piazza, A. (1994) *History and Geography of Human Genes.* Princeton (NJ): Princeton University Press.

Cronk, L. (1999) *The Whole Complex: Culture and the Evolution of Human Behaviour.* Boulder, CO: Westview Press.

Dawkins, R. (1986) *The Blind Watchmaker.* Harlow: Longman.

Klein, R. (1999) *The Human Career,* 2nd edition. Chicago: University of Chicago Press.

Malik, K. (2000) *Man, Beast and Zombie.* London: Weidenfeld and Nicholson.

Ridley, M. (1999) *Genome: The Autobiography of a Species in 23 Chapters.* London: Fourth Estate.

2 Basics of evolutionary theory

CONTENTS

We cannot embark on an evolutionary study of human behaviour without a clear grasp of what evolutionary theory entails. Although perhaps the most elegantly simple theory in the history of science, Darwin's theory of evolution by natural selection has been so often misunderstood and misinterpreted that we consider it essential to review the basic principles in this chapter.

INDIVIDUAL SELECTION AND THE SELFISH GENE

The most important thing to note about the theory of natural selection is that it is concerned with individual survival and not, as is sometimes mistakenly assumed, with the survival of the species. Although individual reproduction inevitably has the effect of perpetuating species, this in itself is not the purpose of reproduction (or evolution). Individuals are selected to behave in their own reproductive interests, and the fate of the group as such is irrelevant to the reproductive decisions of individuals. This must obviously be the case if natural selection is to operate in the way described in Chapter 1: since the whole process is based on the notion of inter-individual competition, any organism that behaved in such a way as to benefit the species or group at some cost to its own reproductive interests would leave fewer descendants than less noble-spirited individuals who just looked after themselves.

Sometimes, however, even the individual is too gross a level to understand the workings of evolution. This is because, although natural selection acts on individual survival and reproductive success, what actually changes over time is the frequency of genes in the population gene pool. Individuals are really transient beings: no matter how long their lifespan, they all die in the end. Genes are the entities that persist and provide continuity over time. Consequently, there are some aspects of evolutionary biology which can be understood much better if we adopt a gene's-eye view of the world, and recognise that the evolutionary process actually consists of genes attempting to promote the survival and reproductive success of the bodies in which they find themselves, rather than vice versa (Dawkins 1979, 1989). Dawkins (1976, 1989; see also Dawkins 1986) made the distinction between 'replicators' and 'vehicles' to get this idea across more clearly (and came in for a lot of misguided abuse as a consequence due to its heavy overtones of genetic determinism: see Malik 2000 for a review). 'Replicators' are the entities (genes) that reproduce themselves and persist through time, while 'vehicles' are the entities (bodies) that the replicators construct to contain themselves and which increase the ability of replicators to reproduce and leave as many descendants as possible.

It is very important to appreciate that when we talk about genes in this way, we are not suggesting that individual genes are consciously striving for such an end; this is simply a shorthand way of speaking about evolutionary processes. What we are actually saying is that genes which code for traits that enable an individual to survive and reproduce effectively are more likely to be represented in the gene pool in succeeding generations than genes coding for traits which aren't so successful in that particular environment.

However, we also need to remember that evolution is always something of a compromise: at any one point in time there are numerous selection pressures acting on different traits in many different ways, with the result that a given adaptation may not always be the perfect solution to the problem in question. The classic example here is that adaptations designed to enhance reproductive capacity are inevitably compromised by those geared toward enhancing survival. For example, a male could have enormously high fitness if he did nothing but mate all day, but his mating activities are likely to be curtailed prematurely if he doesn't spend some time feeding. (The concept of genetic fitness is defined on p. 3.) Equally, if people do not behave strictly according to the dictates of kin selection or sexual selection (see below), it does not necessarily mean that their behaviour is maladaptive. It may just be a consequence of attempting to solve a whole range of adaptive problems all at the same time. Generally speaking, most organisms are jacks-of-all-trades and masters of none and, in this sense, individual organisms, and not genes alone, are the units of selection, since the process of natural selection acts on the organism as a whole and not on genes in isolation.

In a similar vein, the particular set of circumstances that an organism has to cope with during its lifetime can also constrain its ability to behave in a way designed to maximise fitness. After all, you cannot leave all your wealth to your first-born son, for example, if you don't actually have one. If you are prevented from behaving in the optimal manner by circumstances beyond your control, then there is nothing for it but to make the best of a bad job, and see what you can manage as far as getting at

BOX 2.1

Genomic imprinting

In sexually reproducing organisms (such as mammals), individuals receive one set of genes from their mother and one set from their father. In most cases, it doesn't matter which parent provides a particular gene since both are equally active. In the case of 'imprinted' genes, however, the difference is crucial. Imprinted genes carry a biological label that identifies whether a particular gene was inherited from the mother or the father and this information then determines whether or not the gene will be active in the offspring. A gene that is expressed only when inherited from the mother is said to be paternally imprinted (silenced), whereas one that is active when inherited from the father is maternally imprinted.

Over 40 imprinted genes have been discovered so far in mice and humans (see Vines 1997 for a non-technical review; also Barlow 1995). One of the first imprinted genes to be discovered was a gene in mice coding for the manufacture of the hormone that promotes the growth of the embryo (insulin-like growth factor-II, or Igf2) (see Moore and Haig 1991, Haig and Graham 1991). The Igf2 gene is maternally imprinted (paternally active) which means that it is a gene inherited from the father that effectively controls the growth of the embryo in the womb. Another gene, active only when inherited from the mother (paternally imprinted) codes for a 'false receptor' of Igf2 which 'mops up' any excess growth factor and consequently suppresses embryo growth.

This might seem like the wrong-way around: since the baby grows inside the mother's body, shouldn't the mother control growth rates? Moore and Haig (1991) suggested that this odd situation was due to a 'parental tug-of war' over the embryo. While mothers need to produce a large, healthy baby with a good chance of survival, they also have to be careful not to allow the embryo to become so large that it deprives them of nutrients that they could put into the production of other offspring in the future. Fathers are also interested in producing large strong babies but, since most mammals are polygynous and males rarely have more than one offspring with the same female, they are not so concerned with sparing the mother's effort because doing so won't affect their own future reproductive output. Instead, they want their offspring to garner as many resources as possible, even if this has negative consequences for the mother.

Moore and Haig (1991) therefore suggested that male genes evolved the ability to silence maternal genes, since those paternal genes capable of such a feat would have resulted in the production of larger offspring with higher survival probabilities than male genes without the ability to exploit the mother's body. In order to counteract this effect of male genes, females fought back evolutionarily by imprinting the false receptor gene that suppresses growth, so preventing the embryo from exploiting the mother's body to the degree coded for by the paternal genes.

Overt conflict between these two imprinted genes is suggested by the fact that mice who lack both the maternally imprinted growth-promoting gene and the paternally-imprinted growth suppressor gene are of normal size. If these genes were actually needed to ensure normal growth and acted in a complementary fashion, the lack of both genes should be lethal (or at least have a detrimental effect on development). The fact that mice can get by without either suggests that they have nothing to do with growth as such but are the result of 'parental gene warfare' as maternal genes try to keep paternal genes under control.

Other authors, however, have questioned this theory. McVean and Hurst (1997), for example, have shown that imprinted genes evolve very slowly at rates comparable to non-imprinted genes. They state that this is the opposite to what would be predicted if imprinting was the result of an 'arm's race' for control between maternal and paternal genes. Rather than evolving slowly, imprinted genes should undergo rapid rates of evolution in order to be able to counteract each other's effects, in the same way that immune system genes undergo rapid rates of evolution in order to keep pace with genetic changes in the pathogens that attack the body.

BOX 2.1 cont'd

Genomic imprinting

Other genomically imprinted genes have been found to influence brain structure. Keverne et al. (1996a, b), for example, found that paternally imprinted genes were associated with development of the cortex, while maternally imprinted genes were associated with the limbic system. Put crudely, maternal genes appear to code for the parts of the brain associated with higher cognitive functioning, while paternal genes code for more evolutionarily ancient areas like the amygdala and hypothalamus which are concerned with 'emotional' functions. Keverne et al. (1996a, b) suggest that imprinting of this nature may have enabled the rapid expansion of the neocortex seen in primate evolution, and have suggested that certain aspects of primate sociality may provide pointers as to why maternally active genes control the development of the neocortex (see also Skuse 1999: Skuse et al. 1997).

least some offspring into the next generation. After all, something is better than nothing and, with a bit of luck, your descendants may get hold of a winning ticket in the lottery of life next time around, ultimately leading to a high fitness pay-off.

While a gene-centred perspective may seem somewhat reductionist (see **Box 1.2**), especially when dealing with complex multi-cellular animals and their equally complex behaviours, it was only by taking it down to the level of the gene that evolutionary biologists were able to come up with a truly satisfactory explanation of such fundamental puzzles as altruism, parental care and certain aspects of mate choice. In the next three sections, we explain why this is the case.

THE PROBLEM OF ALTRUISM

Altruism can be defined as any act that confers a benefit on the recipient of the act at some cost to the donor. In evolutionary terms, these costs and benefits are measured in terms of reproductive success or, more strictly, fitness (the number of copies of a gene passed on to succeeding generations). As such, the existence of altruism creates something of a problem for the theory of natural selection since, as we've already pointed out, the latter is based on the notion of competition between individuals rather than nóble fellow feeling and self-sacrifice. More specifically, if individuals have been selected to behave in their own reproductive interests, how could a behaviour evolve which entails helping others to increase their reproductive success at the expense of one's own? One would expect any such gene to have been selected out long ago. But even the most cursory glance at the behavioural ecology literature demonstrates that this is not the case. Prairie dogs give alarm calls to warn other individuals about potential predators, although this greatly increases their own risk of being spotted by a predator themselves (Hoogland 1983; see also Sherman 1977, 1980) and vervet monkeys do the same (Cheney and Seyfarth 1990); female lions have often been observed to suckle each other's cubs (see Krebs and Davies 1993) while the women of the Ye'kwana, a Venezuelan tribe of traditional horticulturalists, have been found to care for children who are not their own (Hames 1988) and help to plant crops in other individuals' gardens (Hames 1987).

KIN SELECTION AND HAMILTON'S RULE

One answer to the apparent paradox of altruistic behaviour was provided by Hamilton (1964). He demonstrated, in two classic papers, that fitness benefits can accrue to those who preferentially aid individuals with whom they share genes in common (that is, their relatives or kin). This is because fitness is actually measured in terms of the number of copies of a gene passed on to subsequent generations, rather than just the number of offspring produced. Consequently, individuals can increase their fitness in two ways: they can either do so directly by passing genes on to their own offspring, or they can do so indirectly by aiding the reproduction of other individuals who are likely to carry the same gene. An organism's fitness is thus made up of two components – direct and indirect fitness – which combine to give a measure known as

INCLUSIVE FITNESS

'inclusive fitness'.

Hamilton's solution to the problem of altruism, therefore, was to argue that a gene for altruism could evolve under Darwinian selection if the altruist's behaviour allowed a genetic relative that shared the same gene to reproduce more than it would otherwise have done. One obvious implication of this is that, in order to increase their inclusive fitness, individuals should only behave altruistically toward those with whom they are likely to share a given gene; in other words, they should always prefer to aid close kin over more distant kin, since the chances of sharing a given gene are likely to be higher with close kin.

KIN SELECTION

Not surprisingly, this theory is now referred to as the 'theory of **kin selection**' (Maynard Smith 1964), and can be summarised very neatly in a formula that has

HAMILTON'S RULE

come to be known as **Hamilton's Rule**: a gene for altruism will evolve whenever

$$r B > C$$

where B is the benefit to the recipient of the altruistic act, C is the cost to the actor and r is the coefficient of relatedness between the actor and the beneficiary, where benefit and cost are measured in the same way (namely as the number of extra offspring gained or lost). The coefficient of relatedness is the probability that the two individuals concerned share the same gene by descent from a common ancestor or ancestors (see **Box 2.2**). It is important to note here that the benefits to the recipient must be calculated in terms of *extra* offspring that result from the altruist's actions. A common error is to count all the recipient's offspring and compare that to the altruist's lost offspring; but, in fact, the altruist will gain all those offspring that had already been born as well as all those *that would have been born anyway whether or not he behaves altruistically*.

Altruistic behaviour should only occur when the above inequality is satisfied (that is, when the benefit accruing to the beneficiary, devalued by the probability that the two individuals share the gene in question, is greater than the cost to the altruist). Thus if individuals are unrelated ($r = 0$), then the left-hand side of the formula is reduced to zero, and altruistic behaviour should not occur. Furthermore, Hamilton's rule also demonstrates that the higher the degree of relatedness, the smaller the benefit needs to be in order for altruistic behaviour to be worthwhile. Thus, altruistic acts are expected to occur with greater frequency between close relatives than between more distant relatives because the conditions satisfying Hamilton's Rule will occur with greater frequency between individuals with a high r value.

BOX 2.2

Calculating degrees of relatedness

The degree of relatedness, r, measures the probability that any two individuals share the *same* gene because they inherited it from the same common ancestor. (Note that even in the best textbooks [for example, Wilson 1975] it has sometimes been defined – incorrectly – as the proportion of all genes held in common. Defining it this way can give rise to misleading results.) Calculating r is relatively simple in a species like our own where each individual receives half of its genes from its mother and half from its father. The r value between you and your mother is thus 0.5, and if your father is truly your father (often a moot point, as we will see later) then the r value between the two of you will also be 0.5. Full siblings (those who share the same mother and the same father) also have $r = 0.5$ (they have a probability of $0.5 \times 0.5 = 0.25$ of sharing the same gene by inheriting it from their mother, plus the same chance of inheriting it through their father). However, half-siblings (those who share only a mother or only a father) have $r = 0.25$.

Note that this means that we are as closely related to our full siblings as we are to any offspring we might have: this may mean that, as far as evolution is concerned, it makes no difference whether we help to bring up our own offspring or help bring up our siblings. Either way the fitness gain will be the same. This can often have profound implications for individual reproductive decisions, a point we return to in Chapters 7 and 8.

Using the information given above, it is possible to work out the value of r for any pair of related individuals: you just need to work out the number of reproductive events (births) that separate them in the family tree (or pedigree: **Figure 2.1**) and calculate p^n (where n is the number of reproductive events concerned and p is the **Mendelian** probability of inheriting a particular gene from a given parent at conception – in species like humans that have two sets of chromosomes, $p = 0.5$). If the pair in question is connected through more than one genealogical line (as in the case of full siblings), then you must add together the r values obtained for each line (and in species that live in partially inbreeding communities, as humans do, there may be many more than two such lineages). One final point to remember is that relatedness has a direction: the degree of relatedness between A and B need not be the same as the degree of relatedness between B and A when there is a significant level of in-breeding. When calculating r it is thus necessary to specify the direction, and to make sure that r is calculated the right way around.

So, we now have an explanation for the 'selfless' behaviour of our prairie dogs. Hoogland (1983) found that, far from calling indiscriminately, individual prairie dogs are more likely to alarm call when close relatives are nearby, and least likely to call when there are only unrelated individuals present to hear them (see **Figure 2.2**). Prairie dogs are thus prepared to take a risk if it will increase their inclusive fitness, but tend to keep very quiet if this isn't the case. Female lions within a pride also tend to be closely related, and by suckling their sisters' offspring as well as their own, they are helping to increase their inclusive fitness by promoting the survival of the genes that they share with their nephews and nieces. The same goes for Ye'kwana women: the individuals who devote most time to caring for a mother's offspring are her sisters, aunts, cousins and grandmothers (Hames 1988). However, as the theory of kin selection predicts, there are some interesting differences in the amount of childcare performed by these various categories of relatives, an issue we return to in Chapter 3.

It is important to remember that Hamilton's Rule does not imply that we should be altruistic towards relatives willy-nilly. An important – and frequently overlooked –

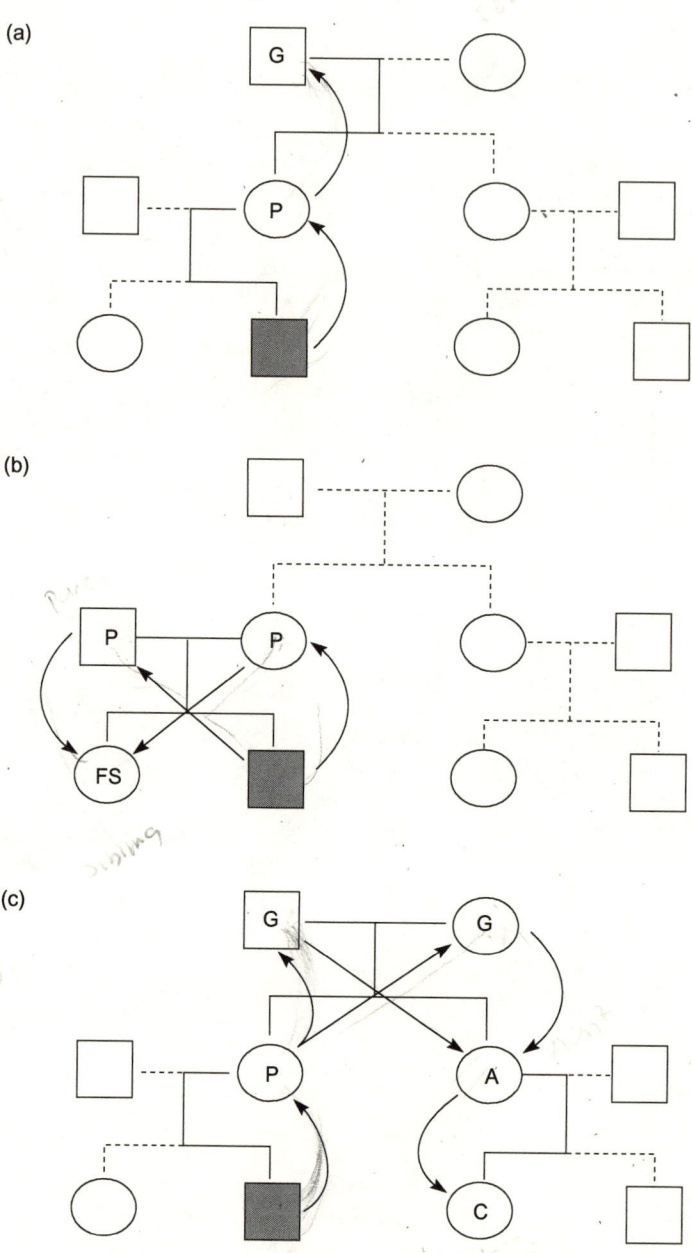

Figure 2.1 Calculating degrees of relatedness between ego (filled square) and (a) a grandparent (G), (b) full siblings (FS) through both parents (P) and (c) a cousin (C) through a parent and an aunt (A) in a three-generation pedigree. Males are indicated by squares, females by circles. The route by which relatedness occurs is shown as a solid line on the pedigree; the reproductive events that contribute to the coefficient of relatedness, r, are indicated by the arrows, with each of these representing $r = 0.5$. In (a), ego's relatedness to a grandparent (G) involves two steps, so $r = 0.5 \times 0.5 = 0.25$. In (b), ego is related to a full sibling (FS) by two steps through each of its parents, so $r = 0.5^2 + 0.5^2 = 0.25 + 0.25 = 0.5$. In (c), relatedness to a first cousin (C) is via both grandparents, so $r = 0.5^4 + 0.5^4 = 0.125$. Full paternity is assumed; if different fathers are involved, then r is half the value given.

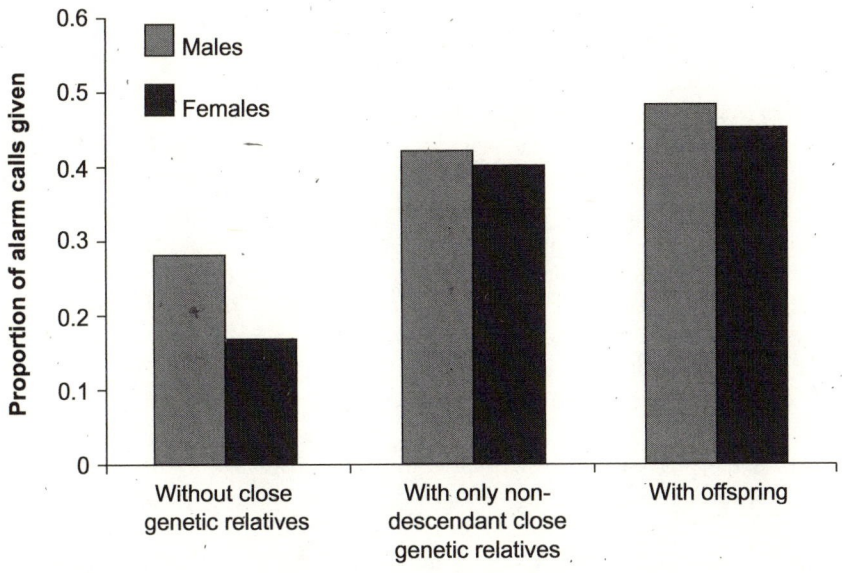

Figure 2.2 Alarm calling by black-tailed prairie dogs in response to a predator stimulus reflects genetic relatedness between caller and audience. Individuals are more likely to alarm call if they have close relatives in the home coterie (social group) than if there are none. Redrawn with permission of the publishers from Hoogland (1983).

feature of all evolutionary analyses is that every decision about how to behave necessarily involves a comparison between at least two alternative courses of action. The decision to favour a relative, for example, involves the actor incurring a cost, and he/she has to evaluate the benefit incurred by favouring that relative against the benefit that *would have been gained* by acting in his/her own direct interest instead. The latter is known as the *opportunity cost* (or, in the economics literature, the *regret*) of opting for the action being considered (in this case, favouring a relative). This contrast between alternative courses of action is built into the Hamilton's Rule inequality, but it is frequently overlooked in most attempts to evaluate the evolutionary significance of behaviour. It is not enough to show that there is a benefit to be gained from acting in a particular way: we need to show that there is a *net* benefit after discounting the opportunity costs of alternative courses of action.

RECIPROCAL ALTRUISM AND THE PRISONER'S DILEMMA

Helping your relatives to spread the genes you share in common is one explanation for the existence of altruism. But what about those instances that involve completely unrelated individuals? For example, unrelated male baboons will often help other males to gain access to fertile females. When a female baboon is in oestrus, a high ranking male in the troop will often form an exclusive 'consortship' with her. However, these males sometimes have trouble maintaining consortships due to fierce competition from rivals. Under such circumstances, a male attempting to wrest access to a female away from her consort will often recruit another male to help him. The recruited male helps in attacks on the consort, while the other male mates with the female (Packer 1977). This looks like true altruism, since the recruited male pays the cost of fighting, but doesn't gain any benefit from mating with the female. However, at a later date, the males may swap roles, and the former beneficiary may return the favour to the former altruist. Lobster fisherman in Maine have been found to trade information on lobster locations in a similar manner (Palmer 1991a), while Yanomamö hunters in Amazonia often share meat brought back from hunting expeditions with unrelated members of their village (Hames 1988). How can we explain the evolution of this kind of altruism in a Darwinian world?

In all these cases, the common element is the trading of benefits between individuals, and this is what led Trivers (1971) to come up with the theory of **reciprocal altruism**. According to this theory, if an individual behaves altruistically but is paid back for their altruistic act at a later date, then both participants will ultimately gain a net benefit. For example, I do your shopping for you today, and you return the favour by washing my car next week. There is a cost to my doing your shopping (I could be doing something more interesting instead), but the benefit to you is larger, and next week the benefit to me of having my car washed will outweigh the cost of you having to do it. So by the end of the week, we'll both have gained.

So far, so good. But there is a problem. What is stopping you from refusing to wash my car? Since there is a delay between me doing the shopping and you washing the car, it's possible for you to cheat on me, and if you decide to do so, there is nothing much I can do about it. The only real way to avoid being cheated is to refuse to go shopping for you in the first place. But then I'll definitely miss out on having my car washed, since there's no favour for you to return. On the other hand, while I won't have gained anything, I won't have lost anything either, and so this might be the safest and most sensible option to choose. So what is the best thing to do? Should I cooperate and take the risk that you will cheat, or refuse altogether – and so risk nothing but gain nothing either? A game theory model known as the Prisoner's Dilemma (see **Box 2.3**, **Table 2.1**) indicates that, when all else is equal, refusing would be the best solution, but if this is the case, then how can a cooperative strategy ever get started?

The answer lies in the frequency with which individuals meet. If individuals only ever encounter each other once, then the best thing to do is cheat. However, if there is a high probability that you are likely to meet the same individuals time and time again, then cooperative strategies prove to be the better option. Under these circumstances, it pays to cooperate on all the early interactions (since the pay-off is larger), and cheat only when the series of interactions reaches its end. If the two individuals don't actually know when the last interaction will be, then cooperation can continue indefinitely.

One of the best strategies to play in a repeated Prisoner's Dilemma is known as *tit-for-tat*. Axelrod (1984) discovered this by organising a computer tournament in which various programmes were pitted against each other in a repeated Prisoner's Dilemma game. *Tit-for-tat* (TFT) was able to out-compete all other strategies in the contest, and once established in a population could not be displaced by any other strategy: it is what's known as an **Evolutionary Stable Strategy** (or **ESS**: see **Box 2.5**). TFT is incredibly simple, having only two rules: cooperate on the first move and then, on each succeeding move, do whatever your opponent did in the previous move. It is a strategy of cooperation based on reciprocity. It works because it is willing to cooperate initially, but has the capacity to retaliate against cheats, and it is 'forgiving': if TFT is defected against, then it will retaliate, but if the opponent then cooperates

RECIPROCAL
ALTRUISM

EVOLUTIONARY
STABLE STRATEGY

BOX 2.3

Prisoner's dilemma

The theoretical model known as the Prisoner's Dilemma is named after the way the problem of cooperation was originally formulated: it dealt with two prisoners being questioned about a crime they had allegedly committed. The only way the police could prosecute them was if one of the prisoners confessed to the crime and incriminated the other. If neither prisoner was prepared to say anything, the best the police could do was to convict them of a minor offence that attracted a much lower penalty. Each prisoner was held in a cell on his own and told by the investigating officer that, if he informed on his partner, he would be granted a pardon and released while the partner would get the full sentence.

Each prisoner has to weigh up the risk of remaining silent (and getting off with a light sentence) against the risk that their partner will inform on them (so that they will be put away for the full sentence, while their partner gets off). The dilemma arises because, although both would do better by keeping quiet, neither can take the risk just in case the other decides to cheat

and inform on them in order to get a lighter punishment. As a result, they both have no choice but to inform on each other.

The Prisoners' Dilemma can be turned into a general model, which can be applied to any situation where cooperation would be of benefit. When applied to evolutionary problems the pay-offs for the different strategies that individuals have available to them are expressed in terms of fitness (the number of descendants gained or lost). The rewards for cooperation versus cheating (or defection) for a Prisoner's Dilemma model are shown in the 'pay-off matrix' in **Table 2.1**. The absolute value allotted to each strategy is arbitrary, but the positioning of the pay-offs in relation to each other is of critical importance. In Prisoner's Dilemma, the pay-offs in the four cells must be in the following order:

$$T > R > P > S$$

And the following inequality must be satisfied:
$$2R > T + S$$

Table 2.1 Pay-off matrix for a Prisoner's Dilemma game

Player A's decision	Player B's decision	
	Cooperate	Defect
Cooperate	R = 3 Reward for mutual cooperation	S = 0 Sucker's payoff
Defect	T = 5 Temptation to defect	P = 1 Punishment for mutual defection

again in the next move, TFT will also cooperate rather than continue to retaliate. These two attributes mean that the strategy very rarely gets into a self-defeating spiral of defection, but can quickly set up long cooperative interactions, especially against another *tit-for-tat* opponent.

Despite its intuitive appeal, some people have questioned whether Prisoner's Dilemma is, in fact, a good model for the evolution of cooperation. This is because, in a Prisoner's Dilemma game, the decisions are made simultaneously during each interaction. They argue that, in real life, reciprocal altruism involves behaviours that occur sequentially: first one individual acts and then the other. In other words, the

BOX 2.4

Other models of cooperation

Although *tit-for-tat* and Prisoners' Dilemma are the most widely known models of cooperation, several others have been developed in recent years. Many of these do not rely on the Prisoner's Dilemma and do not have avoidance of cheating as their basis (see Connor 1995, Noë and Hammerstein 1995). For example, Connor (1995) has argued that certain instances of apparent altruism between non-relatives are better explained by other mechanisms like pseudoreciprocity, kinship deceit or parcelling.

Pseudoreciprocity occurs when an individual, A, performs an altruistic act that benefits another individual, B, but also increases the probability that B will perform an act to benefit itself that incidentally benefits A. Connor (1995) calls this 'investing in by-product benefits', but it might also be described as a form of *mutualism* (Bertram 1982). For example, colonial cliff swallows make food calls which attract conspecifics to insect swarms. It is suggested that birds make these calls because attracting other birds actually increases the caller's foraging efficiency (Brown et al. 1991). This is because the foraging time of the swallows is limited by their ability to track the fast-moving swarms. It seems that the more birds there are, the easier it is for individual birds to track the swarm, so enabling them to feed for longer periods.

Unlike true reciprocal altruism, however, the calling bird is not repaid for its costly act by a similar costly act being performed later by the birds that heard the call, but rather by those birds also later acting in their own selfish interests. The benefit to the caller is a *by-product* of the other birds' selfish actions. This in turn means that, unlike Prisoner's Dilemma, individuals don't have to worry about being cheated on by those they benefit. The only way a calling bird could fail to benefit would be if the other birds didn't feed on the insect swarm – an unlikely occurrence since the other birds would be failing to act in their own best interest. Living in groups as a mutual defence against predators may be another example of the same phenomenon.

Connor (1995) also cites 'kinship deceit' as another mechanism by which non-related individuals may behave altruistically toward each other without involving reciprocity. This model mainly applies to cooperative breeders where individuals other than the parents help to raise the young. Connor and Curry (1995) suggest that in those cases where helpers are unrelated to the individuals they help to rear (that is, where their behaviour can't be explained by kin selection), the helpers do so in order to fool the youngsters that they are, in fact, kin. When these youngsters grow up, they then aid the former helpers because their kin recognition rule-of-thumb – which effectively says something like: 'help those who were nice to you when you were young [. . . because they are likely to be relatives]' – has been led astray. Again, this does not involve a Prisoner's Dilemma, but rather works by subverting kin selection effects.

Finally, Connor (1995) suggests that 'parcelling' cooperation into short bouts can favour cooperative reciprocal solutions to problems, without invoking Prisoner's Dilemma. If individuals require a certain amount of grooming, for example, they can signal this by approaching a potential partner and grooming them for a brief period. The groomed individual has thus received a benefit that it can return, or it can cheat and not groom back. However, if the cheating individual needs further grooming, it would then have to approach another individual and initiate a grooming bout and groom that partner first (with the same risk that this partner will also cheat on it).

Because of this, it always pays to stay with the original partner and continue to groom them. The initial grooming bout and the parcelling of subsequent grooming into small bouts effectively keeps partners out of a Prisoner's Dilemma by always making it more profitable to stay and cooperate than to leave and defect. Connor (1995) was able to show that the grooming patterns of impala and egg-trading by hermaphroditic fish conformed to the expectations of a parcelling model.

BOX 2.5

Evolutionary stable strategies (ESSs)

Maynard Smith and Price (1973; Maynard Smith 1982) defined an evolutionarily stable strategy (ESS) as any behavioural strategy that, once it dominates in a population, cannot be displaced by any alternative strategies that try to invade the population. By strategy, they meant any behaviour or morphological feature that an organism could evolve as a means of solving one of the problems of survival and reproduction that determine its genetic fitness.

The fitness of a particular strategy depends on the balance between the benefits it provides and the costs incurred by acting that way (when both are measured in terms of the numbers of offspring gained and lost). Whenever many individuals compete against each other for a limited resource (whether this resource be food or mates), the costs increase. Eventually, the costs of competing successfully will become so high that some individuals may find it more profitable to opt for a non-competitive strategy. Providing the non-competitive strategy yields at least *some* reproductive benefits, then there will always be a point at which the less competitive strategy is a better option. One of the defining features of ESSs is that they are frequency-dependent: the pay-off to any given strategy depends on how many individuals are playing it.

The idea of frequency-dependent ESSs can be illustrated by a game of life played between two strategies *Hawk* and *Dove*. *Hawks* are aggressive and attack whenever they encounter other individuals; *Doves* always submit. The pay-offs to each strategy when it encounters individuals that play each of the two strategies is shown in **Table 2.2**. Because *Hawks* destroy each other when they meet, a *Hawk*'s pay-off against another *Hawk* is −5 fitness units. However, when a *Hawk* meets a *Dove*, the *Dove* always gives way, so the *Hawk* is able to take all the resource, thus gaining a fitness of +10 while the *Dove* gets a fitness pay-off of 0. When two *Doves* meet, they both back down and thus can share the resource between them (ending up with 2 units each). Once again, the absolute size of the units is unimportant; what matters is the relative order.

When the frequencies of doves and hawks in the population is p and q (where $q = 1 - p$), then the expected pay-offs for each strategy when they encounter opponents drawn at random from the population are:

Pay-off to **DOVE**: $2p + 0(1 - p) = 2p$
Pay-off to **HAWK**: $10p - 5(1 - p) = 15p - 5$

Since the two strategies will be in balance (that is, evolutionary equilibrium) when their pay-offs are equal, we can determine the value of p that will be evolutionarily stable by setting the two pay-offs equal to each other and then solving for p:

$$2p = 15p - 5$$
$$p = 0.385$$

Table 2.2 Pay-off matrix for a *Hawk–Dove* game

Pay-off to:	when encountering	
	Dove	*Hawk*
Dove	+2	0
Hawk	+10	−5

second individual can weigh up what the first individual did (or did not) do and make a decision accordingly. However, Boyd (1995) has argued that the exact patterning of interactions may not be so important, while behaviour is contingent (what one individual does is dependent on what others do) and that there is a time delay between an individual cheating and the other individual retaliating (otherwise the cheater cannot prosper from the deed). If the two players really do behave simultaneously, then mutualism may be a more appropriate explanation for altruism: in this case, both parties benefit simultaneously from each other's altruistic behaviour (Bertram 1982). Cooperating by living in a group so that both individuals benefit from greater protection against predators would be an example.

It is clear that cooperation and reciprocation have played a large role in human social evolution, and there is some evidence to suggest that it has had a profound influence on our psyches. To start with, in order to operate an effective system of reciprocation it is important that individuals are able to recognise each other, remember their past interactions and then behave accordingly. This would obviously select for both a very good facial recognition mechanism (see Chapter 10) and an autobiographical memory capacity (see Chapter 11). Also the ability to detect cheats is very important if reciprocity is going to benefit individuals, and Cosmides (1989) has argued controversially that our ability to solve certain types of problem better than others is directly related to a highly evolved capacity for cheat detection (see Chapters 9 and 10). The actual decisions that humans make in deciding with whom to cooperate are discussed in Chapters 3 and 4.

PARENTAL INVESTMENT AND PARENT–OFFSPRING CONFLICT

The sight of small children throwing tantrums to the exasperation of a frazzled mother is all too familiar, and not just in our own species: young pelicans do the same, as do starlings, zebra, baboons and chimpanzees (Trivers 1974, 1985; see Barrett and Dunbar 1994 for a review). Such conflicts are usually thought to occur over weaning: the point at which a mother ceases to be the only source of food for her offspring, and the offspring are thus forced to start providing for themselves. But why should this process involve such overt conflict? Since parental investment is really just a special case of kin selection, we can use the logic of this theory to provide an explanation of parent–offspring conflict.

Parental investment was defined by Trivers (1974) as 'any investment by the parent in an individual offspring that increases the offspring's chance of surviving . . . [and, in its turn, reproducing] . . . at the cost of the parent's ability to invest in other offspring'. This gives us a clue as to where conflict might have its roots. Since parents are equally related to all their offspring (by $r = 0.5$), they are selected to invest an equal amount in each of them. The individual offspring, however, have quite a different view of things: they are more closely related to themselves (relatedness is $r = 1$) than they are to any of their siblings (even among full siblings relatedness will only be $r = 0.5$). Offspring are thus selected to demand more care for themselves than for their siblings, which is rather more than the parent is selected to provide. Conflicts thus arise over the level of investment that each party considers to be appropriate. As **Figure 2.3** illustrates, parents are selected to continue to invest in their offspring up to the point at which the cost in terms of reduced reproductive success (the more investment in the current infant, the less available for future offspring) starts to outweigh the benefits of increased survival for the current offspring. In other words, as soon as the costs begin to exceed the benefits (that is, $B/C < 1$), parents should stop investing in the current offspring and start investing in the next.

However, at this point, the current offspring would like parental investment to

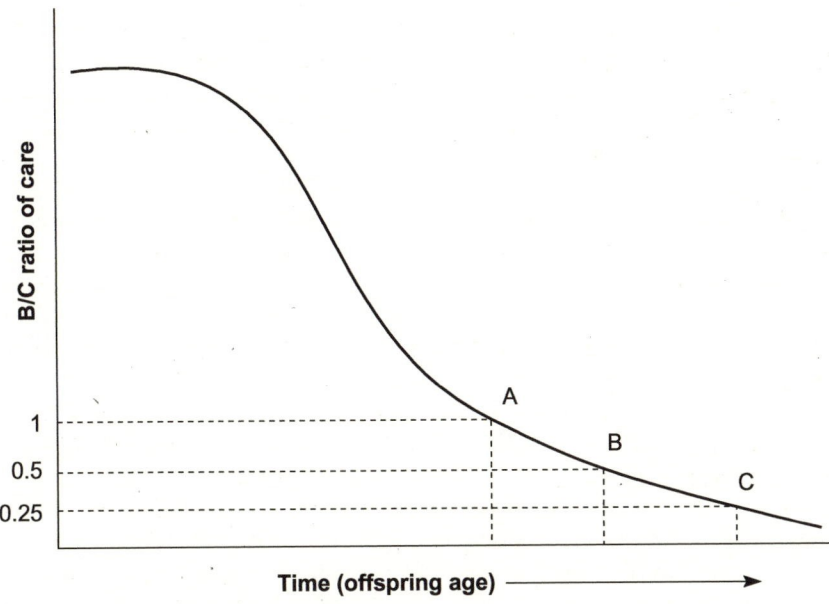

Figure 2.3 Trivers (1974) argued that parental and offspring investment optima differed because of differences in the levels of relatedness between parents and offspring and between siblings. Since a parent is equally related to all its offspring (by $r = 0.5$), it is selected to invest an equal amount in each one. An individual offspring, however, has a different evolutionary view of things: it is more closely related to itself ($r = 1$) than it is to any of its siblings ($r = 0.25$ or $r \pm 0.5$, depending on the mating system and paternity certainty). Offspring are thus selected to demand more care for themselves than for their siblings, which is more than the parent is selected to provide. Conflicts thus arise over the level of investment that each party considers appropriate. At the start of the period of investment, the benefits of investment are high and the costs are low. However, as the infant grows, the marginal value of each unit of investment is worth less (since larger infants need more energy). Parents are selected to invest in their offspring up to the point where the cost in terms of reduced future reproductive success (the more investment in the current infant, the less available for future offspring) starts to outweigh the benefit gained from increased survival for the current offspring. In other words, when the ratio of benefits to costs is less than one, parents should stop investing in their current offspring and begin investing in the next (point A on the graph). However, the current offspring would like parental investment to continue beyond this point because it will continue to benefit until the cost-benefit ratio drops to 0.5 (point B) in the case of full siblings or 0.25 (point C) in the case of half-siblings, reflecting the lower level of relatedness between siblings. After that point, continued demands for investment would lead to a reduction in indirect fitness, since parents would leave fewer siblings with whom the offspring shared its genes. After Trivers (1985).

keep going for a good while longer: being more closely related to itself than to any future sibling, it is selected to demand investment until the cost-benefit ratio drops below 0.5 (or 0.25 if they are half-siblings). After that point, continued demands for investment would lead to a reduction in indirect fitness, since the parent would produce fewer siblings with whom the offspring shared genes. But until that point is reached, an offspring should attempt to garner more care from the parent, since by doing so it will enhance its own direct fitness. Trivers (1974) suggested that weaning conflicts were the behavioural manifestation of this genetic conflict of interests.

Whether parent–offspring conflict in mammals always arises as a consequence of genetic conflict, or whether it is stimulated by more proximate considerations, is still open to debate (see, for example, Altmann 1980; Dunbar 1988; Bateson 1994; Barrett and Dunbar 1994; Barrett et al. 1995; Barrett and Henzi 2000). However, Trivers' model provides a good starting point to investigate this particular facet of parental investment.

PARENTAL INVESTMENT

Since **parental investment (PI)** is necessary to rear offspring successfully, individuals should be sensitive to the most effective ways of managing the way they invest their available reproductive effort and should attempt to balance PI against the potential loss of future offspring production. Given that they are trying to maximise the numbers of genes contributed to some arbitrarily distant generation (that is, the number of descendants they leave), parents should be concerned to ensure that those offspring they rear successfully will reproduce as effectively as they can.

As long ago as the 1930s, the eminent geneticist and statistician R. A. Fisher pointed out that parents should distribute their available PI among their offspring in such a way as to influence their chance of producing (between them) the greatest number of grandchildren for their parents. In most cases, this is achieved by investing equally in both sexes, and the population sex ratio will thus be 50:50 males:females. Fisher demonstrated mathematically that this would necessarily be the case, even though males can often achieve higher reproductive success than females (see below: pp. 37–38). If parents produced only males because they would give them the highest number of grandchildren, a situation would arise in which females became a rare and limiting resource. Under such circumstances, the average female would achieve higher reproductive success than the average male, because females would be able to secure a mate easily whereas the majority of males would never get the opportunity to reproduce. At this point, females would be the best sex to produce, and the sex ratio would become biased in their favour. A sex ratio of 50:50 is maintained by this shift in the relative value of males versus females according to their availability in the population (a process known as **frequency-dependent selection**).

FREQUENCY-DEPENDENT SELECTION

However, if one sex is more costly to produce than the other, then equal investment in the sexes need not result in the production of equal numbers of males and females. Trivers and Willard (1973) suggested that natural selection would favour sex biases in parental investment away from a 50:50 sex ratio whenever the ability of an individual to rear offspring to sexual maturity was dependent on their condition and/ or status. If, for example, male offspring require high levels of investment in early life in order to compete effectively for mates when adult, then parents in good condition would produce higher quality male offspring than parents in poor condition since the former have more of the necessary resources to invest in their offspring. If only high-quality male offspring achieve high reproductive success, then high-quality parents who produce sons will end up with more grandchildren than either low-quality parents who produce sons or high-quality parents who produce daughters (since the variance in female reproductive success is generally much lower than that of males: see **Figure 2.5**). High quality parents should thus bias their investment towards male offspring since this provides them with the greatest return on their PI, whilst low-quality parents should invest in daughters. This type of sex ratio adjustment has been found in a number of species, including our own. In some cases (for example, red deer: Clutton-Brock et al. 1988), adjustment occurs before birth (although we have no idea how this is achieved). In others (including humans: see Voland and Dunbar 1995, Bereczkei and Dunbar 1997), adjustment is made after birth, either through selective infanticide or by giving more resources to the favoured sex (see Chapter 7).

Parental investment is a very important component of human evolutionary adaptation because human babies are born particularly helpless and dependent. In addition, among humans, we find an investment not found anywhere else in the Animal Kingdom; our ability to accumulate material wealth means we can invest more in our children than just milk, food or shelter. Furthermore, in humans, this kind of investment can continue after the parents' own death through wills or bequests. The decisions made by humans regarding their reproductive and parental investment decisions and the way in which children are valued in society are explored in Chapters 7 and 8.

SEXUAL SELECTION

Darwin actually identified two mechanisms by which evolution could occur: in addition to natural selection, he suggested that 'sexual selection' could be a rather potent force (see Darwin 1871). As we explained earlier, natural selection was conceived in terms of selection imposed by the environment, and deals with how individuals solve the problems of everyday survival, such as finding food or avoiding predators. Darwin realised that features like the peacock's tail were a serious disadvantage to an animal in terms of survival. The tail of the peacock, for example, makes it extremely difficult to fly, and so escape quickly from predators. Darwin therefore suggested that the only way such features could evolve was if they conferred an advantage on the animal in terms of mating, which could compensate for the lowered survival probability under natural selection. Traits that make it easier for an organism to secure a mate (or many mates) will increase in the population since, by definition, those individuals that possess such traits will enjoy greater mating success, and hence have higher fitness, than those who do not.

The strength of sexual selection depends on the degree of competition for mates, which in turn can be traced to both the amount of investment that each sex places in offspring and the ratio of sexually receptive females to males (see Krebs and Davies 1993). Among sexually reproducing species, females always invest more than males since eggs are larger and more costly to produce than sperm. In mammals, this situation is exacerbated by internal fertilisation and gestation, which leave females literally 'holding the baby'. Under these circumstances, where females invest most in offspring, males compete amongst themselves for access to females, and this competition increases the potential for sexual selection (Trivers 1972). With regard to the second factor, the fewer sexually receptive females there are around at any one time, the harder males have to compete. This also increases the strength of sexual selection.

Reproductive effort can be divided into mating effort and rearing effort. Because of the way in which sexual reproduction is organised in mammals, females typically emphasize the rearing side of things (**Figure 2.4**). To take humans as our example, no matter how many males a female mates with during the course of her reproductive cycle only one baby will be produced at the end of it. By contrast, in the time it takes a woman to produce a single baby, a man can potentially father hundreds of children since his only input need be an ejaculation of sperm at an opportune moment. Consequently, the amount of variation observed in male reproductive success can be very much greater than that observed in women. Women are limited by their physiology in the number of offspring they can produce during their lifetime, and so the amount of variance among women will always be relatively small. Men aren't limited in this way and, as a result, some can achieve extraordinarily high reproductive success; equally, however, the resulting competition inevitably means that other males will have extraordinarily low (perhaps even zero) reproductive success.

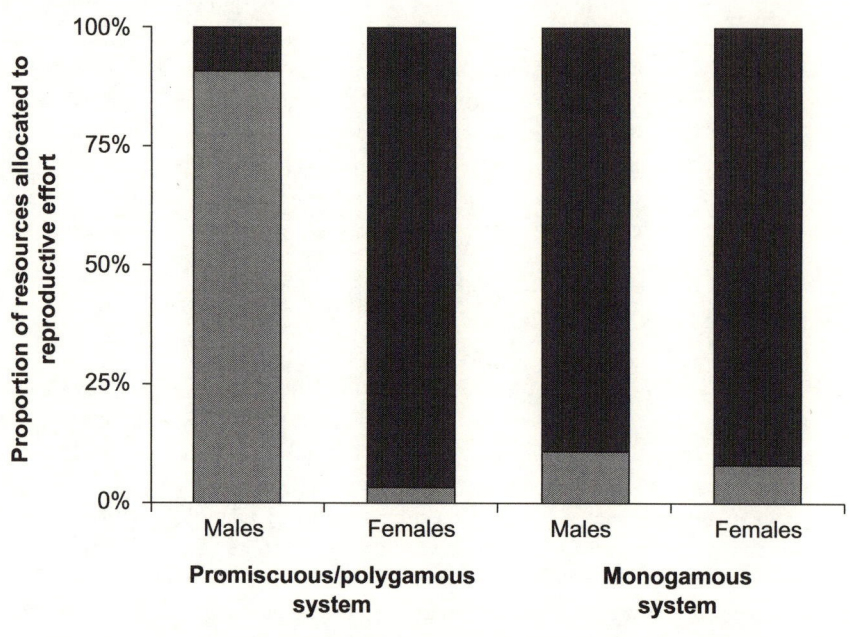

Mating system and relative reproductive effort

Figure 2.4 Sex differences in the allocation of reproductive effort under polygamous and monogamous mating systems for mammals. Reproductive effort can be partitioned into mating effort (grey) and parenting effort (black). In promiscuous or polygamous mating systems, the two sexes pursue complementary strategies, with males investing in competing for mates (mating effort) while females invest in rearing. In monogamous systems, the two sexes' strategies coincide, with more effort devoted to rearing by both sexes. After Krebs and Davies (1993).

Among the Mukogodo of Kenya, for example, there are men who never marry and father children because they cannot afford to pay a 'brideprice' and secure themselves a wife (Cronk 1989). Their reproductive success is zero. By contrast, wealthy men can afford to marry polygynously and may father up to 30 or 40 children in their lifetime (**Figure 2.5**). This difference in the variance between the sexes is known as **Bateman's principle** (Bateman 1948), after the biologist who discovered this difference through experiments with fruit flies. Bateman's Principle has a profound influence not only on mating strategies and sexual selection, but also, as we saw above, on parental investment strategies.

BATEMAN'S
PRINCIPLE

INTER- AND INTRASEXUAL SELECTION

Sexual selection can occur either within one sex (intra-sexual selection) or between the sexes (inter-sexual selection). **Intrasexual selection** occurs when members of one sex compete amongst themselves for access to mates. This leads to the evolution of traits which increase competitive ability, such as the canine teeth of male baboons or the antlers of red deer. Males slug it out and the winner can then claim the female as his reward.

INTRASEXUAL
SELECTION

Intersexual selection (or female choice, as it is often known) occurs when females

INTERSEXUAL
SELECTION

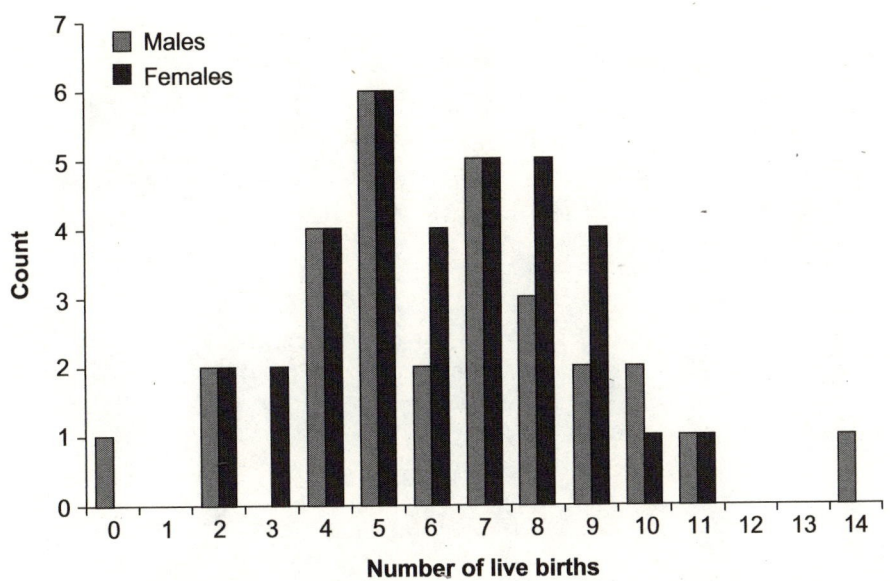

Figure 2.5 The variance in male and female lifetime reproductive output for male and female Aka pygmies. Although both sexes are constrained to having the same mean, the variance in male output exceeds that in females because a male's reproductive rate is not limited by the natural reproductive cycle as in the female. When the variance in the reproductive success of one sex exceeds that for the other sex, the sex with the lesser variance is likely to be more choosy in its willingness to mate because each of its reproductive events represents a more costly commitment. This is referred to as Bateman's Principle, following Bateman's (1948) observation that males can increase their reproductive success by securing matings with more partners, whereas female reproductive success is not increased substantially beyond the first partner with whom they mate. Redrawn from Hewlett (1988).

choose males with particular characteristics because these traits help the female to produce or rear more (or better quality) offspring. In some instances, females choose males on the basis of the resources they have to offer. For example, female North American bullfrogs select their mates on the basis of the quality of the territory they defend, since this is the one factor determining the survival of their eggs that they have control over (Howard 1978a, b). Many birds and insects, on the other hand, select their mates on the basis of the male's ability to provide them with food (Thornhill 1976). Males often have to feed females as part of a courtship ritual. This can make a significant contribution to her eggs and, in the case of birds, can serve as a good indicator of the male's ability to provide food for the young later on (Nisbet 1977).

Females can also choose males on the basis of their personal attributes (the 'good genes' hypothesis). Females select those males who are of high quality (that is, possess good genes) in order to ensure that they produce high quality offspring. There are two ways in which males can be of 'high quality'. The first is by possessing a trait that females find attractive, so increasing the likelihood of securing mates. The trait need not have any survival value for the male; indeed, it can even be positively detrimental. But, so long as it makes males more successful at mating because females show an arbitrary preference for it, sexual selection will promote its evolution.

BOX 2.6

Female choice for exaggerated male traits

Andersson (1982) demonstrated female choice for extravagant and exaggerated male traits in a very elegant experimental study of widow birds. These are small African birds about the size of a sparrow, with males that sport a tail up to 50 cm long. Females, by contrast, have tails only 7 cm in length. In order to test whether females actually did prefer males with longer tails, Andersson cut the tails off some males and stuck the severed bits onto the tails of other males.

This left him with a group of males with tails that were much shorter than average, and another group with tails that were much longer. He then identified two control groups: one was left completely untouched, whilst in the other he cut the males' tails off and then stuck them back on again without altering the length.

At the end of the breeding season, he was able to show that males whose tails had been artificially lengthened had much higher mating success than either short-tailed males or males of the control groups (**Figure 2.6**). Female widow birds much preferred to mate with males who displayed the most exaggerated form of the trait, despite the fact that the trait confers no survival advantage to the male in question.

If both the male trait and female preference are genetically determined, then females who mate with males possessing the preferred trait will go on to produce male offspring who possess the trait and female offspring who express a preference for it (see **Box 2.6, Figure 2.6**). These 'sexy sons' will go on to produce large numbers of offspring (since females will prefer to mate with them), while female offspring will prefer to mate with other sexy males and go on to have successful sons themselves. So merely by mating with a male who possesses a trait which does nothing else but make him attractive to females, a female can achieve high fitness. Fisher (1930) developed this hypothesis during the 1930s and showed how the association between female preference and a male trait could lead to a 'runaway process' whereby females select for ever more exaggerated male traits. Fisher argued that this positive feedback process is the mechanism that produced the magnificent examples of male plumage seen today in peacocks, pheasants and birds of paradise. Selection for the trait by female choice only ceases when the cost in terms of reduced survival probability starts to outweigh the increased mating benefit accruing to males.

The second way females can select for exaggerated traits, like long tails in males, is if these traits act as a signal of the male's genetic quality and allow females to identify those who will provide her with the most robust and resilient offspring. One problem with this hypothesis is that eventually, as a consequence of the increased reproductive success of males with preferred traits – say, long tails – all the variance in the male trait will get 'used up' and selection for tail length will stop. The male gene pool will contain only genes for long tails since genes for shorter tails will have been selected out. All males will therefore have tails of roughly the same length, and females will no longer be able to use tail length as a means of discriminating high-quality genes from low-quality ones.

Hamilton and Zuk (1982) suggested that one way in which variance could be maintained was if individuals were selected for disease resistance. Different pathogens are prevalent in different environments at different times. Genotypes that are selected because they are resistant to one particular disease at one point in time may be selected against in the future because they lack resistance to a new pathogen or a more virulent form of the original one. The genotype favoured in a given environ-

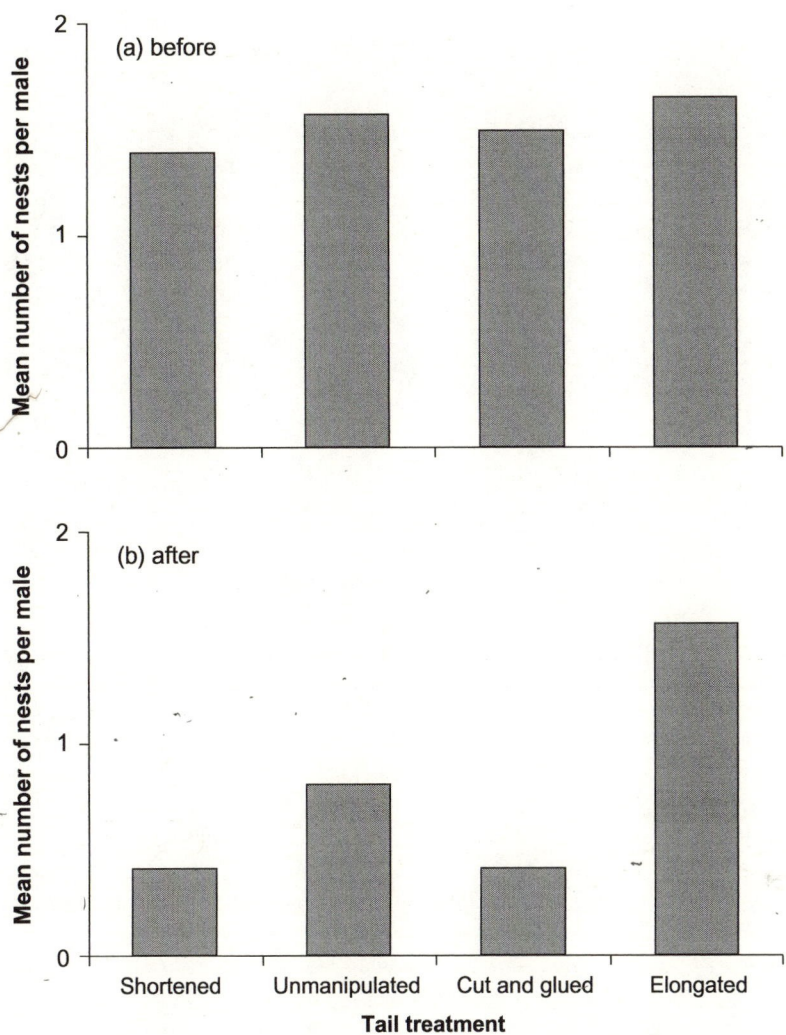

Figure 2.6 Female widowbirds judge a male's quality by the length of his tail. Artificially manipulating the length of a male's tail has dramatic effects on a male's reproductive success: prior to manipulation (a), there was no difference in the number of nests held by different males but, after manipulation (b), males whose tails had been lengthened held significantly more nests than did males whose tails had been shortened or left unaltered (unmanipulated or cut and glued back on again). Redrawn with permission from Andersson (1982).

ment would thus be constantly changing on account of the continual arms race between pathogens and their hosts. In this situation, females would be expected to improve the genetic quality of their offspring by selecting for disease-resistant mates, using exaggerated traits as their cue. The possession of an exaggerated trait requires a male to be vigorous and healthy if the trait is to be displayed to its best advantage, whether this is showy plumage or an elaborate (and exhausting) courtship display. Females can thus assess the health of their partner via the verve with which they are able to perform a courtship dance or the gloss on their tail feathers.

The other model used to explain exaggerated male traits was developed by Zahavi

BOX 2.7

Why do handicaps have to be costly?

Theoretical modelling has shown that handicaps must be costly to produce and maintain in order to stop one individual deceiving others with regard to its quality (Zahavi 1977, Grafen 1990). In this sense, 'costly' means the signal provided by the handicap must involve more effort to produce than is strictly needed to convey the message. This is why males go for overkill when signalling quality to prospective mates by producing highly exaggerated handicaps. If producing a signal of quality like a handicap was cheap, all males would be able to produce a high-quality handicap, but it would bear no relation to the male's underlying genetic quality. Costliness maintains handicap honesty by reducing the marginal value of the potential resource gain to low-quality males.

In other words, if individuals have to spend a disproportionate amount of their energy on producing the trait compared to their expected return on that investment, then only really high quality males are able to pay the price and reap the benefits of an exaggerated trait. Even if low quality males manage to produce an extravagant trait, the high cost means that the quality of the trait itself will be lower than those of high quality males and females are therefore able to discriminate against them.

'HANDICAP
PRINCIPLE'

(1975, Zahavi and Zahavi 1997) and has come to be known as Zahavi's **'Handicap Principle'**. This states that male traits are selected for precisely because they lower male viability and therefore act as a reliable signal of male quality. If a male has managed to survive to sexual maturity despite the cost of the handicap he bears in the form of an exaggerated trait, then clearly he must be a very high quality male and in possession of good genes. Females who choose to mate with such a male will thus gain these good genes for their offspring. Low quality males just can't get away with it because they are unable to bear the cost of possessing an exaggerated trait.

The Hamilton-Zuk hypothesis sits well with this interpretation of things too. If males are going to possess traits that look less than impressive when the male is a bit below par, then this can also be a handicap and act as an honest signal of quality. Only males full of vim and vigour will be able to show off their traits to best advantage, and females will thus be able to select their mates on this basis. In a test of the Hamilton-Zuk hypothesis, Møller (1990) was able to show that parasite resistance was linked to both exaggerated male traits and increased fitness benefits for females that preferred such males. First, Møller (1990) showed that female swallows preferred to mate with males with long tails. Second, he was able to show that parasite resistance in young nestlings was partly genetically inherited from the parents. Then, in an ingenious series of experiments, Møller was able to show that males with longer tails produced offspring with much lower parasite loads than males with short tails. (**Figure 2.7**). In other words, males with longer tails were more resistant to parasites and were able to pass this resistance onto their offspring. Females that used the male trait as a signal of parasite resistance thus produced offspring with greater fitness potential due to lower parasite loads, and in the case of males that would be vigorous enough to grow long tail lengths themselves.

It has been suggested that there are some morphological features of humans that can be attributed to the influence of sexual, rather than natural, selection. Male beards, female breasts (Barber 1995) and even our greatly enlarged brains (Miller 2000) have all been suggested to be the product of sexually selective processes. In Chapter 5, we will be assessing the evidence for and against such suggestions.

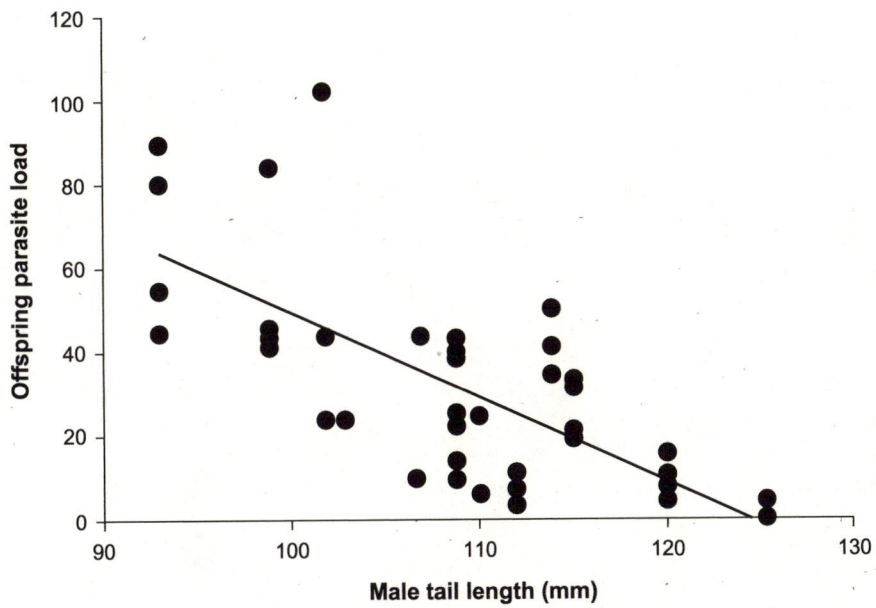

Figure 2.7 Tail length in male barn swallows is used by females as an indicator of fitness potential (indexed by the males' resistance to parasite loads) and this is at least partly inherited by offspring. Møller (1990) was able to demonstrate this by showing that a male's tail length is negatively correlated with the parasite loads of his offspring. Redrawn with permission from Møller (1990).

*

The theory of sexual selection underlines an important point about evolutionary processes in general, namely the fact that individuals are perpetually involved in trade-offs between alternative options as to how they might best maximise their fitness. In long-lived animals like humans, many factors interact to determine an individual's fitness in addition to those issues that arise from kin selection. This mainly arises from the fact that time and energy constraints make it impossible for an individual to optimise its performance on all these variables simultaneously. A male that spent all his time mating would fall prey to predators or starvation, or would be unable to ensure that its mates successfully reared the offspring that it had sired.

Given these constraints, an individual has a large number of alternative (in some case, equally effective) ways of optimising its lifetime reproductive output even if we reduce these many variables to the broad categories of ensuring successful survival, mating and rearing of offspring (Dunbar 1982). If we add to these the options offered by kin selection and, in humans at least, the opportunity to influence the reproduction of one's children through bequests on one's death, it is clear that evolutionary analyses of human behaviour are unlikely to be simple.

This does not, of course, mean that such analyses cannot be undertaken. The note we sound is merely a cautionary one. The main lesson is that we need to beware of declaring that a particular behaviour has no adaptive function on the basis of a simpleminded analysis. Unless and until we have excluded all the alternative opportunities for optimising fitness, we cannot realistically claim that behaviour has no function. In addition, however, there is the less comforting reminder that trade-offs between different components (kin versus non-kin as allies, mating versus parental investment) become an integral feature of our species' behavioural strategies.

In the chapters that follow, we will be attempting to use these principles to explore the adaptiveness of human social, mating and parenting behaviour.

Chapter summary

- Natural selection acts at the individual level, not at the level of the group or the species.

- Although the individual is an appropriate unit of selection, some evolutionary problems, like altruism, can only be understood by taking a 'gene's-eye' view of selection processes. This is because the gene is the ultimate level at which selection acts.

- Hamilton's Rule shows how altruism between kin can evolve. Individuals can increase the representation of their genes in future generations by providing benefits to other individuals with whom they share the same gene in common. This is dependent on the ratio of the costs to the actor relative to the benefits to the recipient and, more importantly, on the closeness of the genetic relationship between them.

- Parental investment is a special form of kin selection. Parents' decisions to invest in their offspring depend on their own and their offspring's future reproductive prospects. Parent–offspring conflict is the behavioural manifestation of a conflict of interests at the genetic level.

- Reciprocal altruism and the repeated Prisoner's Dilemma game have helped to explain the evolution of altruism between unrelated individuals in situations where cheating is likely to be a problem. Individuals that interact repeatedly can form cooperative relationships provided they can (a) recognise each other when they meet and (b) remember the results of their previous interactions. Again, the ratio of costs and benefits to actor and recipient is crucial to whether cooperation will occur.

- As well as natural selection, Darwin identified sexual selection as a means by which particular traits could evolve. This can take place either within (intra-) or between (inter-) the two sexes. Intrasexual competition selects for traits that increase competitive ability such as large body size and weaponry (teeth, horns). Intersexual competition selects for traits that signal a male's quality.

Further reading

Andersson, M. (1994) *Sexual Selection*. Princeton (N.J.): Princeton University Press.

Axelrod, R. (1984) *The Evolution of Cooperation*. New York: Basic Books.

Dawkins, R. (1989) *The Selfish Gene*. (revised and expanded edition). Oxford: Oxford University Press.

Hamilton, W. D. (1996) *Narrow Roads of Gene Land*. Oxford: W. H. Freeman.

Ridley, M. (1993) *The Red Queen: Sex and the Evolution of Human Nature*. London: Viking.

Zahavi, A. and Zahavi, A. (1997). *The Handicap Principle: a Missing Piece of Darwin's Puzzle*. Oxford: Oxford University Press.

Cooperation among kin

As we explained in the previous chapter, an individual's fitness can be enhanced by helping related individuals to survive and reproduce, as well as by producing fit and healthy offspring of its own. In this chapter, we will investigate the various ways – some obvious, some less so – that kin selection operates in human societies. As will become clear, subjects as disparate as childcare practices to Viking sagas can provide us with insights into the behavioural and psychological mechanisms that enable and encourage people to interact favourably with kin.

KIN SELECTION IN HUMANS

Blood is thicker than water, so the old cliché goes. And, if we are to accept that Hamilton's (1964) theory of kin selection applies to humans, then on average we should expect people to behave more altruistically toward their relatives than to unrelated individuals. But does people's actual behaviour measure up to this theoretical expectation? Although it is often claimed otherwise (usually with an obligatory reference to Sidney Carton volunteering to be guillotined in someone else's stead in Charles Dickens's novel *A Tale of Two Cities* (1859)), some rather more scientific treatments of the issue do suggest a strong tendency for the preferential treatment of kin in a very wide range of circumstances.

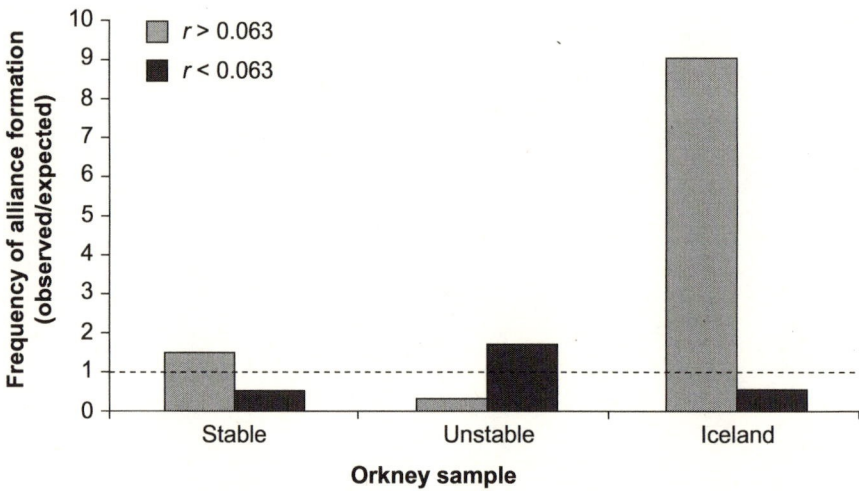

Figure 3.1 Among the Vikings, alliances were formed more often between close relatives (individuals related at least as closely as paternal cousins), and those alliances were less likely to be subsequently broken than alliances between less closely related individuals. The plotted variable is the ratio of the observed frequency of alliances recorded in the sources divided by the number expected on the basis of the number of relatives or non-relatives in the population. Sources: 55 alliances recorded in the *Orkneyinga saga* and 19 recorded in the Icelandic *Njal's Saga* (data from Dunbar et al. 1995).

Both Hames (1987) and Berté (1988), for example, showed that, in South American horticultural tribes, labour services were most likely to be given for free (without expectation of reciprocation in the future) only to relatives, and similar arrangements were reported for Nepalese hill farmers by Panter-Brick (1989). Betzig and Turke (1986) found that, on the Pacific atoll of Ifaluk, food sharing was most common between close relatives. Chagnon and Bugos (1979) showed that, in a filmed axe-fight between two groups of Yanamamö indians, members of the two villages involved tended to side with the protagonist they were most closely related to; even though there were exceptions in that some individuals sided with allies by marriage against more closely related protagonists, biological kinship carried more weight. Among the ten-to-twelfth-century Earls of the Scottish Orkney Isles, Dunbar et al. (1995) found that alliances formed between kin were significantly more stable than those formed between distantly related or unrelated individuals, and were less likely to involve any preconditions (**Figure 3.1**). Smith et al. (1987) showed that handing on wealth to lineal descendants is far more common than giving to less closely related collateral relatives or unrelated individuals – and this may mean excluding spouses in order to benefit children (**Figure 3.2**). Judge (1995) reported similar results for twentieth-century Californian wills. Among the Ye'kwana of Venezuela (see below, **Figure 3.6**), the Efe pygmies (Ivey 2000), black communities in the USA (Stack 1975; Burton 1990) and Hungarian Gypsies (Bereczkei 1998), relatives are significantly more willing to provide care for another woman's child than are unrelated individuals; similarly, a very large sample of Canadian households suggested that individuals were much more likely to provide childcare for relatives than unrelated individuals if they had no children themselves, whereas

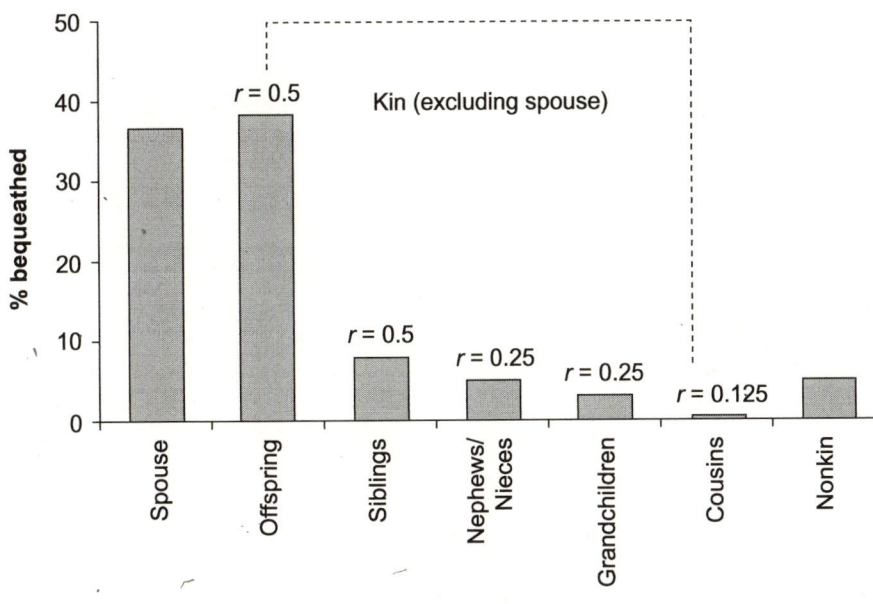

Figure 3.2 Analysis of Canadian wills shows that (with the exception of bequests to spouses) a higher proportion of the estate is given to relatives than to non-relatives, and that more distant relatives receive proportionally less than individuals who are more closely related to the grantee. Source: Smith et al. (1987).

those with children were more willing to engage in (reciprocal) arrangements with non-relatives (Davis and Daly 1997). Essock-Vitale and McGuire (1985) reported, from a study of 300 Los Angeles women, that close relatives were more likely to be sources of help during times of need than unrelated individuals and, in addition, that such help was more likely to be on a strictly reciprocal basis if it was a friend than a relative (see also Hogan and Eggebeen 1995). Finally, a study of contemporary social networks in the UK found that kin account for a significantly higher proportion of network members than would be expected by chance (Dunbar and Spoors 1995), with females tending to have a higher proportion of kin in their networks than males. The latter finding echoes those from a detailed but more qualitative study of the East End slums of London during the 1950s, which showed that kinship played a very important role in female networks, with mothers and daughters and sisters forming mutually supportive alliances (Young and Willmott 1957, Bott 1971).

Such effects may even extend to grouping patterns. Chagnon (1975), for example, demonstrated that, when Yanomamö villages fission, the mean relatedness between members of the daughter villages is higher than that for the parent village, implying that more closely related individuals tended to stay together. Hurd (1985) found a similar effect when Amish church communities underwent fission. Morgan (1979) showed that Inuit (Eskimo) whaling crews have very high coefficients of relatedness (significantly higher than if crews were chosen at random), even though the spoken rules of crew membership were expressed in terms of social kinship (**Table 3.1**).

As impressive as this range of examples may be, however, they offer only circumstantial evidence for kin-bias because they are all based on observational data. Experimental tests have been conspicuously absent. Recently, however, an attempt was made to test whether humans obey Hamilton's Rule using a very simple experimental design (Fieldman et al. submitted). In this study, subjects were asked to perform an isometric skiing exercise in return for a cash reward

Table 3.1 Relatedness of crew in Inuit whaling boats

Relationship to boat captain	r	N
Sons	0.5	15
Brothers	~0.5	10
Nephews	~0.25	7
1st cousins	0.125	1
2nd cousins/grandnephews	0.063	2
Step/adoptive relatives	~0	3
Unrelated	~0	5

Source: Morgan (1979)

BOX 3.1

Rules of thumb and kin recognition

It is important to note here that, in all the examples we describe, individuals are not required to be consciously aware of what they are doing. Individuals who behave in a way that promotes the survival and reproduction of key relatives will leave more descendants than those who do not, regardless of whether they actually understand what they are doing. As far as natural selection is concerned, it does not matter whether you know that the individuals you treat well are your genetic kin, or whether you are just using a rule of thumb that states 'be nice to those you grew up with'. As long as most of the individuals who benefit from your attentions share genes with you, then more of your genes are likely to make it into succeeding generations than will be the case for individuals who do not differentiate between kin and non-kin. Obviously, in many of the cases we discuss, individuals do understand the likely consequences of their actions, but it is not necessary that they do so in order for natural selection to operate.

Kin recognition mechanisms in other mammals are often mediated by smell. The **major histocompatibility complex (MHC)** of mice, for example, is polymorphic (several different versions of the genes coding for this complex are present and stable within a gene pool) and is linked to the production of a distinctive individual odour. Parents and offspring recognise each other on this basis, with female mice able to recognize and preferentially retrieve pups differing only in their MHC but otherwise genetically identical (Yamakazi et al. 2000). It seems that both mother and offspring learn this smell within a few days of birth and then use this cue for recognition thereafter. Other species seem to use visual cues or a combination of visual and odour cues (Arnold 2000, Saidapur and Girish 2000). In humans, there is evidence to suggest that MHC is linked to mate choice preferences with individuals selecting mates with an MHC odour-type different from their own (Wedekind et al. 1995, Wedekind and Füri 1997). Interestingly, a questionnaire-based study by Herz and Cahill (1997) suggests that women rely heavily on smell in choosing mates, whereas men rely more heavily on vision. If humans can use odour cues to identify individuals that are not kin, then they presumably can use the same cues to recognise kin – in addition to the cues provided by growing up in close proximity and learning socially-transmitted kin relationships.

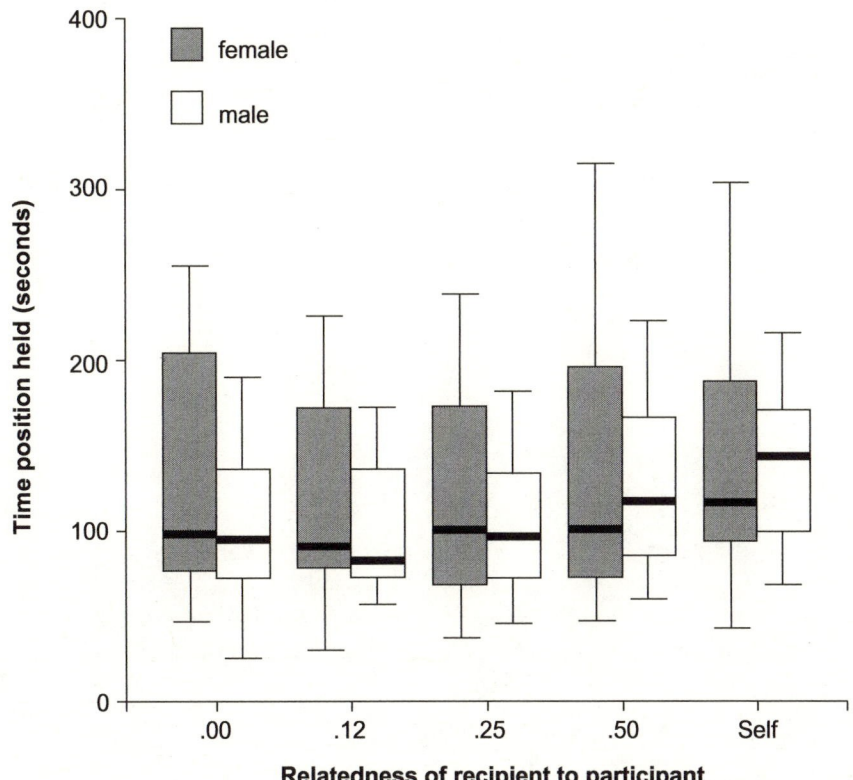

Figure 3.3 An experimental test of Hamilton's Rule in humans. When subjects were paid in proportion to the length of time for which they could hold a painful isometric skiing exercise, the pain they were prepared to sustain was directly related to the relatedness of the person who received the payment obtained by their efforts. The plotted variables are the median, 50 per cent and 95 per cent ranges for the duration the position was held on each trial. Each of 40 subjects ran the trial five times for a different recipient (with trials run in counterbalanced order across subjects). The subject (self) is related to him/herself by r = 1.0. Redrawn from: Fieldman et al. (submitted).

whose value was determined by how long they could maintain the exercise. The exercise (which is designed to strengthen the quad muscles in the thigh) involves adopting a sitting position with the back against a wall: the hips and knees are bent at right angles and take the full weight of the body. This position is reasonably comfortable for about 60 seconds or so, but becomes increasingly painful thereafter.

Subjects were asked to perform this exercise on successive days. On each day, a specific individual was selected to be the recipient of the subject's monetary reward. The recipients varied by relatedness to the subject and included the subject him/herself (relatedness r = 1.0), a parent or sibling (r = 0.5), a grandparent or niece/nephew (r = 0.25), a cousin (r = 0.125) or an unrelated best friend of the same sex (r = 0). On each occasion, the subject knew the identity of the recipient before he/she began that day's trial.

The results were striking and quite unequivocal (**Figure 3.3**). The length of time subjects were prepared to sustain the pain was directly related to the recipient's relatedness to the subject (r_s = 0.284, N = 95, P = 0.006). In effect, they were willing to suffer more for themselves than for their siblings, and more for their siblings than for an unrelated individual. These effects were shown to be independent of the order of the trials, the subject's assessment of the recipient's financial need, the subject's perceived affection for the recipient and the amount of time they had spent together during the previous year. Thus, the results seemed to be driven only by the recipient's relatedness. Immediately after each trial, subjects rated the intensity of the pain they had felt on a 0–100 scale: since this increased linearly with the length of time they had held the position, this suggests that their decision on how long to hold the position was a (semi-)conscious one – they knew how much pain they were prepared to put up with for any given recipient.

BOX 3.2

Adoption: an exception to kin selection?

Although we should not always expect people to behave in accordance with the simple predictions of kin selection, there do seem to be some rather glaring exceptions that go entirely against the predictions of the theory. Adopting and rearing children who are not your own, for example, does not seem like a particularly effective way of enhancing fitness. Marshall Sahlins, an eminent social anthropologist, specifically cited human adoption practices in Polynesia as a cultural behaviour that was inconsistent with the predictions of kin selection, and even went so far as to say that 'there is not a single system of marriage, post-marital residence, family organisation, inter-personal kinship, or common descent in human societies which doesn't set up a different calculus of relatedness and social action than is indicated by the principles of kin selection' (Sahlins 1976).

In many Polynesian societies, as many as one quarter of all children are adopted and raised as members of other families. Sahlins (1976) suggested that, since this falsified the predictions of kin selection theory, it proved that biology was irrelevant to the study of human behaviour. Joan Silk (1980), however, used Sahlin's own data to show that adoption was in fact most common among infertile or post-reproductive couples, that it usually involved (distant) relatives (nieces and nephews with $r = 0.25 - 0.125$ to the adopter), and that (perhaps most significantly) adopted children were often discriminated against in favour of the couple's own biological children in the distribution of the parent's wealth. Adoption transactions in the North American Arctic have been found to show the same pattern (Silk 1987). An analysis of fostering patterns among the pastoralist Herero of southwest Africa revealed that only 16 of 423 fosterings (3.8 per cent) where the relatedness of the foster-parent could be established involved individuals who were unrelated to the child (Pennington and Harpending 1993). Fully 82.5 per cent involved individuals who were related to the child by at least $r = 0.125$ (cousins). Similarly, in a study of a black urban community in Chicago where fosterage was a common practice, Stack (1974) found that only 8 out of 135 fostered children lived with non-kin; foster parents were divided more or less equally between older siblings, maternal aunts, grandparents and other more distant (usually, but not always, maternal) kin. Of a sample of 305 mothers asked whom they would prefer to have their children adopted by in the event of their death, 75 per cent specified maternal kin, and 27 per cent paternal kin; none opted for unrelated individuals (Stack 1974).

The function of adoption in Oceania appears to be the adjustment of extreme family sizes in order to improve the economic viability of the family unit: people with too many children give them up for adoption to families who have too few (Silk 1980). However, it is still far from clear whether this really is an act of kin-selected altruism rather

BOX 3.2 cont'd

Adoption: an exception to kin selection?

than a form of exploitation, and no one has so far attempted to measure the actual effects on the inclusive fitness of the natural parents and/or their children (Turke 1990, Silk 1990).

The one exception to this has been a study by Pennington (1991; see also Pennington and Harpending 1993) who did attempt to assess the impact of fostering on reproductive success among Herero pastoralists in Botswana. 'Fostering' is rather similar to the adoption practices observed in Oceania: children are sent at a young age to live with post-reproductive women, who bring them up and provide all their economic needs. Pennington and Harpending (1993) found no relationship between age-related patterns of fertility and fostering, suggesting that fostering had little effect on the reproductive success of women who fostered out their children. The interpretation of even this result is far from clear cut, however: parents often manipulate the survival and reproductive opportunities of their children in order to maximize their own inclusive fitness (see Chapter 7), and the effects of fostering need not therefore be so direct and obvious. A more sophisticated analysis may be necessary to prove the point conclusively one way or the other. One finding by Pennington and Harpending (1993) does, however, bear on the circumstances of fostering: unmarried Herero mothers were nearly twice as likely as married mothers to foster out their children (usually on getting married). Such women typically left the children of a previous relationship with the maternal grandparents, perhaps suggesting a concern with the risks of discrimination (or even infanticide) by step-fathers.

Trying to work out the potential adaptive consequences of adoption becomes even more confusing when western industrialised nations are being considered. As Silk (1990) notes, adoption in the developed world usually involves unrelated children, prospective parents often expend considerable time and effort to obtain adoptive children and a decline in the number of healthy babies available for adoption means that people are often willing to take on older children

or those with a handicap. None of these point to a simple unitary adaptive explanation for adoption, and it may be that adoption in western society is possibly an instance where modern behaviour has been forced off-track with negative consequences for inclusive fitness. For most of human history we have lived in small, closed groups in which the average degree of relatedness was likely to have been high, and thus kin selection may have favoured a tendency to care for and adopt any orphaned young.

While this sounds potentially plausible, Silk (1990) points out that it cannot explain why people in traditional societies are less likely to adopt non-kin than people in industrialised societies, or why non-human primates, many of whom also live in extremely tight-knit kin groups, show very little tendency to adopt orphaned young (Silk 1990). However, it is true that many women experience a strong urge to have children and, even in modern society, children are highly valued culturally such that parenthood tends to be viewed positively by family and friends. This combination of an in-built biological urge for children accompanied by strong cultural motivations may be so strong that individuals wish to raise children no matter what, even if this makes no sense in fitness terms.

However, even if this is true and ancient psychological mechanisms are led off-track by modern-day opportunities for adoption, it doesn't necessarily mean that this is maladaptive for the individuals involved. We cannot emphasize strongly enough that empirical tests are required in such cases, not unsupported assumptions. While adoption may not enhance the inclusive fitness of adoptive parents (although it may enhance the fitness of the natural parents), neither may it have any adverse effects if the majority of couples who adopt are infertile. Since the direct fitness component of such individuals is necessarily zero, investing in unrelated offspring may not reduce inclusive fitness. If this is so, then adoption may be selectively neutral rather than being maladaptive in present day environments.

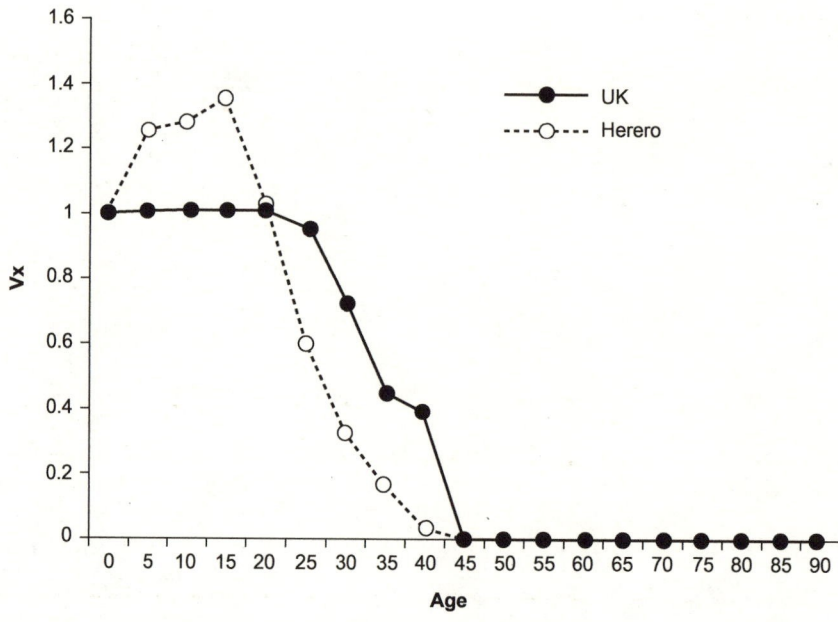

Figure 3.4 Age-specific future reproductive potential (indexed as Fisher's reproductive value, v_x) for women in two modern populations: women in England and Wales (1989–91: filled symbols) and women from the traditional pastoralist Herero of SW Africa (pre-1966: open symbols). Reproductive value is a measure of the number of future offspring of the same sex produced in the remaining part of a woman's lifetime, relative to the mean number produced by all women at birth (which is defined as $v_0 = 1.0$). For present purposes, reproductive value has been calculated assuming a stationary population in both cases (justified in the case of the UK, less so for the Herero). Note that, while the Herero show the classic rising curve between birth and the age of first reproduction, this segment of the curve is much flatter in a contemporary population because of greatly reduced mortality during the pre-puberty years. Sources: UK data: OPCS (1993, 1996); Herero data: Pennington and Harpending (1993).

REPRODUCTIVE VALUE AND KIN SELECTION

Despite the central importance of kinship, simple genetic relatedness will not be the sole determinant of behaviour. For one thing, Hamilton's Rule (see Chapter 2) quite explicitly implies that organisms should be altruistic towards non-relatives (and unhelpful towards relatives) whenever they gain higher inclusive fitness by doing so. More importantly, Hughes (1988) has pointed out that conventional arguments about relatedness invariably assume that all individuals of the same degree of relatedness are equally valuable from an evolutionary point of view. In fact, when the point of the exercise is to maximize fitness (one's genetic contribution to the next generation), they are not all equally valuable. Parents who are post-reproductive may be less valuable than children who are in their prime reproductive years. Hughes argued that the coefficient of relatedness should be adjusted for the individual's *reproductive value*. Reproductive value (RV, or future reproductive potential) is a concept developed by Fisher (1930). It is a measure of an individual's age-specific future likelihood of producing offspring, relative to the average for the population as a whole (**Box 3.3**). In women, RV reaches its peak during the early twenties, and declines steadily thereafter (**Figure 3.4**).

BOX 3.3

Reproductive value

The concept of reproductive value was coined by Fisher (1930) as a measure of an individual's age-specific future reproductive contribution to the species' gene pool. It is a relative measure that compares an individual's future expected reproductive output at a given age against the average lifetime output for all members of the population. The average lifetime reproductive output of all members at birth is taken to be $v = 1$, and age-specific reproductive value is then standardised against this. Because age-specific reproductive rates differ between the two sexes, it is conventional to calculate separate values for each.

Calculating reproductive value at age x, v_x, is complicated by the fact that the growth rate of the population has an effect on v_x:

$$v_x = l_x^{-1} \, e^{rx} \sum_{y=x} e^{-rx} \, l_y \, m_y$$

where l_x is the age-specific survivorship (probability of surviving from birth to age x), m_y is the age-specific fecundity rate (number of same-sex babies produced while age y), r is the Malthusian parameter (annual population growth rate: not to be confused with the coefficient of relatedness!), e is the base of natural logarithms and the summation symbol, Σ, directs us to sum the product that follows it over all age classes from $y = $ age x to $y = $ infinity. It is customary with humans to use 5-year age classes as the time basis. The population growth rate is important because babies born early in a woman's life contribute disproportionately to the future population gene pool when the population is growing, whereas babies born late in her life contribute disproportionately when the population is declining. If the population is stable (neither growing nor declining: $r = 0$), the equation for reproductive value simplifies to:

$$v_x = l_x^{-1} \sum_{y=x} l_y \, m_y$$

Among female mammals in general (including human women), reproductive value starts at $v_0 = 1$ (by definition), increases up to the average age at which females give birth for the first time and then declines steadily into old age (**Figure 3.4**). Note the similarity in the shapes of the curves for the two populations in this sample, despite their very different demographic characteristics. The reason v increases between birth and the age of first reproduction is that some individuals die before puberty: these therefore contribute $v = 0$ to the overall average. Obviously, there are decreasing numbers of these individuals as the age of first reproduction is approached because those that die before age x years do not contribute to the reproductive value of women of older age classes. After age at first birth, a declining proportion of individuals survive to contribute babies to a given age class's gross reproductive value.

Using some rather complicated matrix algebra, Hughes was able to show that the key individuals in any social group were the pubertal cohort (those individuals who had just passed through puberty and were about to reproduce). These individuals had a much greater effect on one's inclusive fitness than any other subset of individuals in the population. He found that, once these focal individuals had been identified, it was possible to predict with surprising precision the clan groupings among a number of New and Old World traditional societies, the extended family groupings of a contemporary Tennessee hillbilly community and even the alliance patterns in Chagnon's Yanomanö axe-fight that we mentioned earlier.

Hughes points out that this preference for the pubertal generation creates a difficulty when it comes to identifying lineages because the pubertal cohort changes each generation. As a result, there is no stable point in the community off which everyone can hang their group membership. Instead, as Hughes notes, we refer backwards to some ancestor from whom we can all claim descent: this ancestor then functions as the fixed point from which to hang the lineage even though in actual fact everyone is using the current pubertal generation when deciding which lineage they should declare allegiance to. Kinship decreases rapidly with each generation back in time (at a rate of roughly 0.5^n, where n is the number of generations to the apical ancestor), and Hughes showed that, because of this, it doesn't really matter who the apical ancestor is, providing he or she is far enough back in time: even if you use the sun or the moon as your apical ancestor, it doesn't affect your calculations about whom you are most closely related to. Hughes' (1988) findings are extremely important because they explain why attributions of common ancestry do not always match up with what we might expect from simple biological considerations. This innovative (if difficult) analysis thus solves rather neatly some hitherto unexplained features of human kinship systems (see Chapter 9).

Similar reasoning regarding the reproductive value of relatives has been used to explain patterns of grief in humans. It is well recognised, for example, that the death of a child is typically more devastating than the death of either a parent or a spouse (Sanders 1980). Crawford et al. (1989) argued that this was a consequence of the higher reproductive value of children. They showed that the amount of grief likely to be experienced (as estimated by male and female subjects) was correlated much more strongly with the reproductive value of the deceased than with their actual ages. Interestingly, however, the results were more strongly correlated with a reproductive value curve calculated for !Kung bushmen (perhaps a better representation of the curve that persisted throughout most of human evolution) than with a curve for a modern day industrial society in British Columbia (compare **Figure 3.4**). However, this study suffers from the drawback that it uses data on estimates of grief supplied by respondents to a questionnaire rather than actual levels of grief experienced by individuals. What people say they do isn't always the same as what they actually do (see **Box 1.3**).

Wang (1996) employed somewhat similar reasoning to investigate the extent to which individual decision-making abilities were influenced by kinship. Subjects were asked which of two alternative medical treatments they would prefer to treat six members of their family who had (hypothetically) caught a potentially fatal disease. One of the possible treatments was deterministic, and ensured the survival of one third of the patients (either the two youngest – for example, offspring or siblings – or the two oldest – for example, parents). The alternative treatment was probabilistic in nature, with a one-third probability that all six patients would be saved.

Interestingly, young (teenage) subjects tended to prefer the probabilistic outcome over the deterministic one (regardless of whether this would save young or old relations), whilst middle-aged subjects preferred the deterministic strategy when it would save young people, but not when it would only save older relatives (**Figure 3.5**). Kin selection effects thus appear to be strongly tempered by age. Wang (1996) explained this by pointing out that the inclusive fitness of individuals who are themselves middle-aged is unlikely to be enhanced by the activity of relatives who are even older, whereas their younger relatives would be expected to contribute quite substantially to their inclusive fitness since these individuals have most of their reproductive lives ahead of them. Teenage subjects, on the other hand, can expect their inclusive fitness to be enhanced by all their relatives at that particular stage in their lives, and they need show no particular preference for one generation over the other. These patterns have also been observed in

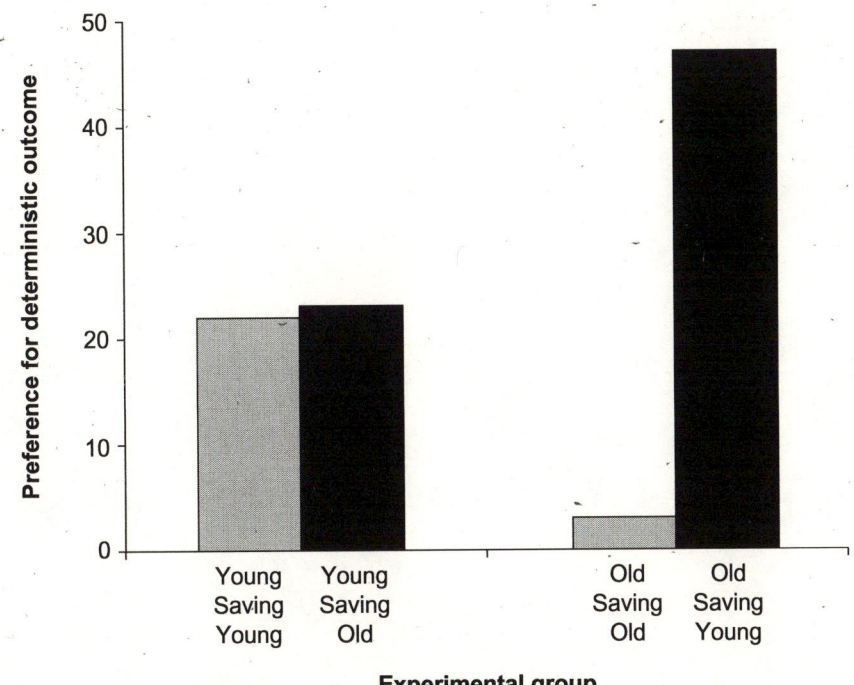

Figure 3.5 When asked whether they would prefer a medical treatment that guaranteed saving the two youngest or the two oldest of six family members (deterministic version) or one that would cure any two individuals drawn at random from this group (probabilistic version), teenage ('young') subjects' preference for the deterministic cure was not influenced by the age of the beneficiaries; but older subjects were significantly more likely to choose the deterministic cure if younger members of the family benefited. Redrawn with permission from Wang (1996).

people's actual behaviour. Essock-Vitale and McGuire (1985), for example, found that investment tended to flow from older to younger kin in their study of 300 Los Angeles women.

Considerations of reproductive value may also account for patterns of childcare among the Ye'kwana of Venezuela. Hames (1988) showed that whilst most direct childcare came from mothers, other female relatives (notably sisters, aunts, cousins and grandmothers) also invested in the care of children under the age of three (Figure 3.6). Despite being related by the same degree, grandmothers were found to invest more in children than aunts. This may be partly explained by a difference in the RV of these females: whilst aunts and cousins are frequently of an age when they could continue to produce their own offspring and thus increase their own direct fitness, grandmothers are non-reproductive, and can only enhance their inclusive fitness indirectly by investing in their grown offspring and their grandchildren, both of whom have a much higher RV.

Grandparental solicitude was also the focus of a study by Euler and Weitzel (1996). Following a similar argument to Hames (1988) and Hill and Hurtado (1997a) regarding the nature of grandparental reproductive strategies, they suggested that grandparents should vary the level of solicitude toward grandchildren according to the sex of the offspring they were helping. Thus, since females put more effort into rearing rather than mating, one should expect maternal grandparents

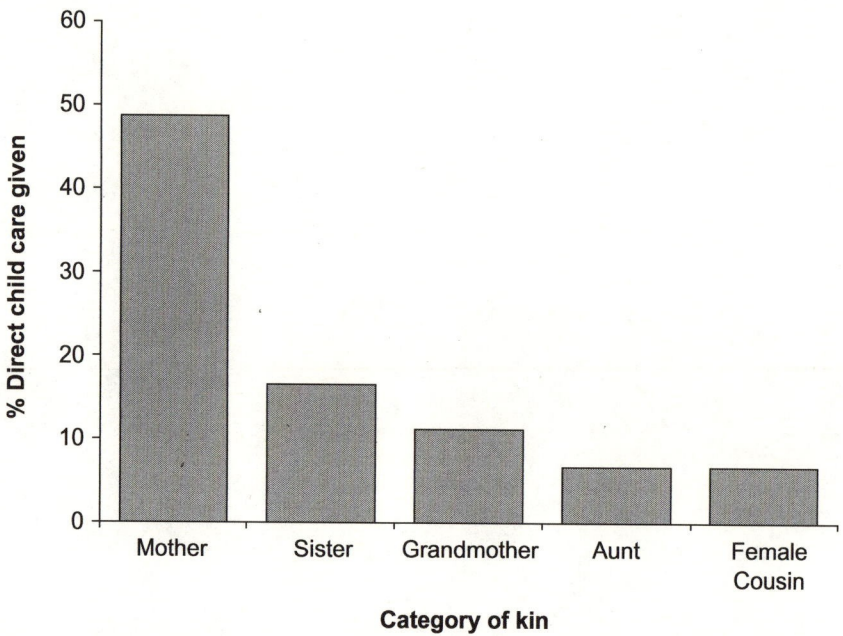

Figure 3.6 The amount of childcare given to infants by Ye'kwana women is a function of both their relatedness to the child and their reproductive value (and hence their own future opportunities for reproduction). This can be seen from the fact that grandmothers (with low reproductive value) are more willing to provide care than aunts, despite the fact that they are equally related to the child. Source: Hames (1988).

to be more solicitous than paternal grandparents, since that is the best way to enhance their daughter's reproductive success. This tendency will then be exacerbated by considerations of paternity certainty. In humans (and other mammals with internal gestation), maternity is always certain, but paternity is not. A man can never be one hundred per cent sure that he is the father of a particular child in the same way that a woman can. Consequently, the degree of grandparental solicitude toward a son's offspring is expected to vary with the degree of paternity certainty a man has. The situation for paternal grandfathers is particularly interesting, since there is not only the chance that his grandchildren may not belong to his son, but there is also a possibility that his son is not his either.

Euler and Weitzel (1996) found that paternal grandparents provided less care than maternal grandparents, and that paternal grandfathers showed the least care of all. Maternal grandmothers provided most care, as might be expected, given reproductive strategy considerations and the fact, that along with the mother herself, she has the highest degree of certainty regarding her degree of relatedness to the child in question (**Figure 3.7**).

KINSHIP, HOMICIDE AND CHILD ABUSE

Daly and Wilson (1988a, b) adopted a rather different approach to the problem of how kin selection influences people's behaviour. Rather than following convention and investigating the occurrence of altruistic behaviour, they used homicide statistics

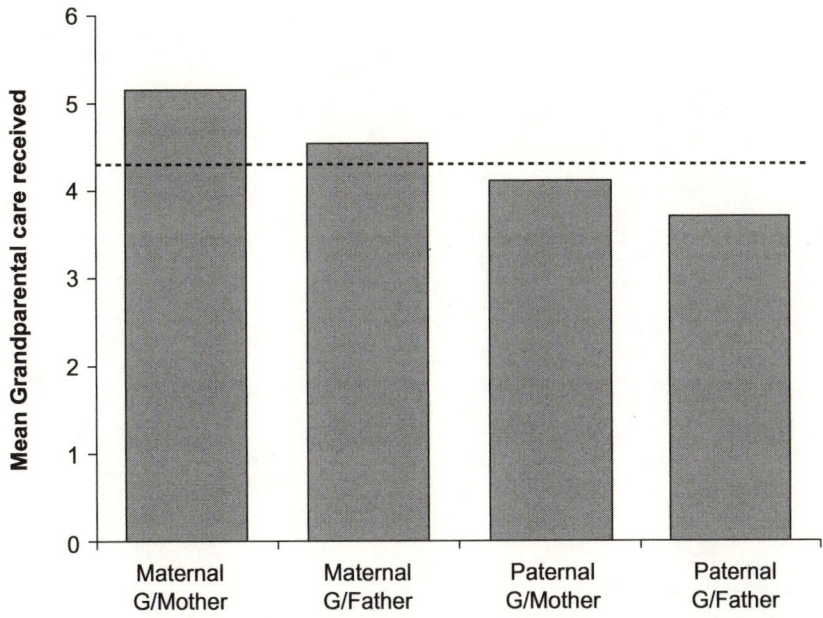

Figure 3.7 Paternity uncertainty is reflected in the amount of care given to a grandchild by maternal and paternal grandparents in a contemporary German sample. Since paternity certainty can never be guaranteed, each additional paternal relationship adds a further degree of uncertainty, with a corresponding drop in care given. The dotted line shows the overall mean for all grandparents. Source: Euler and Weitzel (1996).

to find out who was most likely to come to a sticky end (see **Box 3.4**). They were drawn to this by suggestions in the literature on domestic violence that the 'family is the most violent social group' (Daly and Wilson 1988a). Clearly this doesn't gel particularly well with the idea that you should treat your relatives preferentially in order to enhance your fitness. However, a little delving into the actual crime figures revealed that although 'relatives' constituted one quarter of homicide victims, most of these were in fact spouses rather than blood relatives. In Detroit, 19 per cent of the victims of murders committed during 1972 were related by marriage to their killers, whereas only 6 per cent were related by blood. Similar figures were obtained for Miami where 10 per cent of victims were spouses compared to only 1.8 per cent blood relatives. The prevalence of violence in the home would thus seem to be directed primarily toward non-relatives (**Figure 3.8**). This was confirmed by another analysis which showed that unrelated individuals (whether spouses or not) who were co-resident in the same household as their killer were 11 times more likely to be murdered than co-resident blood relations (Daly and Wilson 1988b).

Daly and Wilson (1988a) also showed that the incidence of infanticide was influenced by ties of kinship. Canadian children living in households with one or more step-parents are 60 times more likely to suffer fatal abuse at the hands of those parents than were same-age children living with their natural genetic parents (**Figure 3.9**). This difference was independent of potential confounds like socio-economic status or the personality characteristics of the abusers. Similar results were obtained from an analysis of child abuse statistics for Canada, England and New Zealand (Daly and Wilson 1981, 1985). In the Canadian sample, for

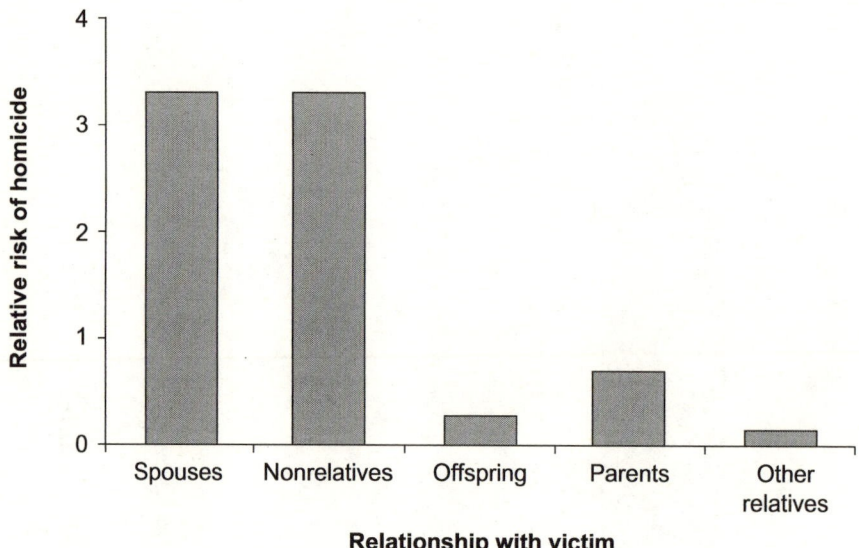

Figure 3.8 Relative to their numbers in the household, non-relatives are disproportionately at risk of being killed by someone they live with than are genetic relatives. This is interpreted as implying that kin selection places a break on going the final step that is absent when an unrelated individual is involved. Source: Canadian homicide data (Daly and Wilson 1988b).

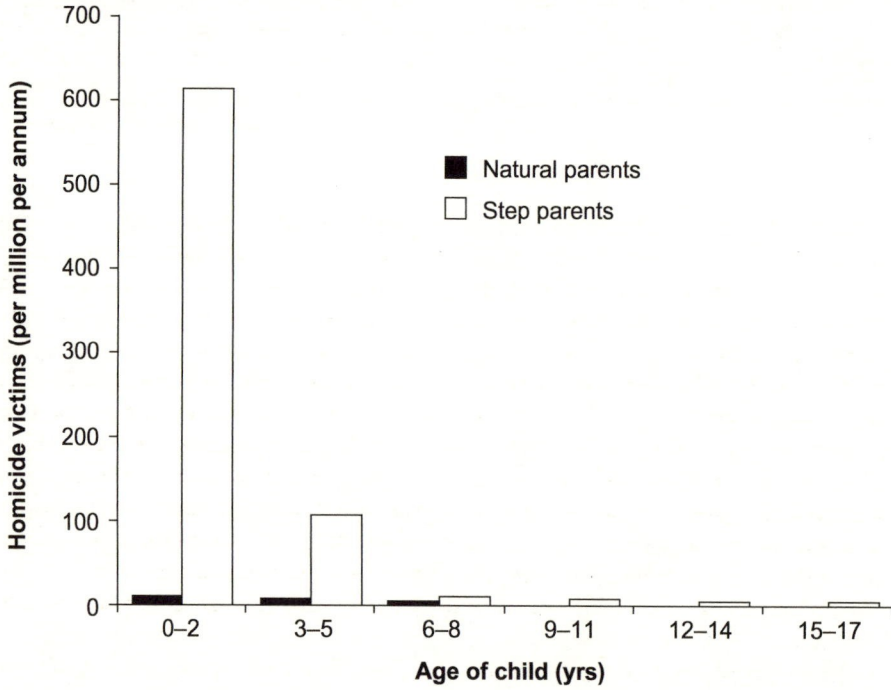

Figure 3.9 Risk of infanticide for Canadian children living in households with either both natural parents or a natural parent plus a step-parent. During the first two years of life, children are 65 times more likely to be killed (usually by the step-parent) when living with a step-parent than when living with both natural parents. Redrawn with permission from Daly and Wilson (1988b).

BOX 3.4

Homicide and infanticide as 'conflict assays'

Daly and Wilson (1988b, 1996) view violence between individuals as a valuable means of investigating the psychological adaptations underlying kinship and sexual bonds. By investigating the circumstances under which violent conflict occurs, it is possible to put together a picture of the underlying mechanisms that function to prevent such conflicts occurring on an everyday basis. Information on rare and unusual events of this nature gives us a key to understanding how links between different factors operate in the normal run of events, similar to the manner in which individuals suffering from unusual and rare damage to the brain can be studied as a means of gaining an insight into the functioning of normal brains.

Data on the most violent assaults of all (those resulting in the murder of an individual) are also methodologically valuable. Since homicide is drastic, final and unambiguous, it does not suffer from problems of quantification associated with non-lethal forms of assault. Non-lethal forms of violence inevitably involve subjective judgements about levels of violence and their impact on vic-

tims and perpetrators: subjectivity of this kind makes it difficult to obtain an objective measure that can be applied across different cases (Daly and Wilson 1996). These kinds of assault are also likely to be under-reported since many victims fail to inform police that they have been attacked (a problem that is especially common in cases of rape), thus inevitably yielding a biased sample. Homicides, by contrast, are more frequently reported since bodies are usually recovered and the seriousness of the crime means that police are likely to become involved. Homicide therefore represents a less biased sample than other forms of assault.

A very important point to remember, however, is that investigating murder from an evolutionary perspective should not be taken to suggest that killing other people has been favoured by selection as an adaptation. Rather, homicide is viewed as the extreme manifestation of conflicts that are usually non-lethal and the emphasis is placed on identifying the psychological mechanisms that underpin such conflicts and those that normally prevent conflicts from escalating to a lethal level.

example, children aged under five years living with one genetic and one step-parent were 25 times more likely to appear on a child abuse register than children of the same age living with both natural parents and three times more likely than children of the same age living in one-parent families with a single natural parent (Daly and Wilson 1985). There was also evidence to suggest that step-parents were highly discriminative when it came to abuse, and would spare their natural children within the same household. Nor, it seems, is this a new phenomenon: in his play *Alcestris* (composed in 438 BC), the great Greek tragedian Euripides has the eponymous heroine beg her husband on her deathbed not to take another wife for fear that her children would suffer at the woman's hands. The wicked stepmother, it seems, has a long history. That step-parents of either sex are a risk is suggested by the fact that, when women acquire a new partner who is not their children's father, it is not uncommon in at least some cultures for children from a previous relationship to be left with the maternal grand-parents (urban black communities in the USA: Stack 1974; Herero: Pennington and Harpending 1993).

These findings suggest that the psychological mechanisms that keep behaviour in check and prevent us from behaving abusively toward our children (however frustrated, angry and fed up with them we may get) may fail to engage when we have no genetic stake in the child in question. Indeed, this seems to be an extension of a general tendency for step-parents to show less benevolent interest in the welfare of unrelated children. Among the Ache hunter-gatherers of Paraguay, for example, 43 per cent of children who were raised by their mother and a step-father died (of various causes) before the age of 15, compared to only 19 per cent of children brought up by both of their natural parents (Hill and Kaplan 1988). In many cases, it was because the step-father deliberately killed them.

Discriminatory behaviour by step-parents may even extend to more subtle aspects of everyday social interaction. Flinn (1988), for example, found that in rural Trinidad step-fathers tended to interact more frequently, and in a more friendly manner, with their genetic children than with step-children living in the same household. These low rates of interaction occurred despite a long duration of co-residence between step-parent and child (often with co-residence since birth), so it wasn't lack of contact and familiarity that could explain these results. In line with these differences, step-children were more likely to be fostered out to other families, and were more likely to emigrate from their natal village at adulthood than were genetic offspring. Most interesting of all from an evolutionary perspective, Flinn (1988) suggested that children raised by step-parents had lower reproductive success than those brought up by two natural parents. This was suggested to be largely a consequence of the increased 'social' benefits that male paternal care could bring to natural children but which were denied to step-children (see for example, Flinn and Low 1986), rather than just a large negative effect of having a step-father. Similar findings suggesting that males discriminate against their step-children in subtle ways have been reported by other studies (Hadza hunter-gatherers: Marlowe 1999a; urban Xhosa: Anderson et al. 1999b).

Given these (often rather alarming) findings, it may seem remarkable that, on the whole, people do actually manage to have peaceful and affectionate relationships with their step-children. Indeed, by no means all studies have reported that step-fathers discriminate against their step-children (for example, the Albuquerque males sample: Anderson et al. 1999a). Daly and Wilson (1988a) suggest that such behaviour by step-parents may actually be part of a reciprocal relationship with the genetic parent rather than with the offspring per se: individuals may be prepared to trade-off greater care of step-children in order to have access to the partner's future reproductive output (the mother's fertility in the case of step-fathers, the father's resources in the case of step-mothers) (see also Waynforth and Dunbar 1995). Males face a trade-off in this respect: excessive discrimination against a mate's offspring from a previous relationship could seriously damage the man's relationship with his spouse. Indeed, there is accumulating evidence to suggest that care-giving and provisioning by step-fathers is geared towards attracting or maintaining sexual access to the mother (Cashdan 1993; Marlowe 1999a; Anderson et al. 1999, Hewlett et al. 2000). Alternatively, at least in the case of males, such tolerance may be due to sexual selection as a result of female choice for benevolent behaviour toward children in general. As yet, neither of these explanations has been subject to rigorous empirical tests. This is important because the vast majority of step-parents do not harm their step-children and this too needs explaining if an evolutionary hypothesis for human behaviour is to be accepted. What is needed now is more detailed data on the costs and benefits of the alternative strategies in order to decide whether or not the men are behaving optimally in their *particular* circumstances.

Just to complicate the picture, studies elsewhere have not always replicated the Canadian findings: Temrin et al. (2000), for example, found that step-parents were not over-represented among child homicides in Sweden between 1975–95. Instead, they identified single-parent families and the psychiatric condition of the parent(s) as the biggest risk factors in childhood mortality in their sample. However, it is clear that there are numerous important contextual differences between Sweden and Canada: childhood homicide rates were nearly 50 times higher in Canada than in Sweden (murder rates for children under 15 years of c. 160 vs 3.4 per million, respectively) and, more importantly, families are economically buffered by a very comprehensive social security system in Sweden (family wealth is a significant risk factor in both child abuse and infanticide: see pp. 181–3). The Swedish results cannot of themselves be taken as disproving an evolutionary explanation, because that would be to leave unexplained the Canadian and other data. A full explanation must show why some behaviours occur in one context but not in another, and we are clearly a long way short of that level of understanding at present.

KINSHIP AND CONTINGENCY

The above findings suggest that non-relatives are likely to come off rather badly when things become potentially violent. However, before comforting yourself with the notion that sheltering within the bosom of your family will ensure a trouble-free life, we should point out that there is an abundance of evidence to suggest that, when the stakes are high enough, individuals can suffer just as badly at the hands of their kin as they can from unrelated individuals. This often surprises people since there is a common misconception that kin selection predicts that relatives should always avoid conflict. What kin selection actually predicts is that altruism should be directed at kin more frequently than non-kin *all else being equal*. Since all else is sometimes far from equal, conflicts among relatives can be expected to break out whenever the benefits of doing so outweigh the associated costs.

Johnson and Johnson (1991) looked at murders among the Earls of Orkney, Icelandic families and the English kings. They found that although close kin were more likely to support each other than be in murderous conflict, there were some interesting exceptions. For example, in the Icelandic sample uncles and nephews tended to support each other, whereas in the Orkney and English king samples murder was extremely frequent. This was suggested to be a consequence of the fact that, in the latter populations, individuals were rich enough to be able to afford paid retainers, and so were less dependent on kin for support when attempting to seize power; the result was strong competition between related individuals. While the frequency of murder was very high between uncle–nephew pairs among co-holders of Viking Earlships (90 per cent, $n = 6$), there were no incidences of murder among half-brothers holding an Earlship in common, even though the degree of relatedness between them is the same as for nephew–uncle pairs (**Table 3.2**). This difference was attributed to an imbalance of power between uncles and nephews compared to half-brothers: while uncles were generally older and wealthier than their nephews and hence more powerful, half-brothers could muster about the same amount of support as each other, and thus lacked any competitive advantage. In such cases, an alliance was seen as

Table 3.2 Relatedness and murder for gain amongst the Earls of Orkney and English monarchs

Relationship	No. of kinship pairs	No. of murders
Earls of Orkney		
Half-brothers	14	0
Uncles – nephews	7	6
English Monarchs		
(1066–1945)		
Fathers – offspring	1	0
Brothers	2	0
Uncles – nephews	4	4
1st cousins	2	2
>1st cousins	8	5

Note: Earls of Orkney: $\chi^2 = 13.26$, df $= 1$, p < 0.001; English monarchs: $\chi^2 = 6.32$, df $= 1$, p < 0.02.

Source: Johnson and Johnson (1991)

preferable to all-out conflict. Similar patterns are seen amongst the English Kings, with murders occurring only between cousins (Queen Elizabeth I ordered the execution of her cousin Mary, Queen of Scots) or between uncles and nephews (the infamous Richard III murdered his nephews, the little princes), but not between father–son pairs or between brothers (**Table 3.2**).

There is, however, another more important reason why we might expect to see conflict between close kin. Hamilton's rule reminds us that the costs to the altruist are as important as the benefits to the recipient of altruistic behaviour. It therefore follows that when individuals gain more by harming a relative than they do by helping that relative, they should opt to harm the relative. In a more detailed analysis of two Viking sagas (the Icelandic *Njal's Saga* and the Scottish *Orkneyinga Saga*), Dunbar et al. (1995) showed that individuals were much more likely to kill blood relatives when the benefit to be gained was high (for instance, the title to land) than when the benefits were low (for instance, a bar-room brawl over an insult) (**Figure 3.10**).

One potential criticism of analyses that use historical documents as their source of data is that the events and motives they describe may be completely fictional and so an inappropriate phenomenon on which to apply real-life evolutionary theory. While this is a fair point, it doesn't necessarily make such analyses any less valid, for if nothing else these analyses show that the mind(s) of the individuals who composed the sagas reflected certain sets of inner beliefs, desires and hopes that seem to be underpinned by evolutionarily driven considerations. Even if the events described in *Njal's Saga* may not actually have happened, the author (and, more importantly perhaps, his audience) nevertheless obviously thought that the behaviour he attributed to his characters was entirely appropriate under the circumstances. Thus, at the very worst, we can use historical documents like these to ask whether the Vikings constructed their mental worlds in accordance with evolutionary theory. The above findings would seem to suggest that they did.

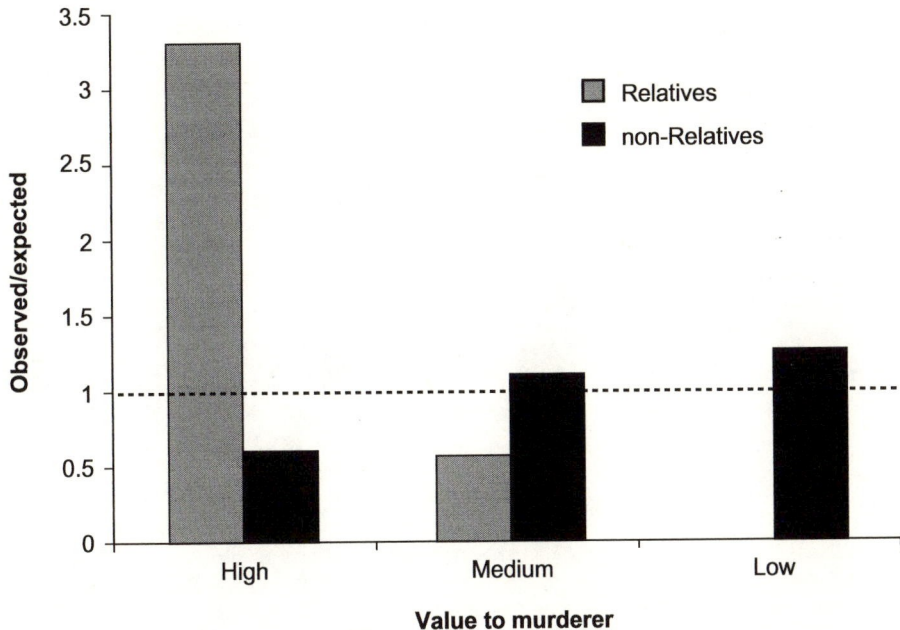

Figure 3.10 Vikings were less likely to murder a relative if the benefit from doing so was trivial (a minor disagreement at a feast) than if the gain was significant (acquisition of land), but no such differentiation was made in the case of non-relatives. The plotted variable is the ratio of observed murders in each category to the number that would be expected if murders occurred in proportion to the number of relatives (paternal relatedness $r = 0.063$) or non-relatives in the population. Source: 48 murders recorded in the *Orkneyinga saga* (data from Dunbar et al. 1995).

Irrespective of whether they are works of fact or fiction, all the findings we have just discussed illustrate another important point, namely the complexity of the decisions that individuals make. In all these examples, individuals do not weigh up the influence of a single factor like kinship and then act accordingly; instead, they appear to evaluate how kinship interacts with the potential costs and benefits of particular courses of action before making their move. It should also be clear how circumstances can constrain the range of options open to individuals: the lack of competitive edge forcing Viking half-brothers into alliances with each other, or infertility leading to the adoption of distant kin in Oceania. As we repeatedly emphasize, this is exactly what one should expect to find if individuals are attempting to make the most of their circumstances despite the odds stacked against them. As dire as the current situation may be, there is always a chance that things will improve for the next generation. Consequently, we need to be very careful when weighing up the various strategies that people employ, since it is all too easy to dismiss behaviour as maladaptive if we are not fully cognisant of all the relevant factors coming into play. Another point worth remembering is that possessing the ability to make the best of a bad job is just as valuable a trait in evolutionary terms as any that help maximise fitness when all is going to plan.

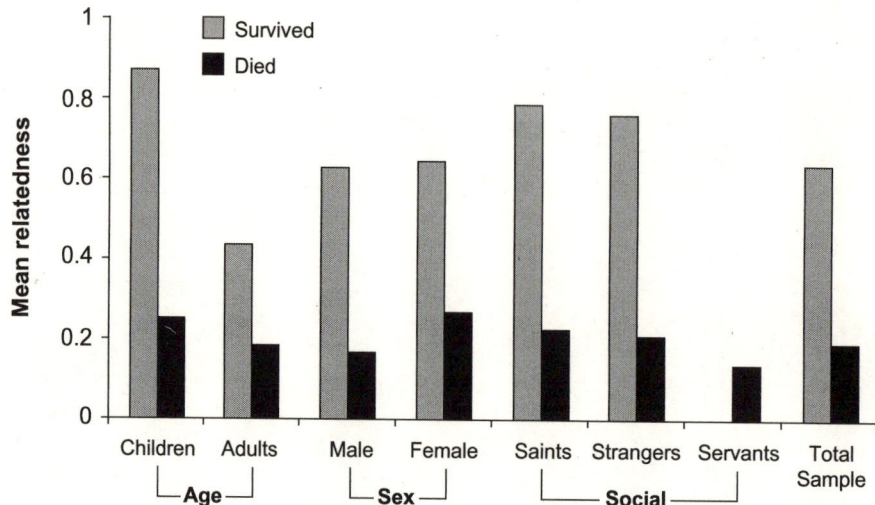

Figure 3.11 During the first winter (1620–21) endured by the original *Mayflower* colonists at Plymouth (Virginia), the risk of dying was influenced by the number of relatives an individual had in the colony. On average, those who survived were significantly more related to the other members of the colony (indexed by the sum of the coefficients of relatedness, *r*, to all other members of the colony) than those who died. Saints were accredited members of the religious community that had organised the expedition; strangers were other individuals who had joined purely for the passage to America; servants accompanied their masters and normally did so without relatives. Based on a total sample of 50 survivors and 53 non-survivors. Source: McCullough and York Barton (1990).

KINSHIP AND HEALTH

Kinship has also been found to influence survival probabilities during crisis periods. McCullough and York Barton (1990) showed that, during their first winter, the *Mayflower* pioneers in the Plymouth colony suffered from severe malnutrition, disease and lack of preparedness. As a direct consequence, 51 per cent of the 103 pioneers died. Except for hired hands and servants (who were in any case few in number and mostly unaccompanied), survivors had a significantly higher mean relatedness to both members of the community as a whole and to other survivors than did those who did not survive (**Figure 3.11**). More interestingly, perhaps, children were found to survive in higher proportion than adults (74.2 per cent vs 37.5 per cent); indeed, if one parent survived, all the children of that family also survived.

Similar patterns have been reported from an analysis of the legendary Donner Party disaster (Grayson 1993). The Donner party set off from Springfield, Illinois, in April 1846, travelling west to a new life in California. The party consisted of 87 people travelling in 20 wagons. The going was tough and rather slow, but largely uneventful – until the party reached the Sierra Nevada mountains. Because of delays along the way, the party hit the mountains in late October just as the winter snows were setting in. As a result, they got stuck, unable to go on and unable to turn back, and were forced to set up camp in order to see the winter through. By the end of winter in late April 1847, 40 of the 87 had died. As with the Plymouth colony, individuals who were members of relatively large kin networks fared better than those who were not. Surviving males travelled with a mean of 8.4 family members, whereas males who didn't survive were in families that averaged 5.7 individuals. Indeed, the individuals who suffered the highest mortality rates were the fit young males travelling alone – the very group one might least have expected to die. Only three of the 15 single men in the party survived. The females who survived were members of families averaging around ten individuals, whereas the one woman who died was travelling with only four other people, suggesting that women survived rather better than males when kin were present. Grayson (1993) suggests that the influence of kin may not have been sufficient to override completely the energy costs associated with the strenuous tasks that male members of the party were expected to perform. The presence of kin also prolonged the lives of those who did eventually die. Males who died early on during the winter were travelling with around five people, whereas those who died later were members of larger families, typically around ten individuals. The presence of a kin network thus provided a high level of life-enhancing support.

Lest this seem to be something peculiar to the historical past, we can point to similar effects even among modern industrialised cultures. That solicitousness for kin may be the key issue here is suggested by two surveys conducted in the city of Haifa during and soon after the 1991 Gulf War: these revealed that people were more solicitous of kin than non-kin in the aftermath of Iraqi Scud missile attacks on the city (Shavit et al. 1994). For example, they were more likely to provide wartime shelter to kin than to friends. During and after missile attacks, they preferentially made phone calls to check on kin and to find out if assistance was needed. In contrast, less costly forms of aid (such as advice on how to seal a room against gas) were more likely to be shared between friends than kin. This is a particularly important study because the researchers were unaware of the theory of kin selection, and their findings were hence not biased by theoretical expectations.

A number of studies of contemporary populations have also shown that there is a positive effect of kin network size on morbidity (especially among children) and longevity (especially for men): children who belong to larger kin groups suffered fewer illnesses than children from smaller families in both industrial Newcastle (England) (Spence 1954) and rural Dominica (Flinn and England 1995; see also Berkman 1984, Werner 1989). More generally, it seems that a large, intense social network significantly enhances health and reduces mortality in adults as well as children (House et al. 1988). Indeed, there is even evidence to suggest that a strong support network can help boost the immune system (Kaplan and Toshima 1990). Strassman and Dunbar (1999) suggest that the prevalence of depression as one of the world's major health problems may be a consequence of the breakdown of kin support networks and the attendant loss of psychological and material security.

*

In this chapter, we have tried to make two important points. First, biological kinship is an important organising force in human relationships: people *are* more generous towards individuals who are genetically closely related to them than they are to those who are less closely related. Second, this tendency to discriminate in favour of kin is not inviolable: we do not behave altruistically towards kin willy-nilly. Rather, kinship is one factor in a complex web of considerations that are weighed up when an individual decides how to behave. The indirect fitness benefits to be gained through collateral kin are set against the benefits to be gained in direct fitness through behaving selfishly. Circumstances dictate the costs and benefits of each strategy, and each occasion has to be evaluated on its own merits. Only when all other confounding factors are held constant (by accident or statistical design) can we expect to obtain clear-cut evidence to support the prediction that individuals will discriminate in favour of biological relatives.

This last point is particularly important, because it reminds us that biological phenomena are complex multivariate systems. Considerable analytical sophistication is required if we are to be able to test evolutionary hypotheses sensibly. Indeed, the history of human sociobiology (and its opponents' attempts to discredit it) is littered with the wreckage of naive attempts to prove or disprove evolutionary hypotheses (see for example Dunbar 1991, Segersträle 2000). In the next chapter, we go on to consider patterns of cooperation and reciprocity between non-kin, where these considerations become even more pertinent.

Chapter summary

■ Many aspects of human behaviour conform to the predictions of Hamilton's Rule, with individuals favouring kin over non-kin and close kin over more distant relatives. This is true for both modern and traditional societies, and for fictional and historical documents.

■ The decisions individuals make are also based on calculations of the potential recipient's reproductive value relative to that of the resource donor. Younger individuals tend to be favoured over older individuals as the recipients of help because these individuals are more likely to make a substantial contribution to the helping individual's inclusive fitness.

■ Apparently maladaptive acts such as adoption and infanticide can also be explained by the theory of kin selection. In many cases, adoption involves taking on distantly related kin, rather than unrelated children, although this does not hold true for modern societies. Infanticide, by contrast, is most commonly perpetrated on infants that are unrelated to the infanticidal individual.

■ The support of a kinship network can have a beneficial effect on health. Individuals are more likely to survive crisis periods if they have kin nearby and the presence of kin appears to reduce susceptibility to disease.

Further reading

Buss, D. (1999). *Evolutionary Psychology*. London: Allyn and Bacon.

Daly, M. and Wilson, M. (1988b). *Homicide*. New York: Aldine de Gruyter.

Hughes, A. (1988). *Evolution and Human Kinship*. Oxford: Oxford University Press.

Ridley, M. (1997). *The Origins of Virtue*. Harmondsworth: Penguin.

Trivers, R. (1985). *Social Evolution*. Menlo Park, CA: Benjamin Cummings.

Wright, R. (1994). *The Moral Animal: Why We Are the Way We Are*. New York: Little Brown.

Reciprocity and sharing

Kinship can clearly explain a large proportion of cooperative behaviours between individuals, but it is nevertheless the case that people also often aid individuals to whom they are not related. In this chapter, we use examples from both traditional hunter-gatherer societies and modern industrialised societies to explore the underlying reasons for the high levels of cooperation which characterise all human societies, and investigate the possibility that humans may be specifically selected for sociality. This latter issue is of great interest at present, since it is possible that the process by which this happened was one of 'cultural selection' rather than genetic selection.

COOPERATION IN HUMANS: A DIFFERENCE IN DEGREE OR KIND?

As we saw in the previous chapter, deciding whether or not to aid one's relatives can be a much more complicated business than we might suppose. In our everyday experience, we tend to leap in and give help to those who need it without a second thought, and it seems difficult to reconcile this with the hard-nosed economic approach that an evolutionary analysis entails. Nevertheless, the data presented in Chapter 3 suggested both that quite subtle cost–benefit trade-offs are made even in those cases that involve kin and that people do tend to behave in a manner consistent with evolutionary theory. As one might expect, the level of decision-making complexity, increases substantially when dealing with cooperation among unrelated individuals, and so do the number of puzzles associated with such behaviour. As we pointed out in Chapter 1, the risk of cheating is ever present when there is a time lag between performing a beneficent act and being repaid. The incidence of cooperation should therefore occur only under certain circumstances, and should reflect an assessment of the particular costs and benefits involved. Once again, however, we are faced with the fact that, in everyday life, we often aid total strangers without hesitation and with no thought of repayment. In other words, we seem to behave in a truly altruistic fashion. Why should this be?

Since, as we showed in Chapter 1, true altruism is extremely unlikely to evolve, it is particularly important to try and understand why it is that humans should apparently display such unusual behaviour. The development of theoretical models other than the iterated Prisoner's Dilemma has provided alternative ways of studying and testing problems of apparent altruism (see **Box 2.5**). Often, it is just the case that our underlying assumptions are wrong, and that one of these different (but equally evolutionary) models is in operation. The food sharing shown by many hunter-gatherer populations is one such example where the adoption of a different analytical paradigm has shed light on some otherwise puzzling aspects of behaviour (see below). What is also apparent is that part of the confusion arises because cooperative behaviour is often placed in a category all of its own, when in reality it may form an integral part of such phenomena as mate choice and parental effort where the strict rules of reciprocity need not necessarily apply. Finally, it is often the case that examples of everyday altruism represent only the very weakest form possible, where the costs of an act are actually so low that they do not impact on the altruist's fitness. Holding open a door for someone and allowing them to go before you is an obvious everyday example. In most cases, one would be intending to go through the door anyway, and the cost of allowing another through first is trivial.

Alternatively, it might be the case that there is indeed something very different about humans with regard to this particular facet of behaviour. Various evolutionary scenarios in which true altruism would be favoured by natural selection have been postulated. However, many other authors have used the notion of inherent altruism as a means of rejecting evolutionary explanations altogether. This latter approach seems to be ultimately self-defeating, since it fails to provide an alternative explanation: if evolution is not responsible for shaping our behaviour, then what is? If we accept that, for most of our evolutionary history, humans have been subject to the same processes of natural and sexual selection as other animals, then it needs to be explained how and why altruism should be so prominent a feature of human nature. The influence of cultural norms has sometimes been suggested as an answer to this problem, but again this is not satisfactory since culture and biology cannot be looked at as independent alternative influences: it is the false dichotomy of the nature–nurture debate writ large.

One particularly promising approach to this issue is to regard human sociality as the product of gene–culture co-evolution, and propose that a form of cultural group selection led to the evolution of the psychological traits that promote cooperative behaviour between group members, regardless of relatedness (Chapter 13). This suggests that there was selection for specific traits that enabled humans to function well

BOX 4.1

Fairness

The concept of 'fairness' lies at the heart of many human social interactions, and has been the subject of considerable interest both from psychologists and economists – see Wagstaff (in press) and Binmore (1994), respectively.

In economics, fairness has often been studied using the *Ultimatum Game* in which a subject is given a sum of money and told they can keep it providing someone else agrees to split it with them. The subject has to make a one-off offer (anything between 0 per cent and 100 per cent of the original sum) and the individual they offer it to must decide to accept or reject the offer; if the response is *reject*, then the money is returned to the experimenter and neither of them gets anything. The subject's task is to decide what offer would maximise his/her gain while at the same time minimising the chances that the person they offer it to turns it down. One might expect individuals to make offers well below 50 per cent (after all, what has the recipient of the offer got to lose by accepting a less than equitable share?). In fact, in hundreds of experiments over several decades, economists have discovered that subjects routinely make offers that are between 45–50 per cent of the total sum.

This surprising outcome has been interpreted as evidence for an innate sense of fairness. However, a recent cross-cultural study carried out on 15 traditional societies all around the world (including the Machiguenga of Bolivia, the Hadza of Tanzania and Zimbabwean farmers) has suggested that equitable (or 'fair') offers of 50 per cent may be characteristic of western societies but not necessarily of traditional societies. In this sample, mean offers within a given society ranged from 15–58 per cent, and were typically much lower than those offered by conventional subjects in developed countries (Douglas 2001).

Neither social complexity nor group size explained the pattern of results. However, a combination of two factors – the importance of cooperation in the way the society makes its living and the level of market integration (the importance of trading in everyday life) – explained 68 per cent of the variance in the results. One interpretation is that, in cases where individuals are exposed to many anonymous individuals in trading relationships (as we are in large-scale industrial societies), simple rules of thumb that emphasize fairness and equity are important mechanisms that allow us to function effectively in a context where we are otherwise always liable to exploitation by unscrupulous individuals whom we are never likely to see again (see Chapter 9). In small-scale societies, in contrast, individuals who live together on a permanent basis are better placed to act in a more equitable (or reciprocal) fashion over a longer time period. What is your good fortune now can be offset by my good fortune another day.

within groups, both to their own individual benefit and, due to the very nature of the adaptations, to the benefit of the group as a whole. The issue of what these adaptations might be, and the impact of cultural evolution on human behaviour, are considered in more detail in Chapters 10, 11 and 13. For now, we will consider patterns of cooperation within human societies and see exactly what an evolutionary analysis has to offer by way of explanation.

RECIPROCITY AND INFORMATION EXCHANGE

Lobster-fishing is a highly competitive business. Because lobsters congregate in large numbers rather unpredictably in time and space, there are only so many lobsters to go around and one man's gain is necessarily another man's loss. Consequently, information on areas of good fishing is extremely valuable. Under such circumstances,

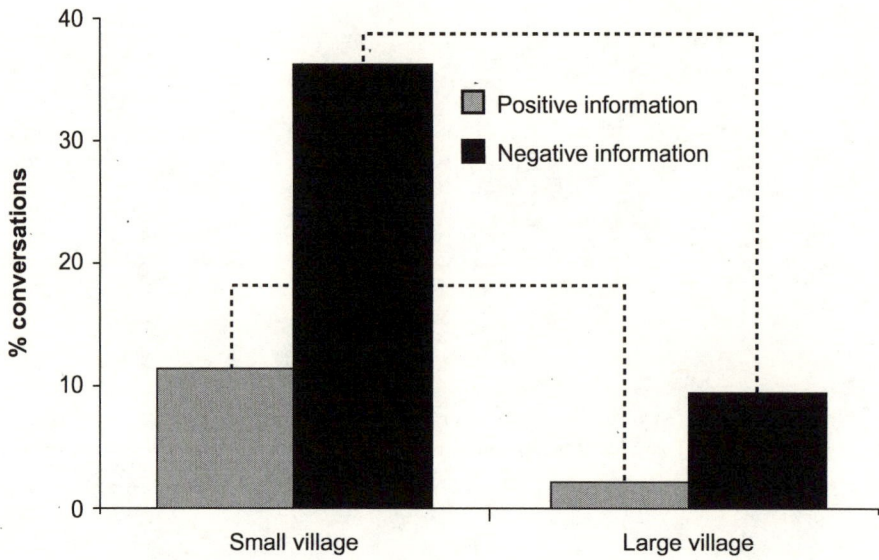

Figure 4.1 Maine lobster fishermen from a small community were significantly more likely to tell each other about the presence (positive information) or absence (negative information) of lobster concentrations than were fishermen from a large community where individuals were less likely to be related to each other or have mutual social obligations. Based on monitored radio conversations between fishing boats. Source: Palmer (1991a).

economic models predict that secrecy regarding the location of catches should be the order of the day. However, Palmer (1991a) found several exceptions to this general tendency. Comparing the frequency of information exchange between a large fishing port and a small fishing village, he found that positive information on catches was more likely to occur in the small village (**Figure 4.1**), and that such information was more frequently communicated by radio, a medium in which there is no control over how many people gain access to the information.

Palmer (1991a) suggested that these differences stemmed from variation in both the degree of relatedness between individuals and in the networks of reciprocal arrangements that existed in the two harbours. The small harbour consisted of a number of families who had fished the area for many years; relatedness tended to be high, and tight networks of relationships existed between unrelated individuals as a consequence of attending the same schools, churches and social clubs. In contrast, the large port contained a small core of fulltime fishermen and a large number of part-timers who fished only during the summer. In addition, the large harbour's fishing territory overlapped with that of a number of other harbours. Thus, only in the small harbour were the conditions appropriate for a system of information-exchange based on either kinship or reciprocity. In the larger harbour, the economic model of secrecy prevailed.

The most interesting finding, however, was the complexity of the reciprocal relationships that existed between fishermen and their families in the smaller harbour. Theoretical arguments advanced by Trivers (1971) and Alexander (1979, 1987) have suggested that altruism need not be reciprocated directly by the person originally assisted, but can be returned indirectly from other individuals or even from society at large. By advertising oneself as an altruist, individuals will be more inclined to act favourably toward you, even if they have not directly benefited from your altruistic

BOX 4.2

Competitive altruism

Alexander (1979) suggested that behaviours such as blood donation could be explained through the desire to be viewed as an altruist by the population at large. Such behaviour could lead to large return benefits for the individuals who perform such acts if other individuals are more willing to aid a known altruist. Performing costly acts in order to gain a good reputation, rather than building up a network of reciprocal obligations, may often be the motivation behind people's behaviour.

Zahavi and Zahavi (1997) and Roberts (1998) have suggested that this kind of 'competitive altruism' may actually be quite common in the animal kingdom, and not something restricted to humans. Zahavi and Zahavi (1997) suggest that Arabian babblers (a small bird species that breeds cooperatively) use altruistic behaviour as a 'handicap' (see p. 42) in a way that can be used as a signal of good genes. In babblers, unrelated 'helpers' as well as parents help to provision young nestlings with food. The helpers compete with each other for the privilege of providing nestlings with the most food. Zahavi and Zahavi (1997) suggests that this might be an attempt by helpers to signal to the breeding female that they are good quality males so that she will select them as mates in the future.

Roberts (1998) makes somewhat similar arguments concerning the occurrence of cooperative behaviour in permanently social groups of animals. He argues that, in such groups, the wider context in which cooperation takes place needs to be con-sidered. Whenever there is competition for social partners, such as when animals need to find grooming partners, individuals may compete to be 'altruistic'. Thus, when individuals cooperate, they are not interacting only with each other, but actively competing against the whole population of potential partners at the same time. Consequently, individuals may forego the benefits of cheating within the narrow context of the cooperating pair in order to maintain an altruistic 'reputation' that has benefits in the wider context.

The theory of **Biological Markets** (Noë and Hammerstein 1995) offers a similar view: here, individuals are envisaged as competing with each other for access to potential allies who offer the 'best value' in the 'market place' that constitutes the social group. This results in good reciprocators preferentially interacting with each other because competition and partner choice enables cheats and exploitative partners to be ostracised. In other words, the 'competitive altruism' of Roberts (1998) and Zahavi and Zahavi (1997) represents a 'market force' that affects an individual's standing (or reputation) in the market place (Noë and Hammerstein 1995, Barrett et al. 1999). As Roberts (1998) points out 'reciprocity is just one way of getting a return on investment in altruism . . . Other mechanisms deserve attention.' These intriguing ideas have relevance for human food-sharing behaviours where male foraging decisions may have more to do with mating benefits than nutritional ones.

acts (see also Leimar and Hammerstein 2001). An altruistic reputation means that individuals are more likely to trust you not to defect, and provide a necessary precursor to embarking on cooperative endeavours. Palmer's (1991a) work suggests that the situation in the small harbour may be an example of the latter. For example, he observed 'one lobsterman's brother helping another lobsterman's neighbour's son to find a job, while another lobsterman repair[ed] a different lobsterman's cousin's car'. He also stated that reciprocal relations may extend across generations, and be inherited by the lobstermen's descendants. Such complex webs may help to explain the relatively free flow of information in this small fishing village. In addition, there may be some long-term benefits in being viewed as an altruist by the population in general, which would further explain why information is so widely exchanged via the radio (see **Box 4.1**).

LABOUR EXCHANGE AND BET HEDGING

Reciprocal altruism has been less widely recorded than kin-directed behaviours among humans (or any other animal for that matter) precisely because of the difficulty of identifying the time frame over which reciprocity operates and whether favours are returned in the same (or different) currency by the original beneficiary or by another individual. There are a few exceptions, like the lobstermen example given above. Another example is provided by Hames's (1979) study of the exchange of labour among the Ye'kwana, in Venezuela. The Ye'kwana frequently help with the planting and harvesting in each others' gardens. Hames (1979) found that kin selection could explain a large proportion of this type of labour exchange: the more closely related a pair of households, the more social interactions they had with each other (**Figure 4.2**) and the more likely they were to help each other in their gardens. However, among households that had no close kin to rely on, these kinds of arrangements were more strictly reciprocal; individuals were more likely to repay in kind the help they had received during the same growing season in which they had themselves received help. Panter-Brick (1989) found similar arrangements among the Tamang hill farmers of Nepal. In addition, the use of flexible, reciprocal relationships between women, between men and women, and within and between households all operating in a tightly knit community, allowed lactating females to participate in the work force in a manner compatible with their childcare commitments.

Among the Ye'kwana, labour exchange is thought to be a way in which individual households deal with the risk of crop failure. Should their own crops fail, individuals can claim food from the gardens in which they assisted with planting. Households that had to rely on reciprocal relationships not only kept a very accurate record of exactly what they had helped to plant and where in case they needed to call favours in, but they also tended to plant bigger gardens of their own. This was thought to be another hedge against crop failure, possibly because reciprocal arrangements tend to be riskier than kin-based ones. Not only are kin less likely to cheat on you in the first place since such relationships tend to be more stable and secure, but even if your relatives do cheat, you may gain sufficient benefit in terms of inclusive fitness to offset this loss.

FOOD SHARING AMONG HUNTER-GATHERERS

Food sharing is even more widespread among hunter-gatherers than among horticulturists. Moreover, it is often highly politicised in terms of both the network of obligations and the protocols involved in offering and receiving food. One of the most widely accepted explanations for food sharing is linked to the risk and uncertainty involved in harvesting natural food crops compared to domestic crops. The foods most likely to be shared are those that are hunted rather than gathered (that is, sources of animal protein). Generally speaking, these resources come in large 'packets' that are acquired sporadically, asynchronously and unpredictably by individuals. The uncertainty associated with procuring such resources may favour sharing in a reciprocal manner. Individuals who fail today can share the food of their neighbours and then return the favour tomorrow (or whenever they get lucky). Individuals can therefore reduce the risks of coming home empty-handed and ensure a more steady supply of these particular foods to their families. Sharing is often a much more efficient alternative to other possible strategies like storing food, which tend to be impractical in

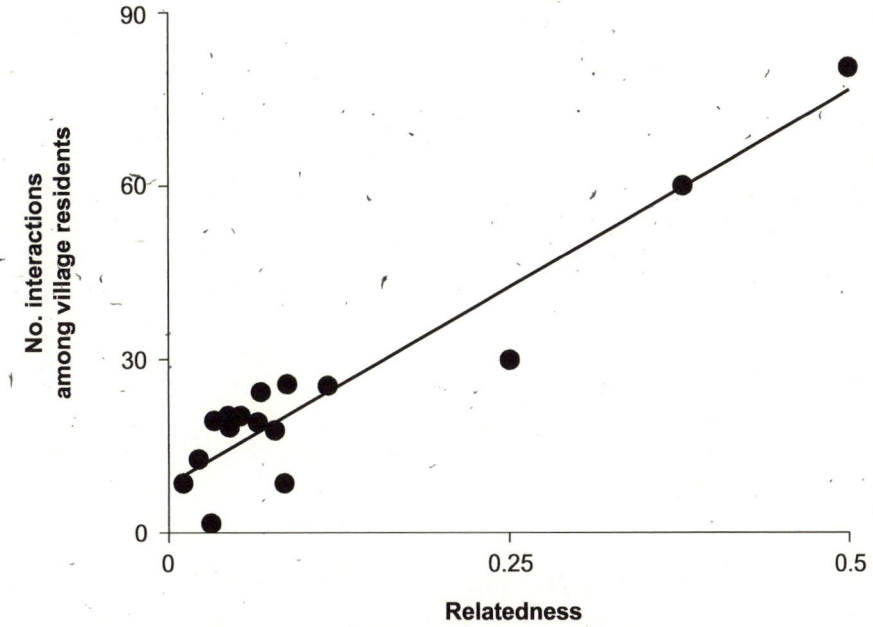

Figure 4.2 In a Ye'kwana village, individuals who were more closely related to each other genetically were more likely to stop and talk than individuals who were less closely related. Genetic relatedness is here measured as the coefficient of relatedness, r (the probability that two individuals share the same allele by descent from a common ancestor). Redrawn with permission from Hames (1979).

most hunter-gatherer societies due both to lack of refrigeration and to the prohibitive time and energy costs involved in preservation methods like smoking. Stored foods also need to be guarded against theft, which introduces an opportunity cost of having to forego other activities as well as substantially reducing mobility – an important consideration for hunter-gatherers.

RISK REDUCTION AND RECIPROCITY

Hames (1990) investigated sharing behaviour among the Yanomamö in relation to the risk reduction hypothesis. The Yanomamö are horticulturists, but men also engage in hunting on a regular basis. The resources obtained by the Yanomamö can be ranked in order of their riskiness to obtain: horticulture is the most reliable and least variable of all the foraging activities (as would be expected given that this is the form of foraging over which individuals have most control), then comes gathering of natural foods in the forest, followed by fishing and, finally, hunting – the most risky of all.

Hames (1990) found a significant correlation between the riskiness of the resource type and the frequency and extent of exchange between households (**Figures 4.3 and 4.4**). Garden foods were shared least, while game was shared widely among a large number of households. Sharing of game therefore reduced the variance in meat consumption for all. However, Hames's (1990) analysis highlights one of the potential problems associated with sharing: successful hunters tended to share more frequently than poor hunters, and therefore experienced a reduction in the average amount of

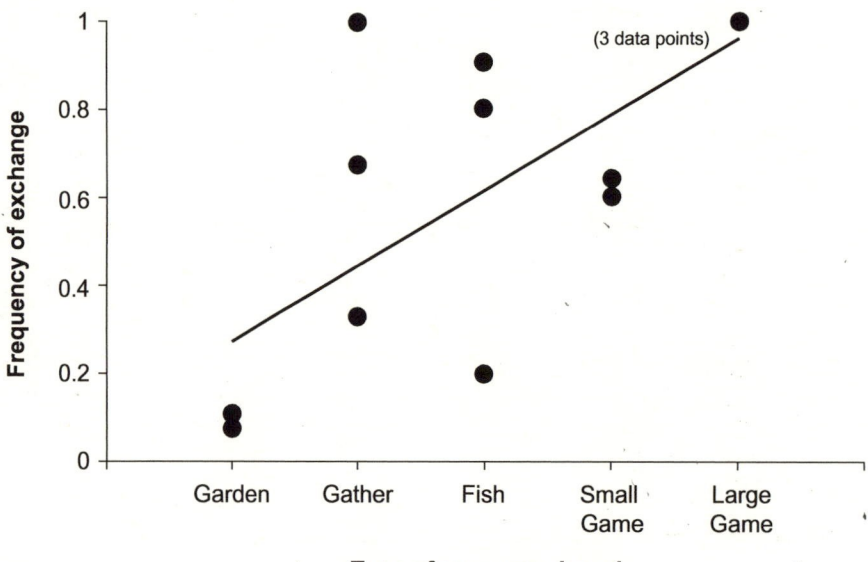

Figure 4.3
Relationship between proportion of food items exchanged and type of food exchanged among the Yanomamö. The products of hunting are more likely to be shared than those that are gathered. Redrawn with permission from Hames (1990).

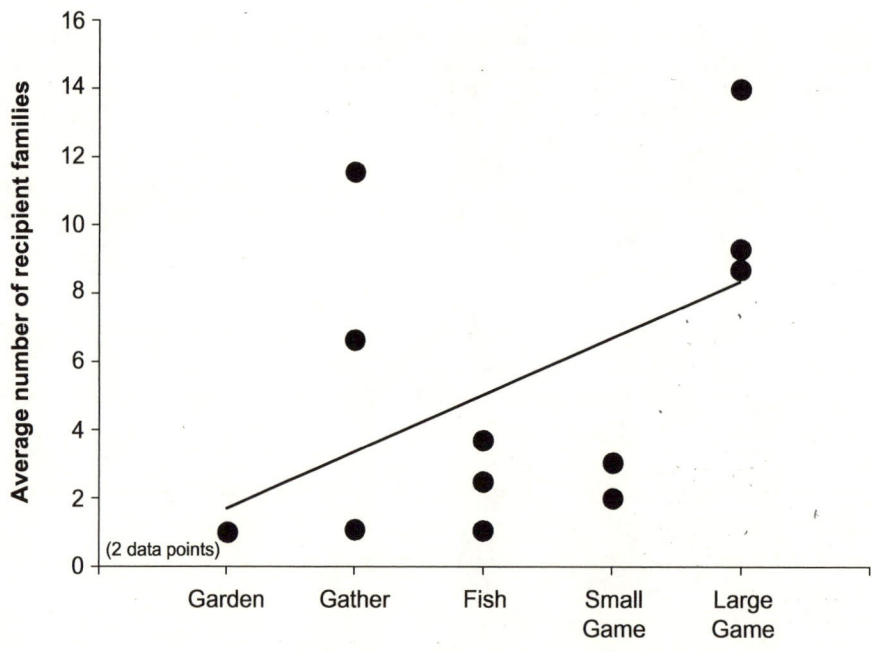

Figure 4.4
Relationship between the average number of families who were recipients of food exchange and the type of food exchanged among the Yanomamö. The products of hunting are exchanged with significantly more families than those that are gathered. Redrawn with permission from Hames (1990).

meat they consumed. Poor hunters, by contrast, managed not only to reduce the variance in their intake (since they ate meat more regularly than if they had relied solely on their own catches) but also achieved a higher level of intake overall. As a consequence, good hunters tended to 'under-produce' and spend less time hunting than poor hunters, which Hames (1990) argued was a strategy for avoiding exploitation. Even so, good hunters still managed to out-produce poor hunters by a considerable margin. Hames (1990) suggested that good hunters might be induced to share because they gained other benefits unrelated to foraging (an idea which has received considerable attention and which we will deal with in more detail below) or that the nature of the resource itself meant that sharing was actually unavoidable.

TOLERATED THEFT

The suggestion that sharing may, in some circumstances, be unavoidable has been formalised as a model of 'Tolerated Theft' by Blurton-Jones (1984). Although food sharing can reduce day-to-day fluctuations in the average amount of resource acquired by individuals within a group, the costs and benefits from any one individual's point of view may differ widely from the group as a whole. A good hunter who produces regularly may actually do worse by sharing his catch, and could often do a lot better by acquiring a more reliable non-shareable resource with which to feed his family. In addition, a very noticeable feature of hunter-gatherer sharing is its tendency to show a pattern of 'sustained one-way flow' (that is, the turn-taking that should characterise true reciprocity is not observed, and good hunters may not actually be repaid for their efforts).

Blurton-Jones (1984) therefore suggested that sharing could arise because the costs of attempting to defend a resource against hungry group mates outweigh the benefits of trying to keep it all to oneself. Blurton-Jones's reasoning is essentially as follows: assuming that individuals acquire food in packets, then as packet size increases the value of the resource to the individual can be represented by a curve of diminishing returns (see **Figure 4.5, Box 4.2**). The first few portions of a resource are worth a great deal to the forager because he is hungry and his need is great (the same holds for the mates and offspring he provisions with food). However, once the possessor of the resource has consumed all that he can, subsequent portions are worth relatively much less to him (their **marginal value** is low). For less successful hunters eagerly eyeing up the carcass, the reverse is true: the remaining portions are still worth a great deal since they haven't yet eaten (in fact, they are worth as much as the first few portions were to the forager who acquired the resource). Consequently, they have a very strong incentive to try and acquire some of the food for themselves, while the possessor of the carcass has very little motivation to defend it against them. If individuals do not differ in their resource-holding potential and if they will all suffer equivalent fitness costs if there is a dispute over the carcass, then the optimal strategy is for the original possessor of the resource to allow other hungry individuals to acquire portions of it until everyone's needs have equalised. Sharing is therefore passive, not active: foragers merely permit individuals to take from them without protest, they 'tolerate' the 'theft' of their resource.

This can lead to a pattern of apparent reciprocity if individuals acquire resource packets asynchronously: those who are tolerated one day will themselves have to tolerate others on the next. However, unlike true reciprocity (where individuals incur a cost now in order to gain a larger benefit in the future, individuals who tolerate the theft of their carcass do not incur a cost by doing so; they share with others because it actually reduces their costs. Moreover, Winterhalder (1996) has shown that tolerated theft can actually increase the total value of a resource packet over and above its initial nutritional value, and that this can increase the likelihood that a resource will be shared. Using a graphical model with the simplest possible case of a two foragers and a single resource, Winterhalder (1996) showed that, because initial portions are worth so much more than subsequent portions, an individual who consumes an initial portion and then gives away a subsequent one (which is, of course, the

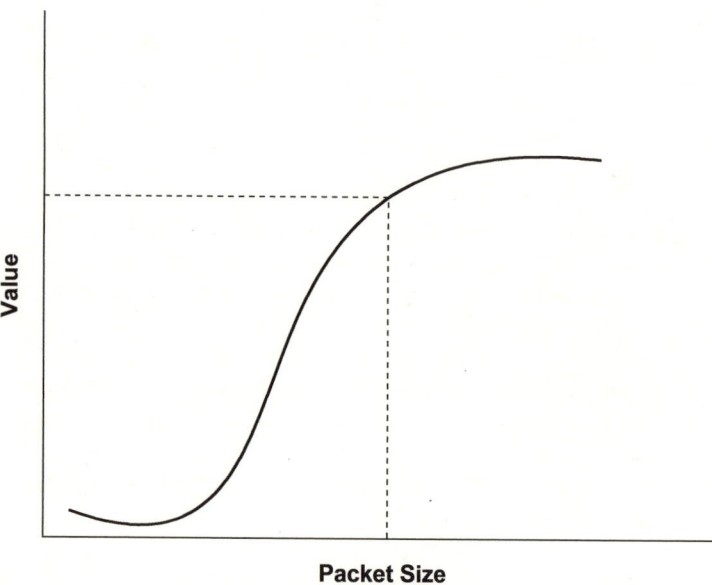

Figure 4.5 The classic S-shaped 'diminishing returns' curve created when a resource has a minimum level below which it offers very limited benefit and an upper limit (or satiation point) beyond which the marginal value of acquiring it rapidly becomes progressively less. The dotted line represents the point beyond which the costs of acquiring a larger packet of the resource no longer produce a commensurate benefit.

taker's first high-value portion) increases the total value of the resource because it now represents the summed total of two high-value portions, rather than one high-value initial portion and a subsequent portion of much lower value. Since the act of sharing in itself increases the relative value of the resource, it becomes even more uneconomic to defend a resource and attempt to avoid sharing. Echoing Blurton-Jones (1984), Winterhalder (1996) suggests that this value enhancement can create the appropriate conditions for the evolution of true reciprocal altruism.

PRODUCERS AND SCROUNGERS

Of course, even with this entirely selfish system the potential for cheating is always present, since some individuals may produce nothing at all and instead rely on getting a share of someone else's resource. While producers (foragers) will occasionally miss the opportunity to obtain a resource from someone else because they are away from camp, a dedicated scrounger will always be around to claim a share. Scroungers will therefore do better on average than producers (and for much less effort), which means that, although in theory everyone does best if everyone produces, in reality a mix of scroungers and producers will be found: it will always pay at least a proportion of individuals to cheat all the time, or for all individuals to cheat at least some of the time (Winterhalder 1996). Scrounging is more likely to occur in large groups than small groups, since the latter can only sustain a very low frequency of cheating before it has a damaging effect on all group members and, obviously, cheats will be more conspicuous in small groups. We explore the issue of group size and the way

BOX 4.3

The marginal value theorem and tolerated theft

Winterhalder's (1996) analyses of tolerated theft were based on a theory taken from the animal optimal foraging literature known as the **marginal value theorem** (Charnov 1976). This was originally formulated to explain how and when animals should leave a particular patch in which they are foraging and move on to a new one. When an animal first encounters a patch, the gains to be made from exploiting it are high. There is lots of food present and it is easy to find. As time passes, however, and the food is depleted, it takes longer and longer to find a food item so that more time is spent searching and less is spent eating.

The rate of energy gain continually drops and can be represented by a graph of diminishing returns (**Figure 4.5**). As time passes, the 'marginal value' of the patch decreases and the slope of the graph levels off. In other words, the amount of food energy obtained represents a smaller and smaller proportion of the amount of energy needed to find it. However, leaving the patch altogether means that time must be spent searching for and travelling to another patch, which might prove even more costly than staying put. The marginal value theorem was developed to predict exactly when an animal should leave a patch in order to ensure that its rate of intake always remained at optimal levels.

Winterhalder (1996) was able to adapt this model for hunter-gatherers because the situation a forager faces is analogous to that of a bird feeding in a patch. When a hunter first obtains a resource, his gains from it are high, and it pays him to keep others away. However, these benefits diminish gradually as the hunter eats his fill. There comes a point at which it pays him to give the resource up to others since the benefit of doing so is greater than the benefit of keeping it. Winterhalder's (1996) analyses are therefore designed to answer the questions: at what point should a hunter tolerate the theft of his resource and to what degree does it pay him to do so?

He also extended his analyses to investigate the factors influencing the likelihood that a hunter would increase the amount of sharing through tolerated theft to see if these matched the observations of sharing seen in hunter-gatherer societies. This is a particularly fine example of how models developed to study animal foraging decisions can be adapted to study those of humans without oversimplifying the situation or making unrealistic assumptions about human behaviour. Winterhalder's (1996) analyses provide an explanation of a human behaviour that is quite complex in economic terms, thus allowing us to identify some crucial factors that determine sharing as well as the impact that cultural factors (for example, conferring high status on good hunters) can have.

that this can encourage 'freeriders' to exploit the better nature of their fellows in more detail in Chapter 9.

Blurton-Jones (1984) suggests that the selfish interests of scroungers can have a marked effect on social strategies. Scroungers have to encourage at least some individuals to produce in order that they might reap the benefits, and they can achieve this by conferring high status on good producers. In this view, producers do not command high status by virtue of their enhanced skills and dominance over others, but rather are granted it by others in an ultimately self-serving manner. Obviously, for such a strategy to persist within groups, high status must have some tangible benefits for producers: a producer who is happy merely to have his ego, rather than his fitness, boosted will not leave large numbers of descendants.

Among the Ache of Paraguay, good hunters are rewarded with quite considerable benefits: both they and their offspring receive preferential treatment within the group, and their legitimate offspring show higher survivorship than those of poor hunters. In addition, good hunters father more illegitimate children, and are more often reported as lovers by the women (Kaplan and Hill 1985b). However, Kaplan and Hill (1985a, b) rejected the tolerated theft model as an explanation for food sharing among the Ache after a detailed examination of sharing patterns. They discovered that good hunters tended to spend more time hunting than poor hunters (that is, they rarely scrounged from others, and therefore overproduced rather than underproduced as Blurton-Jones's model assumed), they did not distribute their kills themselves, and they actually ate less of their own kills than did other individuals. Finally, portions were often saved for those who were absent at the time of distribution. All of these findings seem to go against the predictions of the tolerated theft model, and Kaplan and Hill (1985a) were drawn to the conclusion that something more complicated was occurring.

However, Winterhalder's (1996) more recent analyses have shown that tolerated theft can, in fact, lead to some of the patterns displayed by the Ache. If individuals have access to other resources in addition to the tolerated theft (TT) packets, then sharing of such packets need not be equal between individuals since they are likely to differ in their hunger levels and hence the value they place on the TT resource. For example, if the individual who acquires the TT packet also has the greater amount of non-TT resources, then they will value the TT resource less than individuals who lack both, and will thus be more likely to give the resource packet away. The model suggests that, under most conditions, this can involve as much as half the packet. In addition, as group size grows, the possessor of a resource is forced to give away more of the resource. This is because the more potential takers there are, the more potential high-value first portions there are to be given away, which means that giving away a portion will always be more beneficial to the taker than the costs of defending the portion are to the possessor. The increase in resource value that the act of transfer itself incurs means that it is always more worthwhile to give portions away than to attempt to retain them. In short, individuals may get to consume very little of the resource they acquired.

Winterhalder (1996) also showed that, under conditions where food sharing generated debts of reciprocity in other domains, widespread sharing would confer the greatest benefit on a self-interested producer, even though they may end up consuming relatively little of their own resource. Since the two models yield the same outcome (producers consume less of their own production), it is difficult on this basis alone to distinguish between them. Nevertheless, by considering other aspects of the processes involved, it should be possible to discriminate between the models. Under conditions of tolerated theft, for example, individuals would be expected to attempt to avoid encounters with large numbers of takers, whereas the reciprocity model predicts that producers should actively seek out individuals with whom to share.

Finally, Winterhalder (1996) was able to show that above-average producers might well put up with a high level of unbalanced transfer because they are better off in a group that practises TT than they would be alone. Since giving portions away actually entails a relatively small marginal cost (after all, that's why tolerated theft takes place) it is easily outweighed by the advantage to be gained by being a taker, no matter how rarely this occurs, because being a taker by definition involves gaining very large marginal benefits.

Clearly, determining whether individuals share because the costs of doing so are too high to bear or in order to reduce risk or to gain reciprocal benefits in other domains (or a combination of all three) is a somewhat tricky business. Kaplan and Hill's (1985a) study is one of the most comprehensive attempts to determine the reasons underlying food sharing (see also Kaplan et al. 1990, Kaplan and Hill 1992). Their extensive studies showed that individual's foraging decisions were based on a strategy designed to maximise the rate of energy acquisition and, given this knowledge, they proceeded to look at resource choice and patterns of sharing in more detail. They divided resources into three types: meat, honey and other gathered goods. In all three cases, resource packet size explained a lot of the variance in sharing, and did so more than the standard deviation (SD) of acquisition between families (the standard deviation measures the variance in the quantity of a particular resource that a family obtained and is therefore a measure of risk associated with obtaining those resources).

Kaplan and Hill (1985a) suggested that packet size acted as a more efficient 'rule of thumb' for determining what should be shared, and was better than trying to work out exactly how much each family had caught or gathered over the previous few days or weeks. They also found that sharing food substantially decreased variance in consumption. The mean SD in food acquired across foraging trips by Ache families was 13 243 calories. In other words, the amount of food collected on a trip could vary extremely widely, with some trips producing large 'bonanzas' but others producing very little at all. However, if food sharing was taken into account, this value was reduced to 1945 calories, a reduction of 63 per cent. Furthermore, a simple model investigating the impact of sharing on nutritional status revealed that this was lowest under conditions of no sharing, but that sharing food led to quite large increases in nutritional status. For example, sharing honey increased nutritional status by 20 per cent, sharing meat increased it by 40 per cent, while sharing all foods increased it by 80 per cent (**Figure 4.6**). These findings were supported by their observational data, which revealed that meat was most frequently shared, followed by honey and then all other gathered foods. They also found that, in 52–56 families, sharing substantially increased nutritional status. Food sharing therefore reduced variance in consumption and in so doing increased individual nutritional status.

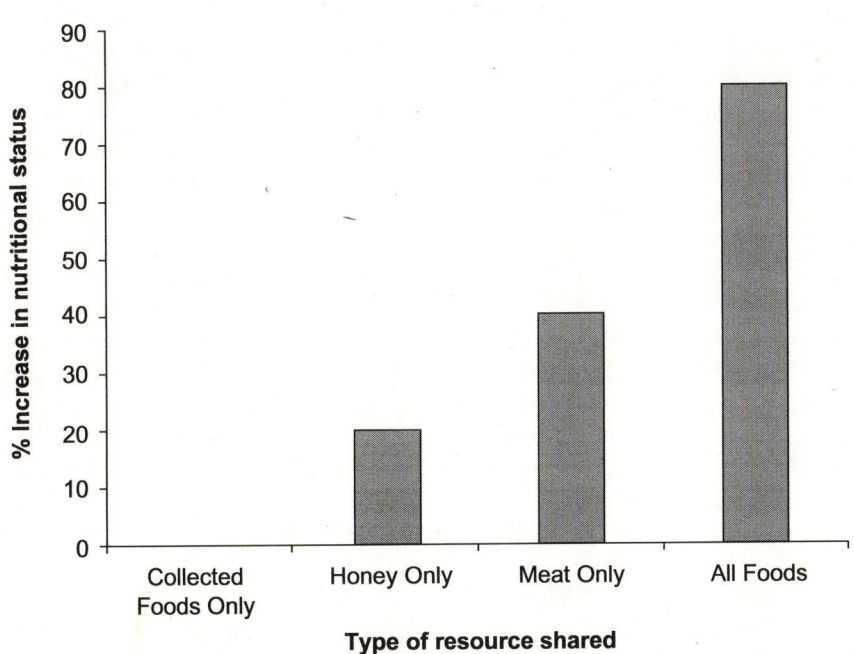

Figure 4.6 Effect of food sharing on nutritional status (measured as the percentage increase in nutritional status relative to that achieved by not sharing) among the Ache for different types of food. Sharing the products of hunting is more beneficial than sharing the products of gathering. Source: Kaplan et al. (1990).

However, some individuals were found to benefit much more than others from food sharing. The overall mean increase in nutritional status through sharing was 80 per cent but the range was very large, varying between −30 per cent to +570 per cent, indicating that some individuals were making enormous gains by sharing, while others were actually losing out. This latter group not only produced more, but were more consistent in their production than others, such that they did worse by sharing their resources than they would have done by consuming them alone. The biggest losers tended to be young single men. Fully 30 per cent of young men on foraging trips produced more calories by themselves than the mean for entire families. Nevertheless, they tended to overproduce consistently and to share more in comparison with other age sex classes.

Men's foraging decisions also differed from those of women's in other ways. Palm starch is a gathered resource that is frequently collected by women, but only rarely by men. In fact, men actually ignore opportunities to collect palm starch in spite of the fact that doing so would increase their return rates by over 1000 calories per hour. Since palm starch is also a very reliable resource, the decision to ignore it not only reduces male return rates, but it actually increases the day-to-day variance in the amount of food produced. Ache men prefer to devote their time to hunting game and collecting honey, both of which show an extreme amount of temporal variability. Hill and Hawkes (1983), for example, estimated that individual men fail to acquire game on 43.5 per cent of all days that they forage. Given that Ache food sharing seems designed to reduce variance, this result is somewhat surprising. Why would men target resources that actually increase the risk of their families going without food?

One explanation could be that meat and honey contain important nutrients (notably protein) unobtainable in gathered foods, and that this increases their value above that calculated on the basis of nutritional value alone (Kaplan et al. 1987, Hill 1988). Also, as Hames (1990) points out, protein cannot be stored in the body, which increases the value of obtaining a steady supply of this particular nutrient. Experimental work has shown that higher protein levels result in enhanced growth rates and body weight (for example, Yoshimura et al. 1982). It is therefore possible that male tendencies to target protein sources in the form of game stem from the increased nutritional returns to be gained. The reason why females target resources like palm starch may be because the demands of childcare limit their ability to hunt, and also because the dangers of hunting pose a larger threat to female reproductive success compared to that of men. Food sharing would then be a means by which individuals ensure that variance in protein intake is reduced to acceptable levels.

THE 'SHOW-OFF' HYPOTHESIS

Hawkes (1990) suggested another reason why men might target risky resources that has nothing to do with the nutritional benefits to be gained. As Hames (1990) alluded to in his study of the Yanomamö and Winterhalder's (1996) model demonstrated, men may be able to use widely shareable resources to obtain benefits in other domains. Kaplan and Hill (1985b) showed that possible reproductive benefits accrued to the hunters with the highest rates of return, and Hawkes (1990, 1991) took this one step further by suggesting that men target risky resources precisely *because* they are risky, not in spite of this fact, and that they do so in order to 'show off' and achieve high status and increased mating opportunities within their foraging bands.

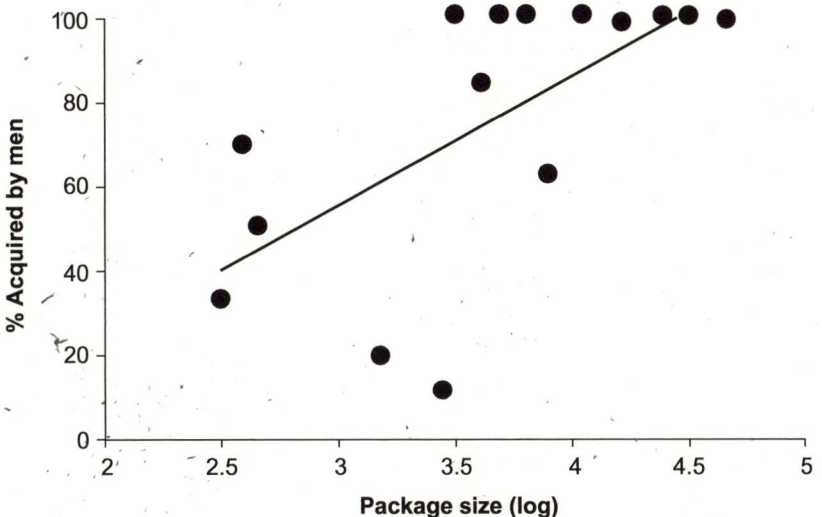

Figure 4.7
Relationship between the percentage of resources acquired by men and the relative size of the resource. Men are more likely to forage for larger resource packages. Redrawn with permission from Hawkes (1991).

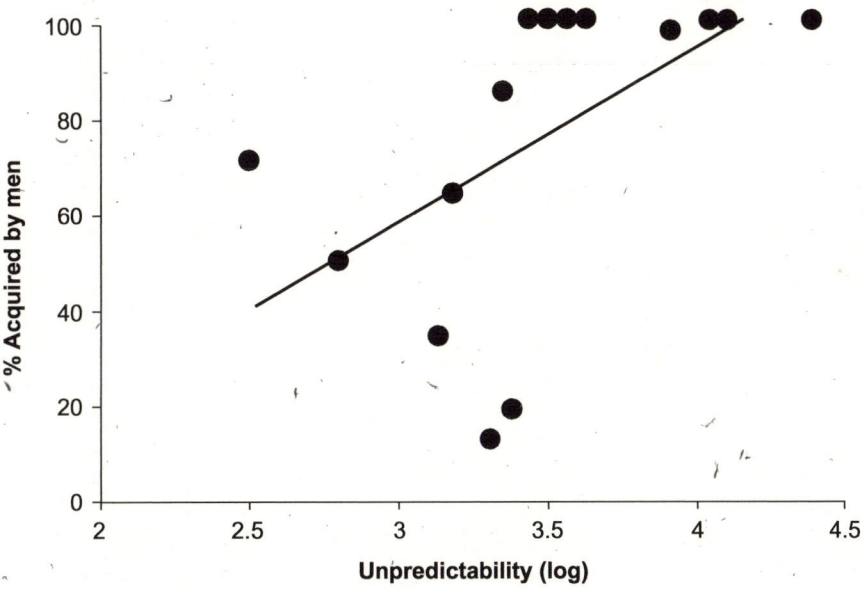

Figure 4.8
Relationship between resources acquired by men and the predictability of the resources targeted (indexed as the log of the variance in quantities acquired). Men are more likely to forage for the more unpredictable resources. Redrawn with permission from Hawkes (1991).

In support of the **'show-off' hypothesis**, Hawkes (1991) presented data showing that men tend to target resources, both hunted and gathered, which come in large packet sizes and are unpredictably acquired (**Figures 4.7 and 4.8**). These are precisely the resources that tend to be most widely shared. Furthermore, Hawkes (1991) was able to show that, when both package size and unpredictability were held constant, the likelihood of a resource being shared outside the nuclear family could be predicted by whether or not it had been acquired by a man. More than 50 per cent of the variance in sharing could be accounted for just from knowing the sex of the acquirer; in contrast, when this variable was held constant, package size explained only 15 per cent of the variance in sharing, and unpredictability of the resource explained nothing at all. All this suggests that, contrary to the common assumption

'SHOW-OFF' HYPOTHESIS

that men forage in order to ensure a regular food supply for their immediate family, men actually forage in order to share (with largely sexual motives in mind).

Hawkes (1991) recognised, however, that this result could have been a consequence of using data from single men in the population who had no family to provision and who therefore biased the sample towards sharing. However, when she repeated her analysis for married men only, the results were the same. Married men were just as likely to show off as single men, and preferentially target resources that would ultimately be shared outside the family. Around 84 per cent of the resources acquired by men with families was consumed by individuals other than the hunter himself and his immediate family (wife and children). When these same resources were acquired by women, only 58 per cent were shared outside of her family.

Men tended to target game resources because their large packet size and sporadic 'bonanza' nature made them particularly suitable for showing off. More interestingly, Hawkes (1991) was also able to show that when men gathered resources, they again tended to show off. Many gathered resources show a curve of diminishing returns over time; the $n + 1^{th}$ unit of resource gathered has a lower value than the nth unit. Hawkes found that men tended to collect larger packet sizes of such resources compared to women, consistent with the view that men reach diminishing returns more slowly since their aim is to share with the whole group, whereas women are only interested in provisioning their immediate family. Men also preferred to gather more 'shareable' resources than women. Hawkes (1991) found a significant positive correlation between the fraction of such foods acquired by men and the fraction of consumptions that occurred outside the household. This last finding is important because it is not confounded by the potential nutritional differences of meat versus carbohydrate products. The differences shown by men and women in this instance relate only to the differences in the shareability of resources, and point to men having very different foraging goals to women.

Hawkes (1991) argues that men 'show off' in order to gain reproductive benefits. In support of this she cites the evidence of Hill and Kaplan (1988) showing that good hunters had both privileged sexual access and better offspring survivorship. However, this evidence is rather circumstantial, and the increased mating opportunities of better hunters could arise as a consequence of another attribute shared by these men. Consequently, it still remains to be demonstrated that showing off directly influences male mating opportunities and status within groups. Hawkes (1991) also makes the very interesting observation that if showing off does occur, then women would wish to live near a show-off and benefit from the extra resources they gain access to thereby, but that they should prefer to be married to a man whose priority is provisioning his family. Most men, on the other hand, should want to be show-offs at least some of the time and gain the extra-marital benefits available. There is thus a conflict of interests between the sexes that is played out in the foraging domain, but which nevertheless reflects the essential conflict of interests between mating versus parental effort.

Bliege-Bird and Bird (1997) also found evidence for showing off among the Merriam Islanders of the Torres Strait. In these islands, turtle meat makes up a considerable portion of the diet, and turtle hunting takes place all year round. However, the risks involved in turtle hunting vary depending on the time of year. During the breeding season, when the turtles are out at sea, hunting is considerably more risky, in terms of both the chance of coming home empty handed and the danger involved during the hunt. At this time of year, only a few young unmarried males hunt, and their catch is widely distributed in the form of feasts. During the nesting season, by contrast, the costs of hunting are low since the turtles come ashore to lay their eggs and are therefore easy both to find and to capture. At this time of year, men of all ages hunt and the majority of meat obtained is confined to sharing within households. As Hawkes's (1990) model predicts, age and marital status influence the level of showing-off exhibited, with young unmarried men willing to take more risks than married men who have provisioning concerns and are less inclined to show off.

Similarly, among the Kubo of New Guinea, men classified as show-offs by Dwyer and Minnegal (1993) obtained more meat per day when resident in their villages than when foraging elsewhere in the area (that is, they show off only when there is likely to be an appropriate audience). Show-offs also tended to travel more frequently between communities and to spend more time away from their spouses, thereby increasing their opportunities for claiming sexual favours in return for providing meat bonanzas.

However, the most critical prediction of the show-off hypothesis – that the reproductive success of show-offs should exceed that of non-show-offs – was not well supported by Dwyer and Minnegal's data. Although the four men identified as show-offs produced more children per ten years of reproductive life than the nine non-show-offs in the sample, the survivorship of non-show-offs' children was twice as high, thereby cancelling out the show-offs apparent advantage (**Figure 4.9**). However, not only is this finding based on a small sample of men, but it includes only those children produced within a recognised union. Since Hawkes' model specifically states that extra-marital matings are what gives show-offs the edge over non-show-offs, Dwyer and Minnegal's (1993) data cannot unequivocally reject the show-off hypothesis. Although Dwyer and Minnegal argue that extra-marital infidelity is probably unlikely in their study population (on the basis of wholly anecdotal evidence, it should be noted), they concede that show-offs may gain sexual favours of a homosexual kind. These favours are obviously not related to reproduction, but they do signal that show-offs may be granted higher status than non-show-offs. Ritual homosexuality of the kind practised in this part of New Guinea is related to the production of masculinity, with younger men performing fellatio on their elders in order to acquire the masculine characteristics of these men. Show-offs accorded sexual favours would therefore be considered to be powerful men. Although their children do not apparently receive special treatment in nutritional terms (in contrast to the Ache), it remains possible that they are also accorded social advantages in later life (at least while their father is alive).

However, we should also bear in the mind the possibility that this behaviour is a cultural phenomenon with no (genetically) adaptive function. As we shall argue in Chapter 13, there is no reason why cultural practices should be adaptive in the genetic sense; indeed, as several authors have shown, it is entirely possible for cultural evolutionary processes to produce behaviour that is entirely maladaptive. Since humans have an evolved propensity to enjoy sex, there is no reason in principle why a culturally instituted behaviour involving the exchange of sexual favours unrelated to procreative function should not evolve; after all, most heterosexual behaviour in western society is now recreational rather than procreational. Using sex as a means of establishing power relations is entirely likely and, given that sexual stimulation is a pleasurable experience, it is likely to be perpetuated because of the immediate proximate rewards it offers.

Dwyer and Minnegal (1993), however, provide an alternative explanation for the show-off phenomenon among Kubo hunters. They argue that showing off is an epiphenomenon of prey preferences and specialisation among hunters. They suggest that the very different modes of hunting required to catch wild pig versus fish result in individuals targeting particular prey types and becoming a specialist in one particular

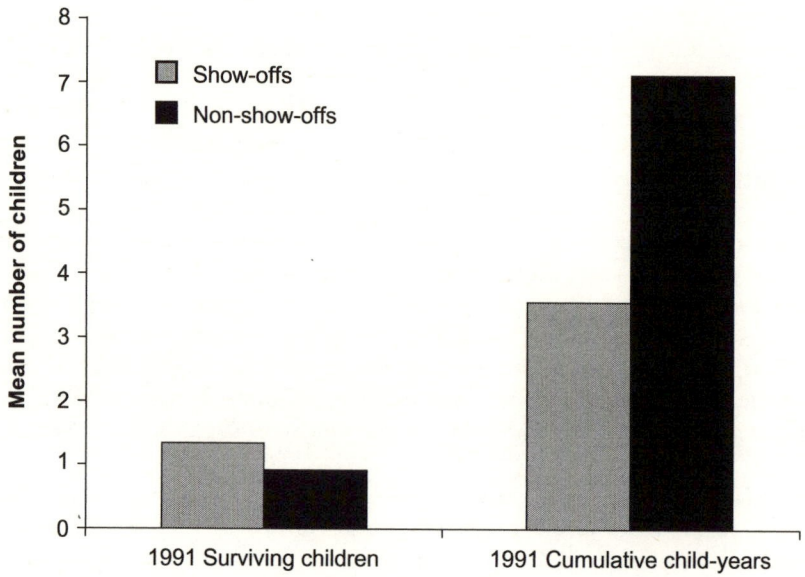

Figure 4.9 Mean number of surviving children and their combined ages (cumulative number of child-years) per ten years of reproductive life for four men defined as 'show-offs' – in Hawkes's (1991) sense – and nine other men among the Kubo of New Guinea. Cumulative number of child years is based on the ages of children alive at the time of data collection and acts as a proxy for child survival. Although show-offs had more children than other males, their survival appears to be poorer. Data are from the more complete 1991 census. Source: Dwyer and Minnegal (1993).

kind of hunting, rather than attempting to master all the available forms. Dwyer and Minnegal (1993) suggest that in order to satisfy protein needs, all faunal types must be exploited and sharing occurs as a means of reducing variance in protein consumption – as suggested by Hames (1990) for the Yanomamö. Individuals who specialise in particular prey types will bring individual benefits by reducing the risks and uncertainties involved in catching a particular prey species, and this will foster the development of trade relationships within a 'biological market' (Noë and Hammerstein 1995) since individuals can exchange commodities to which they have priority of access through increased skill. Showing off therefore arises as an epiphenomenon of target specialisation in the sense that graphical and statistical analyses identify individuals who display characteristics consistent with showing off, but in reality these relationships are abstractions which do not actually exist.

Notwithstanding these potential criticisms, Hawkes has extended the notion of showing off to produce a general model of foraging, which attempts to answer the 'public goods' problem associated with widespread sharing (see **Box 4.4**). Hawkes (1993) argues that the consumption value placed on food resources does not provide sufficient incentive for individuals to continue supplying them. We alluded to this problem earlier in terms of cheating. If resources are of large packet size and both non-defensible and perishable, then they can be considered as 'public goods' available to all who desire them. This being the case, why should any self-interested individual continue to work to obtain such goods? Why not just claim your share of others' efforts?

Hawkes (1993) argues that this other incentive is social attention from others which provides individuals with a larger pool of potential mates and allies. This will often manifest itself in a gender difference since increased mating advantages gener-

BOX 4.4

The tragedy of the commons

The 'Tragedy of the Commons' is a term coined by Hardin (1968) to explain why overgrazing of public rangelands occurs. Hardin (1968) argued that, although reducing pressure on the rangelands would prevent overgrazing and maintain productivity, it would never pay an individual to reduce the pressure by limiting the grazing of his own herd. Since the rangelands are public, a responsible herd-owner has no power to prevent other individuals from overgrazing the land. If other individuals continue to graze their herds with no regard for overgrazing problems, then anyone who limits his herd would pay a cost in terms of reduced food for his animals, but would gain no benefit (in terms of a higher quality area of grazing). The only sensible (that is, economically rational) thing to do under such circumstances is take the same short-term perspective as everyone else and continue to overgraze.

The idea is similar to that of the Prisoner's Dilemma. Although it would pay everyone in the long-term to limit their usage, if there is the risk of other individuals cheating, then it pays everyone to cheat as this provides a better cost–benefit ratio than taking a responsible long-term view. Overfishing of certain fish stocks in the North Sea and elsewhere provides another familiar example of a tragedy of the commons. Individuals can only be deterred from over-exploitation if other incentives (or disincentives, such as fines or imprisonment) that are unrelated to the value of the goods themselves are introduced to keep people's behaviour under control.

The tragedy of the commons argument is based on the broader notion of 'public goods' problems. Public goods (sometimes also known as **common pool resources**) are any resources that cannot be monopolised by or denied to individuals, even when they don't contribute to producing the goods in the first place. Funding the public television service in the UK is a good example of a public goods problem. In order to fund the BBC, everyone who owns a television is supposed to pay a licence fee. But, if enough people are already paying the licence fee and programmes are broadcast, then, as an individual, it doesn't pay to buy a licence, since you are able to watch the programmes for free. As long as there are enough people contributing, then 'freeriders' (see Chapter 9) can get away without contributing. However, if the number of individuals drops below a certain level, then programmes cannot be funded and no one gets any television at all – regardless of whether they are a regular licence payer or not – and, as with the rangeland example, everyone loses out. Large fines are therefore necessary to deter people from avoiding paying their licence fee.

ally accrue mostly to men, while womens' goals are focused on ensuring high family consumption. Hawkes (1993) uses the Ache evidence, plus data from the !Kung and Hadza, to support her argument. In the case of the Hadza, she cites a field experiment in which Hadza men were persuaded to forego big game hunting (shared resources) and rely only on hunting or snaring small game (which are not shared) for food (Hawkes et al. 1991). Although mean earnings per item were lower for small game, success rates were higher (one animal every 3.1 days compared to one animal every 37.0 days when hunting large animals). Interpolating these results into a model of differential foraging benefits suggested that if men were attempting to provision their families with a regular supply of meat (the standard explanation for male hunting behaviour) then they should hunt and trap small animals on a regular basis, instead of ignoring them and targeting large game alone; moreover, they would do best (in terms of minimising the number of days when their families got less than 0.5 kg of meat) if at the same time they could exploit other men's big game hunting (**Table 4.1**). The targeting of widely shared resources by men suggests that social incentives may play a large role in circumventing a public goods problem.

Table 4.1 Probability of obtaining at least 0.5 kg of meat on any given day for Hadza men pursuing alternative hunting strategies, depending on the hunting strategies of the other men in the group

Hunting strategy of individual	Probability of obtaining meat every day	
	Hunting strategy of all other men	
	big game	small game
big game	0.39	0.08
small game	0.69	0.32

Source: Hawkes et al. (1991)

The idea of social incentives therefore presents a strong challenge to the view that hunting is primarily a means of provisioning one's family, and hence a form of paternal investment. (Incidentally, this means that a number of hypotheses concerning the evolution of human pair-bonding, including that of Lovejoy [1981], are also called into question.) It also suggests that risk reduction and reciprocity are not always important determinants of sharing; indeed Hawkes (1993) emphasises that, in all three of the study populations she cites, individuals who never contribute resources to the common pool are nonetheless permitted to consume those supplied by others and that, when individuals request shares in a resource, these demands are framed in terms of 'give to me, since I have none' rather than as 'give to me today, and I will repay you tomorrow'. The social incentives hypothesis can also explain why foraging strategies differ so widely between the sexes as a consequence of the trade-off between consumption and social benefits, since these have different impacts on men and women (for further discussion of the division of labour, see p. 240).

Hawkes' hypothesis (1993) has been quite strongly criticised, although much of this criticism is by social and cultural anthropologists who pitch their arguments at the wrong level of explanation. Eric Alden Smith (1993), though, has made a very valid point in arguing that social attention (if expressed as support in disputes, alliance formation and mating opportunities as Hawkes suggests) actually amounts to delayed reciprocity. Therefore all the arguments that Hawkes levels against the risk reduction hypothesis (where food is exchanged for food) must apply equally to the social incentives model.

This hypothesis clearly needs more rigorous and focused testing before it can be accepted or rejected. But even in its present form, it highlights the difficulty of partitioning human behaviour into convenient categories like 'foraging behaviour' or 'mating behaviour', when it is quite clear that these behaviours interact with each other in quite complicated ways. Indeed, it is worth noting that this form of showing off may extend into other areas of men's activities besides hunting (Bird 1999). In Melanesia, for example, men compete to grow super-large but wholly inedible yams for public display (Kaberry 1971, Weiner 1988).

ARE HUMANS INHERENTLY SELFISH?

Hawkes's model, and indeed all the above findings, suggest that most people behave in a manner consistent with a system of enlightened self-interest, rather than in a truly altruistic fashion. Even such apparently selfless acts as blood donation may operate through a desire to be seen as an altruist by others, rather than merely the desire to help others who are less fortunate.

In an extension of this, Goldberg (1995) asked whether giving money to beggars in the street could be seen as a form of mating display. Goldberg's design was elegantly simple: he watched beggars on a street in Boston/Cambridge (Massachusetts) and recorded who among the passers-by gave them money. He found that lone men were more likely to give to female beggars than to male beggars, and that lone men were more likely to give than those accompanied by a woman. Males did not, it seems, use giving as an opportunity to 'show off' to a current mate (that is, as a display of 'conspicuous wealth'), but more directly as a mating advertisement to a prospective mate. Mulcahy (1999) obtained similar results from a study carried out in England, though he also found that lone women were disproportionately likely to give to male beggars (compared to female beggars). In addition, he interviewed mixed-sex couples after the man had donated and found that, while the man's likelihood of giving was not influenced by his income, it was dependent on the status of the couple's relationship. Men who were at an early stage in a relationship were significantly more likely to give to beggars (especially male beggars) than men who were in a long-established relationship, suggesting that men are sensitive to the fact that giving to beggars may influence their partner's assessment of their potential as mates. In other words, men who are still in the 'assessment' stage of a relationship may wish to convey the impression that they are altruistic and caring.

Some authors have, however, questioned the suggestion that humans are inherently selfish, and claim that people will often behave in an altruistic manner in the absence of any egoistic incentives, whether these are tangible rewards from others or merely internal psychological rewards. One such study is that by Caporael et al. (1989) which appeared to demonstrate the occurrence of altruism under conditions in which all the prerequisites for both kin selection and reciprocity were lacking, thus precluding both of these strategies of self-interest as possible explanations.

The study consisted of a series of experiments in which individuals, placed in groups of nine, were required to make a decision about whether or not to sacrifice personal benefit for the common good of all. Individuals were given $5 and told that they could either contribute it to a common pool or keep it for themselves. They were informed that if enough people donated their money to the pool, then all individuals in the group would be given a bonus of $10, regardless of whether or not they had themselves contributed. Thus, non-contribution would always yield a return of $5 and could potentially accrue $15 if other group members contributed sufficiently. In contrast, contribution would yield either nothing (the individual contributes but not enough other members do so to gain the bonus) or $5.

The potential for kin selection and reciprocity in these experiments was eliminated by ensuring that subjects were always complete strangers, that decisions were made anonymously (precluding punishment of non-contributors) and (initially at least) individuals were prevented from interacting prior to and after decisions were made (to prevent reciprocal arrangements being set up). Despite these restrictive conditions, the results showed that, in three different versions of the experiment, more people than expected contributed to the pool, even though defection (non-contribution) would have been the optimal strategy (**Figure 4.10**). This finding was even more striking when the experimenters allowed the group members to interact with each other prior to taking their decisions (although the actual contribution remained anonymous, so that individuals could still defect with no possibility of being identified). Most striking of all was the fact that in trials where certain individuals were designated

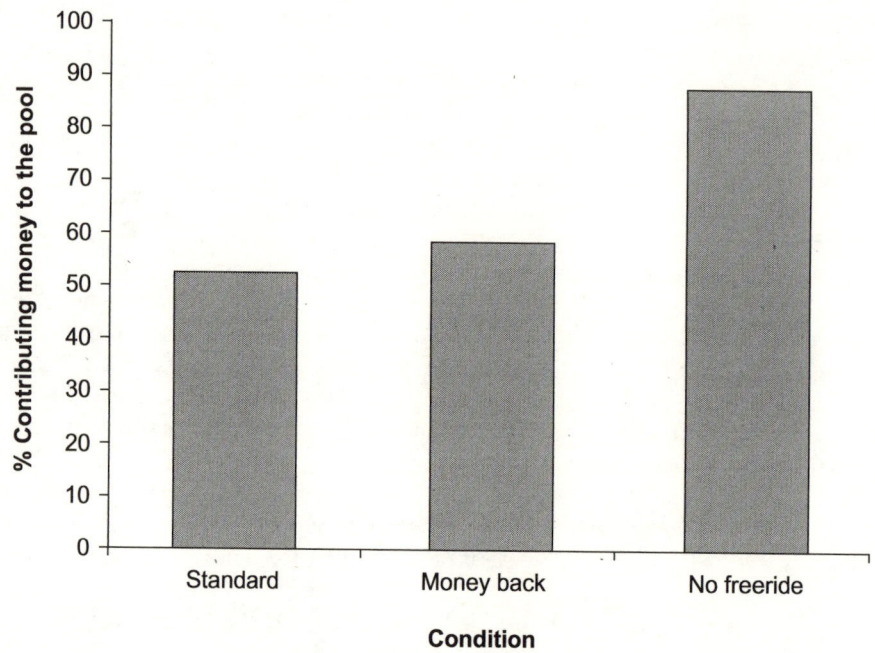

Figure 4.10 Proportion of individuals contributing money to the common pool under three different conditions in Caporeal et al.'s (1989) experimental test of altruistic behaviour. In the 'standard' condition, a specified number of contributions of $5 to the common pool were required in order for all the subjects to receive a monetary bonus of $10; contributors would therefore make a net profit of $5 dollars. However, non-contributing individuals could potentially make $15 dollars by relying on others to contribute to the pool (so that the bonus was provided) while hanging on to their own original $5. In the 'money-back' condition, the contributing subjects were told that they would get their $5 back if too few individuals contributed to the pool. This condition eliminated the fear of losing $5 but freeriders could still make $15 dollars because they kept their $5 while others contributed. In the 'no freeride' condition, the opportunity to freeride was eliminated by setting the maximum amount paid to an individual at $10 so that non-contributors could not make an extra profit compared to contributors. In all three cases, over 50 per cent of individuals contributed their money to the pool even though the rational economic choice was to be a non-contributor. The elimination of potential freeriders had the greatest effect on increasing contributions to the pool. Source: Caporael et al. 1989.

as 'contributors' in order to ensure that a sufficient proportion of the group donated their money to gain the bonus, it was often found that individuals designated as 'non-contributors' nevertheless donated their money to the pool, seemingly to be on the safe side and make sure that the bonus would be forthcoming.

From these results, Caporael et al. (1989) concluded that sociality itself, made salient by designating individuals to groups and allowing discussion between group members, could account for the altruistic behaviour displayed. They argued that, given the benefits of group-living to humans (predator defence, foraging efficiency), selection would favour individuals who were better adapted to group-living, and that this adaptation took the form of a willingness to behave altruistically with no selfish incentives toward other group members. Altruistic behaviour, it was suggested, evolved simply because altruistic individuals were better at group-living, and not necessarily because they received some return benefit from the beneficiary of their altruism or enhanced the spread of their genes by aiding close relatives. The well-documented occurrence of 'in-group biasing' effects, whereby individuals are more willing to aid those perceived as belonging to 'their' group even when this is based on something as superficial as a preference for Klee over Kandinsky, supports this hypothesis (Tajfel et al. 1971).

Unfortunately, Caporael et al. (1989) argue that their findings disprove 'sociobiological egoistic incentive' theories, when in fact they do nothing of the sort. In reality, they support socio-biological explanations rather nicely, and help highlight the fundamental difference between selfish genes and selfish behaviour. Caporael et al. (1989) make the common mistake of assuming that since we have selfish genes, we should therefore behave selfishly whenever we get the chance. But if behaving in an altruistic manner is good for our selfish genes, then such behaviour will be selected. Being nice to other group members, and thus ensuring high foraging success or group cohesion in times of stress (for example, during conflicts with other groups) is not only an egoistic incentive, but also smacks rather heavily of the kind of complex reciprocal altruism that Palmer's (1991a) study of the Maine lobstermen highlighted.

There are two other reasons why Caporael et al. (1989) may have been rather premature in their claim that humans aren't naturally selfish. First, giving people $5 and then asking them to give it away or keep it entails very little cost to the subjects because they can't actually lose: although they may not gain the $10 bonus, neither do they ever find themselves out of pocket (since they leave with what they brought with them). Such an experimental design can't really be used to test for altruistic effects since it doesn't require all that much altruism to give away something that wasn't yours in the first place. Second, the experimental design failed to take account of individuals' reputational concerns. Although other group members were unaware of the choices made, the experimenters knew exactly who was doing what. Individuals may thus have wanted to be seen as doing the right thing by the experimenters (see **Box 1.3**). Similarly, within the groups themselves, individuals may have been reacting to a perceived risk that they would bump into other members of their group again (with its attendant consequences if they had 'cheated'), a perception that would be heightened in the 'discussion' groups. In other words, despite their best efforts, the experimenters were unable to create conditions in which egoistic incentives were entirely eliminated, since they were dealing with people's beliefs about the situation, rather than the true nature of the set-up. Damasio (1994) has suggested that 'somatic markers' (gut feelings that become tied to behavioural outcomes) may play an important role in biasing our decisions in favour of one particular course of action; he argues that this may be particularly so in potentially altruistic contexts because of the associated sense of exaltation that accompanied approval by others on previous occasions. (For a more detailed discussion of somatic markers and the ways in which an individual's beliefs about the world, rather than the actual state of the world, drive behaviour, see Chapter 11.)

Individuals in Caporael et al.'s (1989) experiments may have 'felt' obliged to do the right thing, and might thus have been biased against behaving selfishly because considering such a course of action left them feeling uncomfortable. Caporael et al. (1989) seem to have confused this proximate motivation to behave altruistically with a functional explanation regarding the survival value of altruistic behaviour. Proximate motivations need not be associated with the ultimate function of behaviour in a direct fashion. The fitness consequences of being motivated to behave altruistically may appear way down the line from the point at which the act takes place, in much the same way that an ability to hunt tapirs successfully will eventually translate into high reproductive success for an Ache hunter (via food intake, a lot of

BOX 4.5

How 'selfish' genes lead to non-selfish people

The 'selfishness' of genes has sometimes been taken to imply that individuals will behave selfishly too. But such an inference makes two serious mistakes. First, it assumes that the selfishness of genes has some moral force, when in fact it is just a reminder that the gene (as opposed to the individual, group, population or species) is the proper level at which to evaluate the evolutionary consequences of an action. Second, it ignores the whole point of social strategies of animals: much of what primates, in particular, do is designed to achieve cooperative solutions to problems of mutual interest. Selfish genes therefore commonly produce cooperative individuals. Axelrod (1984) demonstrated that tit-for-tat was evolutionarily stable against all alternative strategies and thus showed how cooperation can evolve in circumstances that appear naturally to favour conflict and contest. Where both players gain a bigger advantage by cooperating and certain conditions are fulfilled (long lifespan to allow repeated 'games' with the same individual, capacity to remember what each opponent did previously, and the ability to learn from reward or punishment), then cooperation will evolve.

One potentially interesting suggestion has been to argue that there is a basic selfishness of individuals derived from the selfishness of genes, but it is the 'bottom line' on human behaviour that is held in check by culture. This is essentially the argument put forward by the seventeenth-century philosopher Thomas Hobbes: the 'beast within us' comes out whenever social controls are removed. The fictional world of William Golding's *Lord of the Flies* (1954) provides a perfect example of this:

a group of shipwrecked schoolboys slowly descends into savagery in the absence of established social norms imposed by adults. The behaviour of troops in occupied territories or during ethnic purges may be a real life example of the same phenomenon.

One can point to various lines of supporting evidence for this proposition. Examples include the fact that all religions emphasize that people have to learn to restrain themselves (because people do not seem to do so naturally) and that moral behaviour commonly ceases when civil authority breaks down (for example, during war or famine). This places considerable emphasis on the importance of culture in determining how people behave (as anthropologists have always insisted), but it does not obviate the fact that behaviour (whatever its origins) still carries fitness consequences. In this case, we might see culture as providing a mechanism for more effective solutions to the problems of survival and reproduction achieved through cooperation in social groups. The reversion to wholly selfish behaviour could then be interpreted as a kind of natural default strategy that we resort to when social life breaks down.

This brings us back to a point we made in Chapter 1 regarding the relationship between moral statements and evolutionary explanations. That we can say that a given ethical proscription evolved for good functional reasons does not necessarily make it right, any more than the fact that we may have evolved to be selfish or cooperative should be taken as a prescription of how we ought to behave now. There is no reason why we can't decide to behave otherwise (for better or worse).

physiology and some reproduction). Caporael et al.'s (1989) study thus reveals a lot about people's willingness to behave cooperatively in groups, but is not really designed to test any functional explanations of such behaviour.

Most of the evidence seems to suggest that while individuals within social groups often act in ways that are only apparently altruistic, it is also true that humans seem superbly adapted to group-living due to a set of psychological adaptations for favouring individuals within the group to which they themselves belong. Richerson and Boyd (1998) term this '**ultra-sociality**' and suggest that the impressive coordination, cooperation and division of labour observed in modern-day western society can be traced back to ancient social instincts combined with modern cultural institutions. Although this level of cooperative behaviour may often conflict with the small-group kin se-

ULTRA-SOCIALITY

lected or reciprocal altruism favoured by our genes, the co-evolution of culture and in-group altruism has proved to be an even more successful strategy than either of these traditional routes since it appears to have achieved the impossible: individuals are capable of true altruism and yet achieve high fitness returns by doing so, not because they have 'overcome' our genes, but because true cooperation was originally to their benefit.

*

In the next chapter, we look at another domain in which individuals with competing interests must somehow cooperate, this time in the mating arena. We have already seen how male hunting practices may actually be a component of male mating strategies within the context of food-sharing, and we now go on to examine the whole range of ways in which individuals select and compete for suitable mates.

■ Humans exhibit many forms of apparently altruistic behaviour which cannot be explained in terms of kin selection. These behaviours range widely in the costs they incur for the altruist: for example, giving money to a street beggar is unlikely to have any effect on the donor's inclusive fitness, whereas sharing the spoils of a long and dangerous hunt is much more likely to do so.

■ Food sharing among hunter-gatherers is usually thought to be an attempt to reduce the risks of going without food. Nutritional status can be increased by sharing and individuals reduce variation in the intake of important foods (like protein sources) through sharing. However, food-sharing patterns do not always conform to the predictions of the reciprocal altruism hypothesis.

■ 'Tolerated theft' can explain many sharing patterns among hunter-gatherers. The size of resources may be too large for a single hunter to consume and may be uneconomic to defend, resulting in food being made available to other individuals. This kind of sharing is passive, rather than active.

■ Male hunters may use hunting to increase their success in the mating domain by 'showing off'. By sharing their catch widely, males can increase the number of extra-marital copulations, improve offspring survivorship and achieve high status within the group – all of which may increase the male's reproductive output.

■ Studies of modern western populations indicate that people may behave non-selfishly, even when motivations such as kin selection benefits or reciprocal altruism benefits are lacking. The proximate mechanism prompting such behaviour seems to be group membership.

■ Psychological mechanisms that predispose humans to behave altruistically in social contexts may have evolved in order to facilitate more effective social living because each individual's fitness ultimately depends on the effectiveness with which groups solve common ecological and other problems. In these circumstances, cultural evolutionary processes may also result in behaviour that is strictly altruistic.

Chapter summary

Further reading

Cashdan, E. (ed.) (1990). *Risk and Uncertainty in Tribal and Peasant Economies*. Boulder, CO: Westview Press.

Ridley, M. (1996). *The Origins of Virtue*. London, Fourth Estate.

Smith, E. A. (1991b). *Inujjuamiut Foraging Strategies: Evolutionary Ecology of an Arctic Hunting Economy*. New York: Aldine de Gruyter.

Smith, E. A. and Winterhalder, B. (1992). *Evolutionary Ecology and Human Behaviour*. New York: Aldine de Gruyter.

Sober, E. and Wilson, D. S. (2001). *Unto Others: The Evolution and Psychology of Unselfish Behaviours*. Cambridge (MA): Harvard University Press.

Zahavi, A. and Zahavi, A. (1997). *The Handicap Principle: a Missing Part of Darwin's Puzzle*. Oxford: Oxford University Press.

Mate choice and sexual selection

CONTENTS

Evolution is concerned not simply with the numbers of offspring an individual produces in a lifetime, but with the numbers of those offspring that successfully reach adulthood and reproduce in their turn. Successful reproduction is thus premised on solving two key problems: one is choosing an appropriate mate, the other is investing appropriately in the offspring that result. The second is concerned with issues of parental investment, and we deal with these in Chapters 7 and 8. Here, we focus on the business of choosing a mate. However, because parental investment looms so large in species like humans, the two components of reproduction cannot be partitioned into discrete sections so easily: issues relating to parental care will inevitably intrude into this chapter.

In exploring human mate choice, there are two general issues to consider. First, there are the general principles that underpin and guide mate choice: these are often considered to be human universals in that, given the nature of the Darwinian process, they apply to everyone. However, individuals' decisions in this, as in every other aspect of real life, are contingent. In other words, no matter how attractive we may find certain individuals, there is no guarantee that they will be attracted to us.

This is because mate choice is almost always a frequency-dependent problem. It is a genuine market place in which individuals make bids and accept negotiated bargains, even though these may often be less than ideal. The second issue is thus the fact that mate choice decisions are necessarily contingent on circumstances.

UNIVERSAL PRINCIPLES OF MATE CHOICE

LIFETIME REPRODUCTIVE SUCCESS

POLYGYNY

Bateman's Principle provides an important starting point for any consideration of mate choice. Bateman (1948) pointed out that, although the mean **lifetime reproductive success (LRS)** of the two sexes must be equal (the collection of individuals that make up one sex cannot produce more offspring than the collection of individuals that make up the other sex), the variance in LRS may nonetheless differ considerably, especially when there is a significant **polygyny** skew (that is, matings are distributed more unevenly in one sex than the other) (**Figure 2.5**). When the variance in LRS of one sex exceeds that of the other sex, the sex with the smaller variance will benefit by being more choosy. In effect, it can afford to exert more pressure on the other sex, which is unable to evolve an effective counter-strategy.

This effect is reinforced when (a) the costs of reproduction are higher for the sex with the lower variance and (b) members of the more variable sex differ significantly in their value as mates. In this respect, individuals can vary in value either because their genetic traits differ or because they differ in their ability to contribute to the rearing process (that is, they differ in the resources they have available or in their parental skills). These two conditions are especially potent in the case of humans because of the extreme costs of the lengthy gestation and long period of lactation needed to grow our big brains (see **Box 6.1**).

The implication of these combined effects is that we can expect human females to be more choosy than human males about whom they are prepared to mate with. It is important not to confuse this choosiness with coyness. Females may behave promiscuously and be adventurous in mating, but still be choosier than the males with whom they mate. A second implication of the enormous costs of human reproduction is that females will tend to select males in terms of their effect on the success with which offspring can be reared. Success in this respect can come in either or both of two respects: the quality of the genes males have to offer (better quality genes presumably mean more successful offspring) and their ability to contribute to childcare (for example, by provisioning). Fundamental evolutionary principles thus suggest that women will select men either on the basis of cues of genetic quality or on the basis of their willingness or ability to contribute to childcare. The latter may involve direct (for example, taking actual physical responsibility for the child) or indirect (for example, providing food for the mother and child) forms of childcare. In hunter-gatherer societies, the most important of the indirect form of investment might involve hunting skills; in agricultural societies, it will usually be ownership of land, while in industrial societies it will typically be wealth and/or status (both being means of purchasing the required provision).

The characteristics of mammalian reproduction stack the odds in favour of females placing a strong emphasis on the successful rearing of offspring. Because gestation and lactation make it difficult for male mammals to contribute directly to the process of rearing, the most effective way that males can influence their reproductive success is to maximise the number of fertilisations they can achieve. Since female mammals cannot easily gestate more than one set of offspring at a time, the most effective way

BOX 5.1

Anisogamy

The biological definition of sex rests on the size of the sex cells (known as 'gametes') that an individual organism produces. Females produce large, immobile gametes known as eggs or ova and males produce small, motile gametes called sperm. Ova are the largest cells found in the human body and can almost be seen with the naked eye. As well as DNA (the genetic material needed to build a new organism), an egg contains nutrients for the growing organism. Sperm by contrast are exceptionally tiny, consisting of little more than the DNA-containing nucleus and an 'engine' made up of mitichondria (the structures that produce energy in a cell) that helps propel the sperm.

Females therefore invest more in an offspring even before fertilisation actually takes place. In mammals, the extra costs of internal gestation and lactation further increase the female's investment relative to the male's. The size of a newborn offspring is many billions of time larger than the male's contribution of a single sperm cell.

These initial differences in investment explain the differences in the mate choice preferences and reproductive investment strategies of males and females. Since females always invest more than males, they should be more choosy since a failed reproductive attempt or a poor quality offspring is more costly for them than it is for males. Since the latter invest only sperm in any reproductive attempt, they can increase their reproductive output by attempting to mate with as many females as possible. Females, on the other hand, cannot produce more offspring by increasing the number of males they mate with because they are limited by the number of large, expensive eggs they produce and the time it takes to produce an offspring.

Human males are, however, more constrained than other mammals because newborn offspring require care from both parents, and this limits a male's ability to find new mates. This means that human males, as well as females, are more choosy than is typical of most mammals and seek mates with high levels of fertility in order to ensure that they produce as many offspring as possible even though they may be limited to a single partner.

---

for a male mammal to maximise his fertilisation rate is to maximise the number of females with whom he mates. When all else is equal, male mammals are thus driven towards an emphasis on the mating side of reproduction. This divergence between the two sexes' reproductive strategies is exacerbated in humans by the length of the reproductive cycle (the interbirth interval or time between one conception and the next), this in turn being a consequence of the prolonged periods of gestation, lactation and post-weaning childcare characteristic of humans.

However, the extent to which human males emphasise mating will be tempered by the extent to which they can contribute to the business of rearing. The extraordinarily high costs of reproduction in humans (as a consequence of having to grow massive brains) means that whenever males can influence the success of rearing, it will pay them to do so (Kleiman 1977, see also Dunbar 1995b). The issue hinges very much on the word 'success'. If males cannot improve their mates' rearing success, then selection will favour males that concentrate on mating at the expense of rearing. If males can contribute to rearing, the extent to which they emphasise mating at the expense of rearing will directly reflect their ability to contribute to rearing.

For example, among the Aka pygmies of central Africa, low-ranking males contribute far more time and effort to childcare than do high-ranking men (Hewlett 1988). Since the latter are able to achieve high mating success by dint of their increased social status, it pays them to concentrate more on mating effort. Investing effort in the successful rearing of offspring is the key to enhancing reproductive success for less socially successful males. Considerations such as these may lead males to place a premium on fertility in those females they do mate with, especially when males are obliged to help with rearing: if males are obliged to stay with a mate for any length of time, they will do best if they choose the most fertile ones.

As a first approximation, then, we can view human mate choice strategies as a simple opposition of interests: females seek to maximise the number of offspring they can rear either by choosing males who can contribute effectively to rearing or choosing males of high quality, while males seek to maximise the number of fertilisations they achieve either by mating with many females or by preferring to mate with the more fertile ones.

In the following subsections, we review the evidence in support of these predictions. In a later section, we will take up the more complex question of how these universal principles can be modified by the contingent circumstances of real life in individual cases.

MATE CHOICE PREFERENCES

Personal advertisements provide a neatly encapsulated vignette of a person's mate choice preferences (see **Box 5.2**). Newspaper advertisements for heterosexual partners placed by women suggest that they place a strong emphasis on one of two kinds of cues: indices of wealth or status or those indicating a willingness to invest in the relationship itself (Kenrick and Keefe 1992, Greenlees and McGrew 1994, Wiederman 1993, Waynforth and Dunbar 1995). In a recent sample of UK middle-class advertisements (**Figure 5.1a**), 24 per cent and 35 per cent listed one or more terms concerned with wealth/status and commitment, respectively. The emphasis placed on these categories has been widely confirmed by analyses of personal advertisements from different parts of the world (Kenrick and Keefe 1992), suggesting that this emphasis is a universal trait. Women's advertisements in the UK sample also mention two other traits as desirable in mates: physical attractiveness (33 per cent of advertisements) and social skills ('charm': 52 per cent). These findings are reinforced by the fact that it is precisely these traits that males tend to advertise in their own advertisements (**Figure 5.1b**). Men appear keen to advertise those traits that they perceive women to be principally interested in. We review some of these results in more detail below; others (notably social skills) we will return to in later chapters.

Wealth may be particularly important in traditional economies simply because it allows women to make higher investments in their offspring. When there is no health care system to provide medical aid during illness, a child's chances of surviving depend crucially on the wealth of the parents (usually inherited or earned by the father). This is well demonstrated both in European historical peasant communities (where farmers have higher infant survival rates than day labourers: Voland 1988) and among contemporary Kipsigis agro-pastoralists living in southern Kenya (where the number of children a woman manages to rear successfully is a linear function of her husband's wealth: Borgerhoff Mulder 1989). For better or for worse, women who are able to gain access to more resources are (a) able to feed their children better (and so reduce their susceptibility to the many diseases that assail young children) and (b) can afford to take them to hospital when they are ill.

Males' interests exhibit a rather different pattern. As might be predicted from evolutionary first principles, men advertising for women place a strong emphasis on cues of physical attractiveness in prospective partners (Buss and Barnes 1986, Buss 1987).

BOX 5.2

Lonely hearts advertisements: methodological considerations

Lonely hearts advertisements provide useful information on people's mate choice preferences since they are explicitly designed for that purpose. However, there have been claims that they may represent a biased or atypical sample (for example, Dardes and Koski 1988). It might be the case that only certain personality types are prepared to advertise for a partner, or that advertisers are individuals who have failed to find a partner by more conventional routes. This would lead to biases if these individuals had choice preferences that were different from the norm, especially if it was these preferences that had resulted in a lack of success in the mating arena in the first place.

However, it is unlikely that advertisers are behaving in an unusual way or that they represent a peculiar subgroup of the population at large. Advertising for mates has become common during the last few decades mainly because traditional mechanisms for finding mates have been lost. Job mobility has meant that young people often find themselves in large anonymous cities where they have no social networks other than those provided by the workplace. Where friends and relatives might have provided introductions to potential mates in small-scale rural communities, economic migrants are strangers who lack access to suitable networks.

In addition, the rising rate of divorce has meant that a significant number of people in their thirties are thrown back into the mate-searching market. At this stage of their life cycle, people commonly lack access to the kinds of social outlets that they exploited when they first married: they stand out as being obviously out of place in the clubs and other venues where they pursued their first (teenage) courtships. For those caught in this predicament, advertising in personal columns has been a godsend. But individuals caught in this predicament are not thereby atypical or unnatural in their mate search habits.

Analyses of advertisements typically involve classifying the words listed by advertisers into a small number of categories. Content analysis of the meanings of terms that appear in advertisements suggests that the number categories is quite small: the most common ones are wealth/status, physical attractiveness, commitment, fidelity, social skills, hobbies/interests and political/religious beliefs (Theissen et al. 1993). **Table 5.1** lists some of the words that commonly appear under these headings. In addition, advertisers often specify a number of other features that, while socially important, are difficult to interpret in evolutionary terms. These include habits such as smoking and drinking, and particular hobbies.

Figure 5.1(b) shows that more than 43 per cent of male heterosexual advertisements in the UK sample mention such cues. By comparison, other trait categories receive much less emphasis. Cues of physical attractiveness have been universally interpreted as indices of fertility in women. Since physical attractiveness changes predictably with age, and age is an uncompromising correlate of fertility in women, such cues provide an indirect measure of a woman's fertility. Women, in turn, seem to recognise this: cues of physical attractiveness dominate their descriptions of themselves (**Figure 5.1a**).

Male and female advertisements also differ in respect of the preferred ages of prospective mates in ways that reinforce these findings. Most advertisements tend to give the exact age of the advertiser and request a partner within a specified age range. If we take the mid-point of the age range requested, we typically find that women seek males who are slightly older than they are, whereas, as men get older, they tend to seek women who are increasingly younger than they are (**Figure 5.2**). It seems that, in western industrial societies, men typically seek women of the about same absolute age (late twenties or early thirties), and this has once again been interpreted as reflecting an attempt to maximise the fertility of partners.

Table 5.1 Key words that typify evolutionarily-valent categories in personal advertisements

Category	Key words
Attractiveness	athletic, attractive, cute, fit, good-looking, healthy, nice body, handsome, hunk, muscular, rugged, tall, well-built; buxom, petite, pretty, shapely, slender, slim
Wealth/status	house-owner, professional, well-off, businessman (-woman), financially independent, solvent, college-educated; terms suggesting lavish lifestyle (holidays abroad, expensive cars, upmarket house locations)
Commitment	kind, (emotionally) stable, mature, dependable, pleasing disposition, good cook, caring, family-oriented, gentle, easy-going, sincere, sensitive, good listener, affectionate, sympathetic, understanding, down-to-earth
Social skills	good sense of humour, amusing, lively, laughs a lot, cheerful, witty
Sexual fidelity	monogamous, faithful, one-man-woman, one-woman-man
Sexual	cuddly, sensual, into fun times

Sources: Buss (1989), Waynforth and Dunbar (1995), Pawlowski and Dunbar (2001)

This contrast in the age preferences of men and women has been confirmed by data from many different cultures (advertisements: Kenrick and Keefe 1992; Greenlees and McGrew 1993; Waynforth and Dunbar 1995; marriage proclamations: Otta et al. 1999). Pérusse (1994), for example, investigated mate choice among contemporary French Canadians and found remarkably similar patterns to those obtained from analyses of personal advertisements. He found that the peaks in the attractiveness of males and females (measured through frequency of simultaneous partners) were at ages 30–39 and 25–29, respectively. These are the same ages for the peaks in market value as assessed from UK personal advertisements (Pawlowski and Dunbar 1999a). Although males universally prefer younger women, the precise age targeted may vary slightly from one socio-economic context to another, either in response to differences in female reproductive schedules (Pawlowski and Dunbar 1999a) or as a function of men's concerns about virginity (and thus paternity certainty: see below).

These differences between the sexes are reinforced by more conventional data sources. Buss (1989) used a questionnaire format to explore mate choice preferences in 37 different cultures drawn from all around the world. He too found that women place a strong emphasis on cues related to wealth and status, while men tend to place an overwhelming emphasis on physical attractiveness. Grammer (1989) has reviewed the extensive social sciences literature on mate choice preferences elicited by interview and questionnaire. The burden of this literature is that women use more cues than men do when evaluating prospective mates. Men typically consider only a single cue (attractiveness) while women consider as many as a dozen different traits, including both social and economic traits. Women appear to make more complex mate choice decisions than men do, and are thus more choosy and demanding of prospective mates. This is confirmed by the personal advertisements data: women list significantly more traits in their demands on prospective mates than men do (**Figure 5.3**). In a staged

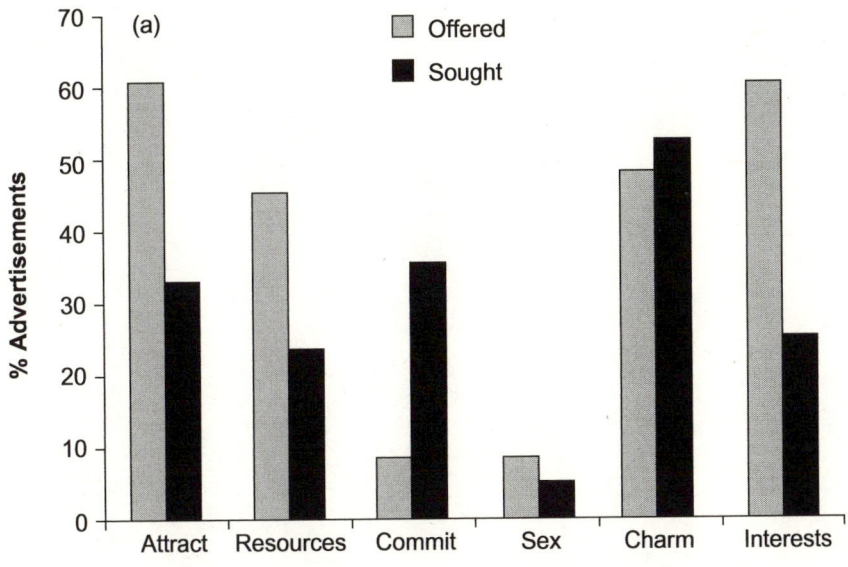

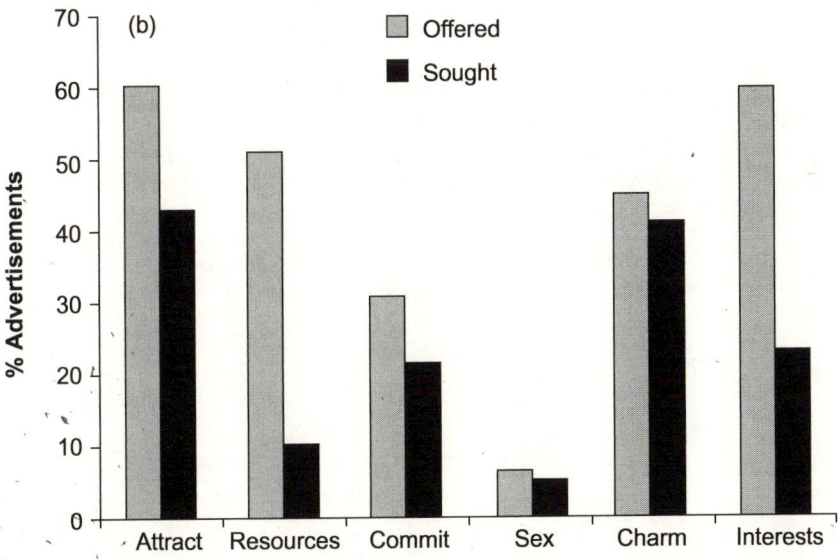

Figure 5.1 Relative frequencies with which (a) women and (b) men offer and seek cues of the six evolutionarily-valent cue categories in personal advertisements. The plotted variable is the percentage of advertisements that mention at least one word in a given cue category. *Attract*: physical attractiveness; *Resources*: cues of wealth or status; *Commit*: commitment to the relationship; *Sex*: cues suggesting sexuality; *Charm*: social skills, humour and so on; *Interests*: social and cultural interests, hobbies and so on. Source: 454 women and 445 men aged 18–59 who advertised in the UK's *Observer* newspaper (see Pawlowski and Dunbar 1999a, 2001).

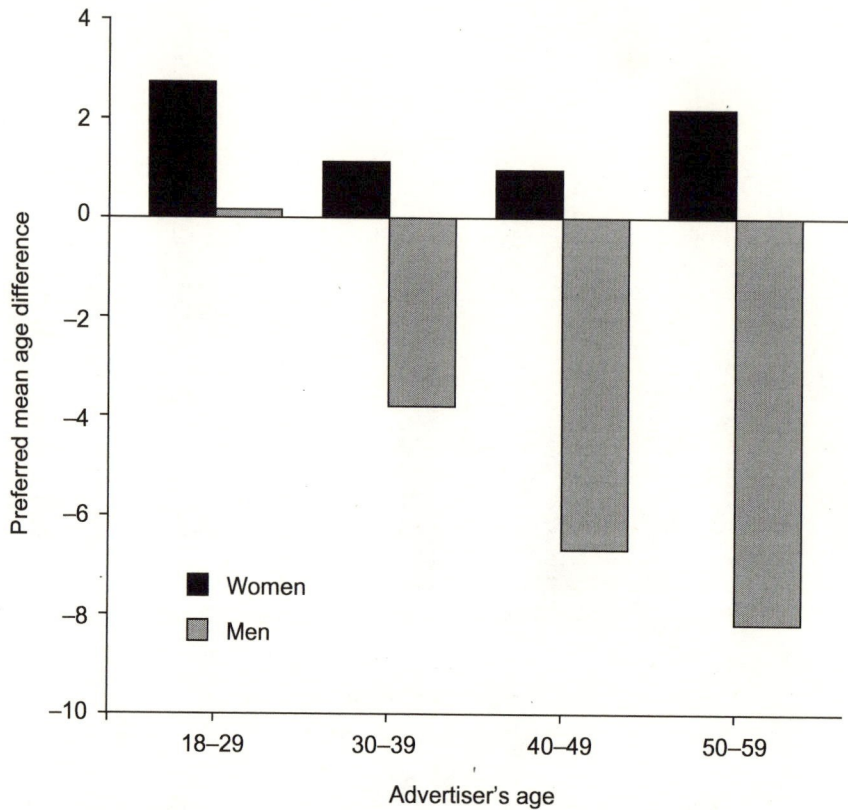

Figure 5.2 Mean difference in age sought in a prospective partner by men and women advertisers of different ages. Women typically prefer men 2–3 years older than they are, irrespective of their own age; men prefer partners that are progressively younger than they are as they get older (men typically target women of about the same absolute age, irrespective of their own age). Source: 454 women and 445 men aged 18–59 who advertised in the UK's *Observer* newspaper (see Pawlowski and Dunbar 1999a, 2001).

encounter between young stranger couples, Grammer et al. (2000) found that the males expressed willingness to date a much wider range of the females they were paired with than did the females.

The contrast between the sexes can also be seen in actual mate choice. Using education as a measure of socio-economic status (education correlates with both family and personal income, as well as with occupational position (Bourdieu and Passeron 1977), Bereczkei and Csanaky (1995) found that Hungarian women preferred mates who had the same or higher levels of education as themselves, and that such couples had significantly more children than did couples where the wife's education level was higher than the husband's. Similarly, most males in the sample preferred and chose younger women as marriage partners, and these males fathered more children than did males who married women older than themselves.

Similar findings emerge from ethnographic studies. Men in many tribal societies prize youth above all in their prospective brides. In many cultures, custom requires the groom (or his family) to pay a brideprice to the family of the bride and this affords us a direct measure of the relative value attached to the women's characteristics. Borgerhoff Mulder (1988a, b), for example, found that Kipsigis are willing to

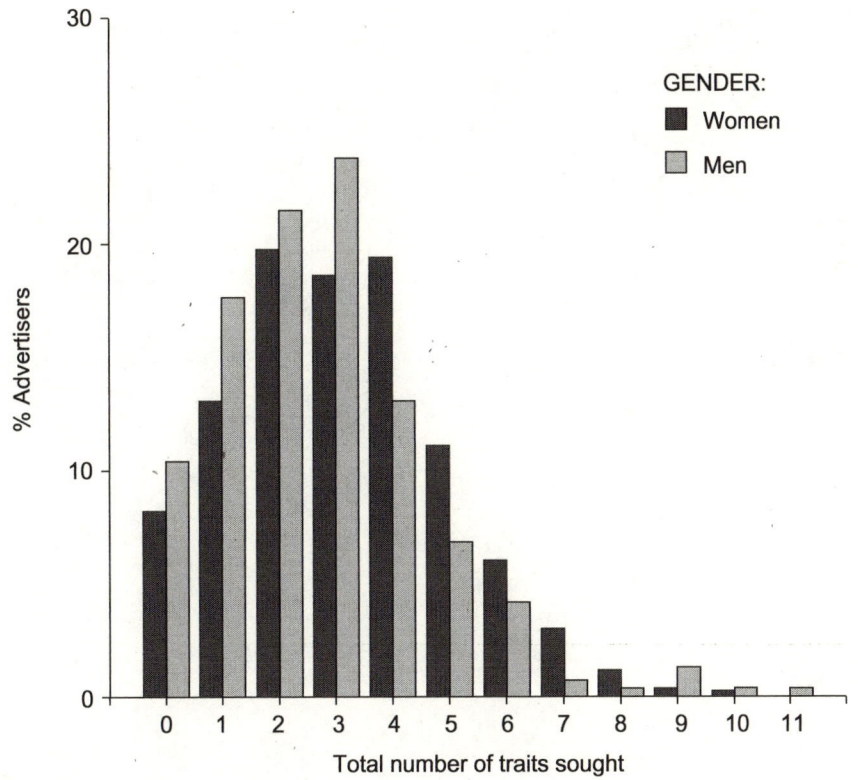

Figure 5.3 Frequencies with which men and women list different numbers of traits as desirable in prospective partners in personal advertisements. Women are significantly more choosy (that is, more demanding) than men. Source: advertisements by 454 women and 445 men aged 18–59 who advertised in the UK's *Observer* newspaper (see Pawlowski and Dunbar 1999a, b, 2001).

pay significantly higher brideprices for younger women, healthier women and (putative) virgins (**Table 5.2**), all of whom are likely to have high reproductive potential. Crippled women and those who had already reproduced attracted significantly lower prices, mainly because they were seen as poor reproductive risks. That a woman's fertility is a primary consideration for men is emphasised by the fact that failure to bear children is one of the principal justifications (or, at least, correlates) of divorce, not only in traditional societies (Betzig 1989) but also in modern western societies (Rasmussen, 1981).

MONOGAMY VERSUS POLYGAMY?

The question as to whether humans are naturally monogamous or polygamous has been a central concern in many discussions of human mating behaviour. Several sources of evidence have been advanced to suggest that men are intrinsically promiscuous while women are inherently monogamous. These have included the claim that women are more faithful than men in their marriages (a claim not overly well supported by the empirical evidence: see, for example, Baker and Bellis 1995).

Table 5.2 Factors that influence Kipsigis brideprices

Trait	Statistic	Value
Age at menarchy	$r_{161} = -0.24$, $P < 0.1$	higher price for younger age at circumcision
Physical condition:		
recent illness	$N = 11$, $P > 0.05$	no effect
plumpness	$N = 11$, $P < 0.05$	higher price for 'plumper' girls
Pregnancy	$t_{14,109} = -2.82$, $P < 0.02$	pregnant women cheaper
Paternity certainty	$t_{19,99} = -2.81$, $P < 0.01$	women with illegitimate children cheaper
Marital distance	$r_{193} = 0.20$, $P < 0.01$	girls from further away cost more (loss of labour to parents)
Affinal connections:		
wealth of in-laws	$r_{140} = 0.00$, $P \approx 1.00$	no effect due to wealth of in-laws
wealth difference	$F_{4,80} = 0.19$, $P > 0.05$	no effect due to wealth difference between parental and affinal families

Note: Degrees of freedom or sample size indicated as subscript numbers against statistic symbol.

Source: Borgerhoff Mulder (1988a, b)

More convincing evidence is provided by the anatomical differences between the sexes: Alexander et al. (1979) pointed out that there is a correlation between sexual size dimorphism and polygyny in many mammalian groups, including primates. Since modern humans are sexually dimorphic in size (by about 8–10 per cent in height and 20–40 per cent in body weight), this implies a mild degree of polygyny in our immediate evolutionary past. However, modern humans are *less* sexually dimorphic than any of our predecessor species (including archaic *Homo sapiens*) (Aiello and Dean 1990, Fleagle 1999). Similar results emerge from a consideration of testis size. Primate species that habitually live in polygynous or promiscuous mating systems have much larger testes relative to body size than species that are obligately or facultatively monogamous **(Figure 5.4)**. Modern humans fall inconclusively between the two regression lines, suggesting a mild level of polygyny has been typical.

The ethnographic data are similarly mixed. Much has been made of the fact that most humans live in monogamous relationships. However, it can equally be pointed out that many more cultures allow polygamy than insist on monogamy; moreover, even in those where monogamy is the legal norm, serial monogamy may be permitted and, more importantly perhaps, individuals of both sexes may entertain lovers while legally married to another individual. Monogamy may thus be a socially enforced state rather than a preferred one. Indeed, the fact that monogamy has to be imposed by legal sanctions in modern societies supports the suggestion that polygamy may be the preferred state: legal enforcement would be unnecessary if humans were naturally monogamous.

It is, however, important to recognise that the sexes may differ in their reproductive strategies. It has been claimed that women pairbond to a particular male in order to obtain resources for child rearing and that, so long as the male continues to

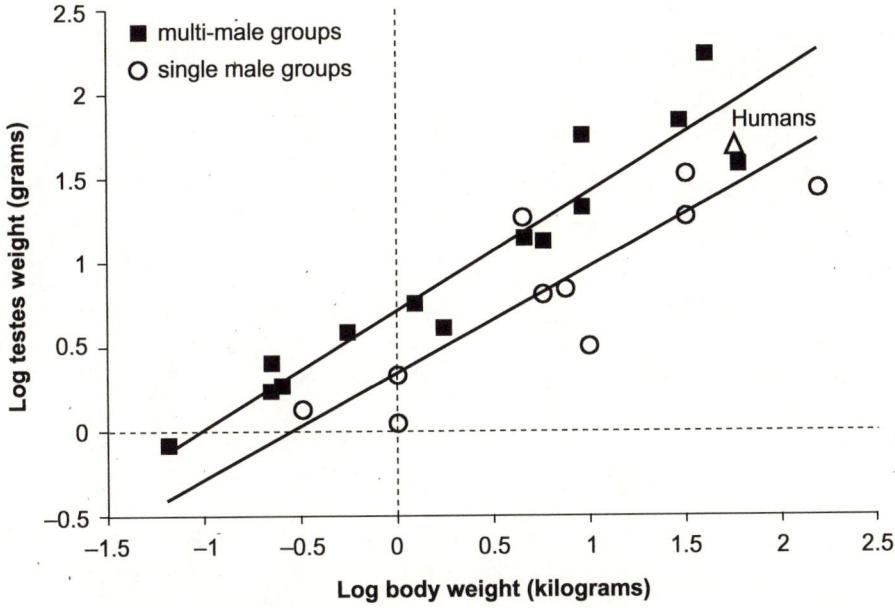

Figure 5.4 Testis size relative to body weight for different primate species. Species that live in more promiscuous mating systems (filled symbols) where males compete with each other over access to individual females have larger testes for body size than species that live in mating systems where males compete only to monopolise a group of females with whom they then have unrestricted mating access – monogamous or harem-based mating systems: open symbols. Humans fall between the regression lines for the two types of mating system, suggesting a limited degree of polygyny is likely to have been typical of our recent ancestors. Redrawn with permission from Harcourt et al. (1981).

provide such resources, women remain relatively less concerned by their males' extra-marital sexual behaviour than men are about their spouses' behaviour. Some evidence in support of this view is provided by sex differences in sexual jealousy. Daly et al. (1982) found that males were more likely than females to express concern about the sexual infidelity of their partners, whereas women were more concerned than men about the emotional and resource investments that their partners make in other women (see also **Box 7.4**). This claim is in turn supported by evidence suggesting that women may be more forgiving of their partners' lapses from fidelity than men are (Daly et al. 1982, Wiederman and Kendall 1999). In effect, women may only become seriously concerned about their partners' infidelity when there are resource implications.

However, it seems that polygamy may have reproductive costs for women, at least in some societies. Chisholm and Burbank (1991) found that, among the Australian Aboriginals of Southeast Arnhem Land, women in monogamous marriages had significantly more live births (means of 6.1 versus 4.6, $p = 0.021$) and significantly more children surviving to five years of age (means of 5.3 versus 3.7, $p = 0.002$) than women in polygamous marriages. These costs do not necessarily reflect reduced male investment in each wife. The polygyny threshold model (**Figure 5.15**) suggests that women ought not to enter into polygamous relationships unless they can be sure that the male can support an extra mate; indeed, there is evidence from the Yanomanö indicating that women in polygamous relationships do not incur any costs in terms

BOX 5.3

Evolution of pairbonding

The evolution of pairbonding in humans remains something of a puzzle. The traditional view (for example, Lovejoy 1981) is that it evolved because females required male help with the high costs of rearing large-brained offspring. Males went out to hunt so as to bring back food to the female at the den. Hawkes's (1990) 'Show-Off' hypothesis raises doubts about the role that male provisioning plays in child-rearing. Nonetheless, the only plausible alternatives are either that the male's contribution is necessary for successful rearing or that the male is engaged in mate-guarding (Brotherton and Manser 1997). A third possibility is that females attach themselves to individual males in order to reduce the levels of harassment that they would otherwise be subject to from males in multimale groups (the 'hired gun' hypothesis: Emlen and Wrege 1986; van Schaik and Dunbar 1990).

The fact that, even in those societies that permit polygamy, polygamous marriages occur only when the male is economically capable of adequately resourcing all his wives offers indirect support for the traditional view because it implies that resourcing is a key consideration for females. Additional support is offered by Foley and Lee's (1991) demonstration that, up to the age of 18 months, the energetic cost of rearing a human infant is about 9 per cent higher than the equivalent cost for a chimpanzee infant (mainly thanks to the difference in brain size). In contrast, the fact that human males go hunting and leave their womenfolk at the mercy of rival males suggests that mate-guarding may not be in the forefront of their minds (but see pp. 337–8). Similarly, the fact that women pairbond with their males is difficult to explain if the male is concerned only with mate-guarding, since the female ought to have no particular concern with the male's problems about paternity certainty (unless, of course, he is willing to offer them resources).

Key and Aiello (2000) used a model based on Prisoner's Dilemma (see **Box 2.3**) to explore the implications of the costs of rearing. The model involved a computerised tournament in which individuals acquired resources by playing Prisoner's Dilemma games repeatedly with each other; once an individual had acquired sufficient resources, it was able to reproduce with an individual of the opposite sex. The results suggested that male–female coalitions emerged readily. When the costs of reproduction were low for males (less than ten per cent of the costs for females), females exploited males, who in turn tended to adopt a weak defector strategy; however, when the costs of reproduction for males were greater than half those of the females, a reciprocal cooperation evolved between the two sexes (**Figure 5.5**).

Key and Aiello interpret this as reflecting the fact that males are drawn into investing in females when doing so allows them to enhance the female's reproductive potential. Key (2000) suggested that the crucial factor was that male investment in parental care allowed the female to reduce her interbirth interval, thereby increasing her fertility rate. A similar explanation based on a game theory analysis was advanced for the evolution of pairbonding in Callitrichid primates: in this case, paternal care enabled the female to quadruple her reproductive rate (Dunbar 1995b).

In contrast to Lovejoy's (1981) assumption, Key's analyses imply that pairbonding (and hence monogamy) arose quite late in human evolution. Her analyses show that it is crucially dependent on the magnitude of the reproductive costs born by the two sexes. Since these are directly related to brain size, and brain size only really entered an exponential phase of increase with late *Homo erectus* or early *Homo sapiens* (Aiello 1996), the evolution of pairbonding is unlikely to have predated this point.

of male provisioning (Hames 1996). A more plausible explanation is, perhaps, that females in multi-female households suffer from reproductive suppression due to stress from female–female conflict (for evidence of this, see p. 152). Some circumstantial evidence to support this suggestion is offered by the fact that sororal polygyny (where the wives share reproductive interests because they are sisters) is more likely to be associated with co-residence in a shared dwelling, whereas non-sororal polygyny is

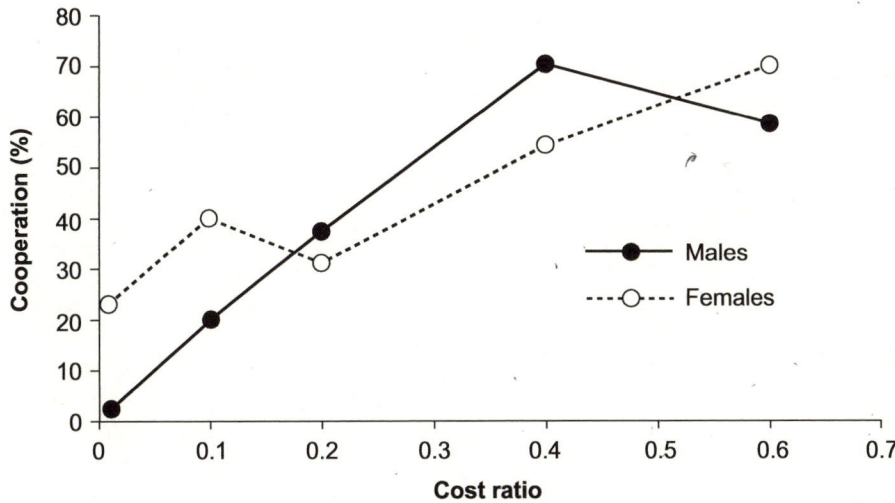

Figure 5.5 In a computer simulation in which the two sexes engage in Prisoner's Dilemma interactions in order to be able to reproduce, the proportion of interactions that are cooperative increases as the ratio of the costs of reproduction for the two sexes tends towards equality. In this simulation, only male costs vary, so a high cost ratio means that male investment in rearing has a significant effect on offspring survival. Redrawn from Key and Aiello (2000).

more likely to involve co-wives living in separate dwellings in order to reduce the social tensions between them (Daly and Wilson 1983). Chisholm and Burbank (1991) also noted that, in their Australian Aboriginal population, sororal polygynists had slightly (but not significantly) higher fecundity and more children surviving to five years of age than non-sororal polygynists.

SEXUALLY SELECTED TRAITS

If one sex exhibits marked preferences for mates that exhibit particular traits, sexual selection should lead to the exaggeration of those traits (or of cues that act as markers for those traits). We first consider physical traits that may be sexually selected cues, and then behaviour patterns that seem to have been subject to similar influences.

PHYSICAL TRAITS

The most obvious case of sexually selected characters in humans concern features such as beards and body shape that differ conspicuously between the sexes. Beards are plainly sexually dimorphic: they emerge at puberty, and do not have any apparent survival value given that they seem not to be necessary for women or children (or even men in some populations) (Barber 1995). If beards are a sexually selected trait, then we would expect that women find men with fuller beards more attractive than clean-shaven men. Barber (1995) reports a number of studies that seem to indicate that men with fuller beards are rated more favourably along a number of dimensions, including being older-looking as well as more masculine, dominant, courageous, mature and confident. Men typically have a deeper voice than women do (men's voices 'break' at puberty, but women's do not). This has long been thought to be a consequence of sexual selection, but its function has been poorly understood since there is no correlation between voice pitch and body size in men. Nevertheless, there is experimental evidence that women find men's voices with harmonics that are closer together (thus yielding a deeper pitch) more attractive; moreover, despite the lack of correlation between body size and pitch, women are able to make accurate judgements about a man's body size (but not his height or age) from spoken vowel sounds alone (Collins 2000). Indeed, the women in the sample strongly agreed in their ratings on all these variables.

It has also been suggested that aspects of male body build, particularly the upper torso, might be sexually selected. The shoulders of men, their upper body musculature, and biceps are all more developed than in women, even when differences in stature are accounted for (Ross and Ward 1982). Using silhouettes as stimuli, a number of studies have shown that females tend to prefer a moderately developed male torso more than extremely muscular physiques, suggesting that runaway directional selection has not occurred (Barber 1995).

Male height also has an effect on perceived social status, with taller men perceived as being of higher status (Barber 1995). There is some suggestion that height is associated with occupational advantage, with taller men receiving better pay and filling higher status positions (Schumacher 1982, Jackson 1992). Taller men are also said to enjoy an advantage in the dating game: they are perceived as more desirable dates (Jackson 1992), and they date more often (Shepherd and Strathman 1989). Taller men also seem to suffer less morbidity (for example, fewer heart attacks: Barker et al. 1990, Herbert et al. 1993, Kee et al. 1997). These differentials seem to be carried through into fitness: Pawlowski et al. (2000) found that Polish men who reproduced were significantly taller than those who did not reproduce, even when education (a proxy for family wealth), age and domicile were excluded as confounding variables (**Figure 5.6**). A regression analysis suggested that, on average, each metre in residual stature (adjusted for age cohort and natal location) was worth one whole offspring over a lifetime:

Number of offspring = 0.769 + 1.096 Residual Stature

$$(r^2 = 0.004, F_{1,3199} = 14.4, P < 0.001)$$

Whether male physique and stature are selected for because they are honest cues of good genes (only good genes are capable of producing tall men despite the less than ideal circumstances in which they develop) remains unclear. Barber (1995) suggests that male musculature and stature might be sexually selected traits given the generally reported female preference for taller, muscular men who, at the same time, are also more intimidating to other men. The fact that extremely tall men and overly muscular men are not attractive to women (and, in the case of the latter, do not appear to enjoy greater reproductive success) suggests that male physique is not necessarily a 'good genes' sexually selected characteristic (Barber 1995).

An alternative possibility is that these traits are cues of resource quality rather than good genes. Since the resources that men have to offer have a significant impact on reproductive success in traditional societies (and even in post-industrial societies), women may learn to prefer cues that are reliable indices of resources. Such learning may occur either through direct experience of life or through cultural transmission of rules of thumb (see Chapter 13). The extensive evidence (summarised above) that

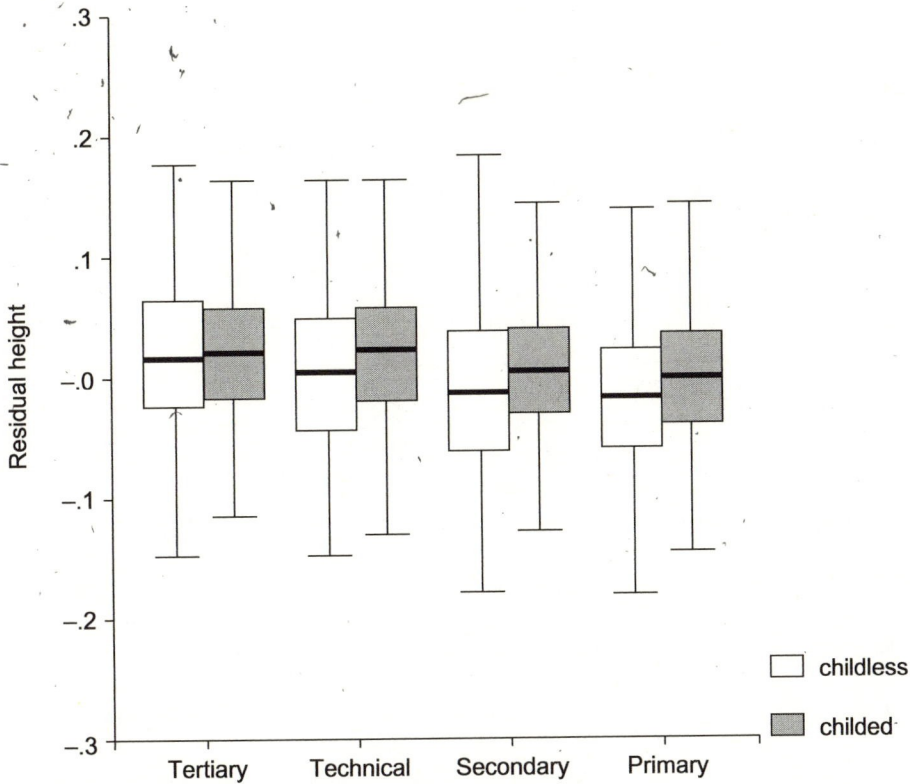

Figure 5.6 In a sample of 4419 Polish men, those who had children were significantly taller (when height was adjusted for age cohort and natal location) than men who had no children, irrespective of family wealth (indexed by educational achievement). The plotted values are the class medians, with 50 per cent and 95 per cent ranges. Data derive from compulsory medicals for men working in the city of Wrocław. Reproduced with permission from Pawlowski et al. (2000).

taller men have greater social and economic success (for whatever *social* reasons that may be) and/or come from families with greater resources (the richness of the rearing environment explains about 80 per cent of the variance in adult stature) provides more than sufficient basis for learning by observation.

The idea that female bodies might have been modified by selection pressure is controversial, not least because intersexual competition acting on females is less common in nature (Barber 1995). Sexually selected traits are typically found in males. Humans present a paradox in this regard: studies that have considered male and female attractiveness have tended to report that it is females that are both more interesting to look at and considered to be physically more attractive when rated by both males and females (Ford and Beach 1951, Jackson 1992). How might this apparent selection on female appearance be explained?

One argument suggests that female attractiveness is a strong predictor of whether or not a woman will marry, as well as the socio-economic status of her husband. The more attractive a woman, the higher is her probability not only of marriage, but also of marriage to a man of relatively high socio-economic status. There is evidence from traditional societies, for example, to show that women married to high-status men have more surviving offspring (for example, Voland 1988; Borgerhoff Mulder

1988c). Women might thus be expected to compete for high-status men who will be able to invest in children. If high status males are discriminating between females, intersexual selection occurs. If males choose females with high reproductive value, then selection for cues that advertise fecundability and reproductive value can be expected. These might include: sufficient fat reserves (convertible to foetal tissue and milk), age (younger women have a higher reproductive value), and general health (possibly indicated by hair and skin quality) (Anderson et al. 1992; Barber 1995).

In western societies, the generally preferred body shape for women is the 'hour-glass' figure, and this body build emphasizes enlarged breasts, buttocks and hips accompanied by a slender waist. Interestingly, there is some suggestion that changes in preferred body shape track economic conditions: Silverstein et al. (1986) have claimed that the bust-to-waist ratio that is seen as ideal by women is inversely cor-related with economic opportunities for women. They argue that women tend to accentuate their feminine morphology when economic prospects are bleak, and sup-press them when employment prospects are good. The **waist-to-hip ratio (WHR)** is controlled by sex hormones and reflects a sex difference in patterns of fat distribu-tion, with the ratio being lower in women of reproductive age than it is in men) (Barber 1995). Not only are women with low waist-to-hip ratios rated by both men and women as being more attractive, healthier and as having a higher reproductive value than women with higher waist-to-hip ratios, but low waist-to-hip ratios are associated with a number of fitness and health indicators (**Box 5.4**).

WAIST-TO-HIP RATIO

Nonetheless, there is considerable variation across cultures in men's preferences for women's body shapes: some cultures prefer slim women, others prefer their women to be plumper. Among the Tuareg of the Sahara, the height of beauty among upper class families borders on the obese and girls are force-fed on milk in order to achieve this (Randall 1995). Anderson et al. (1992) analysed a cross-cultural sample from the Human Area Relations File (a balanced regional sample of cultures) to test be-tween alternative hypotheses for why men's preferences should vary (**Table 5.3**). All but one of these hypotheses (the suggestion that men's preferences for female body shapes are entirely arbitrary and driven by local fashions of the moment) assume that body fat is a determinant of fecundity (at least, in sub-obese women), a claim for which there is considerable evidence (Frisch 1978, 1985).

Anderson et al. (1992) found significant correlations between attitudes towards fatness in women and (a) the reliability of the food supply (fatter when less reliable), (b) latitude (fatter in more equatorial latitudes), (c) relative social dominance of women (thinner when women are more dominant) and (d) the extent to which adolescent sexuality is likely to have adverse consequences for girls if they become pregnant (thinner when both the opportunities for adolescent sex and the costs of pregnancy are both high). The extent to which there are correlations between these variables remains unclear, however.

NEOTENOUS

Facial features of males and females differ quite considerably. Although there are some features that are reportedly attractive in both male and female faces (promi-nent cheek bones, large eyes, wide smile), there are other features that work in opposite directions. For example, small **neotenous** chins are attractive in women whereas large prominent chins (possibly signalling high testosterone titres and social dominance) are attractive in men (Cunningham et al. 1990). Generally speaking, neotenous fea-tures are considered attractive in female faces, while maturity-features are considered attractive in male faces. However, using 'morphed' facial photographs in which an individual's face was digitally enhanced so as to appear more feminine or more mas-culine, Perrett et al. (1998) found that women showed a significant preference for

BOX 5.4

WHR and body mass index

At least during their early reproductive years, human females have a distinctly hour-glass shape (narrow waist, with wider chest and hips) compared to human males (who tend to have a more tubular shape). Singh (1993, 1994, 1995) suggested that this curvaceous shape (which emerges only after puberty in females) was an important attractor in male mate choice because it was a signal of fertility. Fat deposition in the hips (in particular) is regulated by oestrogen in women (Björntorp 1991) and this in turn correlated with a number of indices of fertility and health. When male subjects rated outline female shapes for attractiveness, they generally gave the highest ratings to shapes whose waist-to-hip ratio (WHR) was around 0.7, and the lowest ratings to figures whose shape tended towards WHR = 1. WHR was calculated to be about 0.7 in *Playboy* centrefold models and winners of the Miss America beauty competition.

These findings have been confirmed and extended by a number of independent studies (for example, Henss 1995, Furnham et al. 1997). Singh and Luis (1995) obtained similar results from Chinese-Indonesian subjects (but these had been students at a US university for anything up to five years at the time of the study). The few studies carried out on traditional societies with minimal exposure to western culture typically suggest that WHR may not play so significant a role in men's preferences: in such cases, body mass tends to be a more common basis for choice (Machiguenga: Yu and Shepherd 1998; Hadza: Wetsman and Marlowe 1999). On the basis of these results, Yu and Shepherd argued that the apparently universal preference for low WHR was an artefact of the pervasiveness of western culture.

Other studies have argued that, despite an attempt to partial out the effects of body size, WHR and body mass — or more specifically **body mass index (BMI)**: body mass scaled for height — were confounded in Singh's original line drawings (Tassinary and Hansen 1998). More recently, Tovée et al. (1999) presented evidence to suggest that WHR and BMI are highly correlated in normal women. In addition, they showed that the best predictor of male ratings for the attractiveness of photographs of actual female body shapes (as opposed to outline drawings) was BMI (which explained 74 per cent of the variance in ratings), with preference being a ∩-shaped function of BMI. In contrast, although WHR was negatively related to preference (as Singh had claimed), it only accounted for 2 per cent of the variance in the data.

One possible resolution of this dispute is that males' preferences for female body shapes may be fine-tuned to those traits that are good predictors of female fertility in the local environment. BMI may be a better predictor of fertility in nutritionally stressed populations (that is, traditional horticultural and hunter-gatherer societies) whereas WHR may be a better index in well-nourished populations (Anderson et al. 1992, Wetsman and Marlowe 1999). One likely reason for this is that a mother's body mass is one of the key determinants of infant survival in traditional societies simply because it is a measure of how much spare capacity she has to divert into fetal growth and lactation. Evidence in support of this is given by Pawlowski and Dunbar (submitted). Using a sample of Polish women, they showed that WHR was the better predictor of neonatal weight (a key factor determining infant survival, and hence the mother's fitness) in women weighing over 54 kg, but BMI was the better predictor in lighter women. Women in most traditional societies tend to fall at the lighter end of the weight distribution.

male faces that had been feminised rather than those which had their masculine features enhanced. They suggested that this may be because women are actually seeking a trade-off between good mates and good fathers, and that more feminine faces suggest superior nurturing abilities. Some evidence relating to this is the finding that women's preferences vary with the menstrual cycle: women prefer more masculinised faces at around the time of ovulation, but more feminised faces at other times (Penton-Voak et al. 1999). These findings could be interpreted as suggesting that women

Table 5.3 Hypotheses to explain cross-cultural differences in men's preferred female body shape

Hypothesis	Prediction	Source
1. Food security	Fatness preferred where food supply is unpredictable	Brown and Konner 1987
2. Small but healthy	Small body size is an adaptation to food shortage	Seckler 1980
3. Climate (latitude)	Fatness preferred in colder climates	
4. Cues of fertility	Fatness is a cue of fertility or lactational output	Symons 1987
5. Adaptive reproductive suppression	Women opt for thinner body shapes in order to reduce fertility so as to minimise the risk of unwanted pregnancy	Voland and Voland 1989
6. Battle of the sexes	(4) and (5) are in conflict, and which achieves prominence depends on which sex is socially dominant	
7. Fraternal interest groups (FIGs)	When men's political status is dependent on FIGs, they will de-emphasize women's fertility, and hence ought to prefer thinner women	Paige and Paige 1981
8. Kirche, Küche, Kinder	When pregnancy interferes with women's labour, thinner women should be preferred (because less fecund)	
9. Whims of fashion	Males' preferences are purely arbitrary dictates of fashion	Mazur 1986

Source: Anderson et al. (1992)

prefer cues indicative of good genes when they are likely to conceive, but cues indicative of nurturing and parental investment potential otherwise.

The attractiveness of neotenous female features was nicely demonstrated by Jones (1995) in a cross-cultural study of exaggerated youth indicators in women. He created an index of relative facial neoteny that was based on the difference between an individual's actual age and her predicted age based on measurements of various facial proportions. When the faces were rated for attractiveness by subjects from five different cultural groups, Jones (1995) found considerable support for the idea that female attractiveness involves a substantial neotenous component: that is, ratings of female attractiveness increased as the difference between their predicted and actual ages increased. Females whose predicted age was less than their actual age were considered more attractive (Jones 1995), and this result held across cultures.

Perhaps more intriguing was the next step in his analysis: he measured the faces of models featured on the covers of magazines and then entered these measurements into the regression equations that were used for predicted ages. Jones (1995) found that, when compared to an undergraduate sample, models had significantly more neotenous features: the regression equation between neoteny index and age indicated that the models had predicted ages of about seven years. This is not to suggest that the faces of models match those of seven-years-olds, but rather it illustrates a relatively unexplored issue as to what makes the face of any particular model attractive. Jones (1995) suggests that the faces of models represent a 'supernormal stimulus', an exaggeration of those features that distinguish old from young faces.

It is widely recognised that physical attractiveness declines with age for both males and females, with this decline being more pronounced in females. The beauty industry is premised on a multibillion-dollar quest to retain a youthful appearance. Whatever else physical attractiveness may signal, it seems likely that, at least in women, it signals youthfulness, and thus fertility. This much is implied by the prominence of cues of physical attractiveness in personal advertisements (see pp. 96–7).

However, it remains possible that facial attractiveness in women is not always an entirely honest signal of youth. Underlying this suggestion is the **sensory bias** theory of sexual selection (Jones 1996): this suggests that a preference for facial attractiveness arises incidentally from a preference that is unrelated to mating and this then causes evolution in the opposite sex (Thornhill and Grammer 1999). Female facial neoteny has been cited as a possible example of this: given the male preference for youthfulness in women on account of the strong association between youth and fertility, women with an exaggerated youthful appearance will have had an advantage in female–female competition for mates. The male preference for facial indicators of age-related fecundity is thus seen as a sensory bias that selected for facial neoteny in females (Jones 1996). Thus facial neoteny is a supernormal cue of youth, not a cue of actual phenotypic and/or genetic quality (Thornhill and Grammer 1999).

SENSORY BIAS

FLUCTUATING ASYMMETRY

Fluctuating asymmetry (FA) refers to small random deviations from perfect bilateral symmetry (van Valen 1973; Manning 1995) and is held to be an indicator of 'good genes'. Departures from perfect symmetry are assumed to be the result of environmental stressors (for instance, reduced nutrition, disease, parasitic infections) destabilising those developmental processes under direct genetic control that are responsible for creating our bodily features. Greater symmetry (for instance, in the two sides of the

FLUCTUATING
ASYMMETRY

BOX 5.5

The problem of concealed ovulation

Human females are sexually receptive throughout their reproductive cycle; in contrast, most mammals (and some primates) are receptive only for a few days in each menstrual cycle. Humans also lack external cues of the imminence of ovulation, in striking contrast to the massively swollen perinea of some Old World primates (including our closest relatives, the chimpanzees) (see Domb and Pagel 2001). This apparent absence of any signals for ovulation (termed 'concealed ovulation') has aroused considerable interest because it was at first thought to be a unique feature of human reproduction.

Hypotheses to explain the evolution of concealed ovulation in the human lineage are legion. Among these are that concealed ovulation: (1) promotes paternity certainty (and hence increases the probability of paternal care) because males are forced to continue mating with the female in order to be sure of fertilising her (Alexander and Noonan 1979); (2) promotes paternity confusion (and hence reduces the risk of infanticide) because each of the (many) males that mate with the female has a finite chance of having fertilised her (Hrdy 1981); (3) reduces the risk that human females (because they are sufficiently smart and self-conscious to connect sex with pregnancy) will resort to contraception in order to avoid the anticipated pain of childbirth (Burley 1979).

One of the problems faced by attempts to explain concealed ovulation in humans is the difficulty of testing any hypothesis that ostensibly applies uniquely to one species. However, there is some doubt as to whether the question is in fact being phrased correctly. Rather than assuming that the human condition is phylogenetically derived (that is, a recent evolutionary acquisition), it may be that

what we see in humans is actually the ancestral condition. Phylogenetic analyses of the converse problem (the evolution of conspicuous morphological displays of oestrus such as the swollen paracallosal skin of baboons and chimpanzees) suggest that conspicuous displays have evolved just three times among primates, one of these evolutionary events being in the chimpanzee lineage subsequent to its divergence from its common ancestor with the hominid lineage (Sillén-Tullberg and Møller 1993, Nunn 1999; see also Dixson 1998). In addition, the unrestricted mating pattern found in humans may be only a slightly exaggerated form of normal anthropoid primate reproductive biology: anthropoid primates are characterised by menstrual rather than oestrous cycles in which behaviour is decoupled from the control of underlying hormonal mechanisms (Martin 1990; Dixson 1998). In other words, explanations of human morphology may not be needed.

Nonetheless, Sillén-Tullberg and Møller (1993) show that the evolution of monogamy in primates does not correlate with the evolution of concealed ovulation (thereby excluding Alexander and Noonan's paternity certainty hypothesis); if anything, the evolution of monogamy tends to follow a shift from conspicuous signals to concealed ovulation. Since it seems that conspicuous signals tend to co-evolve with more promiscuous mating systems, it has been argued that these signals are in fact signals of quality designed to entice males to mate with higher quality females (Pagel 1994). There is now evidence to support this claim from comparisons at both the species (Nunn 1999) and the individual levels (Domb and Pagel 2001).

---

face, limb length or any other feature that has a mirror image on each side of the body) thus reflects the presence of 'good' genes, in the sense that they are able to resist environmental insults of this kind.

For this reason, FA is used widely in studies of sexual selection (for example, the tail streamers of male swallows, Møller 1990). The assumption is that, if mate choice results in paired structures or ornaments, then both the size and symmetry of ornaments might function as reliable indicators of mate quality (Manning 1995). Although humans have not evolved conspicuous sexually selected ornaments or structures, FA is nonetheless believed to be important in mate choice. A number of studies have found that facial attractiveness correlates with symmetrical body traits (Thornhill and Gangestad 1993, 1999; Gangestad et al. 1994), and that frequency of female orgasm (which probably results in high sperm retention) correlates with low FA (that is, greater symmetry) in male partners (Thornhill et al. 1995).

Manning (1995) found negative relationships between FA (measured as the mean of several different features) and both weight and height for adult males, suggesting that more symmetric males tended to be heavier and larger, while for females the reverse was true (smaller women tended to be more symmetrical). Manning (1995) argues that his results are consistent with the idea that body weight in humans is a sexually selected character: larger body size and well-developed musculature is important in male-male fighting, with the growth and maintenance of large body size being costly and probably condition-dependent. The converse may be true for women: one of the classic trade-offs in evolutionary biology is that investing in growth comes at the expense of investing in reproduction (Stearns 1992). Smaller body size in women may thus reflect the fact that women withdraw from physical growth earlier than males and so begin to invest in reproduction earlier. This at least seems to be what we observe: women mature earlier than men and usually begin to reproduce before men do.

That FA may have been subject to sexual selection, at least in males, is suggested by two further sets of data obtained by Manning and his co-workers. Manning and Wood (1998) showed that FA in a sample of 90 teenage boys was negatively correlated with self-reported levels of physical aggression (that is, the most symmetrical boys showed the highest aggression), while Manning and Pickup (1998) found a negative correlation between FA and performance among 'professional' middle-distance runners (that is, faster runners were more symmetrical). To the extent that aggression and performance skills have been a factor in inter-male rivalry (or female choice), this implies that the genetic factors involved in FA (in effect, the 'good genes') may have been subject to significant levels of selection. These indices are very likely to reflect testosterone titres and, perhaps, the individual's ability to resist the deleterious effects that high testosterone levels have on the immune system and the body generally. Further evidence for this is provided by Manning and Taylor (2001), who found that low **second-to-fourth digit (2D:4D) ratios** (that is, fingers of radically different length) were correlated with attainment level in both professional footballers and other athletes. (2D:4D ratio typically approximates unity in women, but is significantly lower in men, and the level of divergence from unity is thought to be under the influence of fetal testosterone.) Such effects could be due either to the influence of male–male competition (intrasexual selection) or to female choice (intersexual selection). The fact that professional male musicians also had significantly lower 2D:4D ratios than was typical of men in general (Slumming and Manning 2000) tends to favour the second.

SECOND-TO-FOURTH DIGIT RATIOS

SPERM COMPETITION

Sexual selection theory has made very considerable play out of the fact that intrasexual competition between males may extend beyond the overt business of fighting for fertile females into the less easily observed realms of competition between the sperm of rival males within the female reproductive tract (Parker 1970). We have already noted that relative testis size correlates with the level of polygyny in the mating system (**Figure 5.4**). This comes about because success in sperm competition is a function of how much sperm a male can place in a female's reproductive tract when he mates with her) in those species where mating is promiscuous, males try to overwhelm their opponents by sheer volume.

The theory of sperm competition is well established and thoroughly documented, but virtually all the work has been done on insects and birds; less work has been done on mammals, even though in principle the same processes should apply across all taxa (including humans). Baker and Bellis (1995) have made the strong claim that a number of different sperm types can be differentiated in human semen, and that these subserve different functions. They suggest that some sperm morphotypes are designed specifically to seek and destroy the sperm of rival males (hence the *kamikaze sperm hypothesis*), while others are designed to form a coagulate at the cervix in order to block sperm from a subsequent copulation with another male from entering the ovarian ducts; more importantly, they suggest that only about 10 per cent of sperm are actually designed to fertilise the egg. Although both their claims and their evidence have been disputed by those working in human reproductive physiology (for a summary of this debate, see Birkhead 2000), it is not always clear whether the criticisms of this work are based on biological fact or reproductive biologists' perennial suspicion of (and failure to understand) evolutionary explanations.

EXTRA-PAIR COPULATIONS

Baker and Bellis (1995) are perhaps on firmer ground with the evidence for their claim that (male ejaculate volumes are titrated to take account of the risk of **extra-pair copulations (EPCs)** that their partners might engage in.) They were able to show that ejaculate volume from actual copulations was correlated with the length of time for which the couple were apart, and that this effect was independent of the time since the last copulation (**Figure 5.7**). They argued that this functions to 'top up' the female in order to maximise the probability of conception should ovulation occur, given the likelihood that sperm from rival males may be present. In effect, males attempt to swamp any rival sperm that might be present in their partner's reproductive tract from EPCs. They provide data from a nationwide UK survey to show that (the risk of EPCs increases significantly as the proportion of time that a couple spend in each other's company declines) Although the basis for these data is an unstructured self-selecting survey (and the results have been criticised for this reason), it seems unlikely that a modestly large set of naive subjects could conspire to produce such a consistent set of results.

RISK-TAKING BEHAVIOUR

There are striking sex differences in risk-taking that bear all the hallmarks of being a sexually selected trait. It has long been recognised that, during their teens and twenties, boys are not only more prone than girls of the same age to take risks (driving too fast, trivial competitive altercations and indulging in risky sports and habits such as smoking, unprotected sex and dangerous outdoor activities: Poppen 1995, Clift et al. 1993, Flisher et al. 1993, Harre et al. 1996, Tyler and Lichtenstein 1997, Bruce and Johnson 1994, Powell and Ansic 1997, Howland et al. 1996, Wilson and Daly 1985, Wilson et al. 1996) but they also suffer higher mortality rates from doing so.) Johnson (1996) analysed data for the 676 recipients of Carnegie heroism awards up to 1995 and showed that (a) men outnumbered women 9:1 as medalists but only 3:2 as the recipients of rescue attempts, (b) that women were more likely than men to rescue relatives (20 per cent vs 6 per cent) whereas men were more likely to rescue people they did not know (47 per cent vs 68 per cent) and (c) that these contrasts were even more marked in those cases where the rescuer lost his/her life.

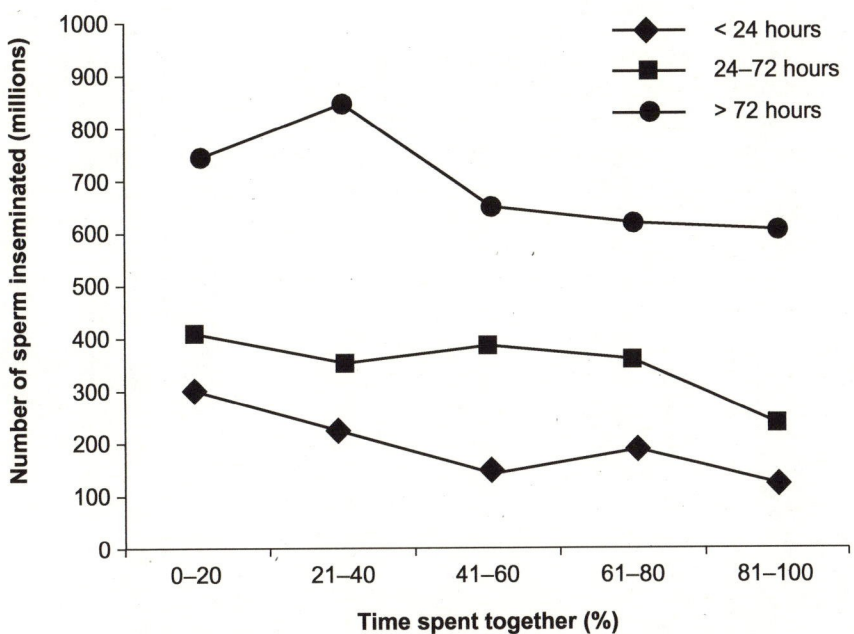

Figure 5.7 Males attempt to reduce the risk of their partner being fertilised by another male by adjusting the number of sperm ejaculated during copulation (estimated from ejaculate volume) as a function of the proportion of time since the last copulation during which the couple were apart (and another male might have mated with his partner). This is not due to the fact that, when a couple has been apart for a long time, the male's sperm store is less likely to have been depleted by copulation, since the relationship holds for three separate intervals since last copulation (< 24 hrs, 24–72 hrs, > 72 hrs). Redrawn from Baker and Bellis (1995).

This pattern seems to be as typical of traditional tribal societies as it is of post-industrial ones. Among the Maasai pastoralists of southern Kenya, young males traditionally achieved manhood (and became eligible to marry) by taking part in a lion hunt. The ultimate accolade (reflected in the attentions of the unmarried girls) was, however, reserved for the one who threw away his spear and acted as the bait for the lion, thereby making it easier for his companions to secure a kill. Honest signals come at a cost, and this behaviour was no exception: by allowing the lion to jump him while protected only by a hide shield, the young warrior often incurred terrible injuries from which death might ensue. But the demonstration of bravery remained beyond doubt.

It has been suggested that risk-taking is a form of competitive advertising: young men who take (and survive) risks are more attractive as mates because their ability to survive these risks is an honest cue of high quality genes. Intra- and intersexual selection conspire to make young males more risk-prone.

That the ability to take (and survive) risks may in fact be seen as an indicator of mate quality by women is suggested by a study carried out by Kelly and Dunbar (2001). They presented subjects with brief vignettes of men that differed in three key respects: risk-taking, the voluntariness of risk-taking and altruism (**Table 5.4**). Subjects were asked to rate these individuals' attractiveness as long-term partners, short-term mates ('one night stands') and friends. Women rated voluntary risk-takers ('heroes') as significantly more attractive as mates than men who undertook risks as part of their jobs (for example, fire-fighters) or altruists ('nice guys') (**Table 5.5**). In

Table 5.4 Vignettes used in Kelly and Dunbar's (2001) study of heroism vs altruism in women's mate choice

Trait	Description
NA, NB, P	*George* lives and works in the city and has a wide circle of acquaintances. He is employed as a financial adviser and when he is not working he likes going to the theatre, restaurants and wine bars.
A, B. V	*Bill* is a quiet but capable man. He's **always ready to lend a hand**, but is unobtrusive about it. He works as a mechanic during the day, which allows him the flexibility he needs as a **key member of the local lifeboat crew.**
NA, B. V	*Frank* likes to be outdoors whenever he can. He is **attracted by danger** and spends many weekends and holidays **rock-climbing**. He has recently taken up **free-fall parachuting.**
A, NB, P	*Charles* **enjoys looking after people.** He works as a **nurse** at present, but is considering training to be a **counsellor** as he thinks the prospects might be better. He is learning to play the clarinet, but doesn't practice as much as he should as he **worries about disturbing the neighbours.**
A, NB, V	*Dennis* lives **next door to his elderly grandmother**, and spends a lot of time keeping her company and **doing jobs for her.** He works in computing and in his spare time enjoys giving dinner parties for his many friends.
NA, NB, V	*Henry* works for a supermarket chain as an administrator. He plays golf, goes to concerts, and enjoys exploring the Internet on his home computer.
NA, B, P	*Edward* is a bit of an **action man.** He is always on the go, and throws himself into his work. He used to be a **steeple jack**, but now works as a **deep-sea diver.** When not working, he goes **jet-skiing** and plays tennis.
A, B, P	*Alec* is the kind of man you can **rely on in a crisis.** He's always getting into and out of **dangerous situations**, usually **on behalf of others.** He works as a **fireman** and has an **award for bravery.** His favourite pastimes include going to the cinema.

Note: A – altruist; NA – non-altruist; B – brave; NB – non-brave; P – professional (altruistic or brave acts are part of the man's job); V – voluntary (altruistic or brave acts are not part of the man's normal job). Bold terms indicate key cues to traits.

Table 5.5 Women's mean ratings of males characterised by different traits

Relationship	Altruist	Non-Altruist	P*	Brave	Non-Brave	P	Volunteer**	Professional	P
One night stand	3.2	2.9	0.02	2.5	3.6	<0.001	2.7	3.3	<0.001
Long-term mate	3.1	3.4	<0.001	2.9	3.6	<0.001	2.9	3.6	<0.001
Friend	2.3	2.8	<0.001	2.3	2.9	<0.001	2.3	2.8	<0.001

Notes: *ANOVA
** Acts of altruism or bravery are part of the male's job ('professional') or not ('bravery')
60 female subjects were asked to rate (on a 1–5 scale) vignettes describing eight males whose descriptions counterbalanced the three traits and their opposites.

Source: Kelly and Dunbar (2001)

contrast, altruists were preferred as long-term mates. Heroism appeared to be functioning as an honest cue of good genes, whereas altruism seemed to be a cue of males' willingness to invest in offspring. Significantly, when male subjects were asked to assess the same vignettes in terms of their attractiveness to women, their responses mirrored the women's responses: men seem to be aware of what it is that women look for in prospective mates.

Other kinds of behaviour that might be interpreted in a similar vein are tattooing (invariably more extensive in males than females) and the surgical modification of the genitalia (which may be both painful and, in the West, expensive). Both are culturally widespread and strongly male-biased (Rowanchilde 1996).

CONDITIONAL MATE CHOICE STRATEGIES

We have so far considered mate choice preferences as absolute universals. This is perfectly reasonable insofar as natural selection can be expected to impose pressure favouring ideal traits. However, the mating arena is a frequency-dependent market: even if we all agree on what constitutes the ideal mate, we cannot all have him or her. Not merely will competition between us for the ideal mate result in most of us failing, but the ideal partner will also have his/her own preferences. The mating marketplace is a two-way process of negotiation, and individuals can be expected to adjust their demands in the light of what mates are realistically available to them. As a result, mate choice is likely to be a conditional strategy that is dependent on the individual's particular circumstances and his/her standing in the market (literally, his or her market value).

Some evidence to suggest that frequency-dependent effects of this kind play an important role in human mate choice is offered by the fact that humans mate assortatively – the coupling of individuals based on their similarity (or occasionally dissimilarity) on one or more physical, psychological or social traits such that 'like prefers like' (Buss 1985). Assortative mating is in fact the human norm, with spouses tending to be more similar to each other on a range of traits (including race, religion, ethnic background, and socio-economic status, as well as a wide range of physical traits such as stature, body build, finger length and so on: Buss 1985) than they are to the population at large. Although assortative mating has often been interpreted as evidence for active choice, it can equally be interpreted as a best-of-a-bad-job strategy: you end up with someone similar to you because you cannot compete effectively for better quality mates.

When individuals cannot offer (or at least perceive themselves as being unable to offer) those traits that members of the opposite sex seek in mates, they must seek alternative strategies if they are to be successful. One best-of-a-bad-job solution to this predicament is to lower one's standards. By being less demanding, you open up the range of potential mates available. Waynforth and Dunbar (1995) demonstrated exactly this effect in US personal advertisements, and these results have been replicated in UK advertisements. (In this case, failure to mention key traits was taken to imply their absence.) Male advertisers who did not mention cues of status/wealth were less demanding (that is, sought fewer traits in a prospective partner) than males who did mention such cues (**Figure 5.8a**). Similarly, women who did not mention cues of physical attractiveness were less demanding than women who did (**Figure 5.8b**).

In addition, Waynforth and Dunbar (1995) found that males who did not offer

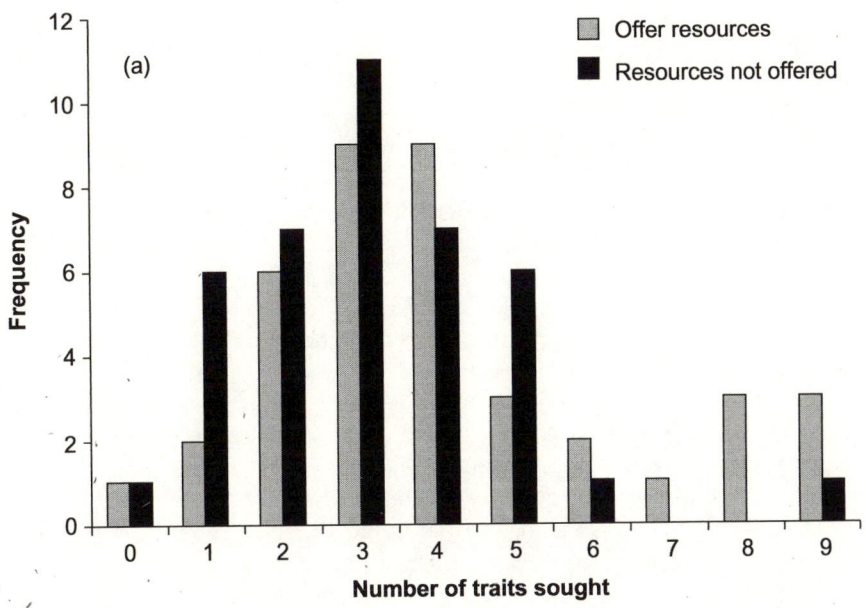

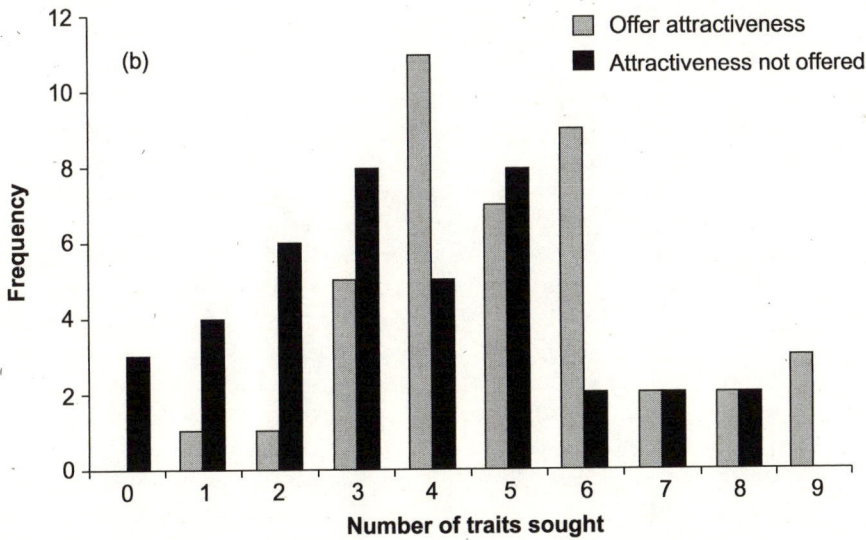

Figure 5.8 Demandingness (indexed as the number of traits sought in a prospective partner in personal advertisements) of (a) men who offer or do not offer resources and (b) women who offer or do not offer cues of attractiveness. Because women demand resources, men who have them to offer can afford to be more demanding. Similarly, because men seek attractiveness, women who can offer this trait can afford to be more demanding. Source: matched advertisements by 40 women and 40 men who advertised in various US newspapers. Source: Waynforth and Dunbar 1995.

resources indicated that they were more willing to accept a woman's children from a previous relationship compared to men who did offer resources: indeed, men with resources to offer commonly stated that they would not accept offspring from a previous relationship. Given males' general reluctance to bear the cost of rearing other men's offspring, the willingness of young males with few resources to offer to accept children is striking, and strongly suggestive of a trade-off: men who recognise that they might compete poorly on the open market seek alternatives that will make them more attractive. This suggestion is reinforced by the fact that women who state that they will be bringing children into the new relationship clearly recognise that they are at a disadvantage: they are significantly less demanding of prospective partners than women who do not have children (or at least do not declare them).

Trade-offs of this kind are probably quite common. Pawlowski and Dunbar (2001) asked subjects to rate the attractiveness of different advertisements that were carefully constructed to balance five different trait categories (attractiveness, resources, commitment, social skills and sexiness). Comparisons of the preferences for advertisements containing different category terms suggested that there were trade-offs between at least two sets of traits, namely resources versus commitment and resources versus social skills.

Cashdan (1993) used a questionnaire-based approach to explore women's mate-searching strategies. Her concern focused on the contrast that Draper and Harpending (1982) drew between 'cads' and 'dads' (the distinction being in terms of the males' willingness to invest in parental care). Cashdan asked subjects to rate agreement/disagreement with a series of statements about mate attraction tactics. She found that women with low expectations of paternal investment were more likely to agree with sexually flagrant tactics (often with a view to obtaining high pre-reproductive investment from males), whereas those who had high expectations of paternal care from prospective mates were more likely to agree with statements emphasizing chastity and fidelity. Male respondents showed a corresponding tendency for those who favoured non-investment to favour flaunting their sexuality to females, while those who favoured investment preferred to emphasize chastity and fidelity. An obvious interpretation of these contrasts is that cads seem to be providing opportunities for good genes, whereas dads provide resources for rearing.

If mating tactics can be adjusted facultatively in the way these findings suggest, then it is likely that their frequencies in the population will depend on simple market forces. Pawlowski and Dunbar (1999a) considered the impact that an individual's 'market value' might have on his or her willingness to make demands of a preferred partner. The market value of a given age-sex cohort was taken to be the proportion of advertisers that requested partners in that age class divided by the proportion of advertisers in that age cohort available in the sample population. Values greater than one indicated an excess of demand over supply, values less than one an excess of supply over demand.

The market value of females peaked in the late twenties, while that for males peaked in their late thirties (see also above). Using population data for the UK, Pawlowski and Dunbar (1999a) were then able to show that female market value was most likely determined by fecundity rather than by reproductive value (**Figure 5.9**), while male market value was most strongly determined by a combination of a man's income and the probability that he will still be married to the female 20 years later (that is, at the end of the period of intense parental investment) (**Figure 5.10**). On the basis of these calculated market values, Pawlowski and Dunbar (1999a) then examined whether individuals were sensitive to their standing in the mating market in terms of how demanding they were of prospective partners (that is, how many

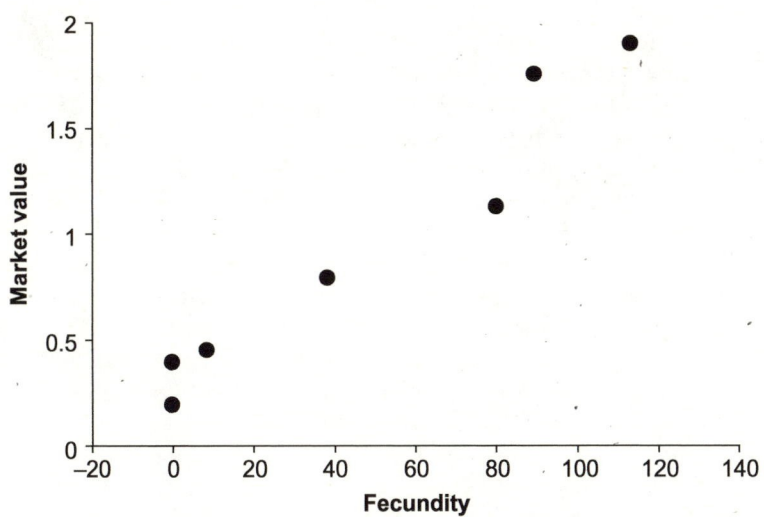

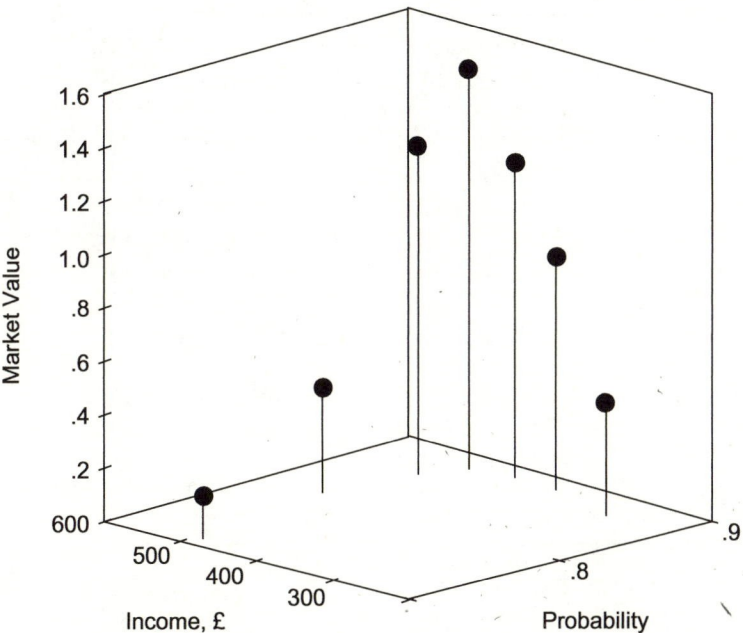

Figure 5.9 Female market value (indexed as the proportion of male advertisers seeking female partners of a given age cohort divided by the proportion of women of that age in the advertising population) is a linear function of female fecundity (number of births per 1000 women per year, determined from the most recent UK government survey data for England and Wales). Each datapoint is a 5-year age cohort (except for the 50–59 year group). Source: 454 women aged 18–59 who advertised in the UK's *Observer* newspaper. Reproduced from Pawlowski and Dunbar (1999a).

Figure 5.10 Male market value (indexed as the proportion of female advertisers of reproductive age seeking male partners of a given age cohort, divided by the proportion of men of that age in the advertising population) is a linear function of both a male's mean weekly income and the probability that he will still be married to the same person in 20 years time. The probability of still being married 20 years after marriage (approximately the end of the period of parental care) is estimated as the age-specific conjoint probability of not dying and not divorcing during the next 20 years. Income and probabilities of death and divorce were determined from the most recent UK national survey data for England and Wales. Each datapoint is a 5-year age cohort (except for the 50–59 year group). Source: 445 men aged 18–59 who advertised in the UK's *Observer* newspaper. Redrawn with permission from Pawlowski and Dunbar (2001).

traits they specified that a partner should have). Overall, with the exception of males (and, to a much lesser extent, females) in the 45–49 age group, there was a significantly positive relationship between market value and how demanding individuals were for both sexes (**Figure 5.11**). Advertisers thus adjust their demands in the light of their own perceived standing in the market place.

An amusing example of how circumstances may lead to the adjustment of mate evaluation is given by Pennebaker (1979). He asked (sober) individuals in three singles bars to rate the members of their own sex and the opposite sex present at three different times during the evening (9 pm, 10.30 pm and midnight). While there was no obvious tendency for the attractiveness of same-sex individuals to change as the evening progressed, members of the opposite sex were judged to be increasingly attractive with time (**Figure 5.12**). Since the two sexes did not differ in their views and it is unlikely that the least attractive individuals are more likely to find partners first, it seems likely that subjects were downgrading their criteria for attractiveness as time passed and the risk of going home alone increased.

Waynforth and Dunbar (1995) argued that mate choice criteria are far from static. Although there are underlying universal trends of the kind that many researchers have assumed, the intrusiveness of these is likely to depend on local economic and demographic conditions. Analyses of advertisements from late nineteenth-century England (Pearce 1982) and Kipsigis bridewealth patterns (Borgerhoff Mulder 1988c) indicate that, in more traditional economies, women place a much higher premium on suitors' abilities to offer resources than do women in contemporary industrial economies, for the very good reason that infant survival is directly related to the resources the woman has available in traditional economies (Krummhörn: Voland 1988; Kipsigis: Borgerhoff Mulder 1988c). Health care and social security arrangements in modern western countries, combined with significant improvements in mean wealth and women's own employment opportunities, have greatly reduced this effect: women no longer need to be quite so concerned about the survival of their children. Men's role as resource providers is thus less important, and a woman may do better to seek males who are willing to invest in the social aspects of childcare. Hence, in modern advertisements, a higher proportion of women seek cues associated with commitment (**Figure 5.1a**). This shift in mate choice preference can be attributed directly to the fact that very significant changes have occurred in women's economic and demographic circumstances over the past half century.

COURTSHIP

Grammer (1989) pointed out that courtship is a process of negotiation. It consists of a series of stages during which a couple become progressively more intimate: at the end of each stage, each individual decides whether to allow the relationship to progress on to the next higher stage of intimacy and commitment or to terminate the relationship. The ultimate endpoint is a commitment to a long-term relationship that is likely to result in the production of children. However, as Grammer (1989) noted, the goals of the intermediate stages may well differ between the sexes in line with the differences in their respective reproductive strategies. That can itself lead to conflict between the sexes. Note, incidentally, that Grammer's framework is applicable only to societies that allow free decisions in courtship, and so excludes those societies or cultures where families arrange marriages.

Courtship begins with advertisement. Physical looks, clothing and behaviour all

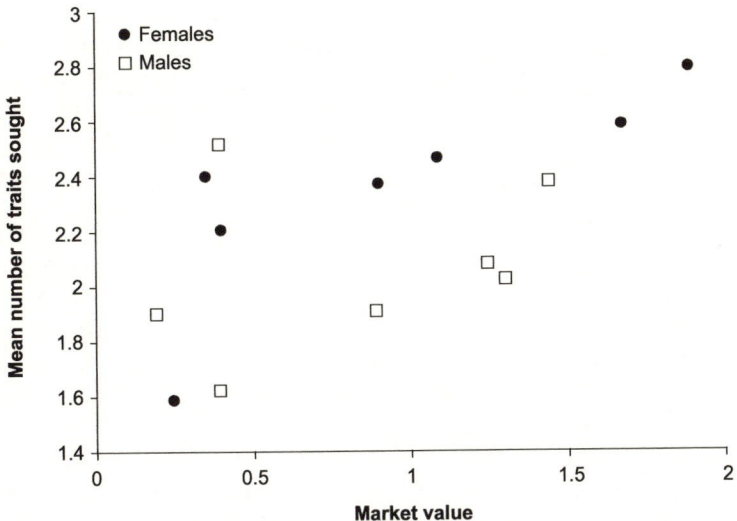

Figure 5.11 The demandingness of males and females (indexed as the mean number of cue terms listed as sought in a prospective partner) is a linear function of market value (defined in legends to Figures 5.10 and 5.11). Each datapoint is a 5-year age cohort (except for the 50–59 year group). Source: 454 women aged 18–59 who advertised in the UK's *Observer* newspaper. Reproduced with permission from Pawlowski and Dunbar (1999a).

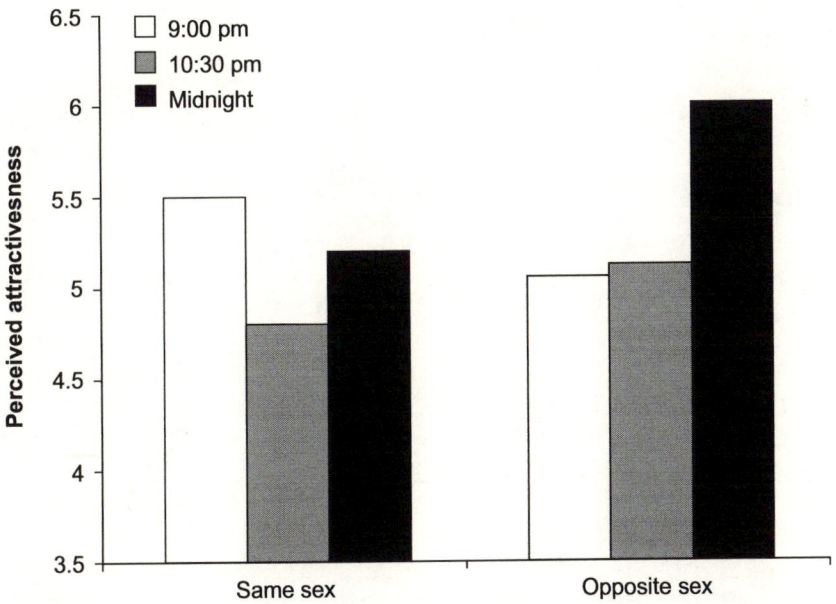

Figure 5.12 The closing time phenomenon: when 103 (sober) men and women in three Virginia bars were asked to rate (on a 1–10 scale) the attractiveness of the other people present at three different times during the evening, their ratings for their own sex remained roughly similar but they claimed that members of the opposite sex were becoming increasingly attractive as closing time approached. The most likely interpretation of these data is that subjects were lowering their standards of what was acceptable in a prospective partner as the risk of going home alone increased. Redrawn from Pennebaker (1991).

combine to signal information to the other individual. The decision to advertise directly to a particular person, or even to approach that person, is at least partly based on the advertiser's assessment of his or her qualities in relation to other same-sex individuals: we are usually cautious about making approaches where we feel they are unlikely to succeed because the other person may not be interested in us (see for example, Duck and Miell 1982).

Once an initial choice has been made, individuals may exhibit quite specific behaviour patterns that indicate interest in the prospective partner. From a study of behaviour in singles' bars, Moore (1985) found that these included prolonged eye contact, swaying the upper body towards the partner while talking, 'accidental' touching and a number of other tactical devices designed to attract attention. Prospective partners use these cues to assess the advertisers' interest in them. Similar results were obtained by Grammer et al. (2000) from a detailed video-analysis of behaviour between stranger couples in a waiting room: the frequency of 'solicitation' behaviours (including elements such as hair-flicks, head tosses and upper body posture) exhibited by the females was correlated with the level of interest in dating the male that they subsequently expressed. Grammer et al. (1998) noted that one important element in signalling interest was the extent to which the woman used rhythmic body movements to mirror the male's behaviour.

More interestingly, Grammer et al. (2000) found that the women's behaviour seemed more designed than the men's was to control the development of the relationship, even though they rarely exhibited explicitly negative behaviour towards the male. At the same time, the women seemed to behave in a more ambivalent way that gave the male few cues as to their real intentions – 'protean' behaviour in the sense of Miller (1997) (see Grammer et al. 1997) – while at the same time probing the male for information that could be used to assess the male's suitability as a mate. In effect, they seem to be trying to keep their options open until they have had a chance to explore the male's suitability in detail. Grammer et al. suggest that this may explain why men are more likely to misinterpret (that is, overestimate) women's level of interest in them. They suggest that, in part, this careful management of how (or whether) a relationship develops may reflect women's greater concern over ensuring the quality of any male they finally choose to mate with.

Static cues such as ornamentation may also play a part in this process. Low (1979) noted that these come in two forms: cues of availability and cues of attraction. The first kind essentially state whether or not the wearer is available as a potential mate. Amongst many others, these include such well known examples as wedding rings, the way the hair is worn (Victorian women would put their hair up after marriage), and, in Tahiti, which ear a flower is tucked behind. In contrast, cues of attraction typically emphasize sexually selected features. These may include the way clothes give prominence to a woman's hour-glass-shaped figure (for example, the use of bustles in Victorian dresses), the use of cosmetics such as rouge and lipstick to highlight features of the face, or the wearing of jewellery to draw attention to key body parts (for example, ankle bracelets, ear rings, necklaces). The use of eye shadow is thought to mimic natural signals of fertility (the skin of the eyelids darken due to increased blood flow during ovulation), while rouge mimics the way the skin flushes when aroused; similarly, the use of digitalis to dilate the pupils of the eyes by the nine-teenth-century prostitutes of Italy seems to reflect the fact that men invariably find women with dilated pupils more attractive (Morris 1977). In the case of males, cues of this kind may involve 'conspicuous consumption' (the purchase or display of expensive cars, clothes, fashion accessories, and so on). Lycett and Dunbar (2000), for example, showed that males are significantly more likely to display and toy with mobile phones as the sex ratio of the group became increasingly male-biased (and the competition for the females in the group increased).

Finally, risk-estimation remains a central issue in the process of courtship: for males, a female of high physical attractiveness carries the risk that another higher ranking male might compete for her, and any investment might thus be in vain (Grammer 1989). For males, risk perception is a function of the quality of the partner and the level of competition for partners. For females, risk perception is predominantly a function of a male's commitment: are there behavioural cues that might indicate philandering? Overall, Grammer (1989) suggests that actual mate choice decisions, and courtship, are in line with ultimate considerations, while the decision to initiate contact is driven primarily by risk-estimation.

TACTICAL USE OF DECEPTION

Since courtship inevitably involves advertising, at least in its early stages, the use of deception as an integral part of the process is inevitable. Pawlowski and Dunbar (1999b) have shown that women older than 35 years tend to suppress reference to their age in personal advertisements (see also Greenlees and McGrew 1994). This appears to be a direct response to the fact that the number of replies received is negatively correlated with female age (Baize and Schroeder 1995), presumably because declining fertility makes older women of less interest to men. By not declaring their age, older women are able to be more demanding of prospective partners than women of the same age who declare their age (Figure 5.13). (Pawlowski and Dunbar were able to estimate the ages of women who did not declare their ages from the ages of the men they sought: it turns out that the age of man sought is highly correlated with the advertiser's actual age.) Pawlowski and Dunbar (1999b) found that the cue profiles of age-withheld advertisers closely matched those of women in their twenties, suggesting that they were attempting to behave like 20-years-olds who are demanding in traits sought but do not have many traits other than their youth to offer.

Although such women risk being 'discovered' once the couple meet, the strategy probably allows women to overcome the initial lack of interest that men would show on the basis of age alone; once the couple has met, the man may be prepared to compromise on his initial requirements and settle for what is on offer (a literal case of the bird in the hand being worth two in the bush). In other words, suppressing explicit information about age may allow women both to stay in the game longer during the initial stages and to exert more control over courtship once they enter the later phases (by being able to decide whether to accept or reject the inquiries they do receive).

Tooke and Camire (1991) found that the use of deception tactics during mate-searching mirrored the mate choice criteria of the opposite sex. In other words, men were more willing to present themselves as more resourceful, trustworthy and sincere than they actually were, while women were more likely to use tactics that enhanced their physical appearance. Corby (1997) asked male and female subjects to rate 76 acts of deception for the frequency that they were used by members of their own sex to attract a prospective mate for either a short- or a long-term relationship. (The 76 acts were elicited from a separate study in which 82 subjects were asked to list five tactics they knew had been used.) Some 59 of these yielded significant differences between either the sexes or the mating context (Table 5.6). These suggest that males are more

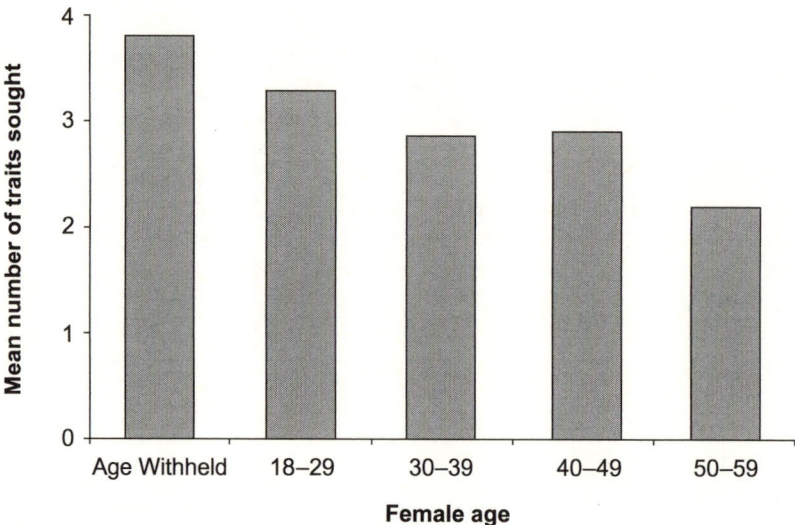

Figure 5.13 Women who suppress information about their age in personal advertisements are more demanding (indexed as the mean number of cue words sought in prospective partners) than those who do not. Because men use age as a proxy for fecundity, older women are less likely to receive replies than younger women. Women who suppress their ages are typically aged 35–50; failing to declare their age allows them to be as demanding as younger women in their twenties. Source: 454 women aged 18–59 who advertised in the UK's *Observer* newspaper. Reproduced with permission from Pawlowski and Dunbar (1999b).

likely to use a variety of bragging tactics whereas females are more likely to enhance their appearance, behave coyly or pay more attention to the prospective mate. Both sexes are also more likely to exaggerate traits in the pursuit of short-term relationships (one night stands) compared to long-term relationships (except for acting coy).

FITNESS CONSEQUENCES OF MATE CHOICE

To what extent do mate selection patterns pay off in terms of fitness consequences? This is not always easy to determine, though there is evidence to suggest that tall men have higher fitness than short men (**Figure 5.6**); in addition, there is evidence that women with more symmetrical features have more children than those with higher FA scores, while, in European populations, those with lower waist-hip ratios have heavier neonates (Pawlowski and Dunbar submitted).

In many traditional societies, as we have seen, women's reproductive success may be as much dependent on the wealth of their husbands as on their own characteristics. Low and Clarke (1991), for example, found that in rural Sweden during the nineteenth century, wealthier women had higher fertility at all ages than did poorer women. However, more detailed analyses often reveal that women's own intrinsic characteristics may also influence their fertility. In a study of three cohorts of Kipsigis women, Borgerhoff Mulder (1988a, b) found a positive relationship between age at menarche and reproductive success (when other factors were held constant). Aside from the trivial fact that women who mature earlier will have relatively longer reproductive lifespans, Borgerhoff Mulder (1988a, b) demonstrated that early-maturing

Table 5.6 Deceptive mating tactics that elicited significant differences in subjects' ratings as likely to be used by members of the opposite sex

| Tactic[c] | Mean rating score[a] for frequency of use | | | | Significant differences[b] | |
| | Male tactics | | Female tactics | | | |
	short[d]	long[d]	short	long	Sex	Context
Spending more money than can be afforded (N=5)	2.83	2.32	2.29	2.16	3	3
Claiming higher past/future achievement than warranted (N=6)	3.38	2.49	2.08	1.95	6	4
Lying about age (N=1)	3.40	2.41	1.60	2.31	1	0
Acting more trustworthy/sincere/kind than normal (N=6)	3.04	2.84	2.89	3.11	5	1
Wearing clothes that enhance appearance (N=7)	1.39	1.32	3.67	3.13	7	2
Enhancing personal appearance (e.g. cosmetics) (N=10)	1.72	1.62	3.49	3.03	9	3
Acting more responsive than normal (N=3)	2.98	2.91	3.07	2.75	3	1
Exploiting other individuals (N=6)	2.95	2.22	2.78	1.91	4	6
Sexual flaunting (N=3)	3.13	2.43	2.64	1.63	2	3
Behaving coyly (N=3)	1.80	2.10	2.96	3.15	3	1
Presenting self in more positive light (N=9)	3.11	2.63	2.78	2.20	4	6

Notes:
a) 65 male and 64 female subjects, rating 76 tactics for frequency of use by own sex on a 5-point scale (1 – never; 5 – very frequent); only those yielding a significant effect by sex or mating context are included in the table.
b) Number of individual tactics whose ratings differed significantly (ANOVA: $P < 0.05$) by sex or by mating context.
c) Tactics categorised by two blind raters using criteria given by Tooke and Camire (1991).
d) prospective short-term or long-term relationship

Source: Corby (1997)

BOX 5.6

Changes in bridewealth among the Kipsigis

Borgerhoff Mulder (1995) has demonstrated that bridewealth payments increased over the half century from the 1930s, but have declined in more recent decades. This decline has been attributed to a combination of a 'flooded market' of marriageable-age women and a general increase in the age at which men marry (due a shortage of land to settle on as well as a need to raise cash for bridewealth). These factors have, according to Borgerhoff Mulder (1995), been paralleled by developments in East Africa in general; relatively high brideprices during the colonial and early independence period were associated with the value of womens' productive labour in an emerging cash-crop economy, as well as with their value in childbearing (given the economic value of children). As wealth earning came to be diversified and less exclusively centred on agricultural productivity, so there was an associated decline in the labour value of women.

At the same time, an awareness of the need to limit family size, as manifested in a dramatic reduction in total fertility rate, meant that womens' reproductive value was in decline. Borgerhoff Mulder (1995) found no evidence to support assortative marriage with respect to wealth: brides and grooms from relatively wealthy families did not intermarry more often than would be expected by chance, although there was a strong tendency for educated individuals to marry one another. However, when marriages between wealthy families did take place, bridewealth was consistently higher compared to other marriages, although this difference has all but disappeared in more recent times. Higher bridewealth payments do persist, however, for women who have received secondary education.

One reason cited by Borgerhoff Mulder (1995) for the decline in bridewealth payments among intermarrying wealthy families is the recognition among the wealthy that 'marrying off' daughters is no longer the primary means toward cattle accumulation. At the same time, there is recognition that demanding too much bridewealth from less wealthy suitors will diminish the wealth-base of the family that the daughter is marrying into, and hence the daughter will suffer in the longer term. Similarly, parent's might 'scare off' a suitable groom by demanding too high a brideprice. On the other hand, the high bridewealth payments demanded for women with secondary education might be explained as an attempt to recoup some of the money expended on educating the daughter: it is the parents who spent money educating their daughters, but the family into which the daughter marries who will gain all the later benefits that derive from education (for example, employment income).

women had significantly more live-births per year than did late-maturing women, and that they also had consistently higher offspring survival (although not significantly so). She also found a negative relationship between parental wealth and menarcheal age, and a positive relationship between parental wealth and reproductive success. In one of the cohorts, the best predictor of a woman's reproductive success was found to be her parents' wealth, even when age at menarche was controlled for.

More importantly, Borgerhoff Mulder (1988a, b) was able to show that these intrinsic characteristics were reflected in the quantities of bridewealth that men were prepared to pay to the woman's parents when purchasing her as a wife. Bridewealth payments were highest for early maturing women, and the association between standardised bridewealth value and age at menarche persisted independently of other factors (**Table 5.2**).

One feature does, however, provide us with a clear basis for testing the hypothesis: since women actively seem to select for wealth (and/or social status) in men, especially in traditional societies, it should in principle be easy to determine whether wealthy men have more offspring than poorer men. We know, for example, that the relationship is in the predicted direction among other non-human species (Clutton-

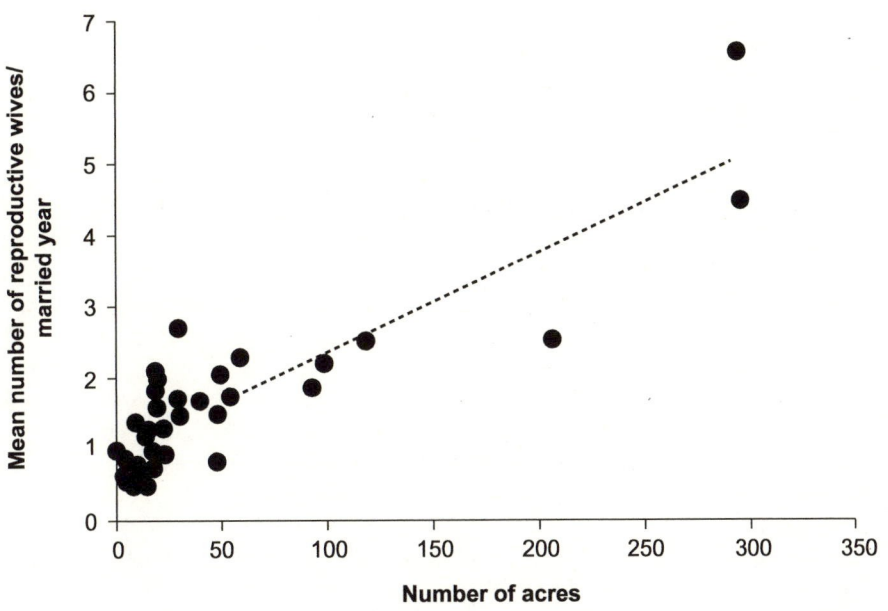

Figure 5.14 Among the Kipsigis of Kenya, a man's lifetime reproductive success correlates positively with the amount of land he owns, mainly because having additional land allows him to marry more wives. Redrawn with permission from Borgerhoff Mulder (1988c).

Brock et al. 1988; Dunbar 1988; Cowlishaw and Dunbar 1991), even if the relationship is sometimes moderated by other factors.

A wide variety of ethnographic and demographic studies provide considerable support for a relationship between status (as measured by wealth or social standing) and lifetime reproductive output in both contemporary and historical human populations. Among the agro-pastoralist Kipsigis, for example, there is a significant positive relationship between a man's wealth and his completed family size. This correlation mainly reflects the fact that wealthy men can afford to purchase more wives (**Figure 5.14**). The same also seems to be true for Turkmen pastoralists in Iran (Irons 1979). Among hunter-gatherers, the more successful hunters have higher social status and also seem to obtain more offspring (Yanomanö: Chagnon 1988; Ache: Hill and Kaplan 1988; Aka pygmies: Hewlett 1988), while social status correlates with completed family size (and specifically total number of sons) among Bakkarwal pastoralists of the western Himalayas (Casimir and Rao 1995). Betzig (1982, 1986) has documented extensive evidence to suggest that, throughout history, wealthy or high-status men have been able to maintain larger numbers of wives (for example, in harems) than their less well off countrymen – especially if they held the reins of political power as despots. The archetypal example is the biblical King Solomon with his innumerable wives and concubines. Social status may also be important in traditional societies: among the Yanomanö, men who have killed other men (*unokai*, literally 'revenge killer') have more wives, 'marry' earlier and have more children than other men (Low 2000).

However, even when monogamy is the prescribed norm, wealthier or higher status men may still have higher reproductive success because the resources they provide enable their wives to manage shorter interbirth intervals or provide their offspring with better chances of surviving to maturity (English feudal nobility: Hill 1999; fishermen/ horticulturalists in Micronesia: Turke and Betzig 1985). Cultural success (as measured by landholding) correlated positively with reproductive success for men in the Ostfriesland region of Germany during the eighteenth and nineteenth centuries (Klindworth and Voland 1995), as well as for men in a rural population in the north of England during the pre-industrial period (1600–1750: Scott and Duncan 1999). In the Krummhörn, this was related to the number of years spent in fecund marriage: wealthier men spent significantly more years in fecund marriage than poorer men because their brides were younger when they married them; in addition, the women married to wealthy men had significantly higher birth rates (as evidenced by shorter interbirth intervals) compared to other women. Moreover, even though infant and child mortality among wealthy families was higher than for the population at large, the reproductive advantage that wealthy families gained from early and fast reproduction persisted. In the English population, it was due mainly to the fact that the wives of wealthier men managed shorter interbirth intervals; that this was a consequence of the husband's wealth and not some intrinsic feature of class-specific fertility is indicated by the fact that both men and women who moved up into a higher or down into a lower class as adults achieved the reproductive rates of their host class rather than that of their natal class (Scott and Duncan 1999). Judge (1995) also found a significant positive correlation between number of surviving offspring and estate size from an analysis of twentieth-century wills recorded in Sacramento County (California).

That these kinds of effects are not confined purely to economic resources such as land holding, herd size or hunting skill is demonstrated by data for the nineteenth-century Mormons. In their case, pure cultural success (indexed by status as a church elder) was positively correlated with reproductive success. From an analysis of genealogical records, Mealey (1985) found that an individual male's status in the Mormon church hierarchy was significantly positively correlated with his fertility (mainly because of the polygynous marriages he contracted), and that this effect was established as early as 25 years of age for most men. As in the case of the Kipsigis (Borgerhoff Mulder 1989), the variance in male fertility was due to the number of wives that a man married, with multiple marriages being most common among men of higher church rank. At the same time, higher ranking males married earlier than did lower ranking males (Mealey 1985).

Mealey (1985) points out an interesting interplay between church rank and wealth, and their respective effects on reproductive success. She notes that the effects of each occurred at different stages in a man's life: wealth had a positive impact when men were seeking their first wives, while rank had a positive effect later in life when polygynous marriages were sought. Mealey (1985) found no relationship between change in wealth and marital status, but she did find that increases in church rank were likely to be followed by a change in marital status: polygynous marriages were more likely to take place *after* a change in rank than they were *before*. When all available variables were entered into a regression model, 24 per cent of the variance in male reproductive success was explained, with the best predictor being social rank at death. Mueller and Mazur (1998) present analogous data from a contemporary sample of US military: these indicate that (some inversions among the higher echelons notwithstanding) individuals of the 1950 graduation class from the prestigious West Point Academy who achieved high career rank (promotion to general) had higher lifetime reproductive success than those who did not.

Of course, reproductive success is not limited to how many children an individual sires. Counting the number of babies born is not necessarily that informative, and any conclusion should take into account the fate of those children: children that die before they can reproduce contribute nothing to their parents' fitness. Voland et al. (1991) and Klindworth and Voland (1995) examined the fate of children (in terms of probabilities of dying unmarried, marrying and remaining in the region, or emigrating from the region), and found that in all cases, the children of wealthy men fared better. Not only did more children of wealthy men get married, but more importantly, the risk of lineage extinction (through failure to raise at least one child to maturity) was only one quarter of the risk that men outside of the wealthy class faced.

EXPLAINING AN EVOLUTIONARY CONUNDRUM

Although many studies provide clear support for a relationship between status and/or wealth and reproductive success, by no means all do so (Vining 1986). In contemporary industrialised societies, the relationship between rank and reproductive success does not appear to hold. High status and wealthy people (royalty, politicians, leaders of industry, sports personalities, and so on) do not seem to have unusually large numbers of children. By contrast, everyone is convinced that the poor breed uncontrollably. In industrialised economies, there appears to be an inverted U-shaped (or perhaps even negative) relationship between rank and reproductive fitness (Vining 1986, Pérusse 1993).

Taken at face value, the apparent absence of the predicted positive relationship between wealth and reproductive success in modern societies undermines what is perhaps the central tenet of evolutionary psychology: that individuals have been selected to maximise inclusive fitness, and that they should exploit all opportunities provided by wealth and status to this end. Are there any proximate reasons to account for the inverse relationship in modern populations?

One possibility is that the evolved relationship between wealth and reproductive success has been derailed because current conditions differ significantly from those of the past. Pérusse (1993) suggests two features of modern environments that might have derailed the expression of this relationship, namely the availability of contraception and legally enforced monogamy. In the case of the former, sexual pleasure becomes decoupled from its reproductive consequences because evolution cannot anticipate such a drastic intervention between a motivation (sex) and its consequence (conception); in the second case, socially imposed monogamy prevents high status males from converting their resources into monopolising the reproduction of large numbers of women.

Pérusse (1993) investigated the relationship between cultural and reproductive success among French Canadians, with particular reference to whether a factor such as contraception prevents expression of the predicted relationship. He assessed cultural success using three indices of social stratification: education, occupation, and income. Reproductive success was assessed using a composite measure derived by combining the number of sexual partners, frequency of coital acts per partner and the probability of conception in unprotected matings. He assumed that, in the absence of contraception, mating success would translate directly into reproductive success.

Pérusse (1993) found no relationship between socio-economic status and 'true fertility' (that is, the total number of known children fathered by each male), suggesting that, in his modern Canadian sample, high status men do not enjoy higher than average reproductive success, even when serial marriages are accounted for. There were, however, strong positive correlations between social status and number of *potential* conceptions for men under 40 years of age, and Pérusse (1993) suggests that this indicates that modern contraception is acting against the predicted relationship between social status and true fertility. Thus, men of high social status appear to have more mating opportunities which, but for contraception, would have resulted in more offspring. (Note, incidentally, that contraception may be driving this relationship: women may be more willing to have short-term relationships and one night stands if they know that there is no risk of an unwanted pregnancy.)

The absence of the relationship for men aged 40 years and older is explained by Pérusse (1993) in terms of monogamy. The vast majority of men at this age (91.3 per cent) were classified as being in a marital-type union; if this implies monogamy, then the absence of a relationship between social status and potential conceptions is to be expected. Men who are in a union will not be as free as younger men to translate their wealth/status into copulation opportunities. The inhibiting effect of monogamous unions on mating opportunities was examined by comparing the relationship between social status and mating success in maritally involved versus uninvolved men. If marriage discourages women from getting involved with married men as well as married men from risking their family life for extra-pair copulations, then we would expect to find a positive relationship between status and number of potential conceptions for uninvolved men, and a much weaker relationship for involved men. This was the case.

More interestingly, perhaps, if the two groups of men (involved and uninvolved) are considered to be alternative mating strategies, it turns out that their frequencies in Pérusse's Quebec population are mirror images of their respective mean pay-offs in terms of numbers of putative conceptions achieved (Dunbar 1993b). This suggests that their frequencies may be held in check by frequency-dependent selection (see **Box 2.5**). The equilibrium frequency is probably maintained by the willingness of the women in the population to mate with uninvolved men. If more women prefer monogamous relationships within which the resources needed for rearing can be guaranteed, there will be fewer mating opportunities for men opting for the uninvolved strategy: with fewer matings divided between the same number of men, the pay-off for this strategy will decline and some men should shift strategy in favour of monogamous relationships.

Overall, Pérusse's (1993) results highlight a very important point. The general absence of a clear relationship between wealth and number of offspring in modern post-industrial societies has been used to question the appropriateness of an evolutionary approach to understanding human behaviour, despite the persistence of the relationship in traditional and historical populations (Vining 1986). Pérusse's (1993) findings show quite clearly that it is at least possible that the relationship might persist in a modern population were it not for two factors that work against it: modern contraception and socially-enforced monogamy.

There are, however, at least three other (and more strongly adaptive) explanations for the apparent downturn in reproductive success among the wealthy in industrialised societies that critiques of the Darwinian approach overlook. One is that, despite monogamous marriages with small numbers of children, the wealthy may still achieve higher lifetime reproductive output than the less well off if their wealth provides them with opportunities for casual or illicit relationships that yield children. In other words, the attractiveness of wealth for women may allow wealthy men to be polygamous even in societies where legally they must be monogamous. This possibility **POLYGYNY THRESHOLD MODEL** derives directly from well established theory in animal behavioural ecology. The **polygyny threshold model** (Verner and Willson 1966, Orians 1969) suggests that a female's preference for monogamy over polygamy will depend on the relative wealth differentials between males in the population (**Figure 5.15**). When wealth differentials are sufficiently large across the population, females should prefer to become the second mate of a wealthy male rather than the sole mate of a poor male, so long as doing so

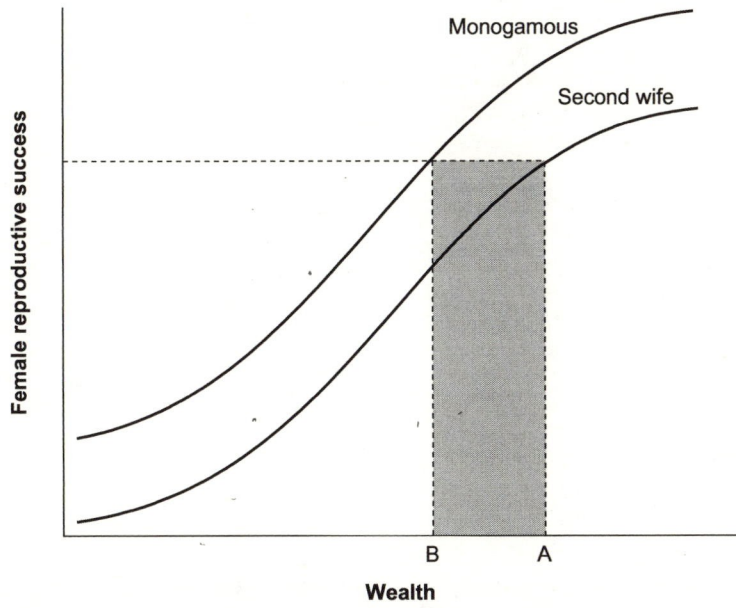

Figure 5.15 The polygyny threshold model. When female's reproductive success is a function of the husband's wealth, a woman will do better by becoming the second wife of a wealthy man (A) than by being the sole wife of a poor man (B). The polygyny threshold is defined by the wealth difference between men (shaded) that is required to make polygyny a preferable strategy for the woman. After Verner and Willson (1967).

yields a larger resource base on which to rear offspring. So long as the man is willing to channel at least *some* of his resources in her direction, it may pay some women to accept a position as the cryptic second wife.

Some evidence that this might be a possibility is provided by the frequency with which high status males engage in extra-pair relationships in all societies, including our own. Among the Ache hunter-gatherers of Paraguay, for example, successful hunters are widely acknowledged as lovers by the women, who attribute a significant number of their offspring to them (Hill and Kaplan 1988).

More generally, there is evidence that extra-pair copulations (EPCs) may be more common at times of high risk of conception. In a survey of British mating habits, Baker and Bellis (1995) found that unprotected extra-pair copulations (EPCs) were more common among women with stable relationships at around the time of ovulation than at other points in the menstrual cycle, even though matings with their principal partner (IPCs, or in-pair copulations) were not (**Figure 5.16**). Since males cannot tell when a human female is ovulating (as is clear from the behaviour of the paired males), Baker and Bellis concluded that the women must be 'choosing' to mate with extra-pair males at the time when they are most likely to conceive, and are therefore biasing the distribution of conceptions against their partners. On the basis of the timing of matings and the probability of conception per copulation, Baker and Bellis (1995) estimated that 7–13 per cent of British children would have genetic fathers that were different to the names on their birth certificates. This estimate seems to be in good agreement with data from paternity exclusion studies from around the world which give a median non-paternity of about 9 per cent (range 1.4–30 per cent: Baker and Bellis 1995). Russell and Wells (1987) reported expressions of emotional closeness to paternal versus maternal relatives that suggested a similar proportion

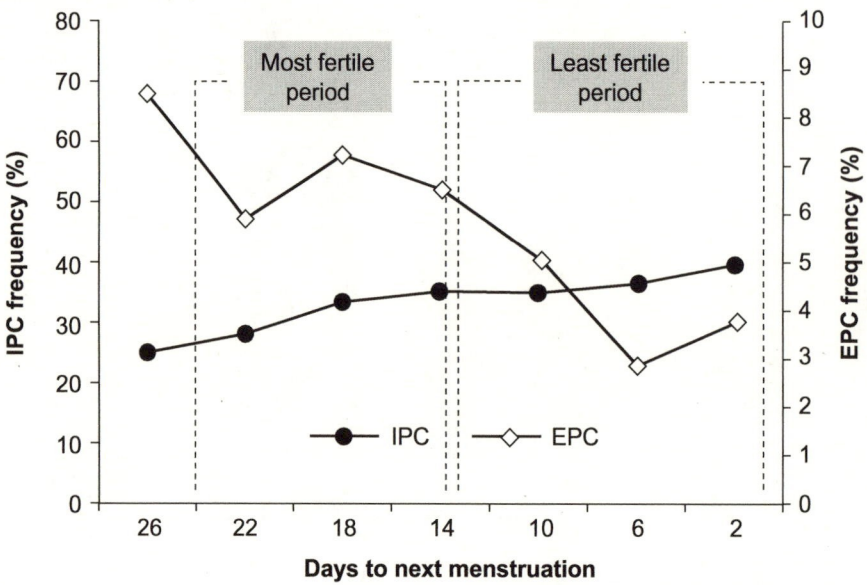

Figure 5.16 Percentage of women in long-term relationships who reported in-pair (IPC) and extra-pair (EPC) copulations on different days of their menstrual cycles. An EPC is defined as an unprotected copulation (that is, one in which no method of contraception was used) with a different male within 4 days of a copulation with the partner male. While IPCs show no variation across the menstrual cycle, EPCs are significantly higher during the fertile (ovulatory) phase of the cycle than during the non-fertile phase. Redrawn with permission from Baker and Bellis (1995).

of non-paternities (assuming that people titrate their emotional bonding to each half of their family roughly in line with their estimates of likely relatedness).

The second possible reason for the apparent lack of reproductive success among the wealthy is more interesting because it suggests that the negative relationship between wealth and reproductive success may actually be genuine. Mace (1998) has argued that, above a certain wealth threshold, the costs of rearing should drive family size down. Her point is that if investment in individual children means that the children will have better social (and hence reproductive) opportunities when they mature and wealth is at the same time fixed, then it will pay to have fewer children so that each benefits from a higher investment. Everything, however, depends on the level of wealth, and Mace's models clearly demonstrate that an inverted-U-shaped relationship between wealth and family size is possible because the poor on the left-hand end of the wealth distribution strive to have many offspring while the rich on the right-hand end have fewer, even though, within each class, number of offspring is positively correlated with wealth (see **Figure 6.12a**). We discuss Mace's model in more detail in Chapter 6 in connection with the demographic transition.

The third explanation is offered by Betzig (1982, 1986). She has argued that the shift from polygamy in traditional societies (especially those with despotic political systems) to monogamy in post-industrial societies is due to the emergence of expertise-based economies. When the wealth of the despot is dependent solely on physical power, then he is able to exercise his demands for a disproportionate share of the women. However, once his wealth and/or power base is dependent on others (and especially technocrats such as skilled administrators, traders, scientists), the latter effectively hold a veto over his ability to exercise his power: in effect, he is forced to share his good reproductive fortune or lose the whole lot through insurrection. Trading down into a constitutional (but still wealthy) monarchy as happened in many European states during the nineteenth century may have been a tactical strategy on the part of despots faced with what amounts to a 'tolerated theft' situation (see **Box 4.3**).

One final twist to this part of the story is the possibility that the high fertility of wealthy men may actually be the product of women's behaviour. That female choice may be an important issue is suggested by the fact that women of all socio-economic classes in the Krummhörn population marry at significantly younger ages if they marry into a higher social class than if they marry into their own natal classes. The fact that the wealthy men of the Krummhörn were in fecund marriages for longer reflects the fact that their wives were younger at marriage than the wives of less wealthy men. This is important in the context of sexual selection since female fertility and reproductive value are a function of age. While the data are not available to assess the correlates of these among Krummhörn women directly, perhaps the more important point is that when women marry early they are more likely to marry wealthy men. In the same way that wealthy Krummhörn men appear to be conforming to the predictions of mate choice theory, so too are the women. Klindworth and Voland (1995) argue that differential male fitness is very much a female-driven phenomenon that results from female reproductive competition where younger females compete for high status wealthy men. The Krummhörn data very much suggest that the women in all the socio-economic classes delayed marriage in the hopes of making a wealthy 'catch'; but if they failed to attract such a mate, they were eventually forced to settle for someone less wealthy in their own social class (the 'best-of-a-bad-job' solution) to avoid being left on the shelf (Voland and Engel 1990). Jane Austen's novels of social life in the eighteenth-century English squirearchy are acute and wonderfully evocative observations of just this process in action.

Even more compelling evidence that women may manipulate the reproductive marketplace to their advantage is offered by the Kipsigis. Kipsigis women produce approximately the number of children that maximises the number of surviving grandchildren irrespective of their wealth (measured in landholding), whereas men in all wealth classes typically produce fewer children than the number that would maximise the number of grandchildren they could have (Borgerhoff Mulder 2000). This suggests two things. First, that women are concerned to optimise the numbers of children they have within the resources they have available (that is, to maximise quality at the expense of quantity), while men would do best by maximising the number of children they have irrespective of quality. Second, that contrary to what one might expect in a strongly patriarchal society with patrilineal inheritance of wealth (in the form of livestock and land), the women win this particular conflict of reproductive interests.

*

As this overview has indicated, there is much more to successful reproduction than choosing an appropriate partner with whom to mate. This is because, in evolutionary terms, rearing offspring to adulthood so that they can in their turn reproduce is the crucial goal. A prospective mate's abilities to rear offspring successfully may thus be as important as mere fertility or the quality of their genes. In the following two chapters, we investigate, first, issues related to reproductive decision-making, such as when to have children and how many to have, and then go on to look at patterns of investment in the children that eventually are produced.

- Human mate choice decisions reflect the essential differences between males and females in terms of maximising fitness; females typically concentrate on the rearing component of reproductive effort while males concentrate on the mating component.

- These differences between the sexes are moderated compared to other mammals by the extreme helplessness of human newborns and the need for high levels of biparental investment.

- Lonely hearts advertisements provide a useful assay of mate choice criteria and show that male and female requirements conform to the expectations of evolutionary theory, with females asking for resources and males seeking cues of physical attractiveness as indicators of fertility.

- Certain physical traits in humans, such as beards in men and low waist-hip ratio and facial neoteny in females, may be the products of sexual selection. Low levels of asymmetry in bilateral characters also appear to be related to levels of attractiveness and may serve as an indicator of good genes.

- Mate choice reflects a trade-off between what an individual requires from a partner and what they can themselves command in the marketplace. Individuals therefore display highly contingent strategies when seeking mates and may undergo lengthy periods of courtship and assessment.

- Data on the fitness consequences of mate selection show that individuals who are more attractive or successful in the mating market achieve higher reproductive success than those who are less successful.

- However, the data from contemporary industrialised societies often show a negative relationship between wealth and reproductive success. The reasons for this may be linked either to a decoupling between mating opportunities and actual conceptions or to the increased costs of rearing children for more wealthy individuals.

Further reading

Baker, R. and Bellis, M. (1995). *Human Sperm Competition: Copulation, Masturbation and Infidelity.* London, Chapman and Hall.

Betzig, L., Borgerhoff-Mulder, M. and Turke, P. (1988). *Human Reproductive Behaviour: a Darwinian Perspective.* Cambridge University Press, Cambridge.

Birkhead, T. (2000). *Promiscuity: an Evolutionary History of Sperm Competition and Sexual Conflict.* London: Faber and Faber.

Buss, D. M. (1994). *The Evolution of Desire: Strategies of Human Mating.* New York: Basic Books.

Cashdan, E. (1996). 'Women's mating strategies'. *Evolutionary Anthropology* 5: 134–43.

Low, B. S. (2000). *Why Sex Matters: a Darwinian Look at Human Behaviour.* Princeton: Princeton University Press.

Life-history constraints and reproductive decisions

CONTENTS

In the last chapter, we dealt with human mating strategies principally in terms of the role that partner choice plays in ensuring high reproductive success. However, the extreme vulnerability of newborn human infants means that successful reproduction, particularly for women, depends on making far more crucial decisions than merely selecting a suitable mate. In this chapter, we investigate how the constraints of our evolutionary heritage combine with environmental variables to determine patterns of offspring production and survival.

OPTIMISING FAMILY SIZE

It is apparent that females in a number of human populations do not maximise their reproductive potential. In modern western populations, women can produce their children with interbirth intervals that can be as low as 12–18 months. Yet, in many traditional societies, intervals of three to four years between offspring are more typical. If the evolutionary goal of female reproductive strategies is to maximise fitness, why should women apparently limit their reproduction in this way?

The key to understanding this apparent paradox is to appreciate that maximising the number of offspring produced does not necessarily maximise an individual's Darwinian fitness. Evolution is interested in grandchildren, not children; if all the offspring that an individual produces die before they are able to reproduce themselves, then that individual's reproductive effort has been in vain. Consequently, it often pays to limit the quantity of offspring produced in order to ensure their quality. Investing a lot of care and effort in a relatively small number of young and guaranteeing that they survive at least until adulthood (assuming that their ability to attract suitable mates is beyond a parent's control) may pay greater dividends than pumping out children at an increased rate and failing to care for them.

This raises the second key issue: the extraordinary costs of successfully rearing a human infant. The unusually large brains that characterise our species bring with them two principal costs. First, brain tissue is extremely expensive to grow and maintain (Aiello and Wheeler 1995), so that the energetic costs of rearing offspring are very heavy. Second, in primates, selection for increased brain size to cope with the demands of social life (see **Box 6.1**) is thought to have selected for an increased juvenile period in order to learn all the skills required to cope with the complexities of life in social groups. Some evidence to support this suggestion was provided by Joffe (1997) who showed that, in primates (including humans), the relative size of the neocortex correlates best with the length of the juvenile period (the interval between weaning and first reproduction). This, of course, is the period of socialisation during which an individual learns all that is necessary for it to function as an effective member of its society. In humans, this gets taken to an extreme: even when they have achieved nutritional independence, human children are still 'culturally' dependent on their caretakers for at least another decade while they learn the skills they need to cope with life in social groups (see for example Blurton-Jones et al. 1999).

Whatever the underlying evolutionary basis for extended childhood, this prolonged dependency imposes considerable demands on the parents who have to provide their dependent offspring with food, shelter and other costly forms of care. Due to internal gestation followed by lactation, the bulk of the nutritional burden, in particular, falls on the mother. However, the extreme helplessness of human infants means that a high level of paternal investment is also needed to ensure offspring survival. More often than not, this occurs through indirect means with males providing resources for the female, rather than by direct care-taking of infants (see Chapter 5). The presence of extended kin networks can also provide a secure base within which children can be brought up effectively (and can help reduce the overall costs of parental investment: see Chapter 7).

One way in which to investigate the trade-off between offspring quantity and quality is to examine patterns of birth spacing. A long interval between births is taken to indicate that the mother is increasing the amount of investment in the current offspring, either because the offspring needs an increased amount of investment in order to survive or because the mother cannot risk overextending herself and putting both her own and the offspring's life at risk. Elucidating the reasons underlying birth spacing can therefore provide us with an insight into both the ultimate causes and proximate motivations as to why women limit their reproductive outputs. We do this by considering two specific examples that represent contrasting conditions in terms of the extent

BOX 6.1

Why do humans have such large brains?

Humans have brains that are six times larger than would be expected for their body size compared to the average primate, while primates as a group have brains that are significantly larger than would be expected for mammals of equivalent size. Conventional explanations have always assumed that brains evolved to process factual information about the world, and that ecological problem-solving provided the impetus for an increase in brain size within the primate order as a whole (for example, Clutton-Brock and Harvey 1980). In humans, tool-making (for hunting) was seen as an added impetus that reinforced this effect (Wynn 1988).

However, given that there are many small-brained species that solve very similar ecological problems to those faced by primates, it is difficult to justify the expense of a large brain in these terms. Consequently, an alternative hypothesis – the so-called **Machiavellian Intelligence (or Social Brain) Hypothesis** of Byrne and Whiten (1988) – has been receiving a great deal of attention. This postulates that the ability to solve complex social problems was the impetus for increased brain size among the primates.

As the name suggests, the ability to manipulate other members of a social group is thought to have selected for increased brain size in primates. Within permanently social primate groups, bonds formed between individuals (through behaviours such as grooming) are used to build strategic alliances that enhance individual reproductive success through the combined effects of kin selection and reciprocal altruism. The initial impetus to service and build valuable social relationships and use them to one's own advantage is thought to have had a ratchet effect on brain size, as the increasingly cunning manipulative strategies used by some individuals selected for equally cunning counter-strategies in others so as to avoid being deceived or manipulated, which in turn selected for individuals with an even greater ability to outwit their opponents.

While the early formulations of the hypothesis tended to focus on the issue of deception and manipulation, it is also apparent that a large amount of computational power is required just to keep track of one's own relationships and those of others in a permanent social group. Although stable through time, primate groups are not static entities; they are subject to constant change as individuals give birth, die, immigrate and emigrate, and, perhaps more importantly, make and break friendships and alliances with each other. In other words, the need for individuals to weigh up their best options with regard to, for example, forming grooming relationships or avoiding trouble from others would be a very strong selective pressure on the brain size of long-lived obligately social animals, even without the ability to manipulate and deceive others. Moreover, it seems clear that the decision-making abilities required to negotiate life in a social group from a purely self-oriented perspective would be a necessary prerequisite for an ability to manipulate the behaviour of others.

There is now considerable evidence to support the Social Brain Hypothesis (see reviews in Barton and Dunbar 1997, Dunbar 1998a). The focus of most of these studies has been on the size of the neocortex, since this is the area of the brain that is particularly characteristic of primates as a whole and which accounts for most of the brain size enlargement that has occurred within primates. Comparative studies across a range of primate taxa (including modern humans) show that relative neocortex size correlates with social group size (**Figure 6.1**), the size of grooming cliques (that is, alliances) (Kudo and Dunbar, in press), males' use of subtle social strategies in mating contests (Pawlowski et al. 1998), frequency of social play (Lewis 2001) and the frequency of tactical deception (Byrne 1995). These various data suggest that the extent to which animals can develop and exploit large numbers of complex social relationships depends closely on the size of the 'computer' they have available to do the necessary calculations.

Figure 6.1 Mean social group size increases with relative neocortex volume (indexed as the ratio of neocortex volume to the volume of the rest of the brain) for New and Old World monkeys (filled symbols) and apes (open symbols). Each datapoint represents a different genus (in most cases, based on data for a single species). The ape genera are (from left to right) gibbons, gorillas and chimpanzees. Source: Dunbar (1992).

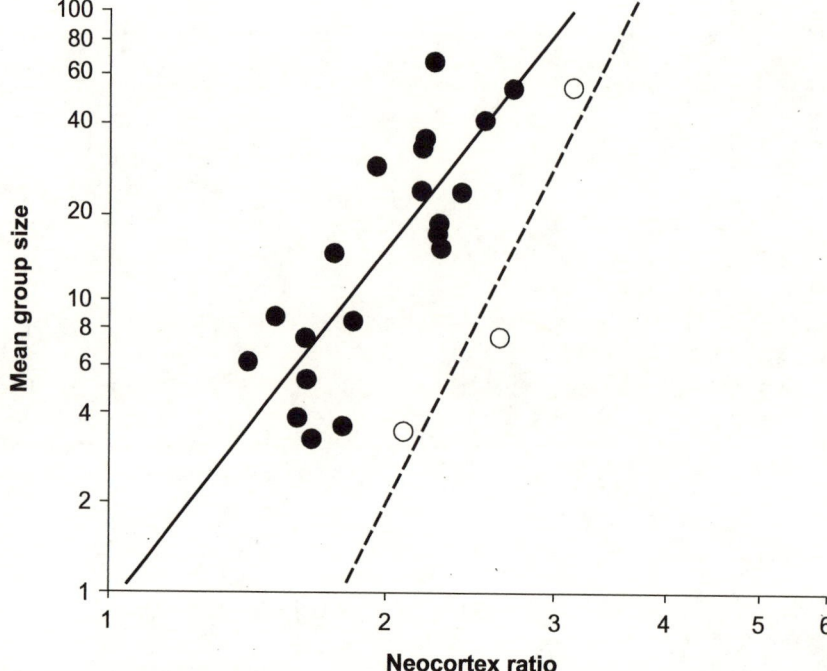

to which material resources (that is, wealth) are necessary for ensuring that offspring can reproduce successfully.

THE !KUNG SAN BACK-LOAD MODEL

The !Kung San hunter-gatherers of Botswana provide one of the classic examples of extended birth spacing as an adaptive strategy. Demographic data collected in the late 1960s by Lee (1979) indicated that the interval between births among the !Kung was around four years, and that completed family sizes (at around 3.8 offspring per female) were relatively low compared to many other human populations. Lee suggested that the need for women to gather large loads of food and carry them long distances under thermally stressful conditions limited their ability to produce offspring.

During the dry season, !Kung women are forced to travel several kilometres from camp under very harsh climatic conditions to collect *mgongo* nuts, their staple diet at that time of the year. During these foraging expeditions they are accompanied by any offspring who are under four years of age. Consequently, a woman not only has to carry all the food required to feed herself and her family, but she also has to carry her offspring. Lee suggested that a shorter interbirth interval would mean that a woman was accompanied by more than one small child at a time, and that this would increase the burden she had to carry to a level (in terms of both physical exertion and heat stress) that would substantially increase the mortality risk of the woman and her offspring.

Using a number of simplifying assumptions, Blurton-Jones and Sibly (1978) calculated the maximum backload a woman would have to carry given a particular **interbirth interval (IBI)**, and then used a computer simulation to investigate how this would

INTERBIRTH INTERVAL

BOX 6.2

Why are human babies born so early?

Primates in general are precocial mammals (their young can move about independently soon after birth); however, humans are a glaring exception to this trend. Human fetuses have a very slow rate of maturation, and are born at a stage reminiscent of more **altricial** mammals (those species whose young are poorly developed at birth, hairless and helpless, with their eyes and ears closed). Compared to other primates, human babies are extremely under-developed at birth: the brain of human newborns is only one quarter of its full size, whereas those of Great Ape infants are already half grown at birth. This means that human infants complete most of their development over an extended period outside of the womb: brain growth is only fully completed at the end of the first year of life. This in turn makes human infants uniquely vulnerable and helpless.

One of the major reasons why human babies are born so early is because of the constraints imposed by the size of the female pelvis. Walking efficiently on two legs requires the pelvis to provide a stable support for the torso; the need to strengthen the pelvis to carry the weight of the upper body has resulted in a rather squat bowl-shaped pelvis that, in turn, restricts the size of the birth canal compared to the elongated pelves of our ape cousins. Giving birth requires a birth canal wide enough to allow the baby's head through (the head is the widest part of a neonate primate's body). The dramatic increase in brain size at a point long after humans had become fully adapted to bipedal locomotion thus created something of a problem. Some sort of

evolutionary compromise was required to deal with the eye-watering prospect of trying to squeeze an ever larger head through an ever smaller hole.

Although the widening of female hips helped to increase the size of the birth canal to a certain extent, this can only occur within very narrow limits before locomotor ability becomes seriously affected. (Even the small difference that exists today between the two sexes substantially reduces female locomotor capacity, and means that women will never be able to run a 100 m sprint as fast as men.) The only other available solution to the problem was to give birth at a stage when the infant's brain was still very small, and could fit through the birth canal more easily – even so, it's still a bit of a squeeze. Incidentally, here we have another good example of the ad hoc nature of evolutionary adaptation: the high frequency of birth complications seen in humans is testimony to the imperfection of this particular design.

Human babies are therefore born prematurely, right on the limit at which it is possible for them to survive outside the womb. This explains why babies born after only six or seven months of pregnancy (the ones we usually think of as being 'premature') experience such problems (including a variety of developmental, cognitive and physical disabilities: see Siegel 1994). Strictly speaking, they really shouldn't survive at all, since they are not actually equipped for life outside the womb. It is only the huge advances in medical science that enable these babies to survive by compensating for their developmental deficiencies.

---

vary across a woman's reproductive lifespan. The results revealed that there was a sharp upward inflection in backload (food plus babies) when IBI was less than four years (**Figure 6.2**), suggesting that, if IBIs were too short, unacceptable levels of work for a woman would result, with a consequent increase in mortality risk for both her and her offspring. In contrast, an IBI of four years resulted in backloads that remained fairly even across the lifespan, while maximising the production of offspring compared to IBIs of five years or more. An IBI of four years would therefore seem to be optimal, as Lee originally suggested.

Blurton-Jones (1986; see also 1987) subsequently tested the backload model using data on !Kung reproductive histories collected by Howell (1979). Specifically, he

Figure 6.2 The average backload carried by a !Kung San woman while foraging (weight of dependent offspring plus weight of gathered produce) declines as the interbirth interval (IBI, measured in years) increases. Short IBIs result in women having to carry two infants for part of the time. Backload reaches an asymptotic minimum at an IBI of 4 years: this represents the IBI that optimises birth rate with respect to backload. Redrawn with permission from Blurton-Jones (1984).

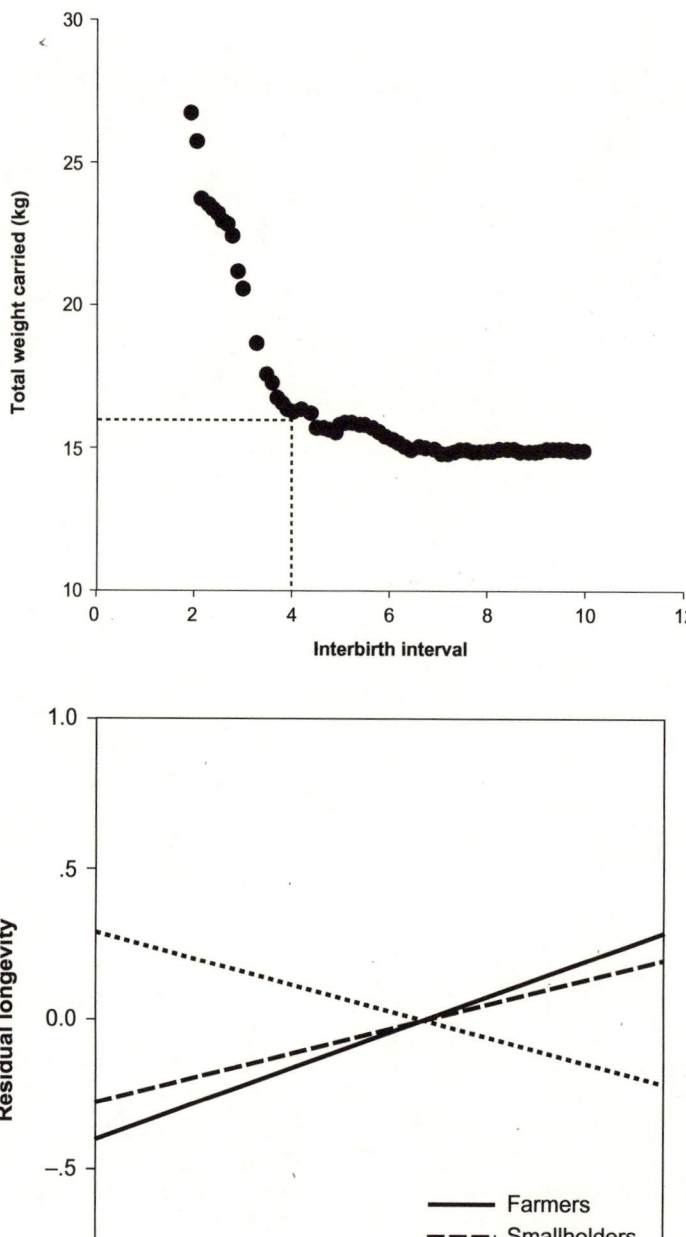

Figure 6.3 Regression lines for residual longevity on residual number of offspring (controlling for duration of fecund marriage) for three socio-economic classes (farmers, smallholders and landless labourers) in the Krummhörn population (NW Germany) between 1720–1870. The costs that reproduction imposed on a woman (indexed as a reduction in lifespan) are felt increasingly strongly as family wealth declines: women from the richer farmer class are largely buffered against these costs. Reproduced from Lycett et al. (2000).

BOX 6.3

Impact of offspring production on parental survival

Rates of offspring production have an impact on parental survivorship as well as on the survivorship of the infants themselves. Westendorp and Kirkwood (1998) found that, among the English aristocracy, the number of children a woman produced was negatively correlated with longevity, while age at first childbirth was positively correlated. Women who placed a lot of energy into childbirth and rearing therefore reduced their lifespans. Interestingly, the same was true for men: the greater the number of children, the shorter the lifespan. This suggests that individuals make a trade-off between continued somatic (bodily) investment which promotes longevity and reproductive investment which diverts resources away from oneself and into one's offspring.

However, other data suggest exactly the opposite: Voland and Engel (1989) and Borgerhoff Mulder (1988c) both reported a positive relationship between fertility and longevity. Only women who have considerably more than average numbers of children show a negative effect of reproduction on longevity (Voland and Engel 1989). Some further evidence which suggests that reproduction per se might not impact negatively on longevity in humans comes from a study which showed that women in the sample who died early

had ceased reproduction earlier in life whereas those who had lived to an especially old age had continued to reproduce into their fifth decade of life (Perls et al. 1997). Perls et al. therefore suggested that the driving force selecting for long lifespan in humans is the need to maximise the time period during which women can bear children. (However, this is at odds with the majority of the evidence which suggests that increased lifespan is associated with a strong grandmother effect (Hawkes et al. 1997; see pp. 164–6) rather than direct reproduction.)

Most recently, Lycett et al. (2000) have shown, using data from the eighteenth- and nineteenth-century German Krummhörn population, that the negative effects of reproduction on longevity are apparent when the confounding effects of economic conditions are removed. For females in each of three socio-economic groups (farmers, smallholders, landless labourers), age at death was negatively correlated with number of children. More importantly, when the duration of fecund marriage was controlled for, the slope of the relationship was itself correlated with wealth (as reflected in socio-economic class) (**Figure 6.3**). Wealth appears to buffer women against the costs of reproduction.

investigated whether child mortality was indeed related to IBI, and then used this to calculate the IBI that yielded the greatest number of surviving teenage offspring. Finally, he compared this to the actual intervals observed in Howell's study population.

The data showed that the observed mortality pattern could be approximated by a curve where mortality was 1/IBI (which approximates the birth rate), while a slightly better fit could be obtained by relating backload and mortality to the same function of IBI, suggesting that, when intervals were too short, increased maternal workloads would reduce offspring survivorship (**Figure 6.4a**). He then showed that an IBI of four years resulted in the highest number of surviving teenagers, and that this matched closely the pattern observed in Howell's data (**Figure 6.4b**).

From this, Blurton-Jones (1986) concluded (a) that !Kung women were optimising their births with quite high precision and (b) that longer IBIs led to more descendants than shorter intervals. !Kung women therefore appeared to be maximising their fitness by limiting their reproductive output to levels appropriate to their ecological circumstances. Blurton-Jones (1984) further suggested that the greatly increased fertility (in terms of the number of living offspring produced in a lifetime) observed among !Kung women who had given up a foraging lifestyle in order to settle at cattleposts

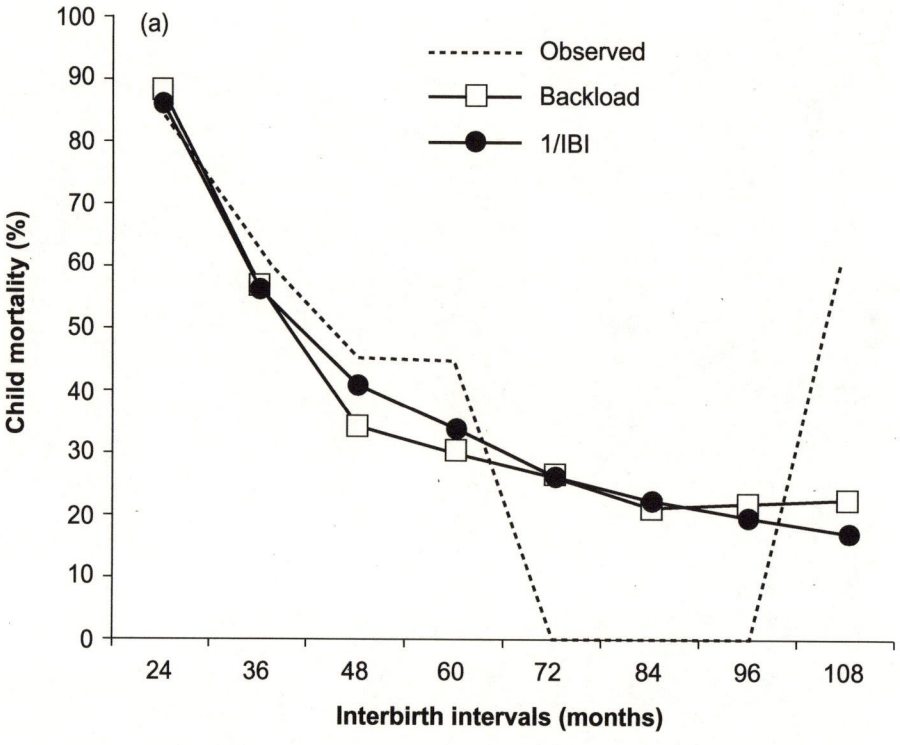

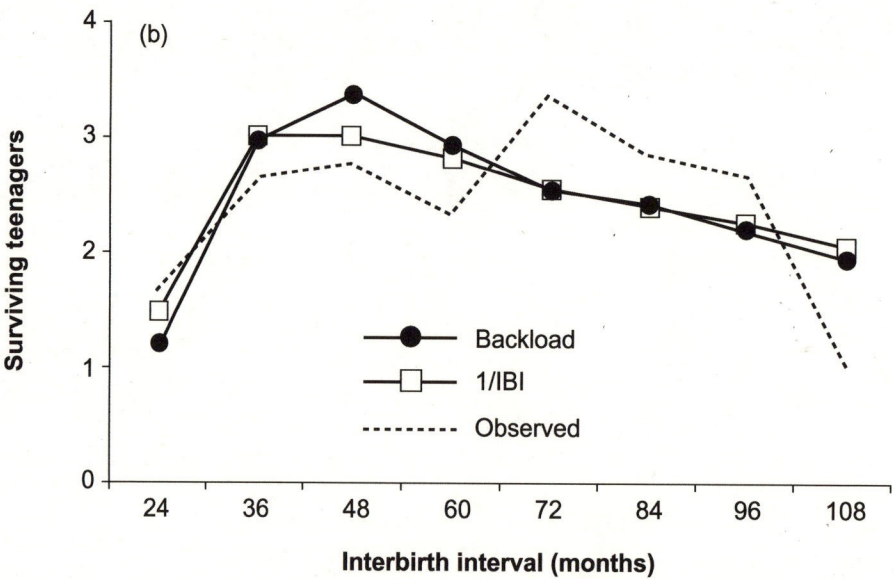

Figure 6.4 Interbirth intervals (IBI) predict (a) childhood mortality and (b) the number of surviving teenagers among the !Kung San. Adding in the additional backload costs of carrying dependent offspring while foraging further improves the fit to the observed data. Redrawn with permission from Blurton-Jones (1986).

could be explained by the fact that the backload constraint had been lifted from them, thus allowing them to produce offspring at more frequent intervals.

Despite the generally good fit of the model to the actual data, there was quite high variability in the length of intervals observed, with a bias toward intervals shorter than four years. Such deviations should, however, be expected since Blurton-Jones's (1986) analysis takes no account of individual female condition and how this might change through time. If a woman is in good condition, for example, she may be able to reduce IBI without any attendant mortality risks, but this will not be picked up by a model that assumes all females are of the same average condition. Houston et al. (1988) have used the statistical technique of **stochastic dynamic programming** (see **Box 6.4**) to show how differences in individual quality could produce variability in the age of menarche (the onset of breeding) and birth rates.

Anderies (1996) later used this technique to investigate whether differences in female quality could explain the variability observed in the !Kung sample. By varying the state parameters of the model, Anderies (1996) could make !Kung women vary in their physical strength and therefore in the size of the load that they were able to carry. These analyses suggested that, if women were weak and in poor health, their optimal IBI would increase to five years, whereas, if they were in good condition and capable of carrying heavier loads, it would decrease to three years. Since real women can be expected to vary in their carrying abilities in exactly this way, differences of this nature could explain the variability observed in IBI.

Equally, Andries found that, if the cost of poor foraging success on child mortality was made to vary, then optimal IBI also varied. When the cost was low (for example, women had access to mechanisms other than foraging to feed their offspring), optimal IBI decreased. When costs were high, on the other hand, it paid women to reduce their reproductive output in order to ensure that their current offspring would not be put at risk; the optimal IBI increased substantially as a result. Anderies' (1996) model therefore supports Blurton-Jones's (1986) suggestion regarding the impact of sedentarism on female fertility, since, by definition, females living at cattleposts are not penalised for lack of foraging success. Including state-dependency in the model, thereby recognising that individual differences between women exist, enabled a more realistic and accurate model of !Kung reproductive behaviour to be constructed.

Given the success of the !Kung backload model, Blurton-Jones et al. (1989; see also Blurton-Jones et al. 1992, 1994) later attempted to generalise the model to other hunter-gatherer populations, such as the Hadza of Tanzania. They found that the costs of children were lower for the Hadza than for the !Kung because, when Hadza women went foraging, they tended to leave offspring behind in camp at a much younger age (around two and a half years); moreover, Hadza weaned their children much sooner than !Kung women did such that, by the age of five, Hadza children can forage almost entirely for themselves. In addition, Hadza children received a substantial quantity of resources from their grandmothers.

Blurton-Jones et al. (1989, 1992) suggested that these reduced costs and lower workloads accounted for the increased fertility and reduced interbirth intervals of the Hadza compared to the !Kung. In one sense, these results support Blurton-Jones' model, since reduced workloads lead to increased fertility among the Hadza. In another sense, however, they don't really test the model directly, since Hadza women may not actually be constrained by backload: they travel shorter distances to find food, leave their children at home, and suffer less heat stress while collecting foods. A constraint on mobility is not the defining feature of Hadza womens' ecology in the same way that it is among the !Kung. Instead, Blurton-Jones et al. (1989) have suggested that time may

STOCHASTIC
DYNAMIC
PROGRAMMING

BOX 6.4

Optimality models and stochastic dynamic programming

Animal behavioural ecology has made wide use of **optimality models** and many of these are easily adapted for use with human data. Winterhalder's (1996) marginal value model (see Chapter 4) is one such example. Optimality models have three main components: a currency (often energy or time), constraints (individuals are always faced with constraints that limit their ability to achieve certain goals) and a decision variable (which defines the problem the model is designed to solve).

Optimality models built with these three components have proved very successful at explaining certain aspects of animal behaviour, but they are not particularly realistic since they take no account of the individual animal's condition or state (for example, how hungry the animal is) and they assume that the world is predictable and constant. Stochastic dynamic programming (SDP) models were developed in order to remedy these deficits and increase the applicability of optimality models to real-life situations (Houston et al. 1988, Mangel and Clark 1988). Like all optimality models, SDP models have a currency, constraints and a decision variable. In addition, they incorporate a measure of the individual's state and allow events to vary stochastically (unpredictably) as they do in the real world. The 'dynamic' aspect of the model refers to the fact that the state of an animal changes depending on what decisions it makes and this in turn feeds back to determine the nature of the decisions it makes in the future.

In an SDP model, the impact of an individual's short-term decision-making on its lifetime fitness is evaluated. An important concept in SDP modelling is the terminal reward function, which is the mathematical relationship between lifetime fitness and the animal's state at the end of the period under study. For example, in an SDP model built by Mace (1996a) to explain the reproductive behaviour of Gabbra pastoralists, the fixed time period over which behaviour was modelled was the 30-year reproductive lifespan and the terminal reward function defined an individual's fitness as a function of its state at the end of this time period.

In order to identify the optimal decision rules that individuals should use, the time period under study is divided into a number of discrete intervals (minutes, days or years as appropriate). In each time period, the animal decides between alternative courses of action, each of which have fitness consequences. For example, in the Gabbra model, individuals made a decision every two years about whether they could afford to have another baby within the constraints of the family's current wealth and the number of children they already had. Because each decision affects the family's patterns of resource use, they influence the subsequent decisions that the individuals make in the next time interval, and so, ultimately, the value of the terminal reward.

SDP models calculate the consequences of alternative actions by working backwards through time. That is, they start at the terminal reward and then reconstruct successive decisions back through time to the beginning of the period. This is computationally easier than going forward and attempting to predict what the long-term future consequences of actions are likely to be. If you start at the beginning and attempt to predict what the best course of action will be in the long-term, a mammoth calculation is necessary to weigh up all the possible alternative decision sequences (or paths) between the start and various possible end points; many of these will be decision paths that die out, so all the calculations up to that point will have been wasted. If, instead, you start at the end point, fewer calculations are needed to determine the optimal course of action because only those decision paths that survive to the end need to be considered.

be the most important constraint for Hadza women: they work longer hours and spend a larger proportion of their time preparing food compared to the !Kung women. The number of hours available in the day may therefore be the critical factor affecting female fertility and mortality schedules, and time should therefore be used as the proper currency in which to measure the costs of Hadza children.

Data from the South American Ache hunter-gatherers provide even more equivocal support for the Blurton-Jones model. In this case, shorter IBIs were found to lead to increased mortality, but the effect was too small to render the observed IBI optimal, and IBIs needed to be shorter in order to optimise rates of offspring production. Hill and Hurtado (1996) suggested that mother's physical size may have had an influence: Ache women are considerably larger than !Kung women which should lead to a slower rate of reproduction. However, Hill and Hurtado (1996) found that heavy Ache women had shorter IBIs than lighter women, suggesting that women who delay maturation to grow bigger ultimately increase their fertility – a suggestion further supported by the fact that faster-growing girls had their first babies sooner than slower-growing girls.

CAMELS, GOATS, BOYS AND GIRLS

In agricultural or pastoralist societies where individuals can acquire and monopolise resources, the reproductive decision-making process becomes more complex. Individuals in these societies not only invest time and energy in their offspring, they must also invest in material goods since these have a strong influence on the future reproductive success of offspring and, hence, ultimately their own fitness. The ability to rear a child to adulthood is only half of the equation here since, in order for offspring to reproduce, they must possess sufficient resources to form their own households (reproductive units). If parents cannot afford to marry off their sons and daughters, then, in fitness terms, they are evolutionary dead-ends. Moreover, unmarried offspring become a drain on the resources of their natal household, further reducing their parents' fitness potential. Prospective parents in these societies therefore have to make a more complicated two-step decision when deciding whether or not to enter the baby-making game. As well as having to decide whether they have sufficient wealth to rear an offspring, they also have to work out whether they will have resources over and above this to provide the appropriate dowry or bridewealth payment when the time comes.

The Gabbra of northern Kenya are a camel-herding people, and household wealth is measured in terms of herd size, particularly the number of female camels present since females determine herd growth rate. There is an intimate feedback between family needs and herd growth rate, since households depend on their camels for milk and, less frequently, meat. If family size increases faster than herd size, then an increasing amount of milk must be taken from the herd, and the growth rate of the herd is reduced. Equally, when immature male and female camels have to be sold or slaughtered to provide food, herd productivity is reduced since those animals are no longer around to breed. When deciding whether to have another baby, parents therefore have to trade-off the probability that they will be able to raise the child and marry it off successfully against the risk that doing so will diminish the family herd and harm the marriage prospects of their existing children.

Using state-dependent dynamic modelling, Mace (1996a) found that, regardless of the cost of marrying off children (as represented by the three different lines in

BOX 6.5

Are !Kung birth rates low by accident rather than design?

A major criticism of the !Kung backload model has been raised by those who question the fundamental assumption that !Kung women display abnormally low fertility compared to other subsistence groups. Pennington (1992) and Harpending (1994) have argued that the !Kung populations which provided all the data for these analyses came from an area of Botswana where overall fertility among all subsistence groups was low due to the prevalence of sexually transmitted diseases (STDs): the low age at last birth of the !Kung women, for example, is a widely recognised signature of pathological sterility.

The low fertility and long IBIs of the !Kung may therefore be a consequence of high levels of infectious secondary sterility, and may not necessarily represent an adaptive response to local ecological conditions. Pennington (1992) suggests that the changes observed among women living at cattleposts owe more to a decrease in mortality as a result of access to medical facilities than to any real increase in fertility.

Pennington claimed that post-reproductive !Kung women who followed a traditional foraging lifestyle in fact had higher completed fertility than post-reproductive women from a sedentary population of Herero pastoralists living in the same area, and that there was no indication that !Kung fertility increased when women settled at cattleposts – if anything, it declined. By contrast, the increased survivorship of offspring born to women leading a sedentary lifestyle may have increased !Kung reproductive success by up to 25 per cent. In other words, the higher fertility of sedentarised women is brought about, not by an increase in the production of offspring (that is, a decrease in IBI), but by the improved survival, compared to bush-living women, of those offspring that are produced.

Pennington (1992) also suggested that high levels of investment in offspring among foraging populations occur not because women use the natural contraceptive effect of lactation to reduce the chance of a new pregnancy, but because their inability to become pregnant as a consequence of a sexually-transmitted disease (STD) means that they do not have to terminate investment in their current offspring due to the imminent arrival of the next infant (pregnancy being a reason for !Kung women to wean their children). In other words, prolonged breastfeeding may be a consequence, rather than a cause, of low fertility – a direction of causality exactly opposite to that assumed by Blurton-Jones (1984).

Blurton-Jones (1994) has countered these arguments by suggesting that, although sterility can indeed explain certain aspects of !Kung reproductive behaviour, his foraging hypothesis is not invalidated since his investigations were carried out on an appropriate subset of women, rather than the whole population. He therefore maintains that among traditional foraging women, long IBI may be adaptive, but concedes that comparisons between cattlepost women and foraging women may be inappropriate given the confounding variables identified by Pennington (1992) and Harpending (1994).

Whatever the real explanation, these analyses do highlight the importance of considering all potential sources of variability when attempting to test adaptive hypotheses. Although the backload model offers an entirely plausible explanation for !Kung behaviour, it is clear that the impact of disease on fecundity can provide an equally valid non-evolutionary explanation. It therefore becomes very important to consider all possible explanations and subject them to rigorous tests before plumping for an adaptive evolutionary account.

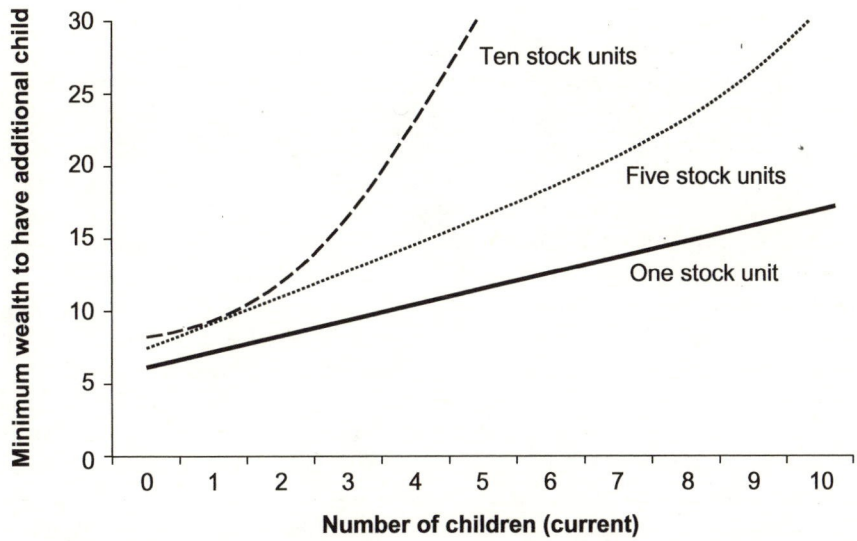

Figure 6.5 Mace's model of the costs of reproduction among the Gabbra of northern Kenya suggests that irrespective of the average cost of marrying off a child (the bridewealth paid out for a son to marry, indexed here as 1, 5 or 10 stock units), the family needs a herd of 7–8 female camels for their first child. Thereafter, the minimum herd size increases with both number of current offspring and the cost of marrying off an offspring. Redrawn with permission from Mace (1996a).

Figure 6.5), the minimum herd size for the first child was around 7–8 female camels; a figure that matched closely the average observed herd size at household formation of 6.75 camels. However, men who married relatively late in life did so with significantly smaller herd sizes. The model revealed that, as the end of the reproductive life span approached, the minimum wealth required to have a baby dropped sharply if the number of existing children was low or zero (as would be the case for unmarried men). This occurred because later in life it became more profitable to take a risk and have a child (even if there is a chance it won't be married off) than to die childless. Men who marry late (and whose reproductive life span is running out) therefore appear willing to take a risk by marrying and start having children as soon as possible rather than wait any longer for herd size to increase to a safer level. Although such men run a higher risk of reproductive failure than more prosperous men, they can do no better since they lack alternative options and are forced to make the best of a bad job.

The Gabbra model also demonstrated how variation in the cost of marrying off children influenced parents' reproductive decision-making. As the cost rises, so the minimum amount of wealth required to have another baby rises increasingly steeply, especially if there are already a number of other children present, as shown by the diverging lines on **Figure 6.5**. This is particularly interesting because these differences in cost mirror real differences in the cost of marrying off male versus female offspring: boys work out to be much more expensive than girls in the Gabbra marriage market. In order to marry off a son, the household needs to provide three camels as a brideprice to the girl's family, plus another six to seven camels to set the son up with his own herd. By contrast, when a daughter is married off, the household only needs to provide a dowry of 15–16 goats. Since one camel is equivalent in value to 10–15 goats, fathers make a net profit when marrying off daughters: they only have to provide the equivalent of one camel, but gain three in the form of brideprice from the groom's family. Marrying off sons is therefore much more costly than marrying off daughters.

Consequently, deciding whether or not to have another baby depends not only on the number of current children but, more critically, on their sex; if too many sons are born early in the parents' reproductive life, they will have to limit drastically the number of children subsequently born in order to allow each son to marry, whereas an equivalent number of daughters would not require reproductive restraint and could even encourage parents to continue trying for a son.

In line with this, Mace (1996a) found that the proportion of households in which a second wife was present (taken as an indication of a man's desire to have more children) depended on the number of children the man had by his first wife (**Figure 6.6**). In all cases where a man had no children during his first marriage, he took another wife, whereas in households with nine or more children, men were never observed to remarry. More importantly still, the number of sons a man had by his first wife had a much greater negative effect on his likelihood of remarrying than either the number of daughters or the current size of his herd, reflecting the high costs of male versus female offspring.

Gabbra men thus behave as if they are attempting to optimise family size and enhance fitness. Remember, this need not be a conscious decision. When making reproductive decisions, individuals may be motivated by much more proximate factors that bear little relation to fitness considerations. However, men who behave in a way that optimises family size will tend to leave more representatives in subsequent generations, and should thus have higher fitness than those who behave differently. Gabbra behaviour appears to be currently adaptive, in that people are behaving in a manner that will maximise the number of descendants in future generations.

SCHEDULING REPRODUCTION

In addition to attempting to manage family size, deciding when to reproduce has important implications. In traditional societies, women give birth at intervals more or less throughout their reproductive lifespan (that is, between the ages of 15 and 45 years), or at least until they reach the point at which they have achieved an optimal family size (see, for example, Pennington and Harpending 1993). However, such a pattern of reproduction is by no means typical of humans in post-industrial societies. The most obvious example is the common tendency for women in western countries to reduce family size to around two offspring which are they produced close together (typically one to three years apart) relatively late after menarchy (typically in the late twenties, sometimes early thirties). This long delay in the onset of reproduction is usually associated with time invested in developing a career. We consider the implications of this pattern of reproduction further in a later section on the demographic transition. Here, we consider a rather different case, namely the phenomenon of teenage motherhood.

As we have noted, the conventional view in most post-industrial nations is that reproduction should be delayed until as late as 10–15 years post-menarche. However, girls from socially disadvantaged backgrounds often become pregnant as young teenagers, a phenomenon that is usually interpreted in terms of ignorance or unintended accident (or occasionally as a deliberate attempt to access social security benefits). An alternative view of this widely recognised social problem is that, so far from being the product of ignorance or indolence, it might in fact be a rational reproductive strategy.

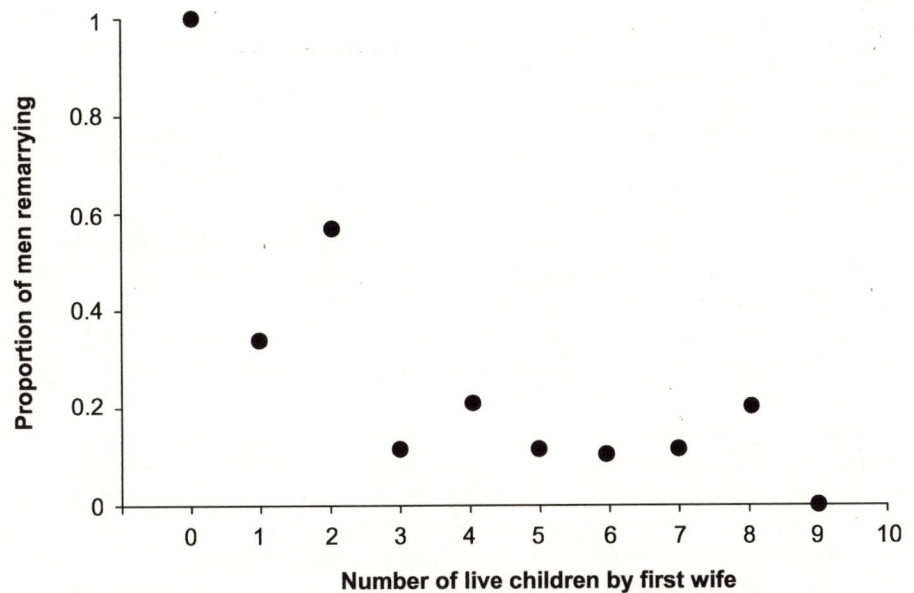

Figure 6.6 Polygyny among the Gabbra is largely driven by men's desire to have children: the proportion of men who remarried is negatively related to the number of offspring produced by the man's first wife. Men with three or more children seldom took on a second wife, and those with nine or more never did so. Redrawn with permission from Mace (1996a).

Geronimus (1987, 1996), for example, has argued that the teenage motherhood syndrome is an adaptive strategy for girls who intuitively appreciate that the future holds very few opportunities for them, either economically, in terms of career opportunities (Geronimus 1987) or in terms of continuing health (the 'weathering' hypothesis, in which poverty is associated with more rapid health deterioration and a relatively early death: Geronimus 1994, 1996). Indeed, she argues, for women in the lower socio-economic strata, job opportunities may be better later in life for mature women who have completed reproduction early and no longer have family constraints. Early versus delayed reproduction should thus be seen as alternative socio-reproductive strategies.

A significant contributory factor here seems to be that, in addition to poor career prospects, young women in these communities also have a very limited choice of good husbands who might remove them from the poverty trap. This partly reflects a tendency for young men to migrate out in search of work (or to seek a marriage partner from a higher socio-economic class in an unusual example of hypergamy), and partly because those males that do remain not only have poor economic prospects but are also likely to die young (Staples 1985). Among black ghetto communities in the USA, for example, mean life expectancy among black males is nearly ten years less than that for white males. Most of the mortality attrition that gives rise to this difference falls on younger men in the 25–45 year age cohorts. Consequently, young women in these communities have an extremely limited choice of high quality mates. The alternative life strategy they then pursue is to reproduce early and fast, in effect going for quantity rather than quality of young. This strategy is possible because they are able to rely on maternal and especially grandmaternal assistance with rearing.

The importance of multi-generation matrilineal households in facilitating this reproductive strategy has been highlighted by Burton (1990) in a study of a black US rural community (see also Stack 1975). It is clear from her observations that early reproduction (sometimes as early as age 14) is crucial to the success of this strategy: by giving birth at around age 15–16, the mother ensures that the grandmother (her own mother, typically aged 30–35) and great-grandmother (typically aged 45–50) are still young enough to provide a combination of household income (usually the grand-mother) and childcare (typically the great-grandmother) in a context where the women themselves have a lower longevity than is typical of better-off women in the USA. A significantly later start would make it unlikely that the child's great-grandmother would be physically able to take on fulltime responsibility for childcare, so allowing the grandmother to provide the economic support for the household.

Flinn (1989) has documented a particularly interesting example of this phenomenon in a study of a rural Trinadadian community. He found that no female became pregnant while co-residing in a household with a 'reproductive female', even though she might be of reproductive age and sexually active. His data revealed that no female who lived with her mother became pregnant until the mother's last-born offspring was at least four years old, while mothers with currently reproducing daughters had ceased reproduction sooner than those without reproducing daughters. Flinn (1989) interpreted this as evidence for the suppression of reproductive overlap between mothers and daughters. Reproducing females increased their own reproductive success through the action of female helpers (women with helpers engaged in less childcare activity themselves, although their children received far more care and attention overall), but at the expense of the helper's own potential reproductive output. Later, as the mother ages, she allows her daughter to take over the reproductive role and herself becomes the helper.

Flinn (1989) suggested that this reproductive suppression effect was the result of a form of parent–offspring conflict between mothers and daughters. As **Figure 6.7** shows, there is a period of potential conflict between mother and daughters regarding their ideal reproductive careers. During this conflict period, a mother would do better by remaining reproductive and obtaining help, but a daughter would do better by starting to reproduce herself. This arises because women are 50 per cent related to their own offspring but only 25 per cent related to their grandchildren. For example, a 36-year-old woman with an **age-specific fertility** of 0.06 infants/year and an 18-year-old daughter (with an age-specific fertility of 0.091) should prefer to remain reproductive because her own direct gain from reproduction ($0.5 \times 0.06 = 0.03$) is greater than that gained by foregoing reproduction in favour of her daughter ($0.25 \times 0.091 = 0.02$), even though the daughter has a higher fertility than the mother. The daughter, however, gains more by becoming reproductive herself ($0.5 \times 0.091 = 0.05$). As we might predict, daughters with reproductive mothers had significantly more agonistic interactions with their mothers than did those with non-reproductive mothers; the highest levels of agonism were found between mothers with daughters in the 18–21 age-group (the point at which it becomes more profitable for daughters to begin reproduction) (**Figure 6.8**).

While extremely interesting, this study raises a number of as yet unanswered questions: why do households typically consist of only one breeding female? Why does competition exist in the first place? One answer is that females initially help because they lack alternative options (for instance, they lack a suitable mating partner) and gain inclusive fitness benefits from helping which makes such a strategy attractive (the standard helpers-at-the-nest explanation), but they are then exploited by reproductive females who try to retain helpers against the helpers' best interests. The mechanism by which reproductive suppression occurs is also unknown, although there is now considerable evidence from the study of primates to suggest that socially stressful conditions can result in the reproductive suppression of low-ranking females (Bowman et al. 1978, Abbott et al. 1986, Wasser and Barash 1983; for review, see Dunbar 1985). In addition, Jasienska and Ellison (1998) have shown that increased workloads can

AGE-SPECIFIC FERTILITY

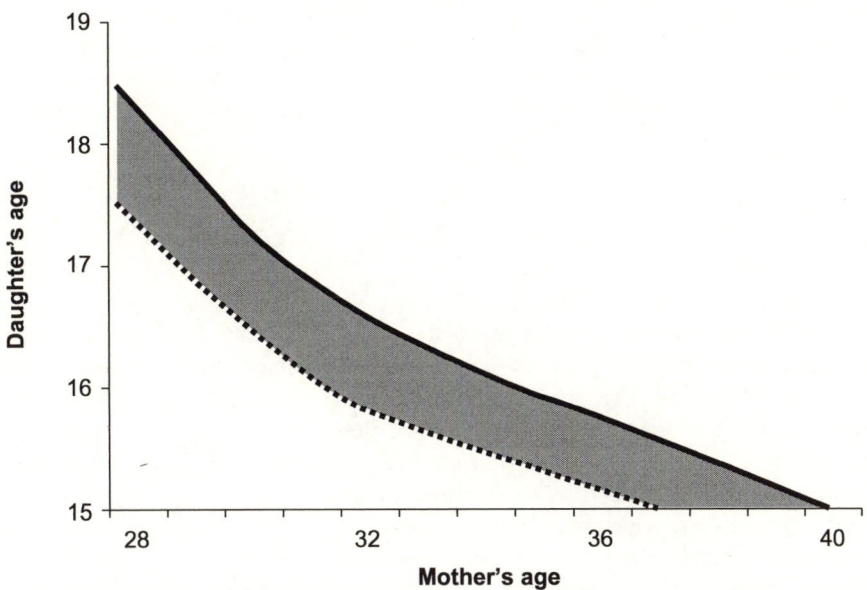

Figure 6.7 When a mother and her daughter live together in the same household in a rural population in Trinidad, only one of them typically reproduces, with mothers ceasing reproduction when their daughters start. The graph shows the relative ages of the two women at which the daughter (lower line) and the mother (upper line) would prefer the mother to cease reproduction and the daughter to start. The switch point for the mother is the age at which the daughter's reproductive value, v_x, first exceeds that of the mother. The shaded area between the two lines represents the period during which the mother's reproductive interests are in conflict with the daughter's. Redrawn with permission from Flinn (1989).

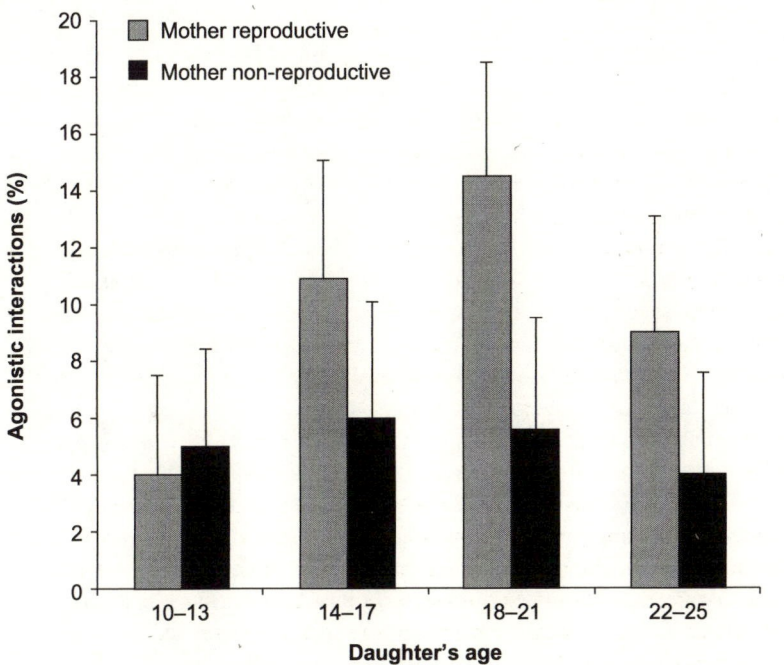

Figure 6.8 The percentage of all household agonistic interactions that take place between a mother and her daughter are significantly higher if the mother is reproductively active than if the daughter is reproductively active. Conflicts are disproportionately more common during the period when the mother's and daughter's interests as to which of them should be the household's reproducer are in conflict (see Figure 6.7). Redrawn with permission from Flinn (1989).

reduce fertility, even when females remain well-nourished and in neutral energy balance. The fact that mothers reduce their workloads when helpers are available, while the helpers themselves have a larger workload, may go some way to explaining the apparently low fertility of helper females. Once again, however, the evidence needed to tie all elements of the hypothesis together is lacking.

FOR LOVE OR MONEY: HERITABLE WEALTH AND DECISION-MAKING

The costs of offspring in terms of the material resources they require as adults, in addition to the high levels of care necessary when they are children, can clearly place a very effective brake on offspring production. This is particularly true of harsh environments, like that occupied by the Gabbra, where resources are generally scarce and low in quality. If we are to give a full account of reproductive decision-making in terms of the quantity/quality trade-off, however, we must also be able to explain the decisions made by individuals living in affluent societies, such as those of the industrialised West, and it is to these that we now turn.

Rogers (1990; see also Harpending and Rogers 1990) specifically set out to see whether, under conditions like those found in the industrialised West, the trade-off between the quantity and quality of offspring was affected by both heritable wealth and the quality of the environment (both important factors in the modern world). Under what circumstances might small completed family sizes lead to high levels of fitness in succeeding generations?

Unlike other animals, heritable wealth changes the nature of the quality/quantity trade-off in humans by extending the value of resource investment into more distant generations. By providing an offspring with more food, a parent bird can increase the quality of an offspring and so increase the numbers of grandchildren that the offspring is capable of producing. But that investment will probably not increase the *quality* of the grandchildren themselves. In humans, however, a large bequest to an offspring may substantially increase the wealth of grandchildren (that is, their quality) which in turn leads to potentially large fitness gains (more great-grandchildren). The standard behavioural ecological method that uses lifetime reproductive success as a proxy for fitness doesn't apply in such cases. Instead, an approach is needed that can account for the effect of bequests made in the present on the reproductive success of individuals in the distant future.

Rogers (1990) used a Leslie matrix approach (a demographic technique for predicting future population size and structure) to build a model that could incorporate the effect of heritable wealth across generations. In this model, individuals could divide their wealth into two components: a *fertility allocation* (used to produce offspring) and a *bequest* (a sum inherited by the offspring). The model set out to discover what the best allocation of total effort between these two categories would be in order to maximise the representation of an individual's genes in succeeding generations. The model laid down some ground rules, which were that at least something had to be allocated to fertility (otherwise no offspring could be produced) and that, in addition to inheriting wealth, individuals could also 'earn' some of their own. The ability to do so depended on the 'quality' of the environment, so a good environment was one in which the earning potential of individual offspring was high.

One of the most striking findings for an 'average' environment was that the relationship between wealth levels and fertility was not directly linear. Fertility (number

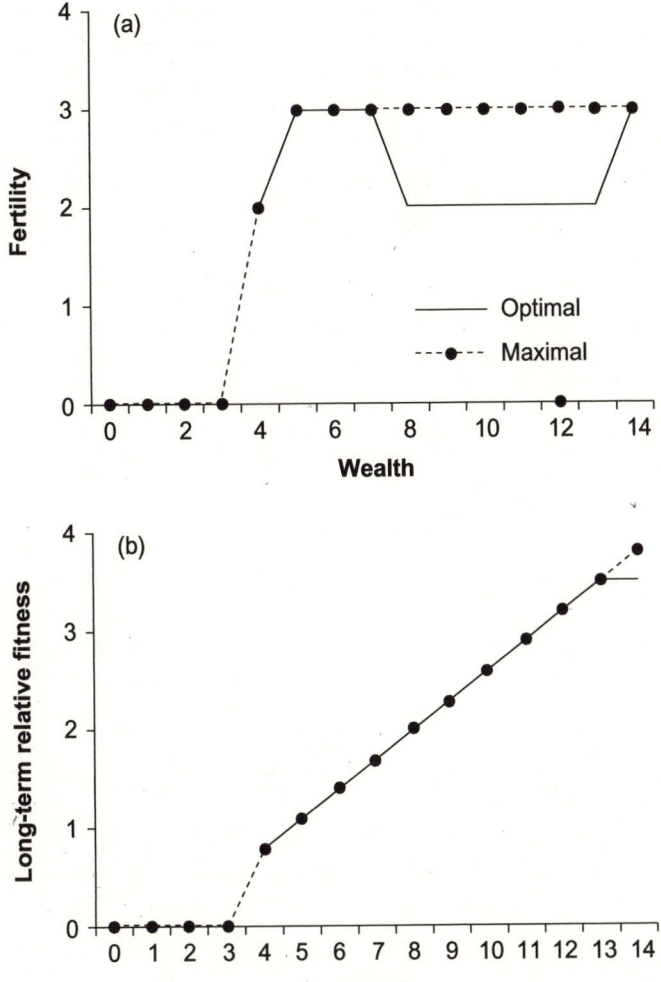

Figure 6.9 The optimal level of fertility (number of offspring produced) is often lower than the maximal number in situations where parental wealth directly affects an offspring's socio-economic opportunities. Rogers' model suggests that under intermediate environmental conditions (a) fertility does not correlate strongly with wealth ($r = 0.508$), but (b) increased wealth is associated with increased long-term fitness (indexed as the number of genes passed on to succeeding generations) in an almost perfect correlation ($r = 0.998$). In other words, it may pay wealthy individuals to favour acquiring wealth over having children so as to be able to invest more in each child. Redrawn with permission from Rogers (1990).

of offspring produced) increased with wealth up to a certain point, but then showed an inconsistent pattern of increase and decrease (**Figure 6.9a**). Consequently, there was only a weak overall correlation between wealth and fertility, although wealth and long-term fitness (the number of genes passed to succeeding generations) were almost perfectly correlated (**Figure 6.9b**). In other words, although wealth increased fitness, this wasn't necessarily associated with the production of large numbers of offspring. A large bequest to a few offspring could have long-term ramifications by ensuring the production of high-quality grandchildren.

These results were thrown into sharper relief when Rogers (1990) adjusted the

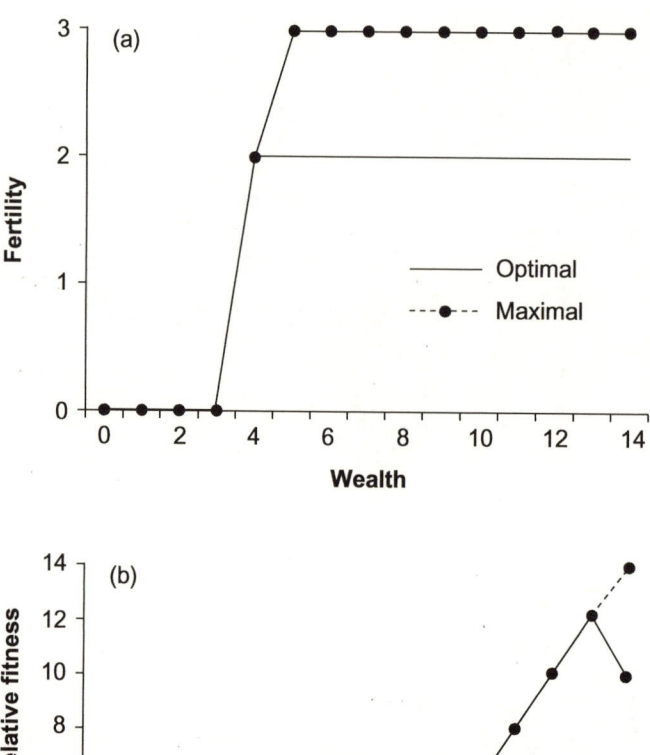

Figure 6.10 In an economically harsher environment than that modelled in Figure 6.9, Rogers' model predicts that (a) individuals should constrain fertility so that the amount of wealth invested in each offspring can be maximised (since offspring success in this environment is dependent on the wealth they inherit from their parents). Optimal fertility is therefore always below maximal fertility over most of the wealth range. This means that there is no relationship at all between wealth and the number of offspring produced ($r = 0.000$), even though (b) long-term fitness continues to correlate highly with wealth ($r = 0.977$). Redrawn with permission from Rogers (1990).

quality of the environment. In poor environments (those in which the ability of offspring to earn wealth was low), inherited wealth became relatively more important to the success of offspring. The optimal strategy therefore was to maximise the amount of wealth inherited by each offspring, which necessarily meant limiting the number produced. Optimal fertility was therefore the same for all wealth classes above a certain level of wealth (individuals below a minimum amount of wealth were assumed to be sterile for the purposes of the model) (**Figure 6.10a**). Consequently, the relationship between wealth and the number of offspring produced fell to zero, but there was nevertheless a high correlation between wealth and fitness (**Figure 6.10b**).

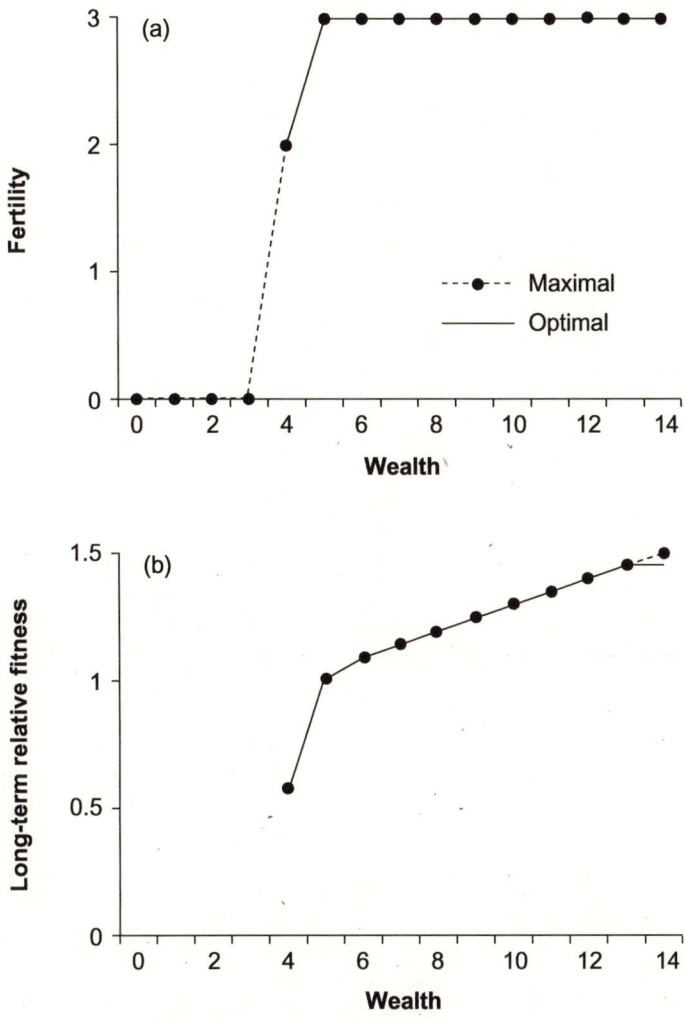

Figure 6.11 In an economically richer environment than that modelled in Figure 6.9, Rogers' model suggests that, when offspring are able to earn their own wealth, (a) it pays parents to reproduce at the maximal rate since parental bequests are not necessary for the offspring's own success. In this case wealth and fertility are highly correlated ($r = 0.838$) but (b) wealth is still better at predicting long-term fitness ($r = 0.931$). Redrawn with permission from Rogers (1990).

Rogers (1990) also investigated what happened in rich environments, where offspring could acquire large amounts of wealth by themselves. Here, the optimal strategy was to maximise the number of offspring produced, and minimise the allocation to bequests. Since offspring could be counted on to ensure the quality of grandchildren through their own earning power, the need for parents to invest large amounts of wealth in offspring diminished. Consequently, wealth and fertility were significantly correlated in this run of the model ($r = 0.838$) (**Figure 6.11a**), but as before, wealth, rather than the number of children, still provided the best proxy for long-term fitness ($r = 0.931$) (**Figure 6.11b**). That is, even in cases where wealthier individuals were able to produce more children, long-term fitness was still tied more closely to wealth itself rather than the number of offspring produced in a generation.

Taken together, the results of these simulations weaken the claim that evolutionary explanations do not apply to modern western populations: Rogers (1990) was able to show that the relationship between wealth and the optimal production of offspring can actually be negative over a large range of wealth categories when individuals are trying to maximise their fitness. It is therefore quite plausible that modern western societies fall into the range where reining back on offspring production in favour of increasing wealth is the optimal strategy. On the other hand, there is, as yet, no empirical evidence to support this model. Although theoretically possible, we cannot actually say whether or not low modern fertility rates are indeed the result of long-term fitness enhancing strategies. What it does mean, however, is that there is no a priori justification for assuming that all modern behaviour is maladaptive because it does not result in high levels of offspring production. (We explore this issue in more detail in the following section.)

What should be clear from these analyses is that people often do behave as if they are attempting to maximise their fitness, and, as Rogers' (1990) analyses indicate, sometimes at a level that is not immediately obvious. For example, since wealth is a strong predictor of long-term fitness, selection may well have favoured economic motivations over reproductive ones. A proximate desire to increase wealth may thus serve the interests of long-term fitness just as well as the desire for children, but this would not be immediately apparent if one takes the narrow view that evolutionary success is a simple function of reproductive success.

THE DEMOGRAPHIC TRANSITION

DEMOGRAPHIC
TRANSITION

Most developed countries have undergone what is known as the 'demographic transition' whereby both mortality and fertility have decreased substantially over the past 150 years or so, resulting in completed family sizes that are typically at or even below replacement rate (two offspring per couple) and a zero (or even negative) population growth rate (Coale and Treadway 1986). These data have often been cited as evidence against socio-biological accounts of human behaviour (for example, Vining 1986), and as such provide an important challenge to evolutionary explanations.

Notice, however, that most authors who have used these findings as evidence against the evolutionary approach have claimed that, in exhibiting the effects of a demographic transition, modern western populations violate the 'central tenet of sociobiology' – that all individuals should attempt to maximise their fitness. It should by now be clear that this argument is somewhat off-beam. (Maximising fitness is not the same thing as maximising the number of offspring produced.) To assume that wealth and cultural success should automatically be associated with the production of vast numbers of offspring without considering socio-economic circumstances is to take a naive and simplistic view of evolution.

A second group of authors have attempted to use the demographic transition to argue that we should not expect to find modern behaviour to be adaptive because the fast pace of cultural change over the last 10 000 years has overwhelmed our abilities to keep up in evolutionary terms (for example, Tooby and Cosmides 1990). However, as we pointed out in Chapter 1, this is an unsupported assumption: there is no reason, in principle, why the behaviour of modern humans should be any less adaptive than that of our ancestors prior to the agricultural revolution.

Harpending and Rogers (1990) extended their model to see whether it could account for the demographic transition. Their approach was based on two assumptions:

that society is stratified into different classes and that there are differences in the relative costs and benefits of downward versus upward mobility of offspring. As before, a model was constructed using the Leslie matrix approach. The class system was assumed to consist of three classes (upper, lower and destitute, the latter being a class in which individuals produce very few offspring, and which therefore acts as a social 'sink').

The mathematics used in the model are complex, but essentially the results showed that, when there is a destitute class in which fitness is extremely low or even zero, lower-class parents would be selected to reduce their fertility greatly and concentrate their resources on a single offspring so as to provide it with the potential to be upwardly mobile. Producing large numbers of offspring who have no alternative but to enter the destitute class through lack of investment boils down to a tremendous waste of effort in terms of long-term fitness.

Once again, low fertility holds the key to long-term fitness, and the existence of a destitute class holds the key to direct selection for low fertility. If we take a look at the actual numbers, the difference is quite impressive: in Harpending and Rogers's model, each upper-class offspring (achieved through upward mobility) produced by a lower class family is 'worth' more in fitness terms than two offspring who stay in the lower class and more than twelve downwardly mobile offspring who enter the destitute class.

So, the crucial question with regard to the demographic transition is: did such a destitute class exist in the period prior to the onset of the demographic transition in the mid-nineteenth century? The historical evidence suggests that this probably was the case. For example, in eighteenth-century France, lower-class men were often removed from the reproductive pool through military service, while lower-class women often migrated to the cities to seek low-status, low-paid work. The levels of infant mortality experienced by these women if they did manage to produce offspring were horrendous: there are records of the streets littered with the corpses of dogs and babies during famine periods. In France, Italy and Spain the frequencies with which babies were abandoned during the eighteenth and early nineteenth centuries were such as to spawn orphanages in very large numbers. Indeed, the frequency with which children (and, incidentally, elderly parents!) were abandoned in France during the eighteenth century correlates strongly ($r_S = 0.81$, $N = 64$ years, $P < 0.001$) with the price of rye (which itself is inversely related to its availability, so providing a direct measure of famine conditions) (**Figure 6.12**). Similarly, the Poor Laws that led to the nineteenth-century workhouse schemes that were described so graphically in many of Charles Dickins's novels were enacted precisely because of the burden that the indigent poor were placing on local ratepayers. In some parts of France, it was not unusual for those in the poorest strata of society to send their children away to work as shepherds or servants as young as seven or eight years of age, taking no further interest in them thereafter (Sabean 1976).

The degree of poverty experienced by the lowest sections of society during the period immediately prior to the onset of the demographic transition in Western Europe was thus clearly of the level required by the Harpending–Rogers model, and those individuals unlucky enough to find themselves in the underclass certainly had little prospect of raising a healthy, happy family. Detailed historical demography studies are now needed in order to test the model more comprehensively.

Somewhat more speculatively, Harpending and Rogers (1990) also suggested that the lack of a demographic transition in Africa could stem from the fact that, until very recently, there was no destitute class that failed to reproduce. The wide avail-

Figure 6.12 The frequency with which babies were abandoned at foundling hospitals in Limoges (France) between 1726 and 1790 (indexed to 100 in 1720) correlates with the price of rye (a well established index of famine conditions: rye becomes more expensive when less of it is produced). Source: Voland (1989).

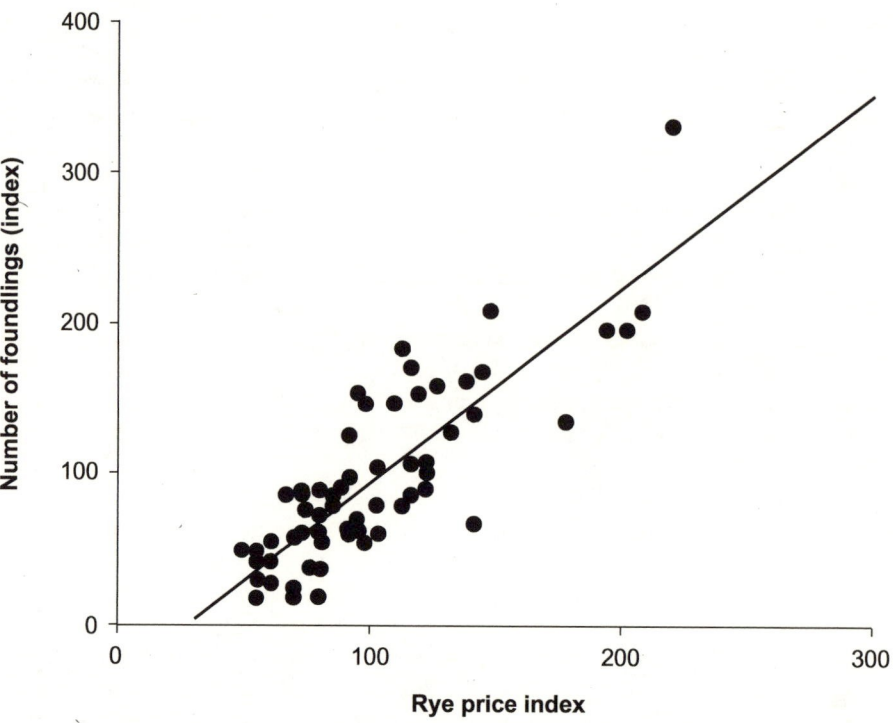

ability of virgin land and the existence of supportive networks of extended kin appears to enable people to support themselves effectively even when poverty levels are high (Draper 1989). In addition, the cultural practice of 'clientship' observed in a number of African societies (in which poor people attach themselves to powerful individuals in order to avoid destitution) also reduces the probability that a reproductively impoverished underclass will emerge within society. Consequently, there may be no disincentive to reproduction in these populations, since the societal stratification does not penalise the very poor in the same way as it did in Europe and America.

Significantly, a similar clientship arrangement existed in the feudal systems characteristic of medieval Europe that bound peasant and lord together in a system of mutual economic dependency. This system had broken down by the end of the seventeenth century when its demise was hastened in western Europe by the improvements of the agricultural revolution that offered landowners considerable financial incentives for becoming economically independent of their rural communities. In some cases, such as the Highlands of Scotland after the 1745 rebellion, political considerations led to social policies deliberately aimed at breaking down the old clan-based loyalties (Prebble 1963). Enclosures of common land, combined with a shift from subsistence to largescale economic farming, set in train conditions for the creation of a landless underclass in the century prior to the demographic transition in Britain.

Turke (1989) has suggested another solution to the puzzle of the demographic transition. He proposed that the development of the nuclear family and the reduction in the size of kin networks prompted the shift to reduced fertility. In circumstances where individuals have high levels of kin assistance when raising offspring, individuals tend to show higher fertility than among nuclear families (see Turke 1988; Bereckzei 1998). Where kin are available to assist in child-rearing, Turke (1989) suggests that the psychological motivation to produce children is increased as the services and resources provided by kin mean that the extra child is viewed in a positive light and is not considered to be an added 'burden' on parents.

Draper (1989) found support for this in her analysis of West African marriage systems where fostering out children to relatives is commonplace and allows parents to produce larger families than would otherwise be the case. Draper (1989) suggested that this influences children's perceptions of resource availability as they grow, with the result that they perceive the child-rearing environment to be much more benign than it would appear to westernised observers, and that this perpetuates the desire for large family size down through the generations. In many cases, the desire for a large family size per se is not the key factor, merely the lack of disincentives against large families in a 'resource-rich' environment.

It is therefore suggested that industrialisation and the rise of the nuclear family, in which the costs of child rearing tend to fall more heavily on the parents lead to reduced fertility by removing the 'safety net' provided by kin. Since decisions about rearing children are made on the basis of having only two adults (or more likely just one, the mother) available, the perceived and actual costs of children are increased, and a reduction in fertility follows as a consequence. Turke (1989) has also suggested that it is precisely the lack of attendant kin that allows parents to focus investment intensely on their offspring. In situations where there are extended family networks, parents are expected to distribute their own resources more widely (as well as receive them); high levels of investment in one's own children are therefore not possible. In effect, the rise of the nuclear family promoted a reduction in family size by removing both the benefit and the cost of extended kin networks.

The notion that increasing the costs of child-rearing influences family size is supported by several other recent analyses. Using state-dependent modelling, Mace (1998) found that, contrary to what most demographic theory suggests, neither decreased child mortality nor greater environmental stability lead to reduced family sizes. Only increasing the costs of children had any significant effect on both optimal fertility and inheritance strategies: as children become more costly to raise and marry off, smaller family sizes become optimal (**Figure 6.13a**). Although the number of sons who gained an inheritance under this fertility schedule was reduced, each of those who did was given relatively more, and therefore achieved higher reproductive success (**Figure 6.13b**). This was at least partly due to the fact that the wealth that each son was given was capable of generating even more wealth. By maintaining the size of the inheritance and not spreading it too thinly, each son and his descendants becomes much more wealthy than would otherwise be the case, and this has positive effects on long-term reproductive success.

Data from the Kipsigis provides empirical support for this finding: men who inherit more capital (land and livestock) become wealthy at a faster rate than those who inherit less capital (Luttbeg et al. 2000). These are somewhat similar to the findings of Rogers' (1990) model, where 'poor' environments lead to an increasing reliance on bequests (and a reduction in offspring production) in order to ensure long-term fitness. The inability of offspring to 'earn' their wealth in Rogers's model is thus analogous to increasing the costs of children in Mace's model. Either way individuals maximise the number of grandchildren produced (Mace) or their long-term fitness (Rogers) by restraining their reproduction in order to ensure that each individual who does inherit gains an increased proportion of the family's wealth, thus ensuring the quality of their own children.

The reduction in offspring number and the concomitant increase in investment to

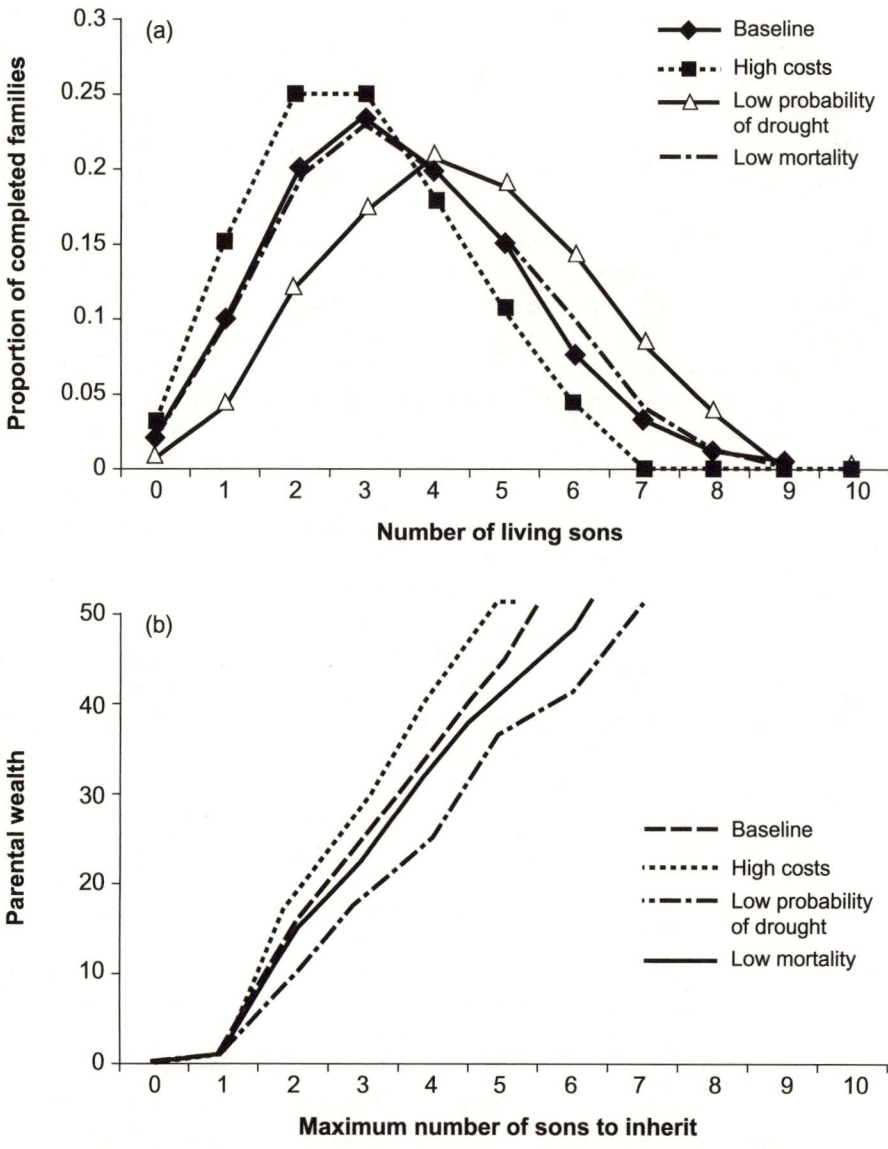

Figure 6.13 Relationship between optimal family size at the end of the parents' lives and household wealth for Gabbra pastoralists, based on the predictions of Mace's (1998) model. (a) Expected number of families with different numbers of sons and (b) the predicted maximum number of sons between whom the family's wealth should be divided, under different rearing conditions. Baseline condition: parameter values based on current environmental conditions experienced by the Gabbra. In the high rearing costs condition, the costs of rearing a child to adulthood are doubled; in the low drought case, the probability of a drought occurring in any given year is half the baseline value; in the low mortality condition, all extrinsic mortality risks are an order of magnitude below their baseline values. In general, the distributions are robust with respect to modest variations in the parameter values: only increasing the costs of rearing children has a significant effect. Redrawn with permission from Mace (1998).

the few offspring that are produced are both features of post-transition societies. Mace (1998) suggests that costs of children can take many forms. Sending children to school can represent a substantial cost even under circumstances where free state education is available. In such cases, the economic value of children to the household is lost because, while they are at school, they are not available as a source of labour – often a major consideration among societies that are at least partly subsistence-based. Indeed, the European pattern of a school year with a long summer holiday is a hangover from the time when offspring labour was essential down on the family farm during summer harvests. If children no longer contribute labour to the household, but continue to require food and shelter (as they obviously do), then they will represent a net cost to the household. It would be very interesting to know parents' motivations for educating their children under these circumstances, and what they perceive the value of education to be, both for their children and for themselves.

Borgerhoff Mulder and her colleagues (see Borgerhoff Mulder 1998; Luttberg et al. 2000) used a stochastic dynamic programming approach to try to understand just what Kipsigis men's reproductive objectives really are. They considered a range of possible strategies that varied between a total preference for maximising the number of children produced and a total preference for maximising wealth. For the Kipsigis, marriage requires a minimum level of wealth (to pay the brideprice) and each child born from a given union requires considerable investment to cover illness, education and (later on perhaps) help with their own marriages. Not surprisingly, perhaps, a comparison of the models' predictions with what Kipsigis men actually did suggested that they give a relative weighting of about 70 per cent to maximising wealth. By doing so, they are then able to maximise the numbers of children (and ultimately grandchildren) produced.

Somewhat analogous findings emerged from a large-scale study of reproductive strategies among contemporary Albuquerque men. Kaplan et al. (1995) argued that a reduction in fertility would be strategically beneficial for wealthier individuals if the value of investing resources in a child are a (positive) nonlinear function of the amount invested. This might happen, for example, because knowledge is cumulative: each additional unit of knowledge or skill acquired provides the platform for an even larger subsequent educational gain or access to much greater socio-economic opportunities (as seems plausibly to be the case in our modern knowledge-based economies: for historical evidence on this, see Lindert 1986). In this case, the opportunity cost of producing an additional child rises disproportionately with each successive child for the better off, whereas they remain relatively constant for the poor whose investment opportunities are limited. Kaplan et al. (1995) were able to show, from a sample of 7107 men, that father's income and length of investment (the total time the father was in the same household as the respondent) correlated positively with both the number of years of education the respondent received and his eventual income, and that the number of sibs the respondent had influenced these two variables negatively. In other words, parental wealth does affect eventual socio-economic status as an adult, while family size detracts from this because that wealth has to be spread more thinly around more offspring. Since the rich can place their offspring into much better conditions in the adult market by investing in them (and because the extent to which they can do this is directly related to the level of investment in each offspring), they have a much greater incentive to reduce family size than do the poor.

THE EVOLUTION OF MENOPAUSE

Limiting the number of children produced during an individual's lifetime can clearly have positive fitness benefits by ensuring that those offspring that are produced receive sufficient resources to give them the best possible chance in life. Similar reasoning has led to speculation that the female menopause – the complete cessation of reproductive functioning – is also an adaptive response to the heavy demands of child-rearing and as such forms an integral part of the human female's reproductive strategies. The so-called 'Grandmother Hypothesis' suggests that, as females age, they may be able to obtain greater fitness benefits through providing assistance to their existing offspring and helping to raise their grandchildren than by attempting to produce more offspring themselves.

The increased risk of mortality for older women has been considered to be the most likely reason for favouring grandchildren over children. If reproductive events place a greater strain on older individuals and they are more likely to die in childbirth (or at least die before their offspring are fully independent), then their offspring are also likely to suffer increased mortality and contribute little to maternal fitness. Reproduction at an advanced age may therefore be wasteful, hence favouring the diversion of reproductive effort to grandchildren.

Using data on the Ache of Paraguay, Hill and Hurtado (1997a, b) have pointed out that, although infants do indeed die if they lose their mothers, the mortality rate of older women is so low (about three per cent per annum for 50 year olds) that very few infants of older mothers would actually have to deal with the situation of maternal death. Having said that, it should be pointed out that this mortality rate is calculated for women who experience menopause and are not reproductively active. If menopause did evolve in response to high maternal mortality rates in older women, then this is exactly the result we should expect to find. Data from the present day where adaptations are already in place cannot be used to test hypotheses concerning the original pressures that selected for a trait.

Hill and Hurtado (1997a, b) recognise that the Ache data can only provide information on whether the grandmother effect maintains menopause in modern populations. Using mathematical modelling, they investigated whether helping by grandmothers had any effect on the fertility of their adult offspring by comparing the reproductive histories of men and women with and without a surviving mother. Their analyses revealed that the benefits of menopause (in terms of producing more grandchildren) only very slightly outweighed the costs (in terms of the loss of direct reproduction). Both men and women with a living mother experienced marginally higher fertility, while grandchildren with a living grandmother had slightly higher survivorship. None of these effects was statistically significant, however.

The problem here is that the measures used by Hill and Hurtado (1997a, b) may not be entirely appropriate for testing the model. First, merely comparing adults with and without a surviving mother does not take account of the fact that individuals whose mothers die may be able to call on other kin for assistance. If this were the case, it would underestimate the grandmother effect, since the analyses assume that individuals without a mother are raising offspring unassisted. Second, the impact of helping may only be critical during times of crisis (drought, disease, food shortage). If these happened frequently enough throughout our evolutionary history, they may have provided a sufficient selection pressure to favour the evolution of menopause. The data collected on the Ache during a relatively stress-free period would not have been able to detect the influence of grandmothers on offspring and grandchild survival during critical periods.

BOX 6.6

Phenotypic correlations

Investigating life-history trade-offs in humans is somewhat limited by the fact that it is only possible to look at phenotypic correlations – that is, comparing the value of one physically expressed trait (for instance, longevity or age at sexual maturity) with another (for instance, fecundity). Phenotypic correlations do not always reveal the trade-offs that exist between traits because there may be a lack of non-adaptive variation in the population. That is, if all individuals are behaving optimally and allocating exactly the right amount of resources to, say, growth rather than reproduction, then there won't be any differences between individuals. If individuals do not vary in their resource allocations, it becomes impossible to measure any trade-off that might exist.

Another reason why trade-offs are not always apparent in phenotypic correlations is that individuals may show adaptive variation in their allocation of resources. That is, individuals may differ in their access to resources and so the optimal allocation to reproduction may also differ. A simple example makes this clear: imagine you are investigating the trade-off between how much people spend on their houses and how much on their cars. You would predict that this relationship would be negative (the more you spend on a house, the less you can spend on your car). However, in most industrialised nations, this relationship is more likely to be significantly positive: the more costly the house, the more expensive the car. This is because only people at the poorer end of the scale actually have to make a trade-off between their house and their car. Individuals at the very wealthy end of the scale, who have increased access to resources, can afford both an expensive house and a very fancy car. Their resources are not limiting in the way that they are for less wealthy individuals. This produces a positive relationship that obscures the extent of the trade-off.

The data from the Ache showing that heavier females give birth for the first time earlier than lighter individuals may be an example of this kind of adaptive variation. Life-history theory predicts that the more an individual allocates to growth, the longer it will take to reach sexual maturity, and so we should expect heavier girls to have their first offspring later than lighter ones. However, if some girls have greater access to resources that allow them to grow faster than those who lack resources, then a positive correlation between heaviness and reproduction will be produced, giving the impression that there is no cost of increased growth in terms of a delay in reproduction.

The only way to identify potential trade-offs when using phenotypic correlations is to experimentally manipulate the population and record the consequences (for example, prevent some females from breeding and then see if they live longer as a consequence). While this enables trade-offs to be measured in non-human animals (see Lessells 1986), this is obviously not possible in humans for ethical reasons. As a consequence, tests of life-history theory using human data need to be treated with caution.

Most attempts to test the Grandmother Hypothesis to date have used an indirect approach based on the predictions of life-history theory. Hill and Hurtado (1997a), for example, argued that life-history theory tells us that if individuals do not have any reproductive value, then there is no pressure to counteract the accumulation of deleterious **alleles** that would lead to senescence and death. Hence, if post-menopausal women are not making any fitness contributions to succeeding generations, they would be expected to age and die faster than men. In fact, they do not. Since Ache men produce approximately 0.15 offspring a year at age 50, whereas women of the same age have terminated all reproduction, life-history theory would seem to suggest, in the absence of any other reasonable explanations for why women do not senesce faster than men, that post-menopausal women must be making a genetic contribution at

ALLELES

least as great as that of men. Since this is not being achieved by direct reproduction, it must occur by aiding close kin (including, obviously, grandchildren).

Blurton-Jones et al. (1999) adopted a different approach, arguing that the long delay to reproductive maturity characteristic of humans can only be made to fit with recent analyses of life-history processes (Charnov 1993) if the gains from waiting longer before beginning to reproduce pay off throughout the entire adulthood, that is, through both child-bearing and grandmothering years. They also suggest that, if the Grandmother Hypothesis is correct, then women should produce babies faster than expected by life-history theory because of the contribution made by grandmothers. It is undoubtedly the case that human IBIs are shorter than those of the Great Apes who often have intervals as long as eight to ten years in the wild (orangutans: Galdikas and Wood 1990); moreover, the human pattern is indeed at odds with the findings of life-history theory which suggest that fecundity decreases as age at maturity increases. Blurton-Jones et al. (1999) suggest that the contribution of grandmothers to children's nutritional needs allows mothers to wean their children earlier and speed up their birth rate. This was also predicted by Anderies's (1996) model: mothers who were not dependent on their own foraging efforts to feed their children would expect to have shorter optimal IBIs.

Although based only on indirect evidence (the fact that women do not senesce faster than men), Hill and Hurtado's (1997a, b) argument suggesting that grandmothers must be making some additional contribution to future generations indicates that there may be an advantage to lengthening the lifespan. Some additional evidence has been adduced for the Hadza: Hawkes et al. (1989) have shown that Hadza grandmothers forage just as efficiently but work slightly longer hours than younger women, with the result that they bring home more food (see also Hawkes et al. 1997). Even more interestingly, the growth of weaned children whose mothers have suckling infants is strongly correlated with their grandmother's (but not the mother's) work hours (Hawkes et al. 1997), suggesting that provisioning by grandmothers is key to children's successful growth and development. Hill and Hurtado (1991) and Blurton-Jones et al. (1994) also found that Ache and !Kung grandmothers, respectively, were efficient and productive foragers.

More recently, however, Sear et al. (2000) used data from a 25-year longitudinal study of nearly 2000 children in rural Gambia to test the hypothesis that grandmothers play an important role in successful rearing. They found that children who had living maternal (but not paternal) grandmothers had significantly better nutritional status (as indexed by weight and height) as well as better survival chances than other children. (The analyses controlled for the child's and parents' ages as well as the parents' own height and weight.) More importantly, these effects were significantly more marked in those cases where the grandmother was non-reproductive (defined as not having given birth in the year prior to the granddaughter's birth). Grandmothers who were having to look after an infant of their own were not a great deal more effective than grandmothers that had died before the subject was born. Similar effects have been reported for Hungarian Gypsies: women who survived past fifty years of age had significantly more grandchildren than women who died earlier (Bereczkei 1998).

Peccei (1995) has offered a slightly different twist on the Grandmother Hypothesis. Her Altriciality-Lifespan hypothesis states that increased encephalisation, and the consequent increase in offspring dependency on mothers, created the need for females to provide prolonged and intensive care well into the juvenile period. Natural selection would therefore have favoured females who became prematurely infertile, since these females would be able to divert all their reproductive effort to their juvenile and subadult offspring, instead of trying to provide care to both older offspring and very young dependent infants. Peccei (1995) is thus arguing for a direct 'Mother Effect', rather than a grandmother effect as such. Peccei (1995) argued that, during the period when menopause most likely evolved, female lifespan would have been too short to allow most women to aid their grandchildren (but see below) and a grandmother effect could not on its own have been selected for. However, an initial mothering effect, which would have increased female survivorship and lifespan, would have allowed a grandmother effect to come into operation, and so helped make menopause an evolutionarily stable strategy.

One final possibility to consider is that menopause is not a specific adaptation in its own right, but merely reflects the operation of other life-history processes. Washburn (1981) and Weiss (1981), for example, have suggested that menopause may simply be an artefact of a relatively recent increase in the length of the human lifespan. Their argument is that the lives of women in the Pleistocene were so 'nasty, brutish and short' that they never reached the age at which modern women undergo menopause.

However, Lancaster and King (1985) reviewed both the archaeological and the ethnographic evidence for the non-adaptive lifespan hypothesis and concluded that women in traditional societies did (and do) live long enough to experience menopause. At least 53 per cent of women across 24 contemporary traditional societies survive to the age of 45, which, if true of conditions during the Plio-Pleistocene, would have given natural selection plenty of material to work on.

Another non-adaptive hypothesis is that menopause reflects somatic senescence exactly analogous to that which occurs later in other body tissues (Wood 1990, 1994). Physiologically, the main cause of menopause is the loss of oocytes (immature eggs). When there are no longer enough viable eggs to be released during the monthly menstrual cycle, reproductive functioning stops. All female mammals show a loss of oocytes as they age, but in species other than ourselves this tends to coincide with general aging processes such that the females tend to die at around the same time that they reach the end of their reproductive lifespan. Since females are born with all the eggs they will ever produce during their lifetimes (unlike males who continually produce new sperm within their testes), Wood (1994) argues that increasing the human female reproductive lifespan beyond its current length would have required too great an investment in reproduction during the period of fetal development. In other words, it is the cost of producing another 20 years worth of eggs during a critical period of pre-natal development, and not the costs of rearing another two to three children, that leads to the early cessation of reproduction in human females.

If this is the case, then it suggests that adaptive explanations of menopause evolution require re-thinking. If the pre-natal costs of increasing the reproductive lifespan are prohibitive, then the phenomenon to be explained is not the early termination of reproduction – since if Wood (1994) is right, reproduction could not continue past this point anyway – but the increase of the post-reproductive lifespan. Women do not stop reproducing relatively early compared to other mammals, rather they die relatively late.

On balance, the weight of the evidence perhaps suggests that long post-reproductive lifespan, rather than menopause, is the adaptation on which the Grandmother Hypothesis should be focused. More studies are now needed, in both modern and traditional societies, to determine in more detail the fitness consequences of grandmaternal investment and to tie this in with observations regarding grandparental solicitude. Blurton-Jones et al. (1999) also suggest that studies of other primate species, where

BOX 6.7

Celibacy and homosexuality

Humans (and, in particular, human males) exhibit two forms of behaviour which, while they may not be unique in the Animal Kingdom, at least require explanation. These are celibacy (voluntary withdrawal from reproduction, sometimes associated with entry into a religious community) and homosexuality (the establishment of same-sex social and mating relationships). Although socially unrelated, these two lifestyles have similar functional consequences in terms of reduced fitness, and thus appear to constitute behaviour that is evolutionarily maladaptive.

Although both have attracted some interest from evolutionary biologists (Alexander 1974, Weinrich 1987, Kirkpatrick 2000) and there has been some (disputed) evidence for a genetic basis to homosexuality (Hammer et al. 1993, see also LeVay 1991), no generally accepted explanation exists for either. There would, however, seem to be four possible evolutionary explanations for non-reproductive lifestyles.

(1) They are functionally maladaptive, and reflect either developmental destabilisation or the fact that our advanced cognitive capacities allow us to make choices about behaviour and lifestyle that are a-Darwinian (we can choose to prioritise proximate rewards over ultimate functional consequences). Such behaviour would simply be the maladaptive tail-end of the normal distribution of behaviour and would not need any further explanation.

(2) They are functionally maladaptive from the individual's point of view, but represent a case of parental manipulation of offspring in order to maximise the parents' inclusive fitness. This may be especially important in cases where fitness is maximised by ensuring lineage survival rather than maximising offspring reproduction in order to reduce the economic pressure placed on the family estate (for more detailed discussion, see Chapter 8). Among the Tibetans, for example, second-born sons commonly join monasteries (Crook and Crook 1988); similarly, the probability that a son will enter the priesthood is significantly higher in American Catholic families with

five or more children than those with fewer (Low 2000). A study in North America also found that homosexual men have significantly more older brothers than heterosexual men (Blanchard and Bogaert 1996).

(3) Celibacy may be an adaptive strategy for individuals if the individuals concerned can invest in the reproduction of their close relatives, so allowing benefits via kin selection. In Medieval and Renaissance Europe, for example, the Church was often an important route by which poor families rose to political and economic prominence (via the coat-tails, as it were, of a sacrificial sibling). Similarly, in one contemporary UK sample, voluntary (and involuntary) non-reproduction by couples was associated with significantly higher investment in nieces and nephews than was the case for couples who had reproduced (Nelson 1998). Judge (1995) found a similar effect in Californian wills: those who died childless were most likely to leave a significant proportion of their estate to siblings or nieces/nephews. However, Kirkpatrick (2000) concludes that the evidence for significant economic contributions to the reproductive effort of relatives by homosexuals is at best equivocal.

(4) Celibacy/homosexuality may simply be a phase that individuals pass through on their way to being reproductively fully active. Such phases may be important in providing either sexual experience or in allowing the individual to concentrate on building an economic base before starting reproduction. The evidence that bisexuality (at least on a lifetime scale, even if not simultaneously) is very much more common than lifelong exclusive homosexuality is persuasive. (That some individuals remain in a non-reproductive phase may simply reflect 'slop' in the ontogenetic mechanisms, and may therefore be uninteresting from an evolutionary perspective.) Baker and Bellis (1995) present evidence suggesting that bisexual women have at least as high fertility as exclusive heterosexuals, and possibly start to reproduce earlier although there are contrary data (for

BOX 6.7 cont'd

Celibacy and homosexuality

example, Essock-Vitale and McGuire 1988). Kirkpatrick (2000) argues that an early homosexual phase may also be associated with same-sex alliance formation, which may in turn have significant benefits in terms of survival. While there may have been advantages to this in terms of mutual help in times of need during hunting expeditions, the available evidence is at best only indirect and the hypothesis needs detailed investigation.

grandmaternal help is absent, may be important in shedding light on the pressures favouring juvenile provisioning as well as those favouring increased brain size and consequent dependency of offspring.

*

The examples in this chapter have illustrated how the issue of offspring production is intimately tied in with the ability to invest effectively both in the present and, more importantly, in the future. In the next chapter, we will look at the issue of parental investment in greater detail, and explore the various ways in which environmental circumstances interact with parental resources and offspring sex to produce characteristic patterns.

■ The heavy cost in terms of time and energy of rearing human offspring is the key determinant of human reproductive strategies. These costs stem from the increased brain size of humans which results in the production of altricial helpless offspring that undergo a long developmental period.

■ Reproduction represents a trade-off between the quality and quantity of offspring. Women in many societies limit the production of future offspring in order to increase the survivorship and prospects of current offspring.

■ Heritable wealth influences reproductive decision-making in humans because of the effect it has on the likelihood that individuals will reproduce themselves as adults. In addition to ensuring that they have sufficient resources to raise offspring, parents also have to ensure that they have enough resources to marry off their offspring and start them off on their own reproductive careers.

■ Theoretical modelling shows that considerations of wealth inheritance can lead to a decoupling between wealth levels and the number of offspring produced, so that high long-term fitness may be associated with low levels of reproductive success in the current generation.

■ Increasing the costs of children may be the factor responsible for the demographic transition in the developed world. A reduction in kin networks and increased costs of child-rearing to the parents may also have contributed to this phenomenon.

Chapter summary

Chapter summary *continued*

- Human females undergo menopause and stop reproducing in middle age, yet live to be much older. The Grandmother Hypothesis suggests that women do so because they can increase their fitness to a greater extent by helping to raise grandchildren than by continuing to produce offspring themselves.

- Other theories suggest that menopause is merely a by-product of the ageing process and does not have any adaptive value. Rather it is the extension of the lifespan beyond child-bearing age that must be explained in evolutionary terms.

Further reading

Charnov, E. (1993). *Life History Invariants*. Oxford: Oxford University Press.

Hill, K. and Hurtado, A. M. (1996). *Ache Life History: the Ecology and Demography of a Foraging People*. New York: Aldine de Gruyter.

Hrdy, S. B. (2000). *Mother Nature*. London: Chatto and Windus.

Kirkwood, T. (2000). *Time of Our Lives*. Oxford: Oxford University Press.

Morbeck, M. E., Galloway, A. and Zihlman, A. I. (1996). *The Evolving Female*. Princeton: Princeton University Press.

Stearns, S. C. (1992). *The Evolution of Life Histories*. Oxford: Oxford University Press.

Parental investment strategies

CHAPTER

7

CONTENTS

In the previous chapter, we discussed the various ways in which fitness can be enhanced by judicious family planning. In this chapter, we investigate the strategies that individuals use to enhance their fitness once they actually become parents, and demonstrate that, within complex human societies, strategies to enhance lineage survival are often of greater importance than those that aim to maximise the number of surviving offspring.

CONFLICT IN THE WOMB

Parental investment by mothers begins long before offspring are born. During the nine months of pregnancy, mothers provide all the nutrition and protection offspring need during the critical months of pre-natal development, and use their own energy reserves to do it. Although it is in both mother's and offspring's interests to ensure that development proceeds as efficiently as possible, nevertheless these interests do not coincide exactly. The differences in genetic relatedness between parents, offspring and siblings means that, in evolutionary terms, offspring should attempt to garner more care than the mother is selected to provide (see Chapter 2). Haig (1998; see also Haig 1993) suggests that these evolutionary conflicts are responsible for some otherwise puzzling features of pregnancy.

Somewhere between 30–75 per cent of pregnancies are spontaneously aborted during the first two weeks following conception. So swiftly does this happen that most women don't even realise that they are pregnant. Haig (1998) suggests that this represents the selective elimination of offspring of low quality by the mother. In order for a pregnancy to be sustained, high levels of the hormone progesterone are required. This is supplied by the corpus luteum (a body formed from the egg follicle in the ovary following the release of the egg). However, luteal progesterone production declines over the two weeks following release of the egg. In order for pregnancy to continue after this point, the embryo itself must release a hormone, human chorionic gonadotrophin (hCG), in order to stop the regression of the corpus luteum and keep progesterone levels high. If there is no embryo, the decline in progesterone production results in the lining of the womb being shed (menstruation). By the same token, if the embryo is not able to produce sufficient hCG, the regression of the corpus luteum is not prevented and the embryo itself is then lost during menstruation.

Haig (1998) suggests that the ability to produce high levels of hCG is an honest signal of quality from the embryo to the mother. Embryos who cannot pass the mother's hormonal 'test' are likely to be those that are abnormal in some way and mothers therefore save themselves the trouble of investing large amounts of energy in low-quality offspring. Haig (1998) also suggests that this can vary according to environmental conditions, either because there is variation in the threshold of hormone needed (that is, more hormone is required when conditions are tough, ensuring that only the best embryos make it through) or the threshold may remain constant, but the ability of the embryo to produce hCG may vary with an equivalent effect.

Haig (1998) also suggests that the high blood pressure often associated with late pregnancy, and which can lead to conditions such as pre-eclampsia (which can cause considerable damage to maternal blood vessels) is another sign of mother–offspring conflict. Over the course of pregnancy there are changes in the maternal circulatory system, designed to accommodate the fact that blood must now flow to the placenta, as well as to all the other maternal organs, in order to nourish the fetus.

As pregnancy progresses and the nutritional needs of the fetus increase, the growth of the placenta allows the fetus to exert more control over maternal physiology. Early in pregnancy, there is often a drop in blood pressure: since an increased rate of flow to the placenta is associated with high blood pressure, this represents an attempt by the mother to reduce the fetus's share of blood to the level appropriate from the mother's perspective. As pregnancy continues, blood pressure rises as the fetus is able to exert greater control and divert more blood to the placenta. This could thus be interpreted as a case of parent–offspring conflict. However, Moore and Haig (1991) have suggested that, because genomic imprinting means that the growth of the embryo and the placenta are under the control of paternal genes (see **Box 2.1**), these phenomena might in fact represent a battle between the maternal and paternal genes, rather than between the mother and offspring per se.

PARENTAL BIASES AND SIBLING RIVALRY

In evolutionary terms, the period of parental investment is probably the most critical in ensuring high fitness returns for individuals. Since evolution is interested in grandchildren rather than children, parents can enhance their fitness by adjusting the level

BOX 7.1

Pregnancy sickness and parent–offspring conflict

Approximately 70 per cent of women suffer nausea and vomiting during the first trimester of pregnancy (based on a meta-analysis of 56 studies: Flaxman and Sherman 2000). Profet (1988, 1992) has suggested that pregnancy (or 'morning') sickness may be an adaptation designed to stop mothers ingesting toxins that may be harmful to the developing embryo, either by causing birth defects (teratogens) or abortion (abortifacients). She argued (a) that the foods most likely to cause sickness or aversion in pregnant women are those that contain high levels of these substances (examples include coffee and many vegetables) and (b) that women show aversion/sickness at the point when the offspring is most vulnerable to toxins (the first trimester of pregnancy – the critical period for organogenesis).

Profet (1992) and Flaxman and Sherman (2000) showed that women who suffered pregnancy sickness had higher pregnancy success rates (lower probabilities of both miscarriage and fetal death) and higher fetal birth weights than those who did not. Results from 20 surveys of food aversion during pregnancy suggested that the most common aversions are to meats (including eggs), strong-tasting vegetables, alcohol and caffeine (Flaxman and Sherman 2000). These authors also noted that, of 27 traditional societies listed in the Human Relations Area File that provide information on pregnancy sickness, meat-free cereal-based diets characterised all seven of those which reported that pregnancy sickness did not occur and none of those where it was common.

Profet (1992) argued that there would have been a strong selection pressure on pregnancy sickness during the Pleistocene in particular as this was the point when humans began to cook their food. This allowed them access to a greater range of foodstuffs (and therefore a wider range of toxins); in addition, the very act of cooking itself can produce new toxins to which pregnant women would have been exposed. On the basis of their more extensive review of the literature, Flaxman and Sherman (2000) suggest that the risk of food-born micro-organisms (more common in meat than vegetable products) may be more problematic for pregnant women because they are immune-suppressed. In contrast, Huxley (2000) noted that, in both humans and animals, pregnancy sickness is associated with reduced maternal energy intake and increased placental weight (see also Pike [2000] who found that Turkana women who suffered pregnancy sickness lost weight). She suggested that reduced energy intake leads to elevated maternal levels of anabolic hormones, insulin and insulin growth factor (IGF-1), which suppress maternal tissue synthesis, thereby allowing nutrients to be directed to the placenta and fetus.

At the moment, it is not clear whether the process is embryo- or mother-driven: does the mother's own body make her sick in order to protect the embryo, or does the embryo itself exert some influence on maternal physiology in order to protect itself. If the latter is true, then this would also qualify as a case of parent–offspring conflict since the embryo would be denying the mother potentially valuable nutrients (in cases of acute pregnancy sickness, women often lose substantial amounts of weight during the first three months).

of investment in offspring in order to maximise the likelihood that they themselves will reproduce and pass on the parent's genes. As ruthless as this sounds, there are many studies of historical and contemporary western societies which show, quite uncontroversially, that parents sometimes manipulate their offspring through biasing investment in favour of one child against another (see below).

Even quite subtle differences in how offspring are treated can have a lasting effect, as evidenced by the fact that children brought up within the same family are more dissimilar to each other than any two unrelated children selected at random. Lalumière et al. (1996a) suggest that parents treat offspring differently according to the traits they display early in life, and this then pushes them along certain developmental trajectories. For example, initial differences in verbal or athletic ability may give one child an advantage over another, with the less skilled child then having to be proficient in another domain in order to acquire an equivalent share of parental attention. Lalumière et al. (1996a) suggest that this could have positive fitness consequences if steering siblings along different developmental trajectories resulted in reduced levels of sibling competition for parental resources, and equipped offspring to compete effectively in different mating environments.

Sulloway (1996) has also suggested that differences between offspring are an expression of sibling rivalry for parental attention. His view, however, emphasises the child's role in all this. From a survey of 120 000 people, Sulloway (1996) discovered that first-born children tend to be more conformist, conservative, responsible and antagonistic compared to later born children, who in turn are increasingly imaginative, flexible and rebellious. Sulloway (1996) argues that first-born children strive to maintain their initial advantage regarding parental attention by developing behavioural traits aimed at maintaining the status quo within the family and sustaining the initial power advantage that allows them to lord it over their younger siblings. Davis (1997), for example, found that first-borns tend to be more status-oriented than children born further down the birth order.

By contrast, younger children have to develop traits that will enable them to gain their share of parental attention despite the best efforts of their older sibs. Since they lack the power to bully them back and aren't as closely aligned with their parents as older children, they develop behavioural traits that allow them to be flexible and imaginative when coping with family tensions. They are therefore more likely to adopt varied and non-conformist strategies in order to achieve their own particular ends.

It seems clear that a combination of these two influences – parental biases and sibling rivalry – is what actually drives the system by a process of positive feedback. It pays siblings to try and differentiate themselves from their older sibs in order to alleviate the negative effects of sibling rivalry, and it then pays parents to encourage and amplify these differences so that each offspring fills his or her own individual niche. The more encouraging the parents, the more the offspring will continue to emphasize their own particular traits, and the more divergent offspring become.

Salmon and Daly (1998), however, suggest something slightly different. They found that first-borns were most likely to name a parent as the person to whom they are closest, while middle born children were especially likely to name friends or siblings and were much less inclined to turn to their parents in distress. Middle-borns were also more likely to be influenced by references to friendship than kinship terminology, again in contrast to first-borns (Salmon 1998). Salmon and Daly (1998) therefore suggest that, rather than using alternative tactics to gain parental attention, offspring born lower down the birth order opt out of the competition altogether, and that they express those traits, such as cooperativeness, that will aid them in forming alliances with non-kin (see also Kidwell 1982).

Regardless of the manner in which it actually serves to promote differences among offspring, differential parental attention is undoubtedly a large influence on the personality traits and values characteristically adopted by children who hold a particular position in the birth order. Pérusse (1993) suggested something rather similar in his study of mating among modern day Quebeçois. In his study, high status males were found to secure more potential conceptions than those of lower status. Pérusse (1993) suggested that the motivation to achieve high status may be a consequence of parental 'teaching biases' which encourage the development of traits likely to lead to high status and therefore high fitness. Many psychological studies have shown that parents play a

BOX 7.2

Teaching biases and peer groups

Harris (1999) suggests that peer groups may be more influential than parents in determining the kinds of personality traits that children display. She argues that, although children may learn a great deal about appropriate and inappropriate behaviour at home, whether or not they retain this information and use it to guide their own behaviour depends on the conditions they encounter outside the home. Consequently, children may behave in one way when they are with their parents but in a completely different way when in the company of their peers.

Other children may also be more influential than parents in moulding children's personalities. Hence, although first-born, middle-born and last-born children appear to slot into particular roles when asked to reflect on their parental relationships, this does not necessarily mean that these roles always dictate their behaviour when they are away from their parents' influence.

Harris (1999) claims that, by the age of two years, children are able to understand which social category they and others fall into (men vs women, children vs grown-ups, girls vs boys) and show a preference for their own social category (girls preferring girls and boys preferring the company of other boys). By the time they are around 4 years old, these preferences have become entrenched and children will preferentially form same-sex groups unless forced to do otherwise by adults. Boys and girls also adopt traits that are characteristic of their sex: boys begin to act tough and hide their weaknesses, whereas girls show little tendency to behave in this manner.

Harris (1999) suggests that these tendencies are not, as Pérusse (1993) and Low (1989) would argue, the result of socialisation by the parents who are behaving in accordance with the values of the culture in which they live. Instead, she argues that it is a child's peer group that dictates what is and is not appropriate behaviour for other children to adopt. This makes a certain amount of sense because, when they become adults, children have to live and operate in a society that consists mainly of their peers rather than their parents: sharing values with their peers may therefore ultimately be more important. In addition, the trade-off between parental and peer group influences on behaviour would allow tests of a number of cultural evolution models (see Chapter 13) which have so far received very little attention.

very active role in encouraging achievement in their children (for example, Child et al. 1958, McClelland and Pilon 1983), often unconsciously, and that such efforts are experienced as rewarding. Pérusse (1993) therefore suggested that parents have evolved psychological mechanisms, expressed as teaching biases, which convey adaptive information to offspring. These teaching biases would make use of acquired information rather than hard-wired genetically coded information, and would thus allow the production of diverse cultural variants. In other words, teaching biases would be dependent on the particular environmental context in which individuals were operating, and would be able to change rapidly in response to changing circumstances.

Low (1989) presents some data in support of this hypothesis. Using cross-cultural data from 93 societies, she showed that boys are taught to show more fortitude, aggression and industriousness in unstratified highly polygynous societies than in societies where the intensity of polygyny is lower. By the same token, in stratified societies where females show high levels of hypergyny (marrying up the social scale) and chastity and obedience are highly desirable wifely traits, older girls are taught to be more sexually restrained and obedient than in less stratified societies, where girls are taught to be more self-reliant (**Figures 7.1** and **7.2**).

Teaching offspring to behave differently according to their sex is just one end of a very long continuum of differential parental investment. Parents can exert a lot more

BOX 7.3

Family environment and future reproductive strategies

Although the ontogeny of adult behaviour has not been the subject of detailed attention by evolutionary psychologists, the possibility that childhood experiences might influence reproductive behaviour in later life has attracted particular interest. In a classic analysis of children's social development, Bowlby (1969) argued that, by the age of seven years, children have internalised key aspects of their family environment, and thereafter treat these as normative. Rutter and Madge (1976), for example, found that the best predictor of marital harmony in a marriage was the marital harmony between their parents, suggesting that at least some aspects of how one should behave in a family context may be learned (see also Thornton 1991).

Building on this suggestion, Belsky et al. (1991) proposed that children use the family environment as a cue to guide their later reproductive behaviour as adults. Children growing up in a stressed environment would tend to regard the future as uncertain, and would opt for early fast reproduction in preference to intensive parental investment. These ideas have been developed and extended by Chisholm (1993, 1996, 1999) who combined Belsky et al.'s original suggestion with the predictions of life-history theory. Life-history theory (see Stearns 1992, Charnov 1993) emphasizes the importance of the trade-off between quantity and quality in reproduction when parents are prevented by circumstances or time constraints from doing both.

Chisholm argued that children who perceive the world as unpredictable (particularly in respect of mortality) should opt for early fast reproduction as the best bet under the circumstances. One counter-intuitive prediction from life-history theory is that, when times are hard and the sources of mortality are beyond their control, parents should *reduce* their investment in current offspring in

order to conceive the next offspring (Dunbar 1988; Chisholm 1996). There is evidence from cats (Bateson et al. 1990), rats (Smith 1991a) and primates (Dunbar 1988) that this is precisely what they do. Chisholm (1993) argued that these effects reflect individuals' attempts to optimise under constraint, and are thus highly adaptive rather than maladaptive.

A number of studies have suggested that, as predicted, girls both undergo menarche and reproduce earlier and more often in households that are single parent (or effectively single parent due to father absence as an economic migrant) and/or economically disadvantaged (the two often go together) (Bereczkei and Csanaky 1996, Wyatt 1990, Surbey 1989, Moffitt et al. 1992, Chisholm 1999). Wilson and Daly (1997), for example, found that the fertility rates for individual Chicago districts correlated positively with their local murder rates, implying that when future survival is unpredictable then the best strategy may be to reproduce as fast as possible.

Chisholm (1999) found that, compared to non-reproducers, young women undergraduates (aged 19–25) who had already reproduced had had significantly poorer relationships with their father, had parents who got on less well with each other, came from lower income families, achieved poorer scholastic grades, and had their first sexual experience significantly earlier. Age at first (consensual) sexual experience correlated with parental education (an index of family wealth), while mean number of sexual partners per year correlated negatively with the quality of the relationship with the father and the reported quality of the parents' relationship with each other. Chisholm interpreted these data as support for the hypothesis that those growing up in stressful environments should reproduce early.

influence than merely deciding the type of information they should impart in order to increase their offspring's chances of reproducing. Evolutionarily speaking, the amount of effort invested in any particular offspring depends on how likely it is that such effort will be translated into grandchildren. Under certain circumstances, this may mean that the optimal parental strategy is to invest nothing at all in a particular offspring once it is born.

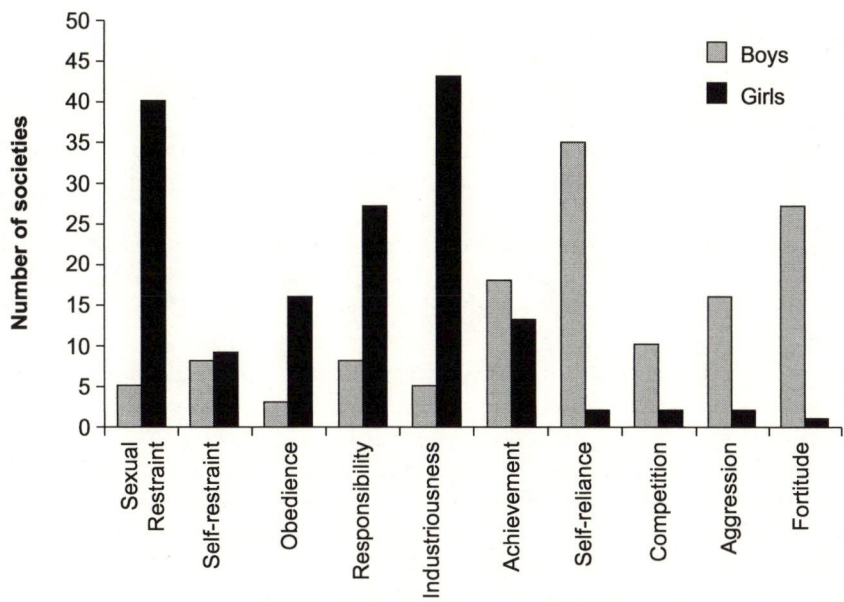

Figure 7.1 Cross-cultural data from a sample of 93 societies suggest that girls are taught to be more sexually restrained and obedient than boys in more societies, whereas boys are more often taught to be competitive and self-reliant. Redrawn from Low (1989).

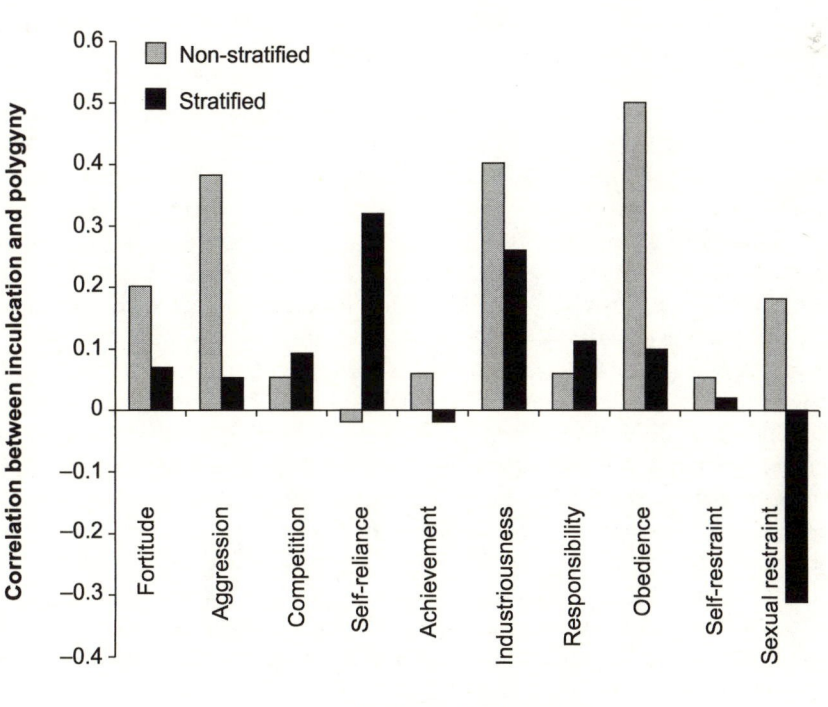

Figure 7.2 Cross-cultural data from a sample of 93 societies suggest that older girls are taught to be sexually restrained and obedient more often in stratified societies (where hypergyny [marriage into higher socio-economic classes] is possible) than in non-stratified societies (where hypergyny is less advantageous). The plotted variable is the Pearson correlation between the extent to which a society inculcates a particular value and the level of polygyny. Redrawn from Low (1989).

INFANTICIDE: SCHEDULING INVESTMENT

Infanticide is obviously a very emotive subject: inevitably, we find it difficult to accept the fact that people can behave in such an apparently cold-hearted way. However, from a scientific point of view, we must be careful not to bring our own prejudices and sensibilities to bear on the behaviour of other individuals in either our own or other cultures. In fact, infanticide is an extremely widespread phenomenon, having been documented from virtually every human society (Daly and Wilson 1981). This is not, of course, to suggest that it is in any sense normative behaviour: were that so, we would not have survived as a species. Rather, the fact that it occurs at all in a Darwinian world in which natural selection reinforces parental solicitude requires an explanation. And its frequency and near-universality implies that any such explanation is unlikely to be that it is merely a maladaptive by-product of organic failure.

The fact that we are designed by natural selection to be solicitous of our offspring raises an obvious problem for any individual that commits infanticide: such behaviour is necessarily in conflict with deeply rooted psychological mechanisms intended to ensure that we invest heavily in our offspring. The solution to this dilemma seems to be that, in most societies, there are mechanisms in place that help to distance parents from their offspring under conditions where mortality (whether natural or imposed) is high. These usually have the form of declaring that babies are not human before some key event has occurred (drawing its first breath, giving its first cry or, in traditional Inuit culture, being named). Such mechanisms provide a brief interval during which abandoning or killing a newborn infant does not count as killing a human, so enabling individuals to cope with what is necessarily a traumatic event.

In addition, as Hrdy (1992) points out, infanticide tends to occur as a last resort; in most cases, individuals who need to drastically reduce parental investment tend to abandon offspring or farm them out to their extended kin. It is only under circumstances when these alternatives are unavailable or impractical that parents resort to infanticide. In late medieval and early modern Europe, for example, infants were frequently abandoned in foundling homes set up for precisely this purpose (Peyronnet 1976, Boswell 1988). Parents would often leave identifying tokens on their offspring, suggesting that they were abandoning their children with at least the hope that they would some day reclaim them. However, as Hrdy suggests, this could equally have been a mechanism for allowing parents to avoid facing up to the fact that, by abandoning their offspring, they were almost certainly condemning it to death: overcrowding and lack of sufficient care raised orphanage mortality rates to horrifying levels (in continental Europe and Russia, they were as high as 53.6 per cent).

This pattern of abandonment is almost unheard of in African societies, both traditional and modern. Here, fostering of offspring to distantly related kin is the most common means by which parents reduce investment in their offspring (see, for example, Pennington and Harpending 1993). For this reason, infanticide is rare in most African societies. This is in sharp contrast to most Amazonian tribal societies where rates of infanticide are often in the order to 12–38 per cent (**Table 7.1**). Fostering is not feasible for hunter-gatherers because most people will face the same stringent conditions as the parents. Similarly, abandonment (in the hope that someone else will raise the child) is not an option because it is unlikely that an abandoned infant would survive for very long in the hostile environment of the Amazonian forest. Hrdy (1992) suggests that, under these more constrained and hostile conditions, the strategy of last resort (infanticide) is the most practical and humane option.

Table 7.1 Percentage of live births terminated by infanticide for nine traditional societies in Africa, Amazonia and New Guinea

Culture and Location	Subsistence type	% infanticide
Efe (Ituri Forest, DRC)	Specialized hunter-gatherer	0
Lese (Ituri Forest, DRC)	Horticulture	0
Datoga (N. Tanzania)	Pastoralism	0
Kipsigis (S.W. Kenya)	Agro-pastoralism	0
San (Kalahari Desert, Botswana)	Hunter-gatherer	1
Mucajai Yanamamö (N. Brazil)	Horticulture and hunting	6
Ache (Paraguay)	Hunter-gatherer	12
Ayoreo (S.W. Bolivia and Paraguay)	Horticulture and foraging	38
Eipo (Highland Central New Guinea)	Horticulture	41

Source: Hrdy (1992)

Daly and Wilson (1981) reviewed the worldwide ethnographic literature on infanticide and concluded that it can generally be attributed to one of three main causes: lack of paternity certainty, poor offspring quality and lack of parental resources. We consider each of these in turn.

PATERNITY UNCERTAINTY

All other things being equal, individual males should only invest in an offspring when they can be sure that the offspring is their own. Investing in someone else's offspring is genetic altruism, and should only evolve when these individuals are genetically related to the investor. The striking asymmetry regarding parental certainty between males and females (internal fertilisation and gestation mean that maternity is always certain, but paternity is never certain) means that males should demand assurances that an offspring is theirs before they are willing to invest.

Being able to guarantee paternity certainty seems to be an especially important consideration in traditional societies. One reflection of this is the insistence that women should be virgins at marriage, an insistence that seems to have resulted in age at marriage being driven down to below the age of puberty in order to ensure that the woman (or rather girl) cannot possibly have conceived before marriage. Dickemann (1979) has argued that the protection of a woman's chastity is the motivating reason behind a host of practices such as the claustration of females (for example, confinement in harems), the imposition of 'modest' dress restrictions (for example, under Islamic law), foot-binding in ancient China and the requirement that women be chaperoned in public. Dickemann (1979) points out that, according to the most ancient Chinese manuals, footbinding was quite explicitly intended to prevent women from straying too far from home, since they could not walk more than a few metres on their badly mutilated feet. More significantly, perhaps, this practice was very much an upper class practice: such families had much to lose and most to gain by the chastity of their daughters prior to marriage and their likely fidelity afterwards. Even more extreme in this regard is genital mutilation of women (cliterodectomy or 'female circumcision'), still widely practised in sub-Saharan Africa where it is explicitly intended to lessen women's sexual interest (and thus prevent them straying).

Strassmann (1992, 1996) has developed a similar explanation for menstrual taboos, whereby intercourse (and sometimes even public appearances) are banned when a woman is menstruating. These taboos occur almost universally and have surprisingly similar forms in quite different parts of the world. Strassmann suggests that these are really designed to allow males to ensure paternity: if the timing of the woman's menses is signalled publicly, it considerably reduces the husband's uncertainty as to when conception might (or has) occurred. Among the Dogon, a West African tribe, women are confined to small huts at the edge of the village while they are menstruating in order prevent them from 'contaminating' (in a ritual sense) the rest of the household. Strassmann (1996) found that 86 per cent of all (endocrinologically confirmed) menses were signalled by menstrual hut visits by the 93 Dogon women in her study, who (with one exception) never visited menstrual huts while pregnant or lactating. Menstrual hut visits thus appear to be (more or less) honest signals of a woman's reproductive state.

All of these practices have been interpreted within the context of 'honour and shame' systems, where the honour and prestige of a lineage is premised on the chastity of its daughters. Daly et al. (1982) noted that, in many societies, failure of a virginity test at wedlock can lead to annulment of the marriage, permanent unmarriageability of the rejected bride, and disgrace for her family. Not unusually in this context, a man's lost honour can be at least partially salvaged by killing the unchaste female (Daly et al. 1982).

Underlying the control of female sexuality is the recognition that women are marketable commodities, that the trade of females through marriage underlies alliance formation while simultaneously serving economic purposes through dowry payments and competition (Dickemann 1981). It has been pointed out that wherever there is significant variance in the resources that are controlled by men, those resources can be converted into control of women, and thus women become resources that must be protected from competitors (Daly et al. 1982).

More direct evidence regarding concerns about paternity certainty is offered by Daly and Wilson (1982) in their study of comments made by parents and grandparents on newborn babies' features in a Canadian maternity ward: grandparents make many more comments about the similarities between the baby's features and the father's than about the mother's ('Hasn't he got his dad's ears/eyes/dimples' and so on). Similar findings have been reported for Mexican infants (Regalski and Gaulin 1993). It seems as though everyone is trying to pressurise the father into investing in the child. One obvious reason why they might want to do this (or at least why the mother and her parents might want the father to be convinced of his paternity) lies in the evidence that step-children suffer much higher rates of abuse and mortality than children who live with both biological parents (see pp. 57–9). Among the Ache, children who have lost their natural fathers have a significantly increased risk of dying before aged 15 years, in most cases because of deliberate infanticide by other males in the group (who explicitly state that they are not willing to contribute to the costs of rearing other men's offspring: Hill and Kaplan 1988).

Interestingly, McLain et al. (2000) showed that, while unrelated individuals asked to judge photographs of newborn babies were more likely to ascribe resemblance to the mother, the mothers themselves stated that the baby was like its father. Tellingly, mothers were more likely to do this in the father's presence than when he was absent, suggesting that the resemblance was possibly more imagined than real and designed to assure men of their paternity. Whether mothers do this consciously, knowing full well that the baby looks nothing like its dad, or whether they deceive themselves into thinking that the baby really does look like its father is unclear. Evolutionary analyses (Pagel 1997) suggest that infants should be selected for 'anonymity' and not show paternal resemblance due to the high potential costs incurred if the putative father is not the biological father (see Chapter 3). The fact that independent judges were more likely to ascribe resemblance to the mother suggests that infants may indeed be concealing their paternity.

The increased risks that illegitimate children run, combined with men's unwillingness to invest in such children, make infanticide by mothers a rational strategy: their interests may be better served by getting on as quickly as possible with a new reproductive attempt. Among the Eipo of New Guinea, for example, an extramarital conception was given as the reason for infanticide in 6 out of 15 observed instances of maternal infanticide (Schiefenhovel 1989). Similarly, Fuster (1984) found that mortality rates in the first year of life were significantly higher for illegitimate births than for legitimate births in rural Galicia (Spain) between 1871 and 1949.

Children raised in hunter-gatherer societies like those of Amazonia and New Guinea are quite heavily dependent on male input for their survival, and the lack of male parental support is therefore a large risk factor for infanticide in these populations. Since paternal investment is much less crucial for the successful raising of offspring in African societies, Hrdy (1992) suggests that this may provide another explanation as to why infanticide rates are so low in Africa. Women, especially in the matrilineal systems of West Africa (systems in which women remain in their natal home on marriage and wealth is passed down to maternal heirs), are self-sufficient horticulturalists and can rely on their extended (female) kin networks for extra childcare. Since male input is so low, men do not risk investing in children who are not their own, and paternity certainty does not feature as a risk factor in these societies.

POOR INFANT QUALITY

Offspring born with a severe physical deformity are likely to be the victims of infanticide, especially in traditional societies where institutional care of the handicapped is rarely available (Daly and Wilson 1984, 1988b). The increased level of care such children require for a low evolutionary pay-off (they are unlikely to reproduce, even if they do manage to survive) means that parents may be better off if they terminate investment in these offspring very early on, and begin investment in a new offspring. Even in our own western societies, handicapped children are more likely to be abused and suffer injury requiring hospital treatment at the hands of their parents than normal healthy children (**Figure 7.3**).

Infants don't have to be physically handicapped in a particularly obvious way to be at risk of infanticide. The ethnographic literature is full of examples of offspring being killed because of abnormal circumstances surrounding their births (often perceived as 'ill omens'). Hill and Ball (1996) collated the reasons given by people in various societies for killing an infant shortly after birth (examples included breech birth – baby born feet first, babies born with teeth, even babies born with red hair in

BOX 7.4

Paternity certainty and sexual jealousy

It has been suggested that humans have evolved certain psychological propensities that function to ensure paternity certainty (Daly et al. 1982). Examples of these include sexual jealousy, the inclination of men to possess and control women, and the use of threat or violence to retain exclusive sexual control of women. Daly et al. (1982) have argued that male sexual jealousy, defined as a state that is aroused by a perceived threat to a valued sexual relationship and which motivates behaviour aimed at countering the threat, is a cross-cultural universal. They suggest that it has motivated legal strictures on female sexual liberty, and that it is a leading motive in homicide and other forms of violence.

In a great many cultures around the world, a wife who has sexual intercourse with another man is regarded as having committed an offence against the *husband*. The wife's infidelity is viewed as a property violation against the husband, and accordingly he is often entitled to damages, violent revenge, or divorce with a refund of any brideprice that was paid (Daly et al. 1982). The marital status of the man with whom a married woman commits adultery is often irrelevant, and the offending male is often treated less harshly than the offending woman. More than this, many older legal codes specified that adultery on the part of a wife was a mitigating circumstance excusing violence on the part of the offended husband, whose behaviour then attracted a lesser sentence (for example, the concept of *crime passionnel* in nineteenth-century French law).

Modern legal codes have mostly repealed such inequalities in adultery, and women are now usually able to petition for divorce on the grounds of a husband's infidelity. Nonetheless, a husband's sexual fidelity is evidently not considered to be as important to a wife as it appears to be in the reverse case. Daly et al.'s (1982) review of the available data indicates that men are far more likely than are women to exercise the right to divorce on the grounds of spousal infidelity. They suggest that men are far less likely than are women to forgive and forget adultery of a spouse. And the overriding reason for this, it is argued, is the doubt created over paternity certainty of both future and existing offspring. What evidence is there to support this interpretation?

Daly et al. (1982) reviewed homicide cases in Detroit, USA and reported that, when crime-specific cases (for example, homicides that occurred within the context of another crime such as armed robbery) are excluded, about 20 per cent of the 'social conflict' homicides on record relate to 'jealousy conflict' and that, of these, the vast majority relate in some way to sexual jealousy. Of these, most cases included a third party, and overwhelmingly the jealous party who committed the killing was a man (**Figure 7.4**). The relatively high incidence of jealousy-induced homicide is not peculiar to Detroit: studies of homicide statistics in both first- and third-world countries reveal a similar pattern – spousal homicides overwhelmingly appear to be related to sexual infidelity, and most usually the infidelity of the wife (Daly et al. 1982).

What do these extreme reactions represent? In the case of infidelity, the husband might (consciously or unconsciously) fear investing in offspring that are not related to him; in the case of infidelity combined with a wife's intent to leave the husband, this represents a loss of exclusive control over the wife's reproduction, and this impacts on the reproductive success of the husband. It should be noted, of course, that only a relatively small fraction of spousal conflicts result in the death of one of the spouses. Far more common is non-fatal spousal beating, and these appear often to be motivated by the same factor. Miller (1980), Whitehurst (1971), Rounsaville (1978) and Hilberman and Munson (1978) all report that the most frequently cited reason for husbands beating their wives is related to jealousy in general, and sexual jealousy in particular. Other studies have, however, reported no sex differences in actual sexual jealousy, and have argued that the supposed 'double standard' in sexual jealousy is principally a male strategy to manage female sexuality (Paul et al. 1996).

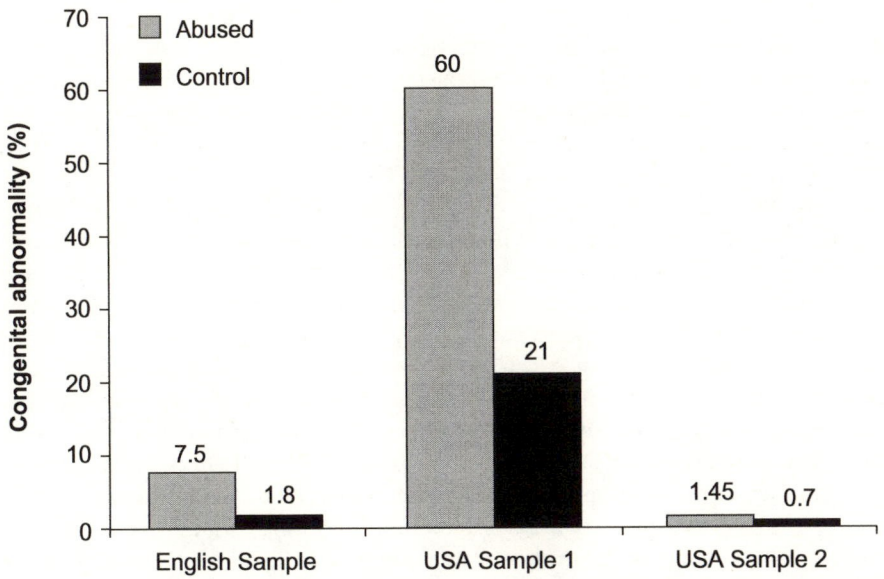

Figure 7.3 Percentage of children admitted to hospital because of injuries resulting from abuse or accidents (the control group) who were physically handicapped in one UK and two USA samples. Handicapped children are more likely to be abused than normal children. Source: Daly and Wilson (1984).

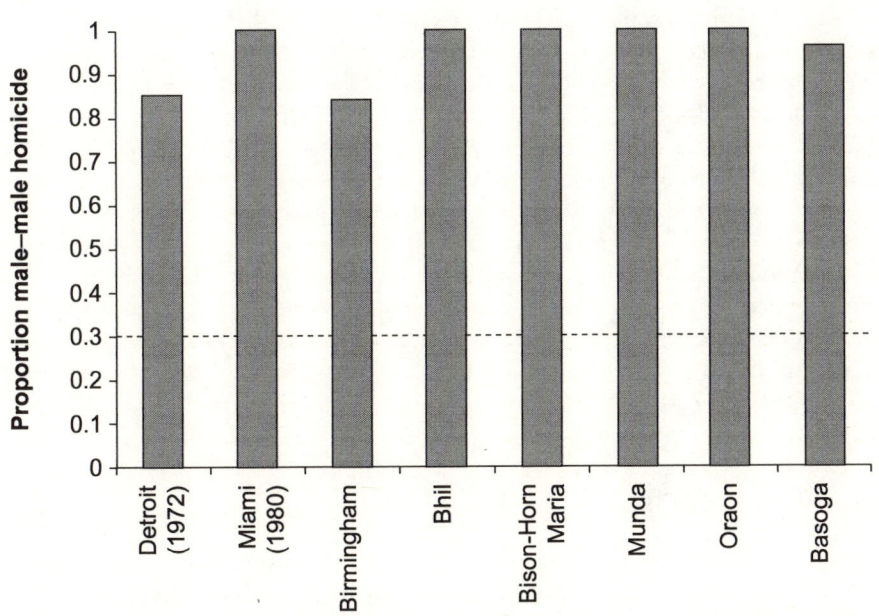

Figure 7.4 For a sample of eight industrialised and traditional societies, the vast majority of homicides arising out of triangular relationships (cases where two men compete over a woman) involve one man killing the other. The dotted line represents the proportion expected if killings were random. Source: Daly and Wilson (1988b).

black-haired populations) and then surveyed the medical literature to see if any of these circumstances were associated with offspring quality. They found that infants produced as a result of abnormal births, or who were perceived as ill omens, tended to display characteristics associated with conditions that increase childhood morbidity and mortality. For example, breech births are often linked to neuromuscular dysfunction and conditions like cerebral palsy (**Table 7.2**). The cultural taboos surrounding unusual births and the cultural prescriptions for ending the lives of infants born in this way may therefore be a proximate mechanism designed to reduce the number of potentially severely disabled children in societies where the necessary facilities to care for them are not available, and where neither their survival nor mating prospects are good.

Twins often suffer a similar fate to physically handicapped offspring (Ball and Hill 1996). Either one or both of a pair of twins is killed soon after birth in a large number of traditional societies; indeed, the practice is clearly very ancient, as is suggested by the story of Romulus and Remus and the foundation of the city of Rome. Twins are often thought to be 'bad omens' and there are many taboos surrounding twin births. The underlying biological basis for this discrimination may be the extra burden that two small babies place on a lactating female. The inability of a female to invest sufficiently in two infants simultaneously means that it may make evolutionary sense to sacrifice at least one of the offspring in order to improve the future reproductive prospects of both offspring and parents. Ball and Hill (1996) found evidence in the medical literature that the mortality rate for twin births is significantly higher than for singletons. Although this was largely as a consequence of the fact that twins tend to be born prematurely, they also found evidence that twinning is associated with a number of negative outcomes, such as an increased incidence of congenital defects (for instance, Klinefelter's and Turner's syndromes, cleft palates, hydrocephaly) plus mental retardation and language disabilities, many of which are also associated with low birth weight (**Table 7.2**). It seems that offspring produced in twin births are often of poor quality compared to singletons, mainly because they have to share the mother's limited resources in the womb. The increased risk of infanticide for twins may reflect this reduction in offspring quality as well as the increased burden placed on the mother.

Gabler and Voland (1994) found comparable fitness costs to twins in an historical peasant community in northeast-Germany. Eighteenth- and nineteenth-century demographic records reveal that both twins and their mothers suffered significantly higher rates of mortality than women who habitually produced singletons (**Table 7.3**). The ultimate fitness of women who produced twins (measured by the total number of grandchildren that survived to age 15 years after all these costs had been factored in) was estimated to be about one third lower than that of women who produced only singletons (**Table 7.3**).

LACK OF PARENTAL RESOURCES

When parents are unable to invest sufficiently in healthy offspring due to unfavourable environmental conditions, it may be optimal to terminate investment as soon as possible, and conserve energy and effort to invest in offspring when conditions improve. Data from an eighteenth to nineteenth-century rural German population show that the children of primiparous widows had an extremely high risk of mortality, that there was an inverse relationship between infant mortality and remarriage, and that such mortality tended to occur before remarriage (Voland 1988). These findings suggest

Table 7.2 Negative outcomes associated with multiple- and breech births

Condition	Syndrome/injury
Multiple pregnancies/births	Klinefelter's syndrome
	Turner's syndrome
	Transfusion syndrome
	Hydrocephaly
	Congenital heart disease
	Hypodontia
	Cleft lip
	Anencephaly
	Mental retardation
	Uterine positional defects
	Minor foot deformities
	Skull asymmetry
	Hydatidiform mole
	Prematurity
	Stillbirth
	Neonatal mortality
Breech presentation/birth	Neuromuscular dysfunction
	Cervicothoracic spinal cord injuries
	Fractures of long bones
	Brachial plexus palsy
	Hip instability
	Sudden Infant Death Syndrome
	Cerebral palsy

Source: Hill and Ball (1996)

Table 7.3 The fitness of twinning in a northeast-German peasant population during the eighteenth and nineteenth centuries

	Twins %	Singletons %	Significance
Maternal mortality (1 year postpartum)	4.8	2.4	$P < 0.05$
Stillbirth rate	8.2	3.0	$P < 0.001$
Child mortality to 1 year	39.9	12.0	$P < 0.001$
Child mortality (1–15 years)	22.6	18.6	NS
Marriage rate	50.2	60.6	$P < 0.05$
Surviving grandchildren*: through sons	0.7	1.4	
through daughters	1.3	1.5	

Note: * Grandchildren that survive to their 15th birthday

Source: Gabler and Voland (1994)

that females were terminating investment in offspring that they could no longer afford after the death of their spouse while at the same time attempting to improve their future reproductive prospects by increasing their attractiveness as a marriage partner to other males (childless women had better chances for remarriage than those with children: Knodel and Lynch 1984, Strakageiersbach and Voland 1990).

Data on modern abortion rates seem to reflect similar considerations. Lycett and Dunbar (1999) have shown that, in England and Wales, rates of abortion were highest among young single women (**Figure 7.5**). They suggest that this is because these women are attempting to postpone reproduction in the expectation that conditions for child-rearing will be more favourable in the future. In line with this suggestion, there was a direct relationship between the age-specific probability of abortion and the age-specific probability of future marriage (as determined from national demographic statistics) (**Figure 7.6**). Furthermore, rates of abortion among single women dropped substantially with age, implying that women switch tactics as they age and begin to value current effort over future effort as the amount of time left to reproduce diminishes: at this point, they opt for single parenthood rather than risk waiting any longer and ending up childless. A somewhat similar pattern was reported for Sweden by Tullberg and Lummaa (2001), although these authors were unable to distinguish married from single women in their sample. Bugos and McCarthy (1984) noted an analogous tendency for infanticide rates to fall among older mothers among the Ayoreo Indians of Bolivia. In contrast, Lycett and Dunbar (1999) found that married women show lower rates of abortion overall compared to single women, but the rate increases considerably as these women age (**Figure 7.5**). Married women do not face the same resource constraints as single women. Their behaviour might reflect an attempt either to reduce the probability of producing a child with congenital defects or to limit family size to a predetermined number.

One other option used to reduce investment in current offspring, but stopping short of the actual destruction of children, was the employment of wet nurses in eighteenth-century Europe. Although it eventually became a widespread phenomenon, it was initially only wealthy families who made use of a hired woman to suckle their offspring (Hrdy 1992). The wet nurse was usually resident with the family, and close attention was paid to ensure that offspring thrived, so that, although parents reduced their personal effort in offspring, they were nevertheless concerned for its welfare.

Hrdy (1992) suggests that heavy pro-natalist pressures from their spouses encouraged women to give their children over to wet nurses so that they could produce children more quickly. The strategy seemed to work: the Duchess of Leicester, for example, gave birth to her first child at the age of sixteen and her twenty-first (and last) child at the age of forty-six. The presence of wet nurses thus allowed wealthy families to achieve spectacularly high reproductive success using a strategy that actually decreased the level of parental investment in each offspring, something that is most unusual for a large slow-growing mammal (Lewis 1986; Hrdy 1992). Although the very act of paying for a wet nurse is obviously a form of parental investment, and only those with wealth and resources could afford to do so, the reduction in personal investment by parents was more than offset by the increased level of reproduction that it allowed.

As might be expected, wet nurses eventually became something of a status symbol since only those with sufficient wealth could afford them. Although over time their use was increasingly emulated by the middle orders, in most cases the less well off individuals could not afford high-quality care for their offspring, and levels of mortality began to increase. The result was an arms' race driven by competition for the best wet nurses, with higher prices paid for the best quality nurses. In addition, children

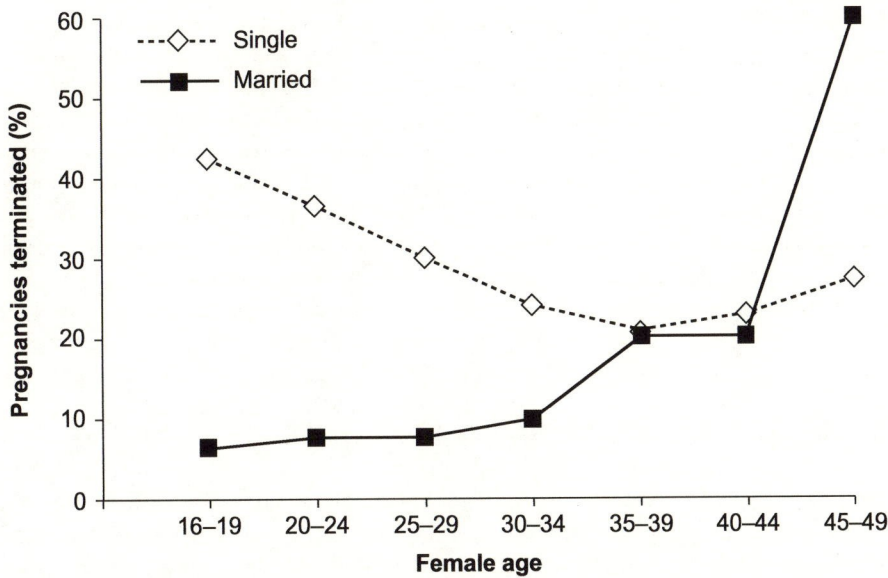

Figure 7.5 Percentage of pregnancies terminated by single and married women of different ages in England and Wales in 1991. Single women were more likely to terminate pregnancies when young, whereas married women were more likely to do so when older. Reproduced from: Lycett and Dunbar (1999).

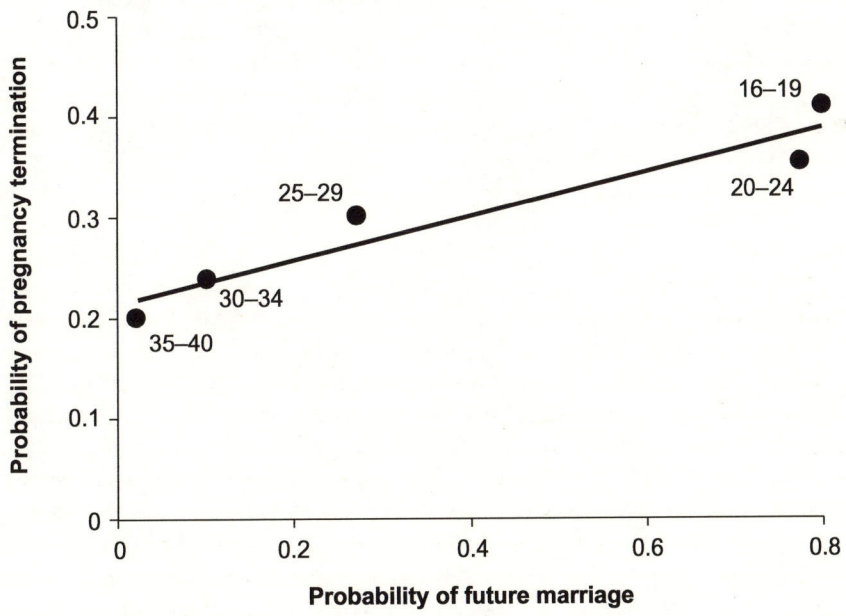

Figure 7.6 Probability of terminating a pregnancy by single women is a linear function of the age-specific likelihood of marrying in the future (based on census data for England and Wales for 1991). Single women who no longer expect to marry are more likely to see a pregnancy through to term, presumably because their opportunities for future reproduction are becoming increasingly limited. Numbers adjacent to data points indicate age cohort (in years). Reproduced from: Lycett and Dunbar (1999).

lower down the birth order tended to receive inferior care: once an 'heir and a spare' had been produced by conventional means, lower born sons were often sent to wet nurses in the country, far from their parents' watchful eye, where the level of care they received could not be closely monitored. The amount of care that parents were prepared to invest in their offspring was therefore related to the reproductive potential of the offspring, often quite consciously (Hrdy 1992). The use of wet nurses was therefore a way to increase reproductive success, but also proved to be an efficient way of reducing investment in children who were likely to be 'surplus to requirements'. Similar effects were reported among the farming families of northwest Germany during the eighteenth and nineteenth centuries: in this case, infant mortality rates for sons with three or more older brothers were double those for earlier born sons (see **Figure 8.5a**).

SELECTIVE INFANTICIDE AND THE SEX RATIO

All the examples we have dealt with so far consider infanticide (or abortion) regardless of infant sex. However, there are a number of good theoretical reasons why infanticide should be highly selective, and practised on one sex only. Trivers and Willard (1973) argued that when (a) one sex has a greater variance in lifetime reproductive success than the other and (b) parents (or more specifically mothers) vary in their physical condition (or resource base), differences in preference for offspring of the two sexes are likely to evolve. If male reproductive success depends on the individual's condition (bigger stronger males gain more mates and/or fertilisations), mothers in good condition who are able to invest heavily in offspring growth will be able to influence the reproductive success of their sons more successfully than mothers in poor condition; they should therefore prefer to have sons. In contrast, mothers in poor condition should prefer daughters because daughters are reproductively less risky (they have lower variance). This general principle is known as the **Trivers–Willard Effect**. We consider three case studies.

TRIVERS–WILLARD EFFECT

RAJPUTS

Dickemann's (1979) review of the historical evidence on infanticide within the Indian caste system showed that female infanticide had been extremely common among the highest castes of the Rajput tribes prior to the twentieth century. Indeed, at one stage, the leading noble family claimed not to have produced a single daughter during the whole of the preceding two centuries. Dickemann argued that the prevalence of female infanticide among the high caste Rajputs was due to the highly stratified nature of society, combined with a strong tendency toward **hypergyny** among females.

HYPERGYNY

In the highest castes (Brahmins, Jhareja, Jetwas and Soomras), daughters had very few marriage (and hence reproductive) options. The rules of hypergyny stated that men could only accept women from a sub-caste equal to or lower then their own. If a female from the highest caste could not find a suitable mate within her own sub-caste, then she was in trouble: there was no higher caste from which she could find a suitable partner. To marry down the social scale was an unacceptable practice that brought shame and disgrace on the entire family, thereby jeopardising the marriage (and fitness) prospects of other family members. Consequently, many of the female infants born into these high castes were killed shortly after birth, and it was explicitly stated that this was because husbands could not be found for them. Why individuals created a society in which such apparently self-defeating behaviour was necessary is, of course, the real issue here, and we will return to this in the later chapters on social structure and cultural evolution.

The reduction in female numbers among the Brahmins can be seen as a very extreme example of the Trivers–Willard effect. Among high-caste Indian families, investment in males paid larger dividends in terms of the production of future grandchildren, and accordingly, parents biased their investment heavily toward males, and invested nothing at all in their daughters. Lower down the social scale, the tendency toward hypergyny meant that daughters usually out-reproduced sons, and parents biased investment toward daughters in such families, and the incidence of female infanticide was considerably reduced. It should be clear, however, that it is the inherent disadvantage that females possess in this particular environment that drives the system; males benefit from extra investment largely because of the lack of opportunity for females, rather than because they require higher levels of investment to compete effectively.

THE MUKOGODO

Infanticide needn't be a deliberate act, and it isn't always females who find themselves the victims. Among the Mukogodo of Kenya, neglect of male offspring occurs frequently, and boys experience much higher levels of mortality as a result (Cronk 1989). In Cronk's study population, there were significantly more female children aged 0–4 years than male children (98 versus 66), a pattern suggestive of parental manipulation of the kind predicted by the Trivers–Willard effect.

The Mukogodo were traditionally hunter-gatherers and only adopted pastoralism as a way of life relatively recently, acquiring the practice from pastoralist groups in the area (mainly Maasai and Samburu). Since they tend to be much poorer than these other groups and are considered to be rather 'low class' by their neighbours, a relatively low brideprice is demanded for Mukogodo women compared to Maasai and Samburu women; this means that the Mukogodo find it relatively easy to secure a husband for their daughters from these tribes. This obviously reduces the number of suitable Mokogodo brides available to Mukogodo men; moreover, the men's problems are compounded by the fact that, due to their poverty and low status, they are unattractive as husbands for women from other tribes and are obliged to pay brideprices well over the normal going rate for Maasai or Samburu women. Many Mukogodo men are therefore unable to find themselves spouses, and have very low reproductive success in consequence. Cronk (1989, 1991) suggests that, in a reversal of the usual situation among polygynous societies, females tend to out-reproduce males among the Mukogodo. In line with the Trivers–Willard hypothesis, Cronk therefore argued that parents should invest more in their daughters than their sons, because daughters would ultimately provide them with more grandchildren.

While it is true that, on average, Mukogodo women do out-reproduce their men, it is clear from Cronk's data (**Figure 7.7**) that the variance in male reproductive success is actually much higher than among females, and that the most successful men do as well as the women. This doesn't necessarily invalidate Cronk's (1989) argument, but it is important to emphasize that, because brides are being exchanged between the various tribes, the Trivers–Willard effect would only operate at the level of the total population of the district, and not within the Mukogodo population alone. The Mukogodo are at the poor end of a socio-economic scale which encompasses all the pastoralist groups, and it is within this whole set of groups that they are competing for

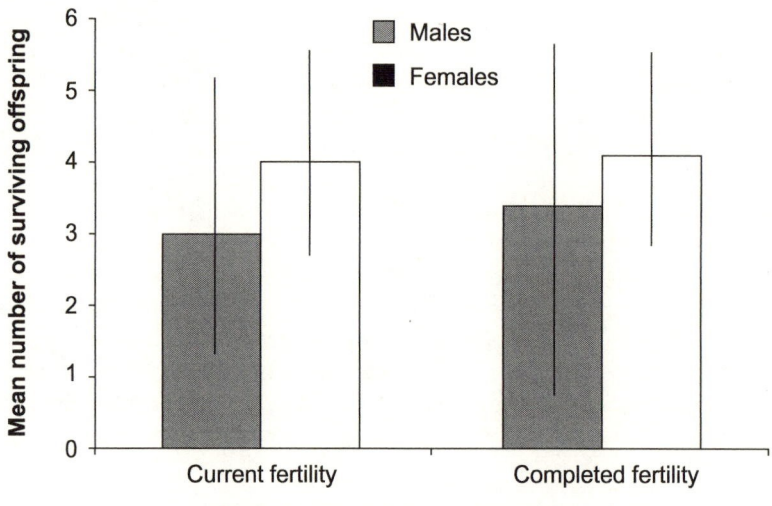

Figure 7.7 Among the Mukogodo of Kenya, the variance in reproductive success is greater for men than for women, even though mean output is higher for women. This is so whether fertility is determined for all individuals (current fertility) or only for those who have completed reproduction. Redrawn with permission from Cronk (1991).

mates. As such, they are expected to invest more in daughters than sons, compared to the non-Mukogodo tribes who are wealthier and able to invest adequate resources in their sons. If these outside groups were not part of the marriage market, then one would expect wealthier Mukogodo families to invest more in sons, given that the polygynous nature of society would allow them to achieve higher fertility than the average female.

Cronk (1989) provides some hints on the processes by which under-investment occur. The frequency with which children were taken to their local mission clinic for medical treatment is an easily quantified measure of parental investment, since this represents a significant cost (medical treatment had to be paid for, and the clinic was at some distance from the Mukogodo camps with walking the only means of getting there). Cronk (1989) found that female offspring were taken to be treated more frequently, and for more minor complaints than male children (**Figure 7.8**). In addition, Cronk (1989) provides anecdotal evidence that severely malnourished male children were often found living in the same family with well-fed healthy girls. He concluded that active neglect of male offspring led to an increased risk of mortality and that this accounted for the female-biased sex ratio in childhood.

If we accept that the differences in solicitousness toward male and female offspring are genuine (and not simply a reflection of some artefact like the fact that Mukogodo girls fall sick more often than boys: see Sieff 1990), then, given the higher completed fertility of women compared to men in Mukogodo society, it would appear that investing more in female offspring does pay fitness dividends to parents. Greater investment may place girls on a higher nutritional plane and, as in the Kipsigis (Borgerhoff Mulder 1988), increase their quality by reducing the age at menarche. This in turn would make them a more attractive proposition as a wife for a high status non-Mukogodo man than would otherwise be the case.

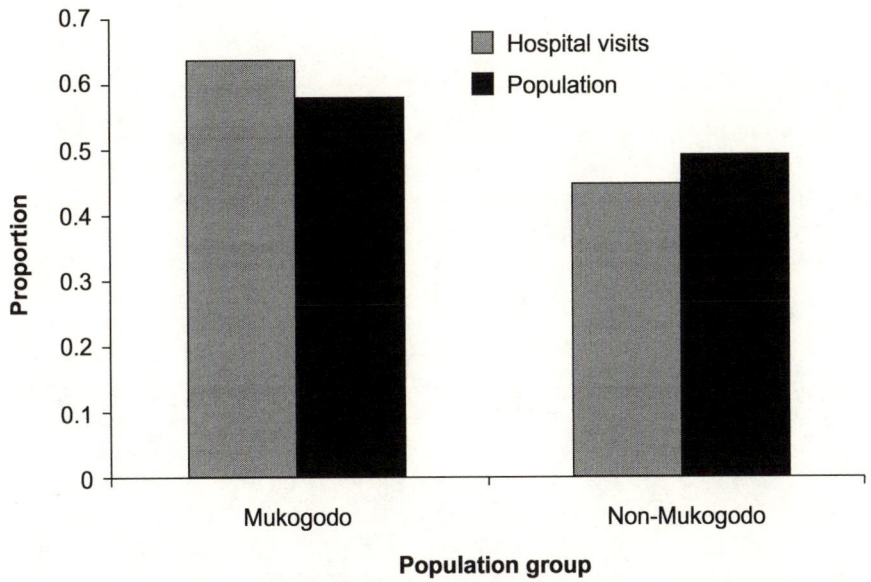

Figure 7.8 Mukogodo take their daughters to hospital more often than they do their sons (indexed here by the proportion of hospital visits that involve daughters as opposed to sons), relative to the proportion of daughters in the population. In contrast, other non-Mukogodo tribes in the neighbourhood take their daughters to hospital *less* often than would be expected given their frequency in the population. This suggests that Mukogodo invest more heavily in their daughters than do other sympatric tribes. Redrawn with permission from Cronk (1991).

INUIT

The previous examples have all emphasized the long-term reproductive benefits that accrue to parents who bias investment in an evolutionarily appropriate manner. However, when environmental conditions are particularly harsh, investment biases may also be important in the short-term in order to ensure the continued survival of the household as a whole. The Inuit (Eskimo) provide a possible example of this.

Nineteenth-century census records suggest that, among the Inuit, childhood sex ratios were highly male-biased (mean childhood sex ratio = 173 males per 100 females), but by adulthood the sex ratio had more or less reached parity (92:100) (see Irwin 1989, Smith and Smith 1994). **Figure 7.9** suggests that the male bias in the childhood sex ratio was correlated with decreasing temperature and increasing latitude (that is, with increased environmental hardship). Since the male-biased sex ratio during childhood was not observed at birth, it seems likely that a considerable degree of post-natal manipulation must have been taking place.

Irwin (1989) argued that because (a) the risks that males incurred while hunting on the ice increased as the environmental conditions deteriorated and (b) women need a male to provision them (due to the poor survival prospects of orphans if the mother died while foraging on the winter ice), the Inuit adjusted the postnatal sex ratio so as to ensure that there were sufficient spare males around to provide for their daughters' future requirements. Female infanticide was posited as the only means by which such a bias could be achieved. Indeed, Smith and Smith (1994) were able to show that the frequencies of female infanticide did correlate negatively with male survivorship (**Figure 7.10**), although actual observations of infanticide were few and far between.

BOX 7.5

Testing the Trivers–Willard hypothesis

Attempts to test the Trivers–Willard effect and other hypotheses of differential parental investment often fail on two critical grounds. First, most studies provide data either on parental biases or on reproductive outcomes and/or the sex ratio, but rarely both together – the Hungarian Gypsy study of Bereczkei and Dunbar (1997) being the one notable exception to date in this respect; as a result, they are unable to demonstrate that the hypothesised fitness consequences actually follow from the relevant behavioural decisions. Second, they invariably fail to test the assumptions on which these hypotheses depend: few studies, for example, present evidence for variance in male versus female reproductive success – the fundamental assumption on which the Trivers–Willard hypothesis is based.

Gaulin and Robbins (1991), for example, tested the Trivers–Willard hypothesis using a large sample from contemporary North American society. They found that individuals of high socio-economic status nursed their sons for significantly longer than their daughters, while the reverse was true for low income households. However, the authors provided no evidence that variance in male reproductive success was greater than that for females in contemporary society, nor that biased investment had different fitness consequences for high and low socio-economic groups. Instead, Gaulin and Robbins (1991) assumed that lifetime reproductive success would differ between the two sexes both because they would have different frequencies of extra-pair copulations and because men are more likely than women to remarry (making modern societies effectively polygynous). However, no evidence was offered in support of either claim (although the latter would be supported by the US demographic statistics: Buckle et al. 1996; Starks and Blackie 2000). While the observation of parental bias is therefore interesting, there is at present, no real proof that this can be considered as an example of the Trivers–Willard effect – although, of course, neither can this yet be excluded as a possible explanation.

Although this might look like an explanation in terms of group selection, it can in fact be interpreted perfectly satisfactorily in terms of parents manipulating their offspring in order to maximise their own fitness. The argument does, however, require families to adhere strictly to a common agreement about the locally ideal sex ratio, and this inevitably opens up opportunities for freeriders to raise more daughters than they should. (We discuss the problem of freeriders in more detail later). Two factors militate against a freerider strategy, however. First, the decision on whether a newborn infant should survive is not made by the parents, but by the community. They do this by naming the child: children that are not named by someone other than the parents are not considered human and are normally left out on the ice. Second, the harsh conditions of life in the Arctic mean that families that overproduce daughters will have extra 'useless' mouths to feed if they cannot marry them off; consequently, the potential costs incurred by freeriders make defaulting on the common plan an unproductive strategy.

In order to circumvent the problems associated with group selection, Smith and Smith (1994) offered a slightly different interpretation of the functional causes involved. They argued that, under the harsh conditions of the high Arctic, the marginal value of boys increases relative to girls, because more effort is required to obtain adequate hunt catches, and because the mortality risks for males are proportionately increased. This in turn means that not only are males relatively more valuable than females, but they are *absolutely* more valuable because an absolutely larger number of male kin is required to ensure household survivorship. Smith and Smith argue that

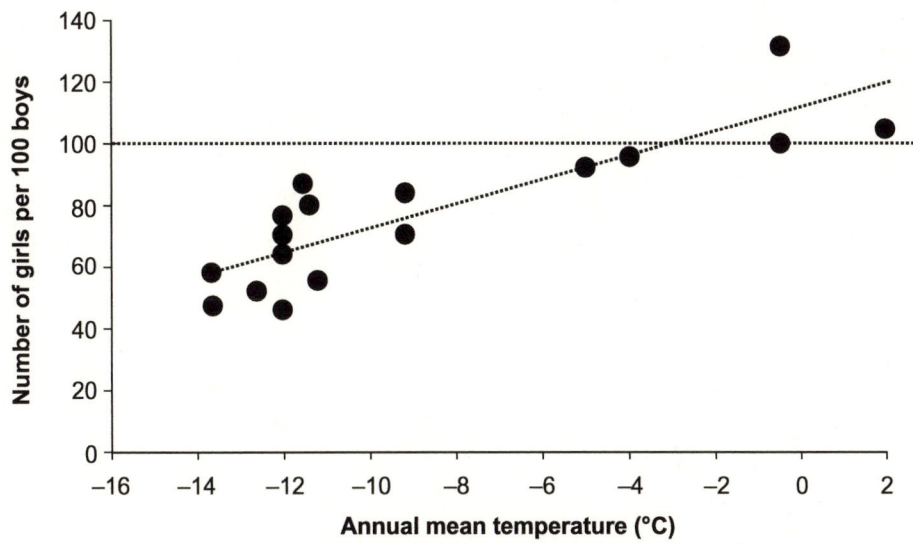

Figure 7.9 Natal sex ratio of historical Inuit populations (indexed as the number of girls per 100 boys) plotted against mean annual temperature. As the climate becomes more extreme, so the infant sex ratio becomes increasingly male biased, despite the fact that the adult sex ratio stays approximately constant. Inuit populations appear to have adjusted the number of surplus male infants through selective female infanticide. Source: Irwin (1989).

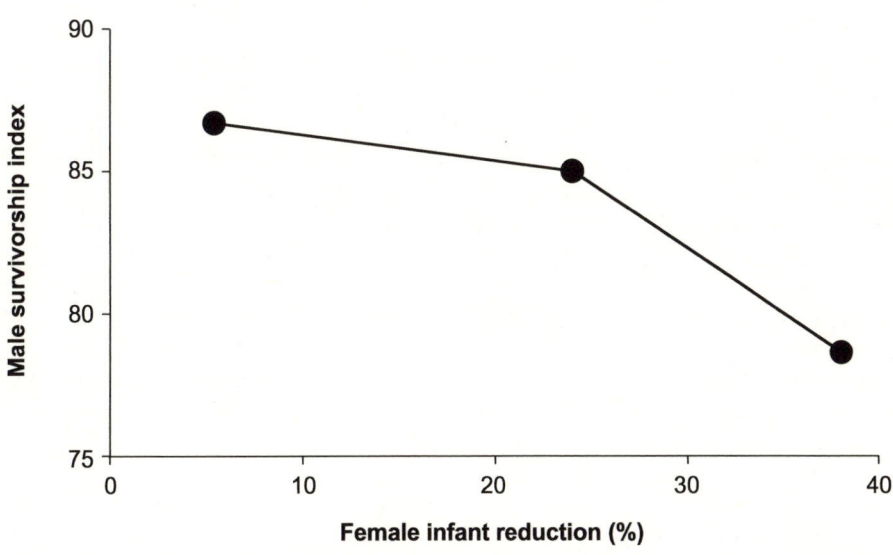

Figure 7.10 Adult male survivorship, indexed as a function of population-specific adult female survivorship (taken to be 100), was a function of the level of female infanticide (indexed as the percentage reduction in female infant numbers relative to males) for three historical Inuit populations. Redrawn with permission from Smith and Smith (1994).

this hypothesis not only explains why infanticide should be female-biased but also why levels of infanticide should vary between populations.

However, Borgerhoff Mulder (1994) has pointed out that, in harsher habitats where males are both increasingly costly (higher mortality rates mean that a greater amount of unproductive effort has to be expended before a son is successfully raised to adulthood, thereby increasing the average cost of rearing a male) and increasingly valuable (for provisioning), individuals might attempt to maximise the number of available males by committing infanticide on female offspring in order to reduce the interval to the next (hopefully male) birth. Committing infanticide and systematically eliminating females may thus have been aimed at ensuring that the first children to mature in any family were male, and thus that the household could be adequately provided for should it increase further in size. A quote from an Inuit informant suggests that such considerations were often a conscious motivation in peoples' reproductive decision-making:

> Life is short . . . parents often consider that they cannot 'afford' to waste several years nursing a girl. We get old so quickly, and so we must be quick and get a son. (Nalnagiaq, quoted in Smith and Smith 1994)

At present, this hypothesis has not been subjected to any rigorous tests (and it is unlikely it ever will, given the incompleteness of these historical data sets). However, it should be possible to estimate the relative pay-offs to families of differing composition and birth order, and then use stochastic dynamic programming to determine whether households with too many females would suffer in terms of future household survival.

CONDITION-DEPENDENT INVESTMENT STRATEGIES

Differential parental investment need not be expressed in such an extreme form as infanticide. In most instances, it is generally enough just to invest more time, effort and wealth in one sex in order to reap the reproductive benefits. This in turn means that sex ratio biases may not always be observed, since differential investment may not actually reduce the survivorship of one sex relative to the other. In such cases, we are not dealing with a Trivers–Willard effect in the strict sense (as an explanation of sex ratio biases), although obviously the underlying rationale is the same: if parents can increase their potential fitness gains by discriminating between offspring, then they may be expected to bias their investment accordingly. We can expect this form of differential investment to be particularly common in those species, like humans, that engage in heavy postnatal investment in their offspring, and especially so when that investment is social in form rather than energetic.

It is also important to realise that there is more than one model available to explain differential investment. Investing more in one sex than in the other need not always be an example of a Trivers–Willard effect, since offspring can also have a direct effect on parental reproductive output in the present, as well as contributing to parental long-term fitness in the future. Under circumstances where offspring of one sex can provide assistance to their parents and directly increase the parents' reproductive output, then it will pay parents to invest substantially more in the sex that helps as this will increase the total number of offspring produced. This is known as the **Local Resource Enhancement (LRE)** model (Silk 1983, Emlen et al. 1986). Since it can often involve older siblings helping to rear subsequent litters, it is also known in the animal literature as the *helpers-at-the-nest* model. By the same token, if one sex actively competes for available resources with its parents (or with its same-sex sib-

LOCAL
RESOURCE
ENHANCEMENT

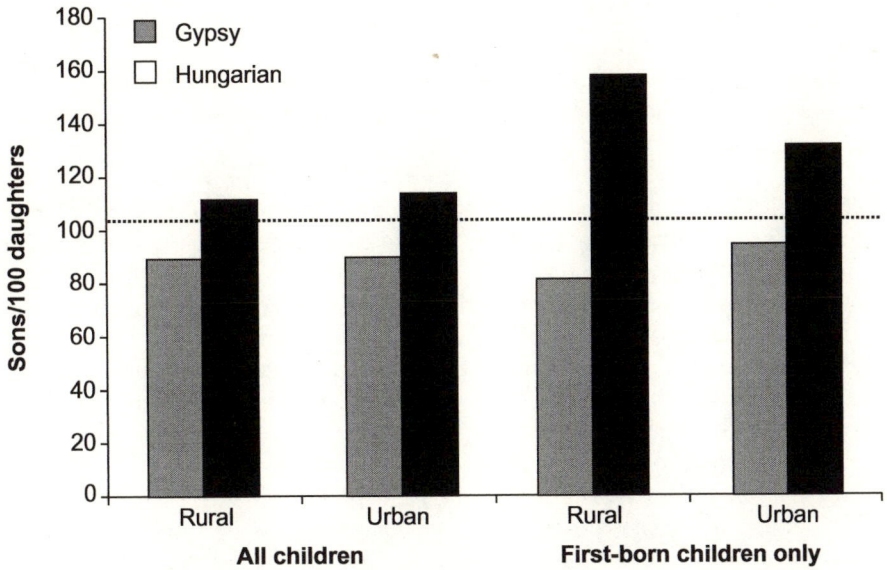

Figure 7.11 In contrast to ethnic Hungarians, sympatric Gypsies have produced consistently female-biased natal sex ratios for more than a century, especially among first-born offspring. Data are given here for two contemporary populations, one rural and one urban, of each ethnic group. Source: Bereczkei and Dunbar (1997).

lings), then it will pay parents to invest less in that sex in order to reduce the level of unproductive competition. This is generally referred to as the **Local Resource Competition** (LRC) model (Clarke 1978, van Schaik and Hrdy 1991).

Both the Local Resource Enhancement and Local Resource Competition hypotheses predict increased investment in one sex, and often these predictions converge with those of the Trivers–Willard hypothesis. This makes it essential to check that all the assumptions of the particular model in question are validated before any tests are made, and that there are a priori reasons to expect such an effect in the population in question. If such checks are not made, it is all too tempting to interpret the results in a post hoc fashion as favouring one's preferred hypothesis. The other complicating factor is that sometimes more than one effect can occur simultaneously within the same population, resulting in similar biases between offspring among different segments of the population but for completely different reasons.

LOCAL RESOURCE COMPETITION

NON-INFANTICIDAL TRIVERS–WILLARD EFFECTS

Like the Mukogodo, Hungarian Gypsy populations show a female-biased sex ratio (Bereczkei and Dunbar 1997). Compared to the native Hungarian populations, Gypsy populations give birth to significantly more daughters than sons (**Figure 7.11**). As in the Mukogodo, Gypsies are at the bottom of the social scale in Hungary: they are generally of a lower socio-economic class and less well-educated than their Hungarian counterparts. Gypsy women, however, are much more likely to marry up the social scale than men and, in doing so, provide their parents with more surviving grandchildren: Gypsy women who marry Hungarian men have infants with higher birth weights, lower mortality rates and lower rates of birth defects than Gypsy women who marry within their own social group.

As predicted by the Trivers–Willard hypothesis, Gypsy parents invested substantially more heavily in their daughters than their sons. Compared to the ethnic Hungarians, Gypsy women were more likely (a) to suckle their first-born daughters for longer than sons (although there was no difference in suckling duration for daughters born lower down the birth order), (b) to abort a subsequent pregnancy after a daughter than after a son (suggesting that women were attempting to continue investment in an offspring of the favoured sex), and (c) to allow their daughters to continue for longer in secondary education (which had to be paid for) (**Figure 7.12**). Even more interestingly, there were marked differences between rural and urban Gypsies in the extent to which daughters were the favoured sex. For each form of investment, urban Gypsies invested more heavily in their daughters relative to their sons than did rural Gypsies. Urban Gypsies have many more opportunities for hypergynous marriages into the ethnic Hungarian population than rural Gypsies, and therefore invest relatively more in their daughters than Gypsies in the rural populations.

This strategy was found to pay off: urban populations tended to produce a higher number of grandchildren through the female line than did rural Gypsies. More importantly, the relative amount of investment placed in sons versus daughters was directly scaled to the expected fitness gains that they provided across the four Gypsy and Hungarian populations (**Figure 7.13**). It is worth noting here that this study is unusual in being the only one to date that provides both evidence of a bias in investment and evidence that this bias pays off in terms of fitness gains.

HELPING AT THE NEST (LRE)

One aspect of the Gypsy study is puzzling, however. The increased investment in daughters over sons for the rural populations is somewhat unexpected, given the reduced probability of hypergynous marriage for rural populations. If exogamy is the only way for females to reap a reproductive benefit, why should parents invest so heavily in their daughters when the likelihood of this occurring is low?

The answer seems to lie in the fact that Local Resource Enhancement is operating in conjunction with a Trivers–Willard effect. Among the Gypsies, daughters are more likely than sons (or Hungarian daughters) to help their parents to take care of children: 72.3 per cent of Gypsy daughters participated in household chores and childcare in households in which at least one other younger sibling was present, compared to only 28.9 per cent of Hungarian girls (Bereczkei and Dunbar, in press). This may explain why first-born daughters are invested in so heavily: with an early-born daughter to act as a helper, a mother can divert her time and energy into producing additional offspring, and so directly increase her own reproductive success.

Among the rural Gypsies, investment in daughters may therefore represent a strategy for increasing the numbers of the parent's offspring rather than a Trivers–Willard effect. Indeed, Bereczkei and Dunbar (in press) found that mothers with first-born daughters tended to have shorter interbirth intervals (IBIs) than those with first-born sons. On average, women with first-born daughters tended to produce their fourth child (if they had one) after only 8.6 years whereas women with first-born sons produced a fourth child after 11.1 years. Women with first-born daughters also continued to reproduce for significantly longer: age at last birth for these women was 37.4 years compared to 34.2 years for women who had produced a son first. As a consequence of

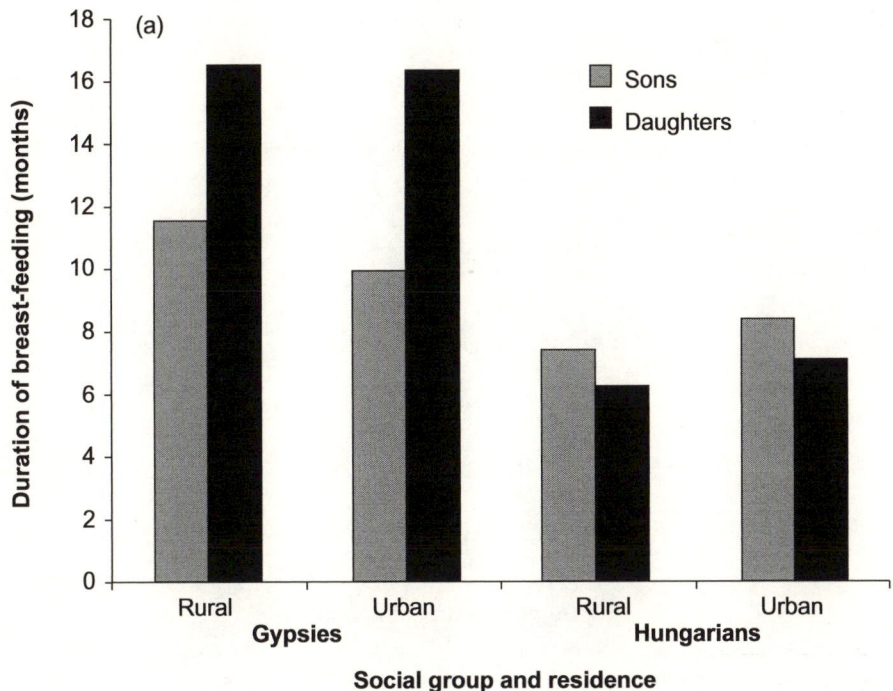

Figure 7.12 Parental investment in terms of (a) the duration of breast-feeding and (b) the duration of secondary education is female-biased among Hungarian Gypsies, but male-biased among ethnic Hungarians of similar economic status. Note that secondary education was not free and thus represents a significant financial cost to parents of low socio-economic class. Source: Bereczkei and Dunbar (1997).

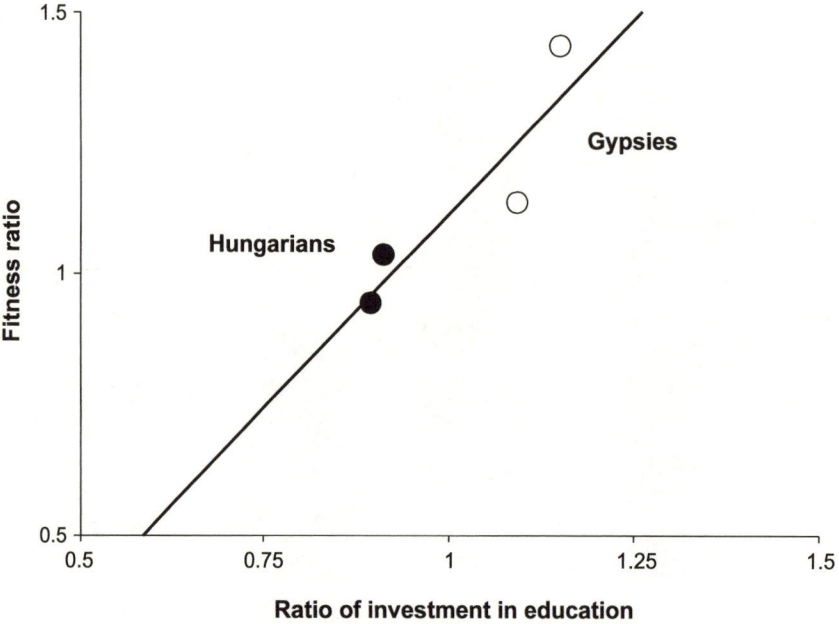

Figure 7.13 Hungarian Gypsies gain higher fitness (indexed by the number of grandchildren) through their daughters than through their sons, whereas ethnic Hungarians gain more through their sons. A plot of the ratio of grandchildren gained through daughters compared to sons is linearly related to the ratio of parental investment in the two sexes (here indexed as the ratio of years spent in secondary education by the two sexes of offspring). Data are shown for two Gypsy populations (one rural, one urban) and two ethnic Hungarian populations (one rural, one urban). Reproduced from Bereczkei and Dunbar (1997).

this reduction in IBI and an increased reproductive lifespan, women with first-born daughters produced an average of 4.4 living offspring whereas women with first-born sons produced an average of only 3.5 living offspring.

Turke (1988) also reported that 'helpers-at-the-nest' had a profound impact on women's reproductive output on the Micronesian atoll of Ifaluk. As in the case of the Gypsies, women who gave birth to a daughter first had higher reproductive success: these women tended to produce on average 6.9 living offspring, compared to only 4.9 for women who bore a son first ($P < 0.05$). The effect was even more striking for women who bore two daughters first: they produced an average of 8.9 surviving offspring compared to 5.1 offspring for women who had two sons (**Figure 7.14**).

As predicted, these differences in fertility tended to appear only once these first-born daughters reached an appropriate age to help their mothers. Early in their reproductive careers, there was little difference in the fertility of Ifaluk women whose first offspring were sons and daughters (2.6 versus 2.7 respectively). Between the ages of 30 and 44, however, women with sons produced on average only 2.7 offspring whereas those who had daughters produced 4.8. Again like the Gypsies, women with first-born daughters also continued to reproduce for longer than those with sons (age at last child: first-born daughter = 40.5 years; first-born son = 33.8).

However, although helping behaviour significantly increased a mother's direct reproductive success, helping to raise her siblings significantly reduced a first-born daughter's own reproductive output. For both the Ifalukese and the Gypsies, there was a signifi-

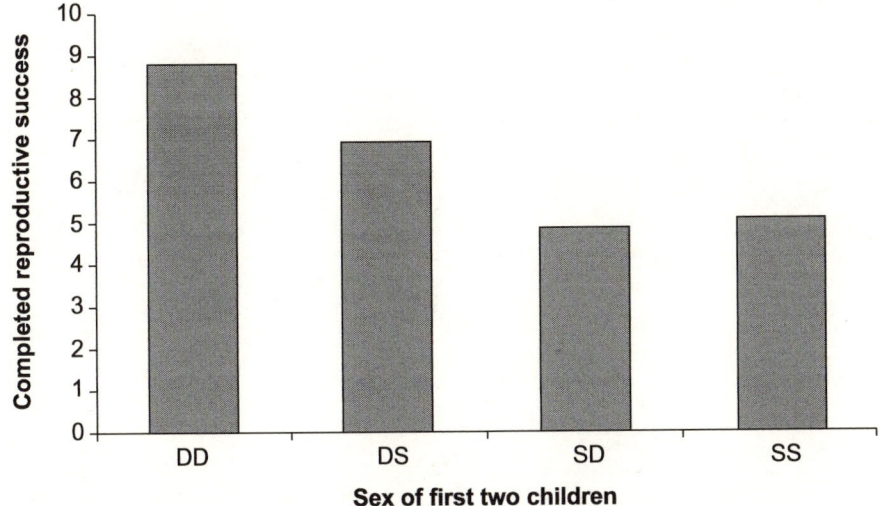

Figure 7.14 On the Pacific atoll of Ifaluk, women who produced daughters (D) as their first-born offspring had more offspring at the end of their reproductive lives than those who produced sons (S) first. A first-born daughter is particularly valuable in allowing the mother to reproduce more often because she can take over the childcare of her younger siblings. Redrawn with permission from Turke (1988).

cant negative relationship between the number of siblings a daughter helped to raise and her own subsequent reproductive output. Unfortunately, it is not known whether daughters' inclusive fitness gains from helping to raise siblings balances out (or even exceeds) the costs incurred in terms of her direct fitness component, or whether helping represents a strategy which is imposed on daughters by mothers against their own best interests (see below).

Voland and Dunbar (1995) provide a possible example of the latter effect among nineteenth-century peasants from the Krummhörn area in northwest Germany. Although the number of surviving siblings had no influence on the age of marriage for either the sons or daughters of the farmer class, a large number of sisters meant a delay in the age at first marriage for the daughters of the landless peasants (the poorest section of society). This suggests that these women may have remained at home to help raise siblings.

The benefits that daughters have for households has also been suggested as an explanation for female-biased investment among the Hutterites, a religious sect found in the northern USA and southern Canada (Margulis et al. 1993). Although there was no overall bias in the sex ratio, Hutterite women suckled their female offspring for longer than males (12.9 months versus 8.3 months respectively) and weaned them significantly later (**Figure 7.15**). Women were therefore found to have significantly longer interbirth intervals following the birth of a daughter compared to a son (14.4 versus 10.3 months, a difference of 40 per cent). Margulis et al. (1993) concluded that this difference in investment could not be related to a Trivers–Willard effect since Hutterites are strictly monogamous and the variance in lifetime reproductive success is similar for men and women. However, children are raised communally on Hutterite farms, and while boys work at tasks that benefit the colony as a whole, girls remain in their natal homes and assist their mothers with domestic chores. The benefit to mothers of investing more in daughters should be clear: female offspring directly aid their mothers in ways that sons cannot. At the proximate level, Margulis et al. (1993) argue that increased investment in infant daughters may facilitate bonding and attachment between mother and daughter, and may therefore be used to perpetuate the strong sex-segregation of societal roles. In contrast to the other examples discussed above, however, it is not known whether this extra investment translates

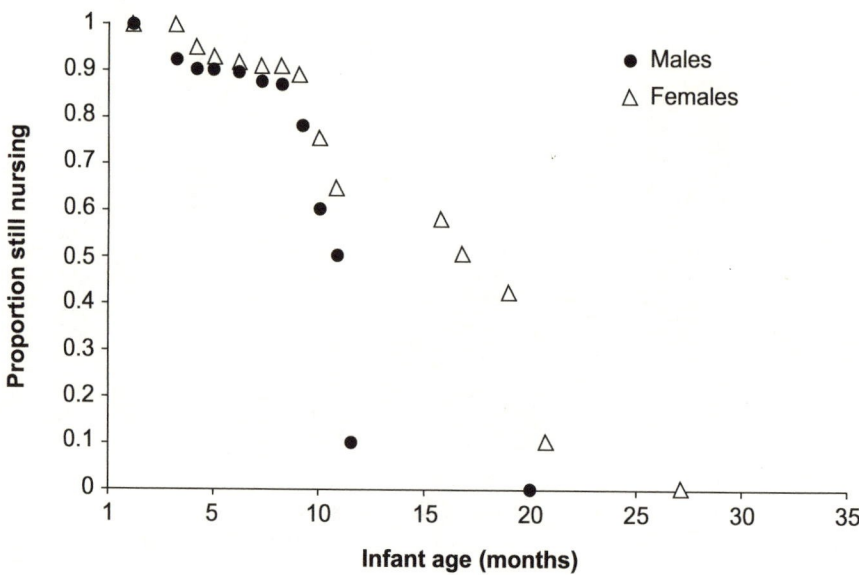

Figure 7.15 Mean duration of breastfeeding among the Hutterites of the northern USA. Female infants are breastfed for significantly longer than are male infants. Redrawn with permission from Margulis et al. (1993).

ultimately into higher reproductive success for women with daughters as predicted by the Local Resource Enhancement model.

REPRODUCTIVE COMPETITION EFFECTS (LRC)

A particularly clear example of this effect is offered by Voland et al. (1997) in an analysis of differential infant mortality rates across six rural populations in northern Germany during one 30-year period in the early nineteenth century. In some of these populations, sons seem to have been favoured, whereas in other areas their sisters were favoured. The extent to which sons were favoured turned out to be a function of the local population growth rate: sons were favoured when populations were increasing, but daughters were favoured in stagnant populations (**Figure 7.16**). A comparison across 30-year time periods between 1720–1870 for two of these populations (Krummhörn in Ostfriesland and Ditfurt in the Harz Mountains some 300 km to the southeast) revealed the same relationship within populations across time. Populations were only able to increase if they had virgin land to expand into, and this placed a premium on sons who could be used to carve out new farms. Once the population had reached its local carrying capacity and all spare virgin land had been put under cultivation, however, additional sons were a liability: even though only one son inherited the farm (usually the youngest), he in effect had to buy out his brothers, and this often necessitated selling part of the land. Hence, in these populations, farmer families attempted to keep the number of sons they had down to two (one to inherit and one as a back-up in case of his death: the so-called heir-and-a-spare strategy) (Voland and Dunbar 1995). Daughters, on the other hand, could always be married off into other families; since their dowries were relatively small compared to the inheritances that sons expected, there was less economic constraint on having lots of daughters. Indeed, detailed analysis of the consequences across generations suggested that farmer families gained significantly in terms both of fitness (number of descendents) and of bilateral land acquisition for the lineage through their daughters (Lycett, Dunbar and Voland, unpublished).

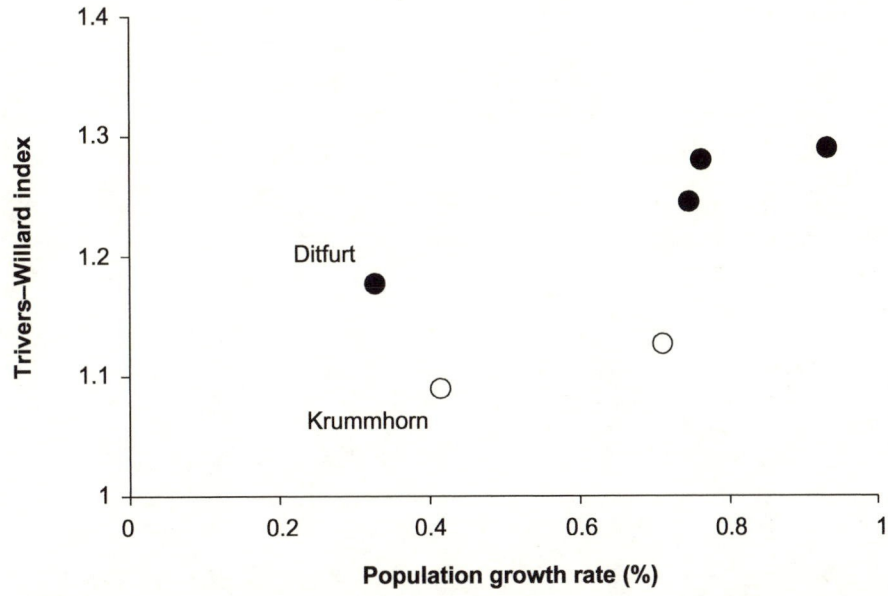

Figure 7.16 Preferential investment in sons (indexed by the ratio of mortality in the first year of life for sons versus daughters among farmer families versus landless labourers) correlates with the rate of population growth for north German rural populations during the mid-nineteenth century. This reflects the fact that sons are advantageous when there is land available for expansion (hence allowing population growth), but daughters are preferred when there is no spare land (hence population stagnation). Data are for the period 1820–70, for four parishes in agriculturally rich areas (filled symbols) and two parishes in poor areas (open symbols). The same correlation can be seen within sites through time for two populations (Dittfurt and Krummhörn) for which there are data spanning the period 1720–1870. Redrawn with permission from Voland et al. (1997).

Scott and Duncan (1999) found a somewhat analogous result in a rural community in the north of England during the pre-industrial period (1600–1800). From an analysis of the pattern of infant mortality, they concluded that the landowning class typically switched their sons from breast-feeding to wet-nursing at two to three months of age, whereas daughters were breast-fed until they were weaned at eight months of age. Survival rates were consequently very much higher for daughters than for sons. In contrast, tradesmen and subsistence farmers breast-fed both sexes for about the same length of time, with weaning at three to seven months; although daughter's survival rates were higher than those for sons (as is to be expected), the difference between the sexes was about half that seen in the wealthiest class. The contrast between the tradesmen and the landed gentry can probably be explained by the difference in opportunity that sons had in these two classes: the landed gentry faced the same problem as the Krummhörn farmers (too many sons created an unnecessary economic risk for the family estate), whereas the tradesmen could generate new filial lineages by simply training more sons in their trade.

Flinn (1989) also found evidence of reproductive competition and suppression among adult female kin occupying the same household in rural Trinidad. As in the Gypsies and Ifalukese, he found that women with co-resident non-reproductive female helpers had larger families than those without. Unlike the Gypsies and Ifalukese, however, there was no significant difference in the reproductive success of women who had daughters first compared to those who had sons first. Furthermore, the causality of the relationship was rather ambiguous, since larger families would be statistically more likely to have a daughter remaining at home than small ones. One possible reason for this apparent contrast might be that, as we suggested in Chapter 6, mother and daughter(s) share the role of reproduction across their combined lifecycle in a multi-generational household.

*

In the next chapter, we consider a peculiarly human form of parental investment: the transmission of heritable resources from parents to adult children, either to enhance their marriage opportunities or to enable them to invest directly in grandchildren.

Chapter summary

■ Conflicts over the level of parental investment begin in the womb. The hormonal maintenance of pregnancy and conditions like pre-eclampsia (high blood pressure) can be viewed as the result of strategies and counter-strategies between mother and embryo (or maternal and paternal genes) for the control of growth. Pregnancy sickness has also been interpreted as an adaptation to prevent women ingesting foods that could be harmful to the embryo.

■ Individual children within families compete for parental attention and personality differences that are dependent on birth order may be the outcome of these effects.

■ Differential investment in offspring is common. The most drastic form of this is infanticide. This is an action of last resort when other means of reducing investment fail. Infanticide is most commonly linked to lack of paternity certainty, poor offspring quality and lack of parental resources.

■ Differential investment by sex is also common in human populations. In many cases, males are the favoured sex because of their greater reproductive potential. Daughters are more usually favoured if helping behaviour can increase reproductive output.

■ Birth order can influence the level of investment by parents. Individuals lower down the birth order often receive less investment than their older siblings due to a reduced reproductive potential and because of competition over resources. Economic as well as reproductive considerations also seem to influence biases of this nature.

Further reading

Altmann, J. (1980). *Baboon Mothers and Infants*. Cambridge (MA): Harvard University Press.

Clutton-Brock, T. H. (1991). *The Evolution of Parental Care*. Princeton University Press.

Foley, R. A. (1995). *Humans Before Humanity*. London: Oxford Blackwell.

Hrdy, S. B. (2000). *Mother Nature*. Chatto and Windus.

Low, B. B. (2000). *Why Sex Matters: a Darwinian Look at Human Behaviour*. Princeton: Princeton University Press.

Rasa, A. E., Vogel, C. and Voland, E. (eds) (1989). *The Sociobiology of Sexual and Reproductive Strategies*. London: Chapman and Hall.

Marriage and inheritance

In the last chapter, we saw how parents manage their reproductive effort in ways that are apparently designed to maximise their fitness (principally, the numbers of grandchildren they produce). The consequence of these decisions might be to delay the arrival of the next offspring, or to alter the survival chances of an existing offspring. In this respect, humans are no different from other animal species that have slow-growing young. Humans, however, differ from all other animals in one particular respect, namely the extent to which they can directly influence the reproductive opportunities of their offspring. They can do this either by providing the resources that their children require in order to be able to marry (by providing dowries, brideprice or land) or by providing inheritances that their offspring can use to invest in their own reproduction.

The notion of marriage as an affirmation of romantic love is a very modern concept, largely peculiar to western culture. For most of human history, the institution of marriage has been a way of safeguarding individual interests by concentrating wealth within families or creating valuable and powerful alliances. Marriages were often arranged by the individuals' parents (or, in cases like the prospective marriages of the English Queen Elizabeth I, were the concern of the entire government); as such, they can be regarded as a form of parental investment. Ultimately, then, marriage systems and inheritance patterns are so intertwined that it is difficult to separate them.

Fully understanding patterns of wealth inheritance and their relation to marriage systems can therefore be a complicated business since it is necessary to disentangle reproductive decisions (which may be consciously or unconsciously made) from conscious strategic efforts to secure wealth or form alliances. Such an investigation is well worth the effort, however, since it provides a very satisfying example of the complex feedback process between biology and culture that characterises so much of human behaviour.

MATRILINEAL VS PATRILINEAL INHERITANCE

Since the agricultural revolution some 10 000 years ago, the great majority of human societies have possessed wealth (usually in the form of land) that is at some point handed on from the parental to the offspring generations. Inheritance systems can take one of two alternative forms: in matrilineal systems, wealth is passed on down the maternal line (sometimes from mother to a son or daughter, but perhaps more typically from a man to his sisters' sons) while in patrilineal systems wealth is passed down the paternal line (usually from father to son). Understanding why societies opt for one form or the other has been a longstanding challenge in anthropology.

POLYGYNY AND WEALTH INHERITANCE

In some cases, the relationship between marriage systems and inheritance practices appears to be straightforward. Hartung (1982) (and more recently Cowlishaw and Mace 1996) used cross-cultural analysis to show that where polygynous marriage systems prevail, wealth tends to be transferred to male descendants. Under a system of polygyny, men are entitled to take more than one wife and their ability to do so is determined by their wealth, since there is usually a brideprice to pay and the man must be able to support his new wife and offspring in addition to his current spouse(s). It therefore makes sense for parents to transfer wealth only to their sons, since wealth has a greater impact on male than female reproductive success (as in the case of the pastoralist Gabbra and the horticultural Kipsigis: see Chapters 5 and 6) and parents gain more grandchildren through the male line.

However, this is not to say that transferring wealth to sons has the same pay-off for mothers as it does for fathers. In polygynous marriage systems, the interests of the mothers and the father as to when and where wealth should be transferred may differ considerably. As far as each mother is concerned, as much wealth as possible should be passed on to her sons to start them off on a successful reproductive trajectory because this will secure her a large number of grandchildren and greatly enhance her inclusive fitness. For the father, however, the calculation is slightly different since he can achieve high fitness directly through his own reproduction as well as indirectly through his sons: wealth that could be used to purchase wives for a man's sons could just as easily be used to purchase more wives for the man himself. Since a man is related to his own children by $r = 0.5$ but to his grandchildren only by $r = 0.25$, his preference will be for investing in wives for himself. Consequently, there is likely to be a conflict of interests between mothers, sons and fathers. Mothers and sons should side against the father over the issue of how the family wealth is spent.

BOX 8.1

Marriage and inheritance: a phylogenetic analysis

Comparative analyses (in which one variable is plotted against another for a large number of species) have been the cornerstone of evolutionary hypothesis-testing ever since Darwin, and considerable use was made of them in the older anthropological literature (for example, Forde 1949). Such analyses are designed to determine whether changes in one variable (say, litter size) correlate with changes in another (say, body size). However, evolutionary biologists have realised that such analyses run serious risk of artificially inflated sample sizes whenever a significant number of species are closely related: the statistical analysis assumes that all the data points are independent, but closely related species are more likely to share traits merely by inheritance from a common ancestor (see Harvey and Pagel 1991).

The solution to this problem has been a set of statistical methods that essentially examine pairwise contrasts in each variable of interest between taxonomically equivalent nodes on a phylogenetic tree (the *method of* **independent contrasts**). Contrasts remove phylogenetic relatedness and yield a set of statistically independent datapoints. When data are in continuous form, contrasts on each variable are plotted against each other and a linear regression set through the origin to establish the relationship between the two variables. Discrete (categorical) data re-

quire a slightly different approach that searches for correlated changes from one state to another within the phylogeny (yielding a contingency table). Details of both methods can be found in Harvey and Pagel (1991).

If all human populations are descendents of an ancestral population (see **Box 1.5**), it is likely that this has occurred by successive fissioning of populations through time. Some cultures/societies will therefore be more closely related to each other than they are to others. This being so, the statistical problems that bedevil comparative analyses at the species level can be expected to apply with equal force to studies that use individual cultures/societies as their data (Mace and Pagel 1994).

To date, only one study has used comparative phylogenetic methods on human data. Cowlishaw and Mace (1996) reanalysed Hartung's (1982) data on the relationship between marriage system and inheritance patterns. With language phylogenies (given by Ruhlen 1987) as a proxy for cultural relatedness, they counted the number of state changes in one or both variables that occurred in the resulting phylogeny. Their results showed that transitions to polygyny are more likely to be associated with shifts to patrilineal inheritance than are transitions to monogamy (**Table 8.1**).

Table 8.1 Phylogenetic contrasts analysis of correlations between changes in marriage system and changes in inheritance pattern in human societies (with cultural relatedness estimated from language phylogenies)

Change in marriage system	Change in inheritance system		Total contrasts
	male-biased	female-biased	
to general polygyny	14 (87%)	2 (13%)	14
to limited polygyny	8 (44%)	10 (56%)	18
to monogamy	5 (25%)	15 (75%)	20

Source: Cowlishaw and Mace (1996)

BOX 8.2

Environmental correlates of polygyny

The ecological determinants of human mating systems are not well understood. White and Burton (1988) noted that polygynous societies tend to be associated with tropical habitats, and argued that this is because these are ecologically richer and more homogenous. Divale and Harris (1976), on the other hand, argued that polygyny was associated with protein stress, as a result of which male competitive abilities would be emphasized; the resulting warfare would lead to a biased adult sex ratio, and hence a surplus of women.

In a more extensive series of analyses of a standardised cross-cultural sample of 93 societies, Low (1988, 1990b) found that the degree of polygyny (indexed in several alternative ways, including maximum harem size and the percentage of men and women who are polygynously married) is negatively correlated with rainfall constancy ($r_s = -0.29$, $P = 0.03$), positively correlated with both rainfall contingency ($r_s = 0.31$, $P = 0.02$) and pathogen stress ($r_s = 0.29$, $P = 0.007$), and positively correlated with the extent to which women contribute to the subsistence base ($r_s = 0.25$, $P = 0.02$), although, since these are simple bivariate correlations, they may not all be independent effects. However, the correlation with pathogen stress at least can be shown to be independent of the correlation with subsistence pattern.

Low (1988, 1990a) has argued that that the crucial driving factor here is the risk of exposure to serious pathogens. These included 18 specific pathogen species (leishmanias, trypanosomes, malarias, schistosomes, filariae and leprosy) coded individually on a three-point scale for severity of incidence in an individual society. Low argued that, when pathogen risk is high, female choice for high quality mates (indexed in terms of pathogen resistance) drives the mating system towards polygyny because women would rather be the second or third wife of a high quality male rather than the sole wife of a poor male. This argument builds on an extension of the classic polygyny threshold model (see **Figure 5.15**) to include Hamilton and Zuk's (1982) model of the impact of pathogen stress on mate choice. Low argues that the quality of offspring produced by parasite-resistant males is the principal advantage that drives this.

Hartung (1982) suggested that this family conflict may actually underlie Freud's 'Oedipus complex'. Freud suggested that sons are sexually attracted to their mothers and that this creates tension and conflict between fathers and sons over potential sexual competition for the mother. Hartung (1982) gave this notion a more plausible twist by suggesting that, in polygynous societies, sources of father–son conflict do not arise over sexual access to the mother but over who should be the most successful polygamist. Mothers get involved in this not because they are the object of sexual competition, but because their interests coincide with their sons' and conflict with those of their husbands.

Hartung's (1982) suggestion highlights the value of adopting a longer term perspective by recognising that females can also achieve very high fitness within systems which seem to favour only male interests. That is, if one considers only the first offspring generation, the variance in female reproductive success within polygynous systems will be low compared to that of males. However, in subsequent generations, the variance in female reproductive success will actually begin to approach the variance in male reproductive success because some women will have produced highly successful polygynous sons who provide their mothers with an extraordinary number of grandchildren. These differences would then be amplified over the generations until there was little difference between male and female reproductive variance. Polygyny may therefore be just as much in women's reproductive interests as in men's, at least providing the women are in a position to ensure that their sons will be successful reproductively.

MATRILINEARITY: ARE SISTERS DOING IT FOR THEMSELVES?

It has long been suggested that matrilinearity is associated with low paternity certainty: men invest in their sisters' children because doing so guarantees that the children who receive his bequest will actually be related to him. Hartung (1985) showed, from a cross-cultural sample, that matrilineal inheritance was indeed typically associated with moderate to low paternity certainty (17 out of 20 societies), whereas only 5 out of 50 patrilineal societies were classified as having a low probability of paternity. These data fit well with the paternity hypothesis.

However, as Hartung (1985) points out, it actually requires a very low probability of paternity indeed for a man to have, on average, a higher degree of relatedness to his nephews than to his own sons. Since a nephew is related to the man by $r = 0.25$, paternity certainty will have to be less than 50 per cent (that is, only half the offspring of the man's wives were sired by him, such that $r = 0.5 \times 0.5 = 0.25$) for investing in his nephews to be worth his while. In other words, unless levels of promiscuity are very high, he is still, statistically speaking, likely to be more closely related to his own sons than he is to his sister's sons, and investing in his sons would always be a worthwhile risk. Indeed, the situation is likely to be worse than this because this analysis assumes that the man and his sister are full sibs. Given the high level of promiscuity assumed in this example, we would need to adjust his relatedness to his nephew by the same amount (50 per cent) to allow for his mother's promiscuity since she will be just as promiscuous as his wives. This would give a relatedness to a nephew under 50 per cent paternity uncertainty of $r_{PC=50\%} = 0.25 \times 0.5 = 0.125$. In fact, unless the man has knowledge about actual paternity, paternity certainty would have to be zero for a matrilineal strategy to be worth his while. What this means is that matrilinearity is actually very hard to justify if only the advantage to men is considered.

The paternity hypothesis can be strengthened if the advantage to females is also considered. Although there is no such thing as maternity uncertainty, there is, of course, grandmaternal uncertainty through the male line: a woman can never be sure that her son actually fathered any of his children. (We touched on this in Chapter 3 with our discussion of grandparental solicitude.) As paternity uncertainty increases, the probabilistic relatedness of a woman to her son's son (patrilineal heirs) diminishes substantially, whereas low paternity certainty has no effect at all on her relatedness to her daughter's sons (her matrilineal heirs) (Hartung 1985).

Matrilinearity thereby ensures that wealth is transferred only to those males with whom a female is likely to be related, that is to her son and son's sister's sons (that is, her daughter's sons), but not to her son's sons. **Figure 8.1** shows the relative advantage of matrilineal over patrilineal inheritance (in terms of passing wealth on to the heirs to which an individual is most closely related) for a man and a woman separately over successive generations assuming that paternity certainty is $PC = 75$ per cent. Matrilineal inheritance benefits women (greater relatedness to matrilineal than patrilineal heirs) from the very first generation onwards, and this is true as long as $PC < 100$ per cent. Men, however, are less related to their matrilineal heirs in both the first and second heir generations, and do not begin to benefit from matrilinearity until generation four. Overall, men do not benefit widely from matrilinearity unless

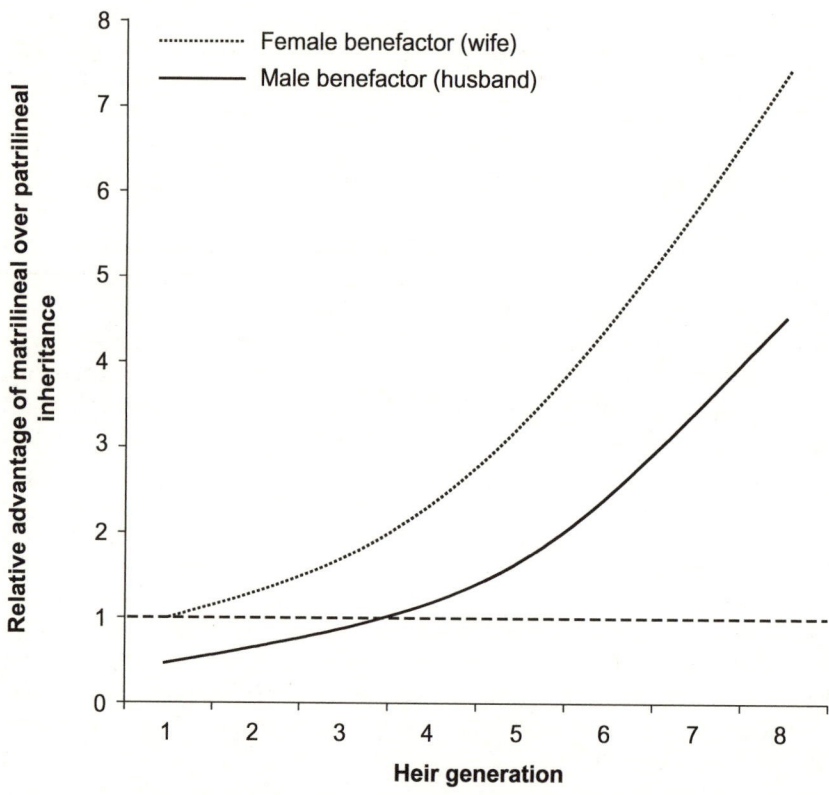

Figure 8.1 Estimated advantage to male and female benefactors (in terms of the numbers of descendents produced in various subsequent generations) when inheritance is matrilineal compared to that when inheritance is patrilineal under conditions of 75 per cent paternity certainty (that is, there is only a probability of 0.75 that a man sired any given child that his wife produces). Under these conditions, matrilineal inheritance is always more advantageous for the women than it is for the men; men do not begin to benefit from matrilinearity until the third or fourth generation. Redrawn with permission from Hartung (1985).

paternity certainty is at extremely low values (*PC* < 46 per cent), whereas it is the ideal strategy for women whenever *PC* < 100 per cent. Matrilinearity therefore allows females to associate their wealth with their biological heirs – and thereby gain all the attendant fitness benefits – much more successfully than it allows males to do so.

However, there is still a problem with this analysis, since it seems unlikely that matrilinearity evolved through its benefits to women, given that men invariably control wealth in most societies. As Hartung (1985) points out, one would expect the transfer of resources to be directed to patrilineal heirs since it is men who control the distribution of inherited wealth and patrilinearity benefits male fitness to a greater extent than matrilinearity does. However, as well as being fathers, men are also sons, brothers and uncles, and in this capacity matrilinearity is advantageous. In addition, matrilinearity is also advantageous for every man's mother, sister, nephew and father. Since the collective advantage to all these relatives outweighs the disadvantage that matrilinearity brings to individual men, the combined pressure of these individuals may make it difficult for men to circumvent these collective interests. Moreover, once matrilineal inheritance has become established, it is highly resistant to change and is likely to persist despite its disadvantages to individual males' long-term fitness.

An alternative explanation is that men do not assess their probabilistic relatedness in the way the models suggest, but rather use a deterministic rule of thumb that leads them to invest in their matrilineal kin whenever they feel that the probability of paternity is even slightly less than one. After all, it is quite possible under such circumstances for a man not to be the father of any of his putative children. If this were the case, then an individual who passed his wealth to 'his' male offspring would not achieve much in the way of fitness. In other words, men may be selected to be risk-averse with regard to paternity certainty and make their decision based on even the slightest perceived risk of paternity uncertainty, not just whether they are more or less likely to be related to their matrilineal kin.

Mace and Holden (1999) offer a slightly different version of this argument. They note that matrilineal descent is most common among farming/horticultural societies, which are heavily dependent on women's labour, and that it is significantly negatively associated with pastoralism (cattle herding). Matrilinearity is therefore associated with situations in which male earning power is limited and the key resources are produced by women, implying that women actually control resource distribution. However, men can and do farm within matrilineal systems, and they also wield a lot of political power. Mace and Holden (1999) therefore suggest that, in matrilineal systems, men do not *want* to control resources, not that they are prevented from doing so. They suggest that this is because, under matrilinearity, men may do better by employing a 'roving' strategy, and opportunistically taking advantage of mating and marriage opportunities as and when they arise. Maintaining control of resources and acting as family head would not allow them to adopt such a strategy. Consequently, matrilinearity is most prevalent where ecological conditions preclude subsistence modes like pastoralism that allow men to accumulate large amounts of personal wealth. Since women under these same conditions are largely self-sufficient, the costs of desertion by a male partner are not high, and the fitness benefits of matrilinearity mean that such a system is even more beneficial for women. Thus, men are prevented by circumstance from following the strategy which is most optimal for them, and resort instead to the one that offers the best returns given their constraints, while women are actually engaged in a strategy that is ideal from their own reproductive perspective.

┃ RESOURCE COMPETITION AND LINEAGE SURVIVAL

The examples discussed so far suggest that cultural rules regulating marriage systems have, at their base, sound biological reasons linked to maximising long-term fitness. However, maximising reproductive success per se may not be the best strategy once heritable resources become available. As we saw in Chapter 6, the impact of resources on future reproductive opportunities may be so great that fitness may be more effectively maximised by ensuring the integrity of the resource base rather than by simply maximising the numbers of children one produces. In effect, parents are forced into a new game in which fitness is maximised by ensuring lineage survival. The evolutionary significance of this is brought into sharp focus by the observation that even in modern post-industrial populations, the risks of lineage extinction may be as high as 20–50 per cent (Keyfitz 1977). When this happens, both parents and offspring may be forced to adopt strategies to ensure lineage survival that may appear to be reproductively disadvantageous if the history of the situation is not considered.

Biasing investment toward offspring of one sex is often extremely marked in situations where heritable wealth is involved, and again provides a situation where parents can

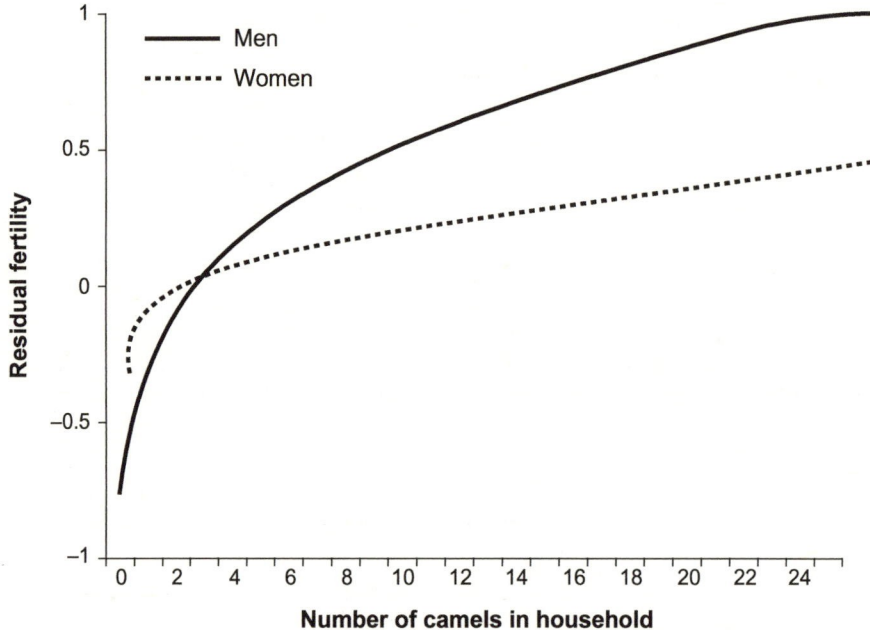

Figure 8.2 Among the Gabbra pastoralists of northern Kenya, residual fertility (corrected for the man's age) is related to household wealth (indexed as the number of camels). However, the effect of wealth is greater for men than for women, probably because a woman's fertility is ultimately limited by the biologically constrained rate at which she can produce offspring whereas richer men can increase their fertility by using their wealth to buy another wife. Note that in very poor households, the relative fertility of men is lower than that of females. Redrawn with permission from Mace (1996b).

achieve potentially high reproductive rewards by judicious allocation of resources, without any overall bias in the production of one sex over the other. In Chapter 6, we saw how the Gabbra pastoralists base their reproductive decisions on estimates of the future costs of offspring according to their sex. Now we can examine how these varying costs determine overall patterns of investment within families.

A PASTORALIST EXAMPLE

Mace (1996b) found that, among the Gabbra, wealth had a positive effect on the reproductive success of both men and women, but as in most other human populations, the effect was greater for men. As the wealth of the household increased, the marginal impact on female fertility declined, probably due to a combination of the fact that there is a biological upper limit on the number of offspring a woman is able to produce and because at the very wealthy end of the spectrum, women are likely to be sharing resources with more than one wife. By contrast, men's fertility continued to increase steadily with increasing wealth, with very rich men out-reproducing rich women by a considerable margin. However, very poor men actually did considerably worse than very poor women (**Figure 8.2**).

These differences in the variance of male and female reproductive success according to the level of resources available are the conditions conducive to a Trivers–Willard effect, and it might be expected that, despite a general tendency for male-biased

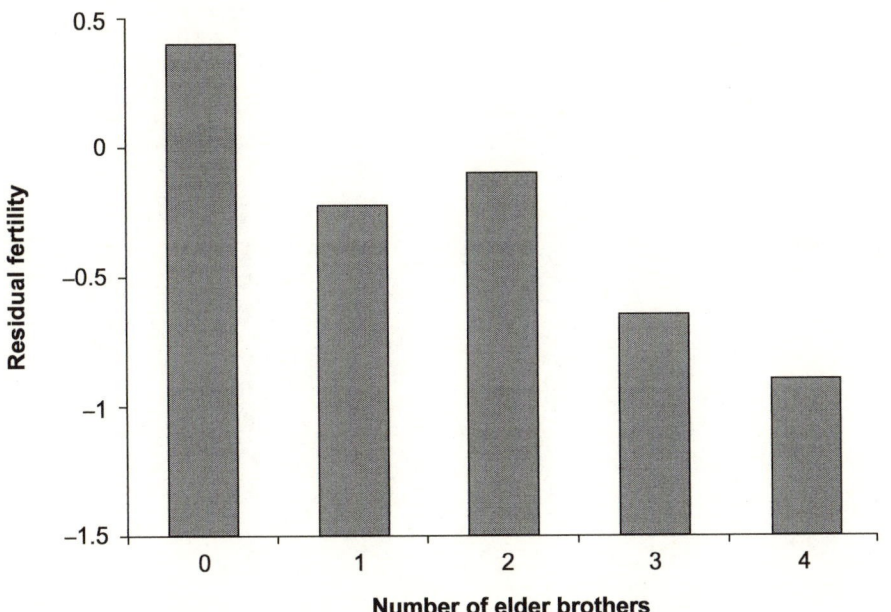

Figure 8.3 Mean residual fertility (corrected for age) of a Gabbra man as a function of the number of older brothers he has. A late-born son is likely to produce fewer offspring. For further details, see Figure 8.2. Redrawn with permission from Mace (1996b).

inheritance within the Gabbra, poor families would actually invest more in their female children. There was, however, no indication of such an effect in Mace's (1996b) data: both wealthy and poor households showed a slightly male-biased sex ratio and no evidence of any patterns of sex-biased investment. Mace (1996b) points out that marked investment biases are not really to be expected in a society where the labour of children of both sexes is valuable (both boys and girls act as herders from the age of about nine onwards) and female offspring provide a net source of income in the form of brideprice. This latter point is in fact the key to the Gabbra system: since the cost of marrying off sons and daughters varies widely, parents can bias the level of investment in offspring within and between sexes not by denying them basic care and resources as children, but by varying the number of animals passed on to a child at the time of its marriage. (However, Mace and Sear [1997] did find that interbirth intervals following male children tended to be longer than those after female children and that this was especially true for first-borns, suggesting that mothers tended to invest more care in children with a higher reproductive potential.)

Mace (1996b) was able to show that married men with elder brothers tended to have lower fertility than those without elder brothers (that is, first-born sons). As **Figure 8.3** shows, the magnitude of this disadvantage is quite substantial, but Mace suggests that even this is an underestimate since it doesn't include brothers who were never married, or those who left the district through lack of reproductive opportunity. By contrast, and as might be expected, the number of sisters a man had influenced his reproductive success in a positive manner, although the effect was rather small.

The decline in fertility with increasing birth order was interpreted by Mace (1996b) as representing a strategic decision on the part of fathers as to how they partitioned their herd among their sons. Since male reproductive success is very sensitive to wealth, it may make more sense from a long-term fitness perspective to endow one or a few sons with a large amount of wealth, and ensure that they will be reproductively successful, than to allocate resources evenly between all sons, all of whom will then be relatively poor and unsuccessful. Small herds are not viable for pure pastoralists, and household

Figure 8.4 The marriage opportunities of a Gabbra man are strongly influenced by his birth rank. (a) Mean age at first marriage and (b) mean number of camels in his herd at marriage are a function of the number of older brothers a man has: the lower down the birth order he is, the later he is likely to marry and the fewer camels his father is likely to have from which to provide him with a 'starter' herd. Redrawn with permission from Mace (1996b).

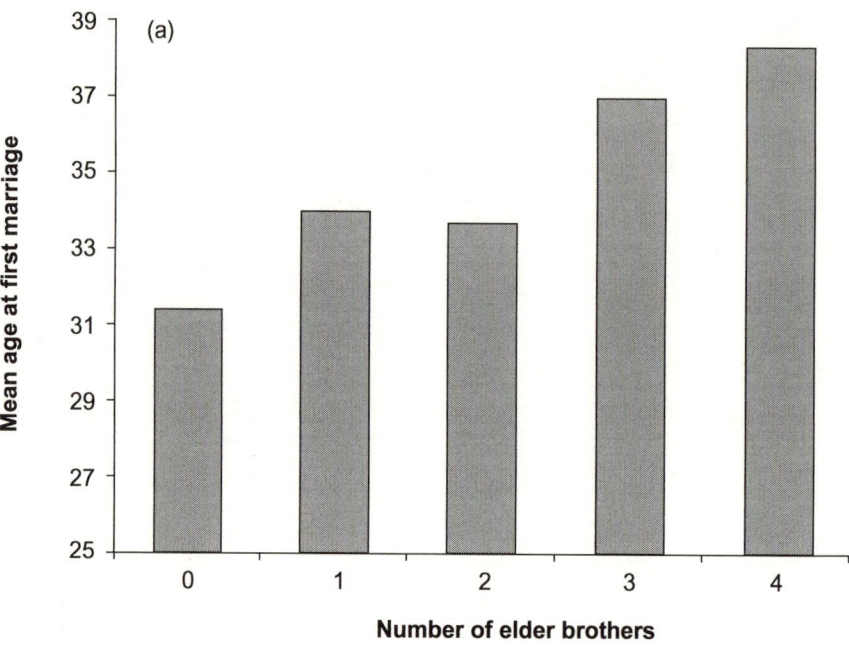

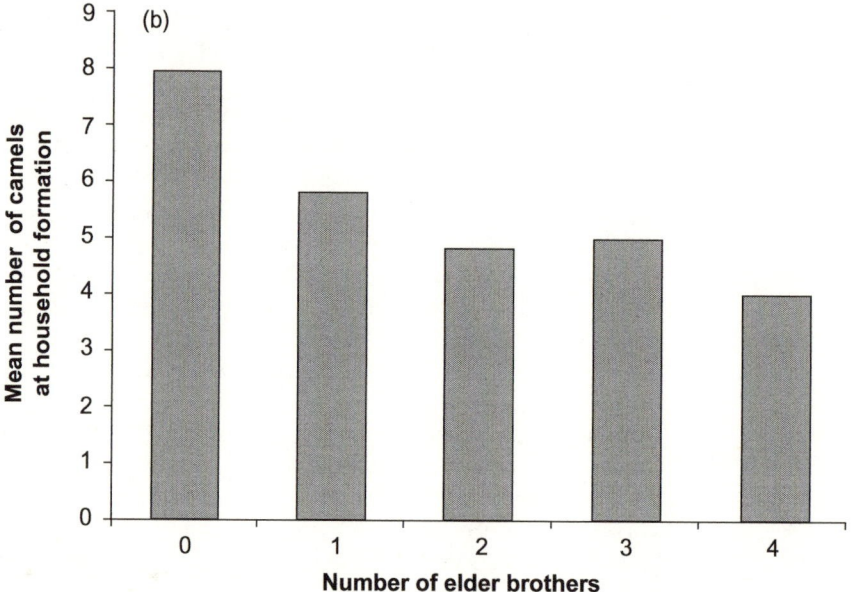

survivorship is threatened when herds fall below a critical minimum size. Fathers therefore do better by providing a small number of sons with viable herds, and they achieve this by investing in early-born sons. First-born sons, as well as being ready to begin their reproductive careers earlier, are also more effective at controlling younger brothers and coercing them into herd management than are younger sons.

Sons born lower down the birth order tend to marry later (**Figure 8.4a**), and when they do, they establish smaller herds (**Figure 8.4b**). First-born sons therefore have a marked advantage over their younger brothers, but the fact that most of these brothers

BOX 8.3

Wealth-dependent herd management

The Gabbra pastoralists of northern Kenya provide evidence that humans might be trying to maximise lineage survival and not simply short-term gains in terms of numbers of children produced. Using dynamic programming models, Mace (1993) analysed the decisions that Gabbra made about managing the fertility of their sheep in order to determine which breeding strategy maximised the chances of household survival over a ten-year period, given the risk that one in four years would be a drought (famine) year with low sheep fertility and high mortality and that ewe mortality is at all times a function of birth rate (but nearly an order of magnitude higher during drought years than during non-drought years). The parameter values for the model were based on actual sheep demographic data from nine herds. The key to these analyses is that the variable being optimised is the probability of household (or lineage) survival in the long term.

The analyses suggested that reducing sheep breeding rates minimises the household's risk of falling destitute (having no sheep during any given six-month period over a ten-year cycle). However, this strategy is optimal only when herd size exceeds a certain threshold. The precise herd size at which it pays to reduce breeding rates depends on the household consumption of sheep

(normally surplus males) for their own subsistence. When ten sheep are removed from the herd each year, the minimum herd size at which reduced breeding rates is commensurate with a 95 per cent chance of not falling destitute over the next ten years is 34 sheep, but it is 125 when as many as 40 sheep are eaten each year.

When herd sizes are below this value, it does not pay to reduce breeding rates because the risk of not producing enough lambs to maintain herd size increases. Although poor households run a risk of overtaxing their ewes by breeding them at maximum rates, nonetheless failure to do so increases the risk that they will not survive a famine year. Households with smaller flocks are obliged to maximise short-term goals (at the expense of increased short-term risk) rather than long-term goals in order to have any chance of survival.

The optimal strategy thus depends on household wealth (as measured in terms of herd size), with wealthy and poor households pursuing diametrically opposite strategies. Since this is exactly what the Gabbra appeared to be doing, it seems that their ultimate goals are to maximise lineage survival rather than simply maximise short-term birth rates.

do eventually marry means that the Gabbra do not employ a strict system of **primogeniture** (first-born son inheritance). Younger sons might have to wait longer but they marry in the end, and it may be the case that households have to allow enough time for the herd to increase after each marriage to provide a sufficient endowment for each son. Parents may therefore be managing their investments very wisely, and not permitting sons to marry with a herd size that will not be viable in the long term. The ability to bias investment in offspring by manipulating wealth provides humans with a much finer tool than the somewhat heavy-handed sex ratio bias envisaged by the original Trivers–Willard effect.

PRIMOGENITURE

AGRICULTURAL SYSTEMS

A similar effect of birth order on child mortality rates and reproductive success is found among a number of historical populations with agricultural-based economies. Low (1991), for example, found an effect exactly equivalent to the Gabbra in nineteenth-

Figure 8.5 In the Krummhörn during the eighteenth and nineteenth centuries, the chances of surviving infancy decline as the number of living same-sex siblings increases for (a) the sons and (b) the daughters of a landed farmer, but not for the offspring of a landless labourer. Note that farmers' daughters only begin to suffer higher mortality if they have three or more sisters. Redrawn with permission from Voland and Dunbar (1995).

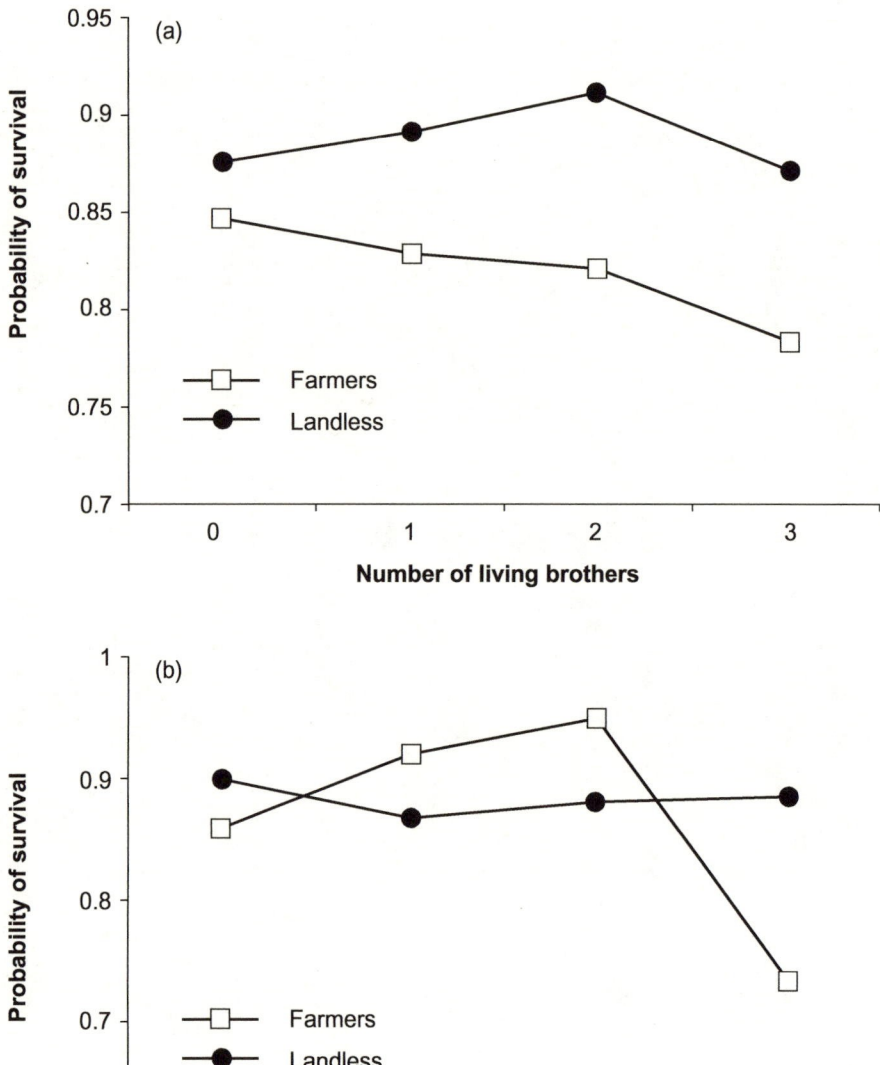

century Sweden: here, elder brothers were found to depress the reproductive success of their younger male siblings. Similarly, Voland and Dunbar (1995) found that male infant mortality among eighteenth–nineteenth-century Krummhörn farmers was strongly influenced by birth order: the greater the number of brothers in the family, the higher the mortality rate experienced by male infants (**Figure 8.5a**). An infant with 3 or more brothers had a 20 per cent chance of dying before his first birthday compared to 12.5 per cent for the population as a whole. A similar pattern was found in female mortality rates: with three of more living sisters, infant female mortality increased to approximately 27 per cent, compared to only 5–10 per cent with 1–2 living sisters (**Figure 8.5b**). By contrast, male and female mortality rates were unaffected by birth order among the landless labourers (the poorest section of the population who owned no heritable land).

Voland and Dunbar (1995) suggested that competition among siblings for parental resources was responsible for these patterns. The geographical location of Krummhörn (surrounded by the North Sea on three sides and by inaccessible moorlands on the fourth) meant that farmers with large families could not expand the area they occupied in ways that would allow them to provide an inheritance for all their sons. Dividing the existing landholding evenly between heirs was unfeasible, since after only a very few generations the original area would be subdivided into units that would no longer be economically viable. In order to sustain the size of the landholding through time and maximise the duration of tenure, the Krummhörn farmers operated a system of **ultimogeniture** in which the youngest son inherited all his father's land. Consequently, resource competition between siblings was strong and, as in the Gabbra, tended to have a large impact on reproductive opportunities. (Note that elsewhere in Germany during the same period, communities that could expand into virgin land did not restrict the numbers of sons they had, but rather tended to favour sons at the expense of daughters [Voland et al. 1997]. This suggests that the Krummhörn farmers' behaviour was a facultative response to their circumstances rather than a characteristic of contemporary German culture.) More generally, the practice of limiting land inheritance in order to avoid the dispersal of the family land holding in successive generations – even to the extent of preventing more than one son and daughter marrying in each generation (the so-called 'stem-family system') – was a widespread practice in northern Europe (Hajnal 1965). Indeed, this practice continued to be prevalent in rural areas of Ireland as late as the mid-twentieth century: there was, for example, a significant negative relationship between the number of celibate sons and the size of the farm during the first half of the twentieth century (Strassmann and Clarke 1998).

Whilst men and women of the landless class were equally likely to marry regardless of the number of brothers and sisters they had, a farmer's son's probability of marriage was negatively correlated with the number of surviving brothers. With three or more brothers, a man had only a 30 per cent chance of marrying compared to a 50 per cent chance among the landless (**Figure 8.6a**). This reduction in reproductive opportunities for farmers' sons high up the birth order seemed to increase their likelihood of emigrating away from the area. Farmers' sons had higher overall rates of emigration than the sons of the landless, but farmers' sons with three or more brothers show the highest rates of all. Similarly, three or more sisters drastically reduced the likelihood of marriage for farmers' daughters. Whilst an only surviving daughter was almost certain to find herself a spouse (a 72 per cent chance of getting married), the youngest of four daughters had less than a 60 per cent chance of getting hitched (**Figure 8.6b**).

An obvious interpretation of these data is that wealthy farmers were attempting to accumulate and concentrate wealth by limiting the number of heirs produced, rather than attempting to maximise the number of grandchildren. Although Krummhörn custom dictated that the youngest son inherited the family farm, he nonetheless had to compensate his brothers with modest financial settlements and provide his sisters with dowries, and it was sometimes necessary to sell off part of the farm to meet these obligations. Consequently, the more offspring the parents had, the more rapid was the dispersal of family wealth. Differential investment may therefore have been linked to economic rather than reproductive motivations. The Krummhörn thus seems to provide us with another clear example of the fact that humans often maximise their fitness by maximising wealth rather than simply maximising numbers of offspring

ULTIMOGENITURE

Figure 8.6 In the historical Krummhörn population, the likelihood of marriage declines with the number of same-sex siblings for both (a) sons and (b) daughters of farmers, but not for the offspring of landless labourers. It seems that farmers attempted to manage family size carefully in order to avoid having to split up their estates. Reproduced with permission from Voland and Dunbar (1995).

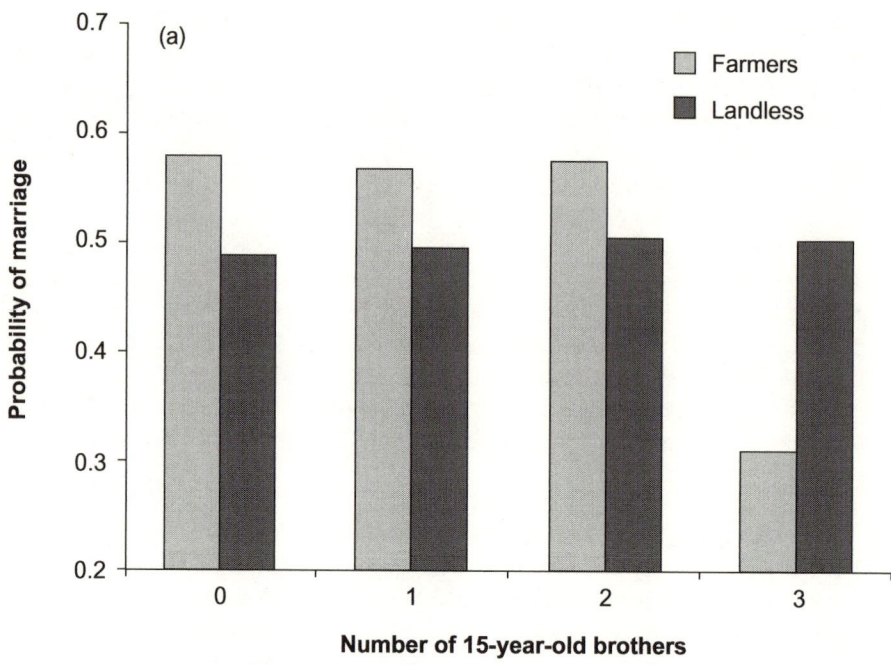

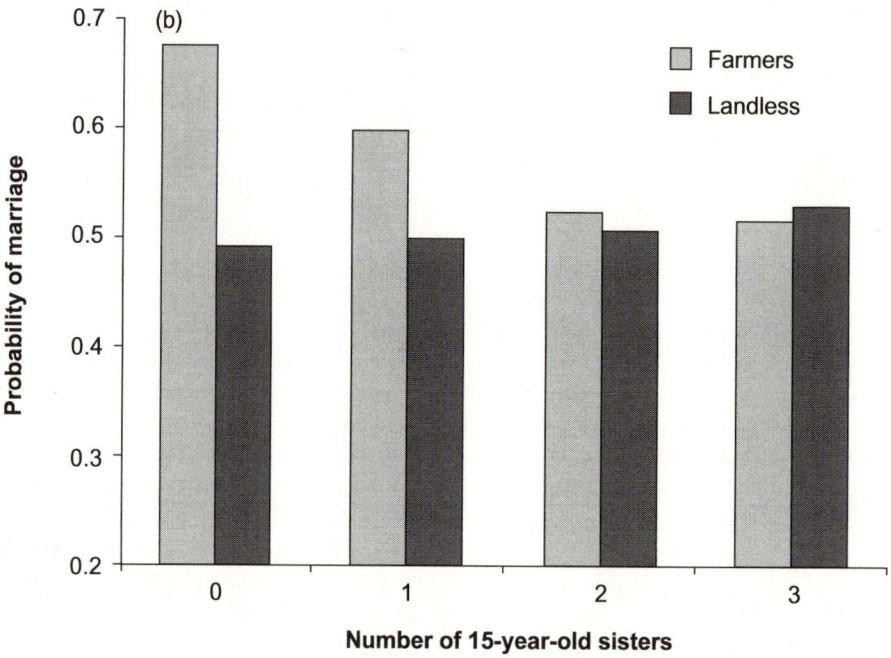

produced (see Chapter 6) and that lineage survival forms a crucial component of this strategy.

In line with this, Klindworth and Voland (1995) found that 'elite' Krummhörn men (the men with the highest absolute wealth) were no more likely to raise a son to adulthood than the non-elite population, and that up to a quarter of elite men did not succeed in raising a son at all (**Table 8.2**). However, only four out of 60 elite

Table 8.2 Probability of lineage extinction among elite and non-elite men of the Krummhörn (NW Germany) for marriages contracted between 1760–1810, using two measures of lineage extinction

Extinction criterion[a]	Elite men[b]	Non-elite men	P
At least one surviving 15-year-old offspring:			
son	0.233	0.260	NS
daughter	0.067	0.262	0.001
either sex	0.033	0.124	0.050
At least one married offspring:			
son	0.333	0.459	NS
daughter	0.133	0.415	0.0001
either sex	0.067	0.239	0.010
Sample size	60	954	

a) Probability that a man will leave at least one offspring on his death
b) Men who owned at least 25 grasen (or leased at least 50 grasen) of land (1 grasen = 0.37 ha), thus entitling them to vote and hold political or church office within the parish.

Source: Klindworth and Voland (1995)

men died without leaving behind at least one married daughter. In other words, elite men were more likely to increase their long-term fitness through the female rather than the male line, despite the operation of a patrilineal system of inheritance. The investment strategy shown by Krummhörn farmers would thus seem to be an attempt to get the best of both worlds: reducing the number of inheriting males ensured the concentration of wealth within the lineage, while increasing the number of female offspring who could marry into other land-owning families ensured the long-term fitness of the farmers by preventing lineage extinction. Since women inherited only half the amount that their brothers were given, increasing the production of daughters did not reduce family wealth quite so much, and conforms to Fisher's sex ratio principle (see Chapter 2) which states that parents should produce more of the cheaper sex in order to equalise investment.

Boone's (1988) study of the Portuguese nobility during the fifteenth and sixteenth centuries provides a particularly nice example of the interplay between cultural rules and reproductive behaviour. Portuguese noble lineages were organised around a series of primary lineages of ducal status and their associated lower ranking lineages which were tied to a smaller estate or lower titles. Boone (1988) compared high- and low-ranking lineages to investigate patterns of differential parental investment so as to assess the influence these patterns had on lineage survival through time. The men of high-ranking lineages tended to have higher reproductive success than those of low-ranking lineages since they were more likely to have more illegitimate children, more likely to marry and, most importantly, were more likely to have more than one marriage during their lives. By contrast, there was little difference between women when they were compared on the basis of their natal status. However, when women were grouped according to the status of their husbands, those with high-ranking husbands had higher reproductive success.

Conversely, investment in female offspring was much more pronounced among the lower nobility than among the primary titled lineages. In a manner similar to that observed in the Indian caste system (see p. 188), high-ranking females faced stiff competition for husbands since they had a restricted set of high-ranking men available to choose from, whereas low-ranking women could marry up the social hierarchy as well as within their own class. Excess (that is, unmarriageable) daughters thus accumulated among the primary lineages and these were often cloistered in convents. These women were known as 'Brides of Christ' and were quite explicitly distinguished from nuns with 'proper' vocations; indeed, unlike conventional nuns, they could withdraw from the contemplative life and return to society if their services as more conventional brides were required (Duby 1997, E. Hill 1999). In effect, nunneries became storehouses where excess (usually younger) daughters could be kept secure and chaste relatively cheaply.

While excess daughters were sent to nunneries, the surplus sons were often sent off to war. Boone (1988) has shown that a son's probability of dying in warfare increased with birth rank. The family could not afford to provide these males with estates capable of attracting a marriage partner of high status; later-born sons therefore had little choice but to opt for 'a best-of-a-bad-job' strategy in the hope that, by accumulating spoils of war, they could eventually return with sufficient wealth to attract a wife.

Boone (1988) has pointed out that, during this period, the Portuguese nobility began to face the same kind of problem encountered by the Krummhörn farmers. During the early medieval period, they had practised a form of partible inheritance in which all sons shared in the inheritance of the family estate. However, once the conquest of the Arab-dominated southern parts of the Iberian peninsula had been completed (around the middle of the thirteenth century), it was no longer possible for the nobility to expand their estates by conquest and sequestration. As a result, the subsequent division of estates in successive generations began to threaten lineages' economic and political power (and hence survival). Primogeniture (inheritance by the eldest son) was seen as the answer to this problem and rapidly became the norm after about 1250.

A similar shift to patriliny and primogeniture had been adopted about two centuries earlier in France and England, for essentially the same reasons (Howell 1976). In central France, primogeniture was implicit rather than explicit, since all sons formally inherited an equal share of their father's estate; however, only the oldest son was allowed to marry and produce legitimate heirs who then inherited their uncles' shares of the family estate, so that the inheritance system was effectively a rather complex form of primogeniture (Duby 1977). A similar practice occurred among the Venetian nobility in the early modern period: in many cases, only one brother married while the others typically joined the Church (or remained celibate) and in due course left their inheritances to their nephews (Cooper 1976). That partitioned estates ultimately resulted in the downfall of wealthy families is amply attested to in the European medieval and early modern historical record (Ladurie 1976, Thirsk 1976, Cooper 1976). It thus seems that this is yet another case of subtle parental manipulation of offsprings' life opportunities in order to maximise the parents' own fitness.

Obviously, with a system of primogeniture, birth order effects become exaggerated. As we saw in the case of the Gabbra, being a first-born son can have a highly significant influence on a man's reproductive success even in a system where all sons eventually inherit. Among the Portuguese nobility, birth order also had a large effect on patterns of migration and mortality (**Figure 8.7**). Younger sons were more likely to

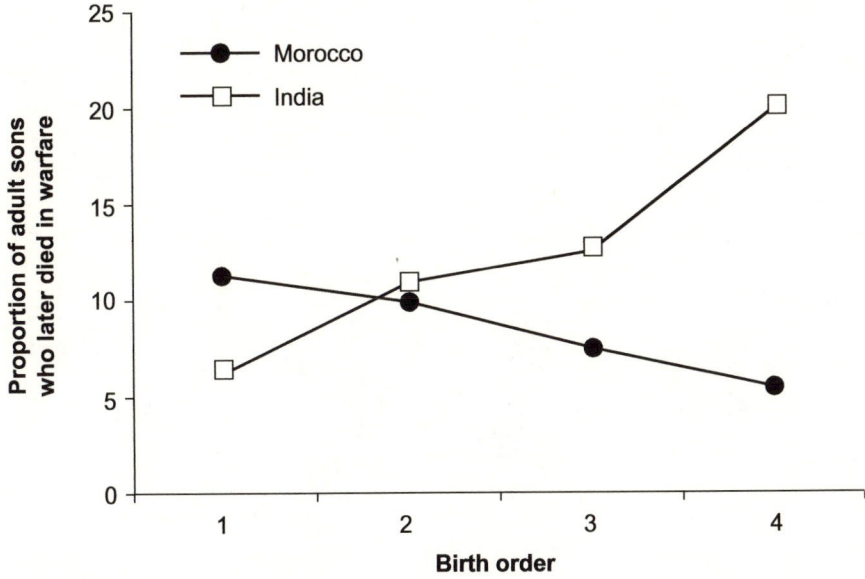

Figure 8.7 Among the fourteenth–fifteenth-century Portuguese nobility, the probability of a son dying on a colonial expedition far from home (for example, India) increased with birth rank. Early-born sons typically died nearer home (for example, Morrocco). Redrawn from Boone (1988).

be sent away to make war in distant lands, and were more likely to die there as a consequence. Older sons who were due to inherit the family estate tended to participate only briefly in military campaigns in the Iberian peninsula and its immediate environs prior to establishing themselves on the family estates in Portugal (where, of course, they subsequently died). These career differences obviously affected the probability of marriage and, as one would expect, resulted in highly significant differences in reproductive success according to birth order (**Figure 8.8**).

Birth order also had a negative effect on the reproductive success of women. As in the Krummhörn, the older daughters of the Portuguese nobility achieved higher reproductive success than their younger siblings. Boone (1988) suggested that this reflects the level of dowry competition experienced in the population, with families often able to afford dowries only for the oldest daughter. Dowry competition itself was another consequence of the switch to primogeniture, since this drastically reduced the number of eligible bachelors and substantially increased the 'market value' of the few who were available. Dowries became inducements to persuade the most sought after families to take on daughters as wives for their sons. Indeed, dowries entered into something of an arms' race during the early modern period throughout much of southern Europe, to the point where both the Church and the various monarchies became concerned about the ruinous consequences for high-ranking families (Cooper 1976).

The strategies employed by the ruling elite to ensure lineage survival created intense competition between families at a societal level. The period covered by Boone's (1988) sample represents the point where Portuguese colonial expansion reached its height, and Boone has argued that the two phenomena are causally related. Over this 200-year period, the proportion of war deaths among the nobility increased from 40–75 per cent as increasing numbers of men were forced to join the military because

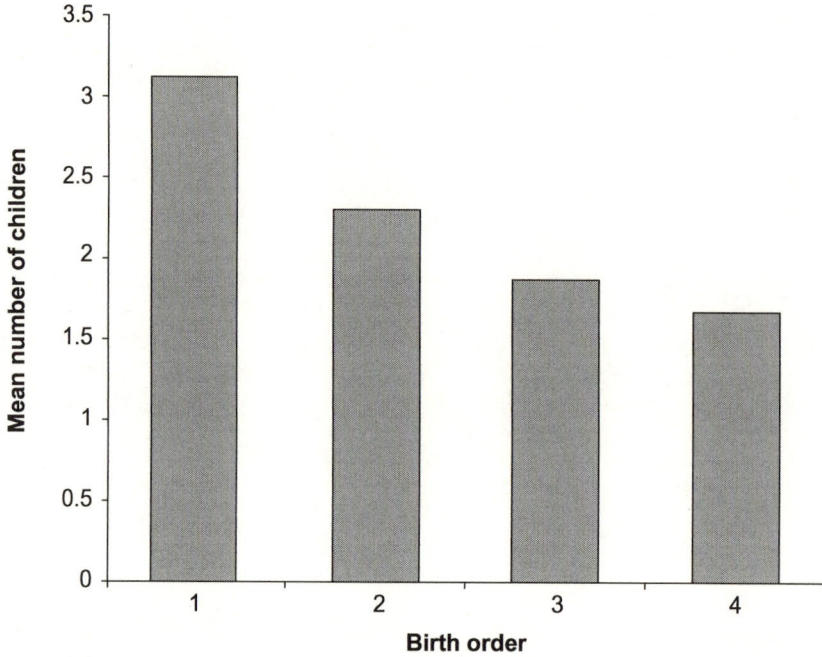

Figure 8.8 Reproductive success declined with birth rank among the sons of the late medieval Portuguese nobility, partly because of an increased likelihood of being killed on campaign and partly because of reduced chances of marriage. Source: Boone (1988).

they could not inherit. (Up to 25 per cent of daughters were placed in convents as a direct consequence of the combined effects of increased competition for titles by males and the reduction in the number of marriageable men.) Boone (1988) has argued that the pressure created by a growing pool of disaffected landless younger sons was directly responsible for the initiation of the Portuguese explorations and the resulting colonial episode – with the explicit encouragement of the monarchy who saw this as a useful way of diffusing a growing social problem.

Primogeniture and ultimogeniture are enormously successful as systems for maintaining an individual family's fortune and stabilising the process of succession, but they tend to have a negative impact on individual reproductive success because of the competition engendered at a societal level. This means that, somewhat counter-intuitively from an evolutionary point of view, culturally enforced rules which have at base a biological origin – reproductive decision-making and parental investment – ultimately serve to reduce some individual's reproductive output. This may seem maladaptive (and indeed there is no reason why cultural processes should have to be adaptive: see Chapter 13), but it is very important to remember that within these culturally-imposed constraints, individuals are still attempting to optimise fitness by getting the best return they can on their biological investment under current circumstances.

A rather nice example of this is provided by the French nobility in the mid nineteenth century. In 1809, Napolean Bonaparte promulgated the *Code Civil*, one of whose clauses banned primogeniture. Napolean had hoped that, by forcing the old French aristocracy to divide their estates equally amongst all their children, their economic (and hence political) power would soon be broken. However, he reckoned without human ingenuity and the instinctive urge to ensure lineage survival: within two generations, the French nobility had reduced family size to two and instituted the practice of marrying cousins. This ensured that, although the actual land units changed from one generation to the next, the size of the land holding did not.

Boone's (1988) study makes it particularly clear that cultural practices impinge on the biological decisions made by individuals in ways that are not always open to simple-minded evolutionary analysis. Culture cannot be disregarded as an explanatory variable within the biological domain because it is an integral part of that domain. The cultural rules and regulatory systems that exist in a society have been created by individuals motivated to do so for specific reasons. Whether these reasons are ultimately adaptive is another question, but it remains the case that the capacity to create systems of rules and laws regulating individual behaviour stems from the ability of humans to decide on a particular course of action given the potential costs and benefits, and that ability is, at base, biological in origin.

KEEPING IT IN THE FAMILY: INCEST AND MARRIAGE RULES

Until recently, most authorities believed that the existence of the 'incest taboo' was the mechanism by which potentially deleterious inbreeding was prevented in human populations. With Freudian analysis at its height during the early decades of the twentieth century, it was generally supposed that close kin were frequently desired as mating partners and that, without a specific injunction against it, incest would be rife. Incest taboos were therefore necessary to 'keep all humans from ruining themselves'.

Thornhill (1991) argued that, in fact, such taboos ought to be unnecessary because humans already possess a built-in psychological mechanism (the Westermarck effect: see **Box 8.4**) to promote incest avoidance. Yet, despite this, 44 per cent of societies have taboos that specifically prohibit sex between close kin, a value that, as Betzig (1991) points out, is rather low from a Freudian standpoint yet a little too high from an evolutionary one.

Thornhill (1991) suggested that the solution to this apparent anomaly lay in the fact that, in reality, incest taboos have very little to do with sex between close relatives and everything to do with protecting property, paternity and status. Using a cross-cultural sample of traditional societies from all of the world's main geographical regions, she identified three sets of incest rules. As well as true incest rules regulating sexual relations between close kin, Thornhill found another set concerning distant kin (those with $r < 0.25$) which she suggested represent an attempt to regulate wealth concentration and lineage survival since marrying within the same extended family ensures that wealth remains within a particular lineage and can increase that family's power base. The third set of rules were concerned with the relations of **affinal kin** (for example, a woman and her father-in-law) and may more properly be viewed as designed to prevent a particular form of adultery. These, Thornhill argued, represent attempts to protect male paternity and prevent the disruption of within-family alliances by avoiding sexual tensions.

AFFINAL KIN

Thornhill (1991) found that rules prohibiting sexual relations between affinal kin differed between **patrilocal** societies (where a woman lives with her husband and his relatives on marriage) and **matrilocal** societies (where the woman and her husband live with her relatives). Rules were much more extensive in patrilocal societies, where men tend to leave their wealth to their wives' sons, and where paternity certainty is

PATRILOCAL
MATRILOCAL

BOX 8.4

The Westermarck effect

The Westermarck effect refers to an observed lack of sexual interest between two individuals who were raised together as children (Westermarck 1891). It appears to be a psychological mechanism triggered by living in close physical proximity during a critical period during the first five years of life. Many animal species show equivalent effects to those seen in humans (Bischof 1972, Kuester et al. 1994). In the majority of cases, individuals raised together will be close kin, so the effect of this mechanism is to lessen the chance of potentially deleterious close inbreeding (incest) between individuals. This seems to be a very old human adaptation that probably extends back into our more distant mammalian past.

Several sources of evidence provide support for the proposed mechanism. Shepher (1971), for example, reported that children from Israeli *kibbutzim* who had been raised communally in nursery groups rather than with their biological parents in their natal homes preferred to marry outside the kibbutz rather than marry each other (as their parents had intended). Bevc and Silverman (2000) reported that, in a sample of 170 Canadians, prolonged childhood separation was significantly more likely to be associated with incestuous post-childhood sexual activity involv-

ing genital intercourse than was the case for children who grew up together.

However, the most detailed study of the Westermarck effect to date has been of so-called 'minor marriages' in Taiwan (Wolf and Huang 1980, Wolf 1995). Minor marriages are arranged marriages in which mothers give their daughter, shortly after birth, to other families, who then raise these baby girls as wives for their young sons. Because the future bride and bridegroom are brought up in intimate contact with each other, the subsequent marriage is not a great success, reproductively speaking. Minor marriages have a fertility that is 31 per cent lower than marriages between individuals that were not raised in intimate contact, and a divorce rate that is three times higher (Wolf and Huang 1980).

Using carefully controlled statistical analyses, Wolf and Huang (1980) were able to show that these differences were not the result of the effects of adoption, the health of adopted daughters-in-law, age at marriage, poverty or the low prestige of such marriages. The only factor that could explain the difference was the Westermarck effect producing a maladaptive outcome among individuals in these marriages (see also Irons 1998).

thus likely to be of greater concern for men. In addition, because patrilocal residence involves groups of related men living together with their associated wives, this form of residence arrangement presents men with greater opportunities for gaining sexual access to other men's wives than do matrilocal societies. Covert sexual relations between individuals could therefore be relatively easy to achieve. In line with this, patrilocal societies were thus much more likely to extend incest taboos to include affinal (that is, non-biologically related) kin than were matrilocal societies (**Figure 8.9a**); moreover, the punishment for infringing these rules was much harsher in patrilocal societies (**Figure 8.9b**).

Rules prohibiting marriage between distant kin were also more extensive in societies with a high degree of social stratification. In such societies, it is not in the interests of the ruling elite (that is, those at the top of the social hierarchy) to allow marriages to take place within lineages because this allows lineages to concentrate wealth and power in a way that may ultimately threaten the status of the elite. Rules prohibiting inbreeding were therefore both more common and involved much harsher penalties in highly stratified societies than in non-stratified societies. The diversity in punishment was also very telling: whereas men of the unstratified Trumai society were

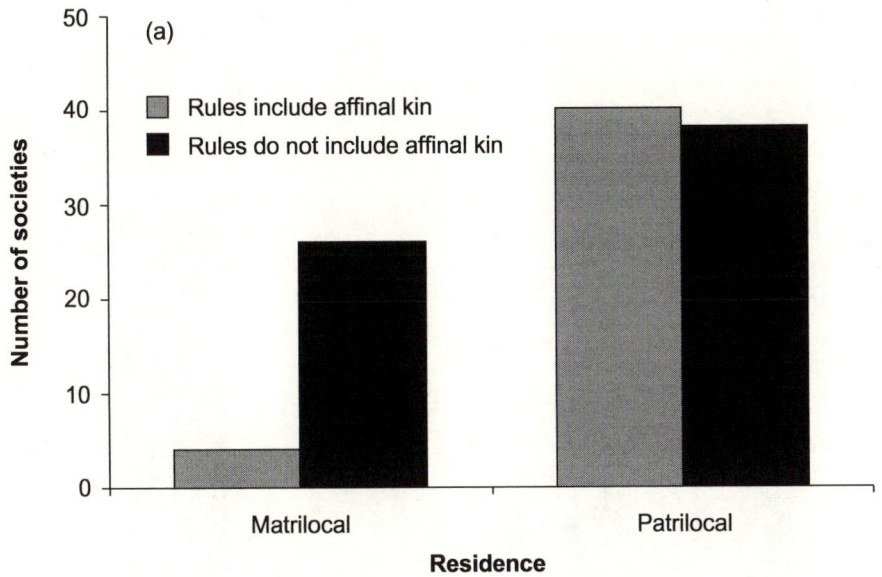

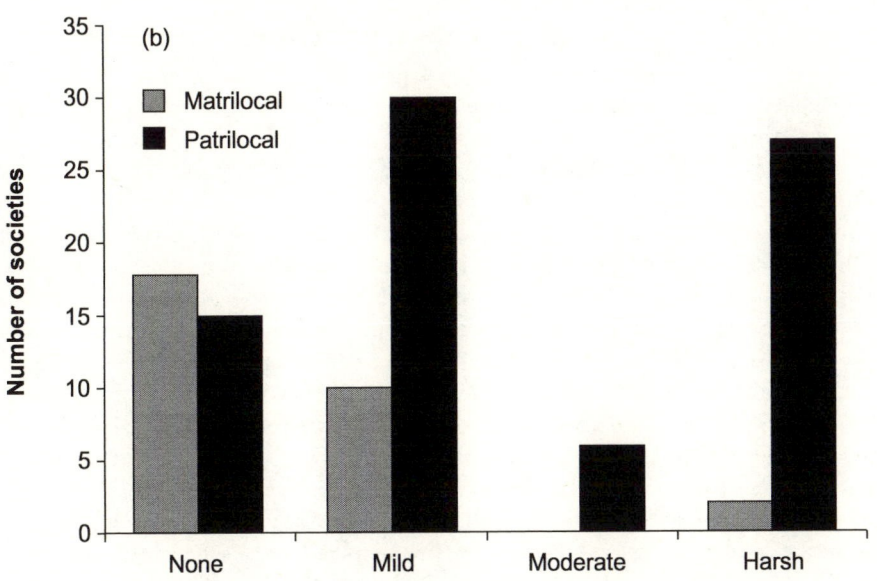

Figure 8.9
Prohibitions against marrying close relatives are (a) more likely to include affinal (that is, genetically unrelated) kin and (b) the punishments for infringements are likely to be harsher in patrilocal societies than in matrilocal ones. Source: Thornhill (1991).

merely 'frowned upon' if they married their sister's daughter (the only prohibited relationship in Trumai society), Inca men had their eyes gouged out if they engaged in sexual relationships with their aunts, nieces, cousins, godmothers or any other female 'relative'.

As one might expect, these rules often didn't apply to those who made and enforced them. The ruling elite in society were more concerned than anyone about preserving lineage wealth and power and, as such, were highly prone to break the rules in order to further their own interests, safe in the knowledge that there was no risk of punishment. Thus, among the more highly stratified societies, inbreeding rules

Figure 8.10 The extent to which inbreeding rules were equitably applied varied as a function of the level of stratification in a given society. With increasing social stratification, inbreeding rules are applied less equitably. Redrawn from Thornhill (1991).

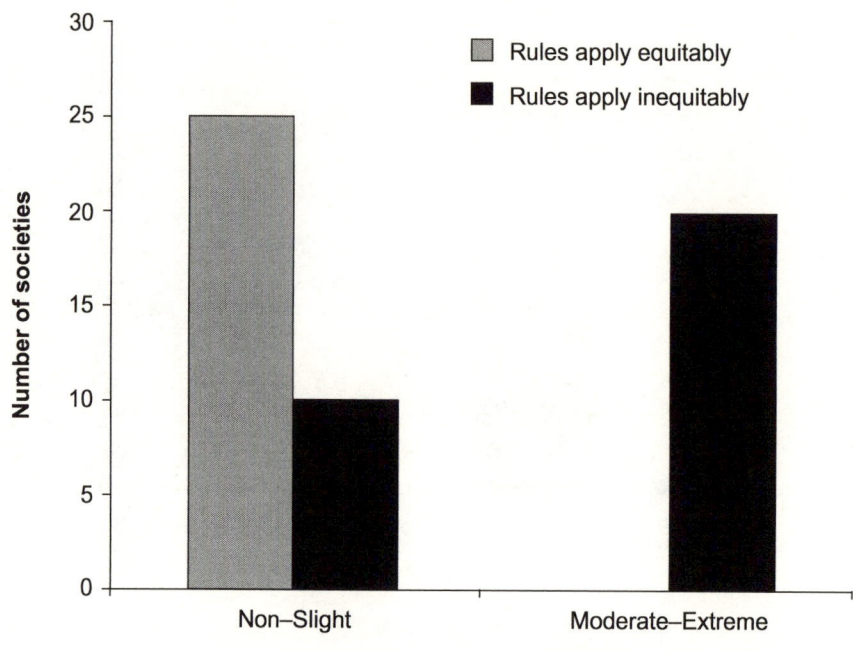

were applied less than equitably (**Figure 8.10**). Among the Incas and Ancient Egyptians, for example, royalty could even go so far as to marry their own siblings (although it's important to remember that there may have been a difference between royal sibling marriage and sibling sexual relations).

Taken together, these findings highlight the fact that, when individuals are constrained to behave in certain ways by the existence of certain rules, it is not society itself which somehow imposes these constraints but rather individual members of that society. This means that such constraints will only be tolerated if there are at least some individuals who perceive them to be in their long-term interests. If individuals have the ability to break free from these constraints, they generally do so, regardless of what 'society' has to say about such behaviour (see also Chapter 13). The flouting of incest rules by the Inca royal family may be one of the more extreme examples of this, but there are many others, like the changes observed in Kipsigis bridewealth payments over time (see Chapter 5) or the Krummhörn farmers forcing their daughter's husbands to take on their family name in order to ensure lineage survival (Ohling 1963).

▎TIBETAN POLYANDRY

POLYANDRY

Another striking example of the way in which cultural rules help serve the interests of lineage survival concerns polyandrous systems of marriage (those in which a single woman has more than one husband at the same time). Although found in a number of areas throughout the world, **polyandry** is rather uncommon among humans because of the fundamental differences between male and female reproductive potential. Unlike polygyny, where a man can make huge gains by having lots of wives who have to share him, it doesn't make good reproductive sense for a man to share his wife since women cannot increase their birth rates proportionally by having additional spouses.

BOX 8.5

Incest and exogamy

In most of the anthropological literature, rules of incest and rules governing exogamy (marrying outside one's natal group) seem to be confounded: the latter are often assumed to be equivalent to the former, and are presumed to be designed for the same reason (to reduce inbreeding). Lévi-Strauss (1969), for example, argued that once culturally imposed incest avoidance rules had been established, exogamous marriage could then be used as a means of forging alliances between groups via the exchange of women, leading to the complex cooperative societies that characterise humans as we know them. The cultural imposition of incest taboos was therefore regarded as the key feature that helped raise humans above other animals in terms of their cultural and social complexity.

However, van den Berghe (1980) pointed out that this can't be quite right, for while incest taboos inevitably reduce the level of inbreeding, exogamous marriage practices very frequently do the opposite. For example, there are literally hundreds of marriage systems that have rules regulating cross-cousin marriage (of which the Yanomamö Indians of Venezuela are one: see Chapter 9). Under these rules, individuals are allowed to marry certain types of cousin (for example, their mother's brother's daughter or their father's sister's son) but not others (their father's

brother's daughter or mother's sister's son). As should be apparent, such arbitrary rules have little to do with incest: your mother's brother's daughter is just as closely related to you as your father's brother's daughter, and the level of inbreeding produced would be equivalent. Consequently, the net effect of these rules in small-scale societies is to increase the level of inbreeding by encouraging high levels of first and second cousin marriage.

Van den Berghe (1980) argued that the need to form valuable and strong alliances between lineages would provide sufficient reason for the development of particular marriage rules, even in the absence of specific incest taboos. Since exchanging women between clans is an efficient means of ensuring that long-term reciprocal and trustworthy alliances can be formed, all that is needed is for a rule to be established that prohibits a man from marrying a woman from the same side of his family. As Lévi-Strauss pointed out, the simplest way to get the system going is to have a rule specifying that only cross-cousin marriages are acceptable. This has the dual function of keeping the clan fairly tight-knit by ensuring that the individuals involved are at least distantly related, but also allows for the formation of bonds between the two distinct lineages which makes for further increases in levels of harmony and cooperation.

Although there is, as yet, no definitive explanation for the evolution of polyandry, there is at least some agreement that polyandrous systems tend to represent a unique solution to a particular type of environmental challenge. Most evolutionary biologists contend that polyandry is practised when men are limited in their ability to support wives and offspring by severe ecological and/or economic constraints.

Polyandry is particularly common among the Tibetans where families typically live on large agricultural farms of low productivity which simultaneously have a high demand for labour (Crook and Crook 1988). The poor productivity of this high altitude habitat means that conservation of the family farm over time is a crucial strategy if lineage fitness is to be maximised, since dissipation of the farm among inheriting sons in future generations would soon whittle it down to a point where economic survival (and hence reproduction) would become impossible. The Tibetans' solution to this problem was fraternal polyandry: a set of brothers would marry the same spouse and raise children as a single family under the same roof. This limited the number of individuals produced by a family since there was only one marriage and one reproducing female per generation (the '**monomarital principle**': see Goldstein 1976) while, at the same

MONOMARITAL
PRINCIPLE

time, ensuring an adequate and cooperative workforce. This effect was further enhanced by the fact that marriage opportunities for daughters tended to be limited under such a system, with the result that daughters often remained unmarried and lived on the farm providing additional (often virtually slave) labour.

Goldstein (1976) proposed that polyandry represented a 'homeostatic' adjustment of group size to resources under conditions when economic opportunities other than agriculture were limited. The 'latent function' of the system – restraint of population growth to ensure that people did not outstrip their resources – was achieved through the conscious individual goals of avoiding division of land and concentrating wealth within families. Notice that this need not be a group-selectionist argument if individual goals to ensure that lineage survival and prosperity lie at the heart of it rather than acts that function for the good of the group as a whole.

Goldstein (1976) found that, under polyandry, 31 per cent of potentially reproductive women did not bear children, and calculated that a switch to monogamy for the whole population would result in a 16 per cent increase in population size. So, as in the case of the Portuguese nobility and Krummhörn farmers, the established marriage and inheritance system tended to reduce the reproductive success of the population in general, but represents an optimal strategy given the constraints imposed by the local economy. However, unlike the Portuguese, where concerns for lineage survival increased levels of competition between and within families, the constraints placed on Tibetan farmers were more strongly ecological in origin. Switching to a monogamous system would have been unfeasible because the attendant risks of land shortage, over-population and individual survival were more intrusive in a habitat of low carrying capacity.

Polyandry therefore functions in the same way as the extreme forms of primogeniture found among the Portuguese nobility: it limits the number of marriages that take place per generation, and therefore limits the number of individuals who can put strain on the system. Unlike the Portuguese, however, non-inheriting sons are not a burden, but are essential for the continued existence of the farm and therefore the family. Consequently, these individuals must be encouraged to stay, and it seems that this was achieved by permitting them to marry the eldest brother's wife, thus giving them a stake in the future of the farm through the children born into the family.

Although these concerns for lineage survival and farm prosperity would have been enough to maintain the system and account for the success of a polyandrous strategy, Crook and Crook (1988) were in fact able to show, from long-term ethnographic data for their study area in Ladakh in the Himalayas, that females in the grandparental generation gained significant fitness benefits through polyandry compared to their monogamous counterparts (**Figure 8.11**). More importantly, Crook and Crook (1988) calculated that, since the mean family size of all monogamous marriages was 3.1 children, polyandrous men would have to produce 4.4 offspring in total in order to achieve the same inclusive fitness as they would have done in a monogamous marriage. Since the mean completed family size of polyandrous households was 5.2 offspring, they concluded that males did not suffer any fitness losses from polyandry and may even have made a marginal gain.

Recently, however, Smith (1998) has questioned some of the assumptions in these calculations. Using the same approach as Crook and Crook (1988), he calculated the family size that would be needed to give the senior brother in a polyandrous marriage the same inclusive fitness as he would have experienced had he been the sole monogamous heir. According to Smith's (1998) calculations, which dealt with all forms of polyandry separately rather than averaging across them as the Crooks' had done, the senior husband within a diandrous (two-husband) marriage was actually doing

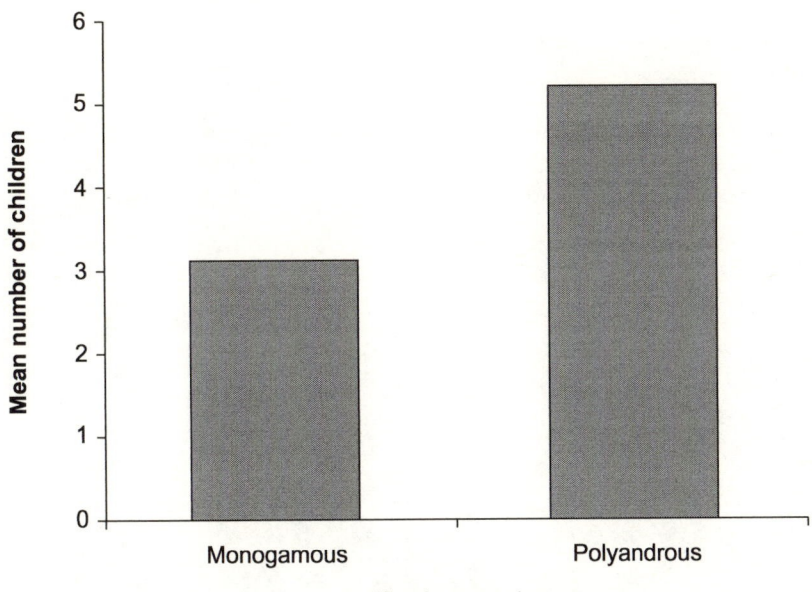

8.11 Completed family size for monogamous and polyandrous households in the Tibetan population in Ladhak (India). Source: Crook and Crook (1988).

worse and sacrificing inclusive fitness (assuming that paternity was equal across brothers) compared to a monogamous strategy.

Furthermore, when Smith (1988) relaxed the Crooks' rather stringent assumption that junior brothers who did not join a polyandrous marriage failed to produce any offspring at all (an assumption that produced the marginal fitness gain of polyandry over monogamy in the Crooks' analysis), he was able to show that senior brothers were sacrificing their inclusive fitness under almost all levels of polyandry found in the Crooks' original data set. Only tetrandous marriages (five-husband) produced more children through polyandry than a senior brother could have produced under monogamy (although this was based on a sample of only one household).

The most interesting part of Smith's analyses, however, lies in his recognition that the costs and benefits of polyandry as a reproductive (rather than lineage survival) strategy depend critically on each individual's standing within the family. For example, the Crooks' calculations were based on the notion that a man either became polyandrous or was the sole heir of a farm. In reality, however, only the eldest of a set of brothers would be able to achieve this. All the others would be choosing between polyandry and an option that was unlikely to yield the same reproductive rewards as inheriting the family farm.

Smith (1998) used a modelling approach known as the 'member–joiner' game (see **Box 8.6**) to see how the value of polyandry could vary between individuals. Essentially, a member–joiner game calculates (1) the benefits to an individual of joining a group of a particular size as opposed to pursuing an alternative strategy and (2) the costs to those individuals who are already members of the group of adding another person. To keep things relatively simple, he dealt only with the interests of brothers and their parents, and did not consider the pay-off to women or grandparents.

Using the observed values obtained from the Crooks' data set, Smith (1998) assumed that the opportunity cost of polyandry versus monogamy for the elder brother would be 3.75 offspring (which is the number of offspring they would potentially sacrifice by allowing a brother to join them in a polyandrous marriage). If we further assume that the junior brother sacrifices 0.54 offspring by joining a polyandrous marriage (this being

BOX 8.6

The member–joiner game

In order to calculate the reproductive advantages of polyandry, Smith (1998) assumed that men prefer marital arrangements that increase their inclusive fitness. He then used Hamilton's Rule (see Chapter 2) to calculate the effect that a polyandrous marriage would have for individuals' inclusive fitness. Hamilton's Rule states that a net gain in inclusive fitness is produced if the effect on kin, once devalued by the degree of relatedness, is greater than the effect on the individual.

Smith (1998) then laid out his model as follows: in the simplest case of two brothers, the senior brother can expect m offspring if he marries monogamously, while his younger brother will average e offspring if he is excluded from the polyandrous marriage. If the two brothers share a wife and jointly inherit the farm, they will jointly father p offspring. Assuming that polyandrous co-husbands have equal paternity, the optimal rule for an existing group member (senior brother) is to include his younger brother as a co-husband as long as:

$$\tfrac{1}{2}\,(p + rp) > m + re$$

where r is the co-efficient of relatedness. In other words, he should tolerate an additional brother in the marriage if the inclusive fitness gains are greater within the polyandrous marriage than the gains he would make if he married monogamously and his brother had to find a wife of his own.

For a joiner, the optimal preference is that he should join if:

$$\tfrac{1}{2}\,(p + rp) > e + rm$$

which is just a mirror image of the member's rule. That is, he should only join a polyandrous marriage if both his direct fitness gains from monogamy and the inclusive fitness gains via his brother are lower than the gains that can be expected within the polyandrous marriage.

It is possible to extend this analysis to take account of any number of brothers by adding two more variables: n for the total number of brothers and h for the number who are co-husbands,

where $h \leq n$. This gives what Smith calls a 'slightly messier' form of the formula for members:

$$(p_h/h) + [(rp_h/h)(h - 1)] + re(n - h) > [p_{h-1}/(h - 1)] + [rp_{h-1}/(h - 1)]\,(h - 2) + re(n - h + 1)$$

The left-hand side of the inequality gives the fitness gain to an existing member (that is, the oldest brother) if that h^{th} brother is included in the marriage, while the right-hand side gives his fitness gains if that brother is excluded. On the left, the three terms correspond to the following (from left to right):

1. The expected number of children fathered by the member (if the hth brother is included as co-husband);
2. The expected number of children fathered by the $(h - 1)$ co-husbands, devalued by r;
3. The expected number of children fathered by the $(n - h)$ brothers who are excluded from the polyandrous marriage and marry monogamously or remain bachelors, also devalued by r.

On the right-hand side, the terms refer to:

1. The expected number of the member's children if the hth brother is excluded;
2. The expected number of children fathered by $(h - 2)$ co-husbands, devalued by r;
3. The expected number of children fathered by $(n - h + 1)$ brothers who marry monogamously or remain bachelors, devalued by r.

The rule for brothers who are potential joiners of a polyandrous marriage is that they should join if:

$$(p_h/h) + [(rp_h/h)(h - 1)] + re(n - h) > e + rp_{h-1} + re(n - h)$$

As above, the terms on each side of the equation refer to the fitness gains made through the potential joiner's own reproduction, that of his polyandrous brothers and that of his monogamous brothers.

the average number of offspring that non-polyandrous junior brothers managed to produce), then the oldest brother (as an existing member) would lose inclusive fitness from polyandry compared to monogamy for all cases except tetrandry (although we've already explained why this may not be a robust finding). For younger brothers (as joiners), by contrast, polyandry resulted in a gain in fitness at all levels of polyandry. These findings are not unexpected, given Smith's other analyses, but he considered them to be puzzling in the sense that, within polyandrous societies, elder brothers actively encourage their younger sibs to join them in marriage and it is the younger brothers who tend to leave polyandrous marriages (see Levine and Silk 1997, and below).

Smith (1998) offers two possible explanations for this apparent contradiction. First, he suggested that perhaps the fitness loss to members was exceeded by the fitness gain to joiners so that, on average, a cohort of brothers would be reproductively more successful under polyandry than otherwise. In other words, it is the fraternal cohort's parents who benefit most in inclusive fitness terms, even though individual brothers may experience a fitness reduction. In support of this, Smith (1998) found that polyandry produced a greater number of grandchildren for each cohort size of brothers than did a strategy whereby brothers were excluded from the marriage. However, this doesn't explain why elder brothers should be actively encouraging polyandry, when it is their parents that benefit most.

Smith's alternative explanation was to accept the Crooks' original suggestion that elder brothers tended to father a disproportionate number of children within polyandrous marriages, and hypothesize that this paternity skew increased the value of polyandry to these individuals. According to tradition in Ladakh, paternity was ascribed equally to all co-husbands within a polyandrous marriage. In reality, there was often a wide disparity in age between the eldest and the youngest brother at the time of marriage, with the latter often being no more than a child and, for obvious reasons, unlikely to engage in sexual relations with the wife. It therefore seems likely that the eldest brothers achieved much higher fitness than their siblings. Furthermore, Crook and Crook (1988) noted that the odds were often further increased in favour of the oldest brother because the second oldest was usually sent to a monastery. The most frequent birth order position among the monks in two of the monasteries in the Crooks' study area was second place. It therefore seems likely that the eldest brother fathered most offspring and that the tradition of ascribing equal paternity to all the husbands was a device to persuade the younger brothers that they had an equal stake in the future and therefore should remain working on the farm (compare Daly and Wilson's [1982] study on the family resemblance of newborns which we discussed in section 7.3). Smith (1998) calculated that a skew of 69 per cent (based on the data in Levine and Silk 1997) could make polyandry advantageous for elder brothers, and that even a paternity skew as high as 98 per cent would still make polyandry worthwhile for younger brothers, given their otherwise limited reproductive options.

However, one option that Smith (1998) doesn't consider is that current reproductive strategies are of less importance than long-term lineage survival. The prime mover here is the need to preserve the integrity, and therefore the economic viability, of the farm under harsh ecological conditions, and not high offspring output. Most analyses which consider the fitness outcomes of polyandry tend to forget that the *raison d'être* of polyandry is to limit the number of heirs produced in the interests of long-term farm and lineage survival.

In the conventional literature, polyandry is considered to be something of a puzzle, something that makes no sense unless it can be shown to have positive reproductive consequences for all involved. By now, however, it should be obvious that, when individuals face competition for resources and are limited by the constraints of their particular environment, strategies aimed at enhancing long-term fitness are rarely associated with high reproductive success. Consequently, elder brothers may have had strong economic, rather than reproductive, motivations to encourage their younger brothers to join their marriages and remain on their farms, and these economic motivations may ultimately have had significant long-term fitness benefits (see Chapter 4; Rogers 1990).

This brings in another important point which is that, if polyandry is a means of ensuring long-term survival of families and their farms, then, strictly speaking, analyses investigating the success of polyandry should model the long-term economic and lineage survival consequences of families that pursue a polyandrous strategy against families who pursue a monogamous strategy, not the fitness of individuals within one particular family type. Attempting to explain the apparent puzzle of polyandry by showing that it could have fitness benefits for all the brothers involved, younger as well as older, somewhat obscures the point that such individuals are constrained into the system by their relatives' and their own concerns for lineage survival and wealth concentration.

When viewed from this perspective, it becomes immediately apparent that younger brothers in polyandrous systems are actually in a much better position than younger sons among the Portuguese nobility or the Krummhörn farmers who were shipped off to the army or never married at all. As Mace (1997) points out, they face similar constraints on their reproductive success to the younger sons of Gabbra pastoralists: among the Gabbra, the lower down the birth order a son was, the lower his reproductive success (see pp. 211–12). A younger brother in a fraternal polyandrous marriage may do worse than his older brothers or a monogamous man, but that does not make polyandry any more 'puzzling' from a reproductive perspective than the Gabbra system of inheritance or the extreme primogeniture practised by the Portuguese nobility. Younger brothers do worse in *all* these systems. The fact that a polyandrous man may gain some direct fitness benefits from his marriage (and will definitely gain some inclusive fitness benefits) actually makes polyandry seem much more sensible than the strictly monogamous systems operating in Portugal and eighteenth-century Germany which prevented younger sons from gaining any fitness benefits at all.

The notion that men are forced into polyandrous marriages by the ecological constraints imposed on the system is supported by the fact that, in the Crooks' study area, the advent of improved employment opportunities after the 1960s resulted in a substantial reduction in the frequency of polyandrous marriages. As soon as younger brothers were able to become economically independent (for example, though government employment or the tourist trade), they abandoned their polyandrous way of life to make their own independent monogamous marriages (although they often continued to live on the family farm). Even though polyandry was firmly entrenched culturally – and indeed, was highly valued by those living under that system – there was little hesitation over changing as soon as the opportunity arose.

Levine and Silk (1997), however, have argued against such an interpretation for the breakdown of polyandrous marriages, at least among the Nyinba, another polyandrous Tibetan population living in Nepal. They predicted that if the dissolution of a household was dependent on economic opportunities, then households with larger landholdings should be more prone to partition than smaller farms, as the breakaway co-husbands would be able to secure themselves larger individual landholdings and so maintain economic independence. However, they found that the size of the family farm had no consistent impact on whether the household partitioned (one or more co-husbands leaving the marriage to form a new household) or became conjoint (co-husbands within a polyandrous marriage marrying another wife, so that the union became **polygynandrous** – several wives and several husbands). While households that became conjoint were often significantly larger than those that remained intact, those that partitioned altogether were not significantly larger than those that remained intact.

While this does suggest that economic considerations need not have an impact on individual decisions to dissolve a polyandrous marriage, Levine and Silk (1997) do not take account of economic opportunities outside of agriculture. If individuals can gain alternative employment, then the size of the family landholding need not be of such critical importance to the decision to leave a polyandrous household. If the farm can remain viable due to wealth flowing into it via alternative sources of income, then there is no reason to expect the size of the farm before partition to have any impact on individuals' decisions.

Just such an explanation may account for findings from another Tibetan population. Haddix (2001) found that, in contrast to the Nyinba, increased wealth did tend to increase the probability of marital breakdown among the Karnali Tibetans. However, the diversity of the household's economy was also a very important factor, such that the interaction between wealth and diversity was a better predictor of marital dissolution than either effect alone. The Karnali divide their time between agriculture, trade and livestock herding and it is important to have some involvement in each of these spheres in order for households to remain viable. Multivariate models showed that households with low wealth but high economic diversity were least likely to split, whereas marital breakdown became increasingly likely in wealthy, low diversity households. These results support the idea that men's decisions to leave a polyandrous marriage are based on more subtle aspects of household viability than just the total amount of resources available. Economic diversity is enhanced within a polyandrous set-up because each brother can specialise in a particular economic sphere and control that aspect of the household's income. This may help to reduce tensions between brothers and increase household harmony, as well as enhancing long-term viability. For households in which there is low economic diversity, all brothers work in the same sphere and the benefit of polyandry may thus be reduced, since the management of the system can be undertaken by one individual alone. When households of this kind are sufficiently wealthy, they are thus more likely to split. A simple measure of household wealth does not take into account the complexities of Tibetan subsistence practices.

To be fair, Levine and Silk's (1997) paper is more geared toward understanding the proximate factors influencing the breakdown of polyandrous marriages now that prevailing economic and ecological conditions have eased the constraints on this system. Under such circumstances, individual reproductive decisions should begin to take precedence over previous concerns for wealth concentration and lineage survival. In line with this, individuals who played an active role in the partitioning of households tended to be significantly younger than their wives compared to their co-husbands. On average, active partitioners tended to be six years younger than their wives (not that these women were particularly old, however: the wife's average age at partition was 35 years) and, when they remarried, tended to find partners who were, on average, approximately four years younger than themselves. By contrast, men who remained in stable polyandrous marriages tended to be the same age as their wife. In line with this, birth order was a very good predictor of the role that men were likely to play in the dissolution of polyandrous marriages: only seven per cent of first-born sons were active partitioners, whereas 37–39 per cent of third-born sons were involved in the dissolution of a marriage.

These findings appear to be directly related to the reproductive opportunities experienced by men within households: first-born brothers have a distinct advantage with 67 per cent of all first-born children within a marriage attributed to them, although second-born children were generally attributed to the second-born brother. (Note that 'attributed' is the operative word here: what women tell their husbands need not necessarily be what they actually know to be the case.) Overall, birth order was negatively related to the number of children a man produced during marriage for intact households. Levine and Silk (1997) were also able to show that in households which had become conjoint or had partitioned, the eldest co-husband had produced more children than his younger counterparts. Active partitioners tended to have fathered only 0.5 offspring in their original marriage compared to 1.8 offspring for men in intact polyandrous households. With remarriage however, their rate of reproduction was found to increase threefold, supporting the notion that lack of opportunity rather than infertility was the factor responsible for the low reproduction of active partitioners.

However, there is also a cultural sanction on husbands who leave a wife with whom they have had sexual relations, and they are expected to compensate her with a share of the family's heirloom jewellery. Any offspring of a former marriage must also be provided for by leaving a share of the co-husband's property behind for them. Consequently, men who have it in mind to leave a polyandrous marriage may deliberately avoid having sex with their current wife in order to avoid these costs. The decision to partition could therefore be a cause, rather than a consequence, of low fertility among such men.

Interestingly, mean relatedness did not vary significantly among households that remained polyandrous compared to those that became conjoint or partitioned. Levine and Silk (1997) predicted that households with higher levels of kinship should be more stable because of the inclusive fitness benefits men gain by sharing a wife with their brothers. The fact that no such effect was found suggests that men value their own direct reproductive contributions over those of their brothers (as evolutionary theory in fact predicts they should) and/or that proximate considerations relating to sexual access and sexual jealousy are able to override the kin-selective advantages of fraternal polyandry.

This is an extremely important point, since it needs to be made very clear that individuals are not selected for polyandry per se (as some authors seem to think must be the case if we are to accept that polyandry is adaptive). Rather, individuals have been selected to adopt strategies that are optimal given the constraints under which they are forced to operate. The fact that individuals are able to dissolve marriages and embark on new relationships with greater reproductive advantages suggests that the ecological constraints placed on the system today are not as great as previously, and that polyandrous societies are undergoing exactly the changes that one would predict if individuals were now able to consider their own reproductive success, as well as the original concerns for lineage survival and wealth concentration.

The lack of stability within some polyandrous marriages is also very interesting from the female perspective – something that is often neglected in studies of polyandry. As Voland (1997) very rightly points out, females are just as much reproductive opportunists as men, and one would expect them to play an active role in protecting their own reproductive interests. He suggests that, since women do better reproductively out of polyandrous marriages than monogamously married women do, one would predict female strategies designed to stabilise polyandrous marriages as far as possible so that they remain intact. He suggests that women should not attempt to exclude younger husbands from sexual and reproductive opportunities, and that paternity at-

BOX 8.7

Reproductive rivalry and the risk of fission

Levine and Silk (1997) found that the size of the fraternal cohort had a significant influence on the probability of partitioning. Only 10 per cent of two-brother households partitioned whereas 60 per cent of four-brother households split. As male reproductive competition increased within a household, the more likely it was that younger brothers would be active partitioners. The neighbouring Karnali population, on the other hand, showed no effect of male cohort size on marital breakdown (Haddix 2001). Haddix attributes this to the fact that, while the Karnali use a per capita inheritance system (sons inherit an equal portion of the family estate), the Nyinba use a per stirpes system where each son gets a proportion of the estate that would have been received by his father. So, if one brother in a two-husband polyandrous marriage has three sons, they will each inherit one third of their father's share; if the other brother only has one son, then this offspring will inherit the whole of his father's share. That is, the latter will inherit half of the total family estate, whereas each of the other three sons will only inherit one sixth of the estate. Possibly as a consequence of this, the Nyinba are meticulous about tracking the paternity of children, whereas among the Karnali paternity is attributed to the eldest brother alone (which is again different to the Ladakh population where a child's paternity was attributed to all the males in the household).

Haddix (2001) states that the tradition of attributing all children to the eldest brother probably reflects the reality of the situation, with the eldest brother actually fathering the majority of offspring.

Haddix (2001) therefore suggests that the difference in the effect of fraternal cohort size between the Karnali and the Nyinba reflects the way in which paternity is distributed among husbands. In the Nyinba, all men have at least some chance of fathering offspring and attempt to do so. However, the greater the number of brothers there are in the household, the more competition there is to father children and the less likely it is that any one brother will be successful. Hence, larger cohorts are more likely to undergo marital breakdown. Among the Karnali Tibetans, on the other hand, younger brothers are extremely unlikely to father any children at all (and any child known to be the offspring of a younger brother is a subject of great interest and gossip in the community) so that it doesn't matter how many other brothers there are; a male with one older brother suffers the same lack of opportunity as one with three brothers. Consequently, the size of the cohort is not a factor in Karnali men's decisions.

Why these differences in inheritance systems and paternity attribution exist between populations has not yet been answered, but it may relate to differences in the ecological circumstances between them. One might predict, for example, that relatively harsher environments would have systems where paternity is ascribed either to all men or only one of them, and all children receive an equal inheritance, as a means of keeping competition between men to a minimum, thus promoting household stability under conditions where household breakdown could spell economic disaster.

tributions should reflect not only the mother's 'best knowledge' but also her strategic concerns regarding household stability. She should use her 'knowledge' in the most effective way to bind potential 'active partitioners' to the marriage and prevent its dissolution. However, this may require both men and women to behave in a far more rational manner than is actually possible, since emotions such as love and jealousy (which have evolved for very good reasons) cannot always be controlled, and it may be that women inevitably form a strong attachment to one of their husbands, thereby rendering them less than willing to behave in a perfectly optimal manner. Nevertheless, a study investigating female decision-making and reproductive strategies within a polyandrous context would be extremely interesting in terms of determining the limits of women's ability to control their own reproductive destinies.

*

In this chapter, we have investigated a few examples in some depth with the aim of demonstrating how cultural rules interact and feed back on biological tendencies and reproductive strategies. In the next chapter, we take an even broader perspective and consider how individual decisions in combination with ecological conditions influence some of the larger-scale emergent properties of social systems.

Chapter summary

- Cultural practices dictating the distribution of wealth to descendants show patterns generally consistent with evolutionary theory. However, it is clear that cultural practices place constraints on individual behaviour that must be overcome.

- Polygyny is associated with patrilineal inheritance whereby a man's sons receive an inheritance. Patrilineal societies also show high levels of paternity certainty. Matrilineal inheritance patterns are associated with societies where paternity certainty is lower. In such societies, men may be pursuing a 'roving strategy' under conditions where they cannot monopolise resources.

- In some circumstances, individuals may be more concerned with long-term lineage survival in the face of outside competition for resources. Data from medieval Portugal and hill farmers in Tibet are consistent with this interpretation.

- Although incest rules are frequently interpreted as strictures against in-breeding, detailed analysis shows that they are actually closely related to protecting property rights, paternity and social status, rather than being concerned with limiting sexual access per se.

- Cultural rules interact with and feed back on biological tendencies, but do not necessarily constrict individual decision-making in a maladaptive way. Changes in ecological circumstances can lead to changes in cultural practices, as evidenced by a reduction in the frequency of polyandrous marriages among Tibetans as alternative economic opportunities became available despite the fact that polyandry was valued by society and conferred status on the individuals involved.

Further reading

Betzig, L. (1986). *Despotism and Differential Reproduction: a Darwinian View of History*. Hawthorne: Aldine de Gruyter.

Clutton-Brock, T. H. (ed) (1988). *Reproductive Success: Studies in Individual Variation in Contrasting Breeding Systems*. Chicago, University of Chicago Press.

Crook, J., and Osmaston, H. D. (eds) (1994). *Himalayan Buddhist Villages*. Bristol, University of Bristol Press.

Goody, J. (1976). *Production and Reproduction: a Comparative Study of Domestic Domains*. Cambridge, Cambridge University Press.

Goody, J., Thirsk, J. and Thompson, E. P. (eds) (1976). *Family and Inheritance: Rural Society in Western Europe 1200–1800*. Cambridge: Cambridge University Press.

The individual in society

Humans are, above all, social animals. That sociality represents part of our primate heritage, a heritage characterised by a peculiarly intense form of sociality not seen in any other group of animals. Social solutions to the problems of survival and reproduction represent the primates' unique evolutionary adaptation. However, social life of the intensity that primates exhibit comes at a price: if groups are to function effectively as social solutions to the problems of ecology and demography, individuals cannot pursue their personal interests while completely ignoring the interests of those with whom they live. We saw in Chapters 2 and 5 how the two sexes have different strategic interests in respect of the business of reproduction. But just as the sexes have divergent interests, so too do individuals: Chapter 3 reminded us that each individual has a unique constellation of genetic relationships whose interests bear on his or her inclusive fitness, and these give each of us a slightly different perspective on the set of individuals with whom we happen to live. Moreover, each of us is unique and brings to a given situation a particular history, with its unique set of abilities, advantages and disadvantages. Thus we have our own peculiar interests at stake: what suits me need not suit you, even if you are my closest relative.

The dilemma created by sociality at any level – and exacerbated in the intense sociality of primates – lies in the conflict between short- and long-term interests. We are social precisely because group-living allows us to solve certain problems of survival and reproduction more effectively than we could do on our own. But for a group of individuals to coexist stably over time – and, indeed, to function as an effective coalition – we are forced to compromise on our immediate personal interests in order to allow our prospective companions to have a stake in the group. Groups and coalitions are strategies whose pay-offs are long-term, and to gain that long-term benefit we have to be prepared to incur a short-term cost in terms of our immediate personal interests.

As will have become clear from the discussions in Chapter 4, any strategy based on reciprocity runs the risk that short-term interests may lead some individuals to renege on the implicit agreements that hold a group together. Freeriding (taking the benefits of social life without paying the costs) becomes an increasing temptation. That in turn destabilises the delicate balance on which the coalition depends, threatening to undermine the foundations of the group itself. The threat that this engenders and the ways in which we seek to defuse its effects are the focus of this chapter. First, however, we consider two aspects of human sociality that are unique, namely kinship naming and the division of labour.

KINSHIP AND KINSHIP-NAMING

In Chapter 3, we considered the implications of genetic kinship in some detail. We argued that, in general, people behave rather differently toward kin than toward non-kin, and suggested that this is related to issues concerning inclusive fitness. In the examples that were presented, close examination of cooperative behaviours in a range of societies living under different subsistence regimes in a variety of habitats emphasized the fact that people do seem to place biological kin in a special category in their daily interactions. Being able to do so rests on an important feature of human behaviour: the ability to name and classify individuals by their degrees of relatedness.

Kinship classification is a universal feature of human social behaviour (Fox 1979, Keesing 1975), but it is rather more than simply a means of labelling who-is-who in relation to everyone else (in the sense of 'he is my brother's wife's father'). In a social milieu, kinship naming carries with it important implicit cues as to how one should behave towards another person, which may also have implications in terms of whom one can marry, whom one can call on for favours or support and whom one can expect to inherit from. The complexities of kinship classification across the broad range of human societies have sown discord and confusion among socio-cultural anthropologists for more than a century, not least because, although some patterns emerge, it has not been obvious why cultures should differ in their kinship terminologies (Keesing 1975). One thing they are all agreed on, however, is that social (or classificatory) kinship has nothing at all to do with biological (or genetic) kinship (for a trenchant example, see Sahlins 1976). Anthropologists, for example, often draw an important distinction between *genitor* (biological father) and *pater* (social or classificatory father) (see for example Barnard 1994), arguing that the latter often carries greater significance for humans.

That kinship terminologies do reflect something more than arbitrary naming devices is suggested by the fact that, despite the great variety of languages in the world, all of the world's kinship terminologies show sufficient similarities to be grouped into a handful of major types. There are several ways of doing this, one being in terms of how close male relatives are categorised: (1) those (like Hawaiian) that use the same term for father, father's brother and mother's brother, (2) those (like Iroquois) that distinguish between paternal and maternal relatives among these relatives, (3) those (like Eskimo) that distinguish father from either parents' brothers and (4) those (like the Sudanese Nuer) that give a different name to each type of relative (see Barnard 1994). In Eskimo kinship systems, no distinction is made between different types of cousins (the offspring of mother's brother or a father's sister), but other kinship systems (such as the Hawaiian and Dravidian) do make such a distinction. In Crow kinship systems, members of different generations may be classified together: a woman's brother and her own son are treated as exactly equivalent, and the same applies to a mother's brother's children (cousins in English-speaking cultures) and a brother's offspring (our nieces and nephews). **Table 9.1** illustrates some of these contrasts.

English follows Eskimo terminology in which many types of relative are lumped together by sex and generation – or, at least, has done so ever since the Normans introduced the French term *cousin* into the language: prior to that, it followed a Sudanese structure in which relatives in this particular class were specified by their exact relationships (for example, *mother's brother's son*) as is still the case in traditional Scots Gaelic.

Rather than indicating that such cultural traditions arise from a few sources, this clustering of kinship terminologies suggests that kin-naming patterns are drawn from a limited set of alternative possibilities that may reflect subtle aspects of the mating system (Hughes 1988). Although there is no simple correlation between kinship naming

Table 9.1 Kinship terminology for some representative classification systems

Category	Eskimo*	Hawaiian	Crow	Omaha
Father	father	father	father	father
Mother	mother	mother	mother	mother
Brother	brother	brother	brother	brother
Sister	sister	sister	sister	sister
Mother's brother	uncle	father	... mother's brother ...	
Mother's sister	aunt	mother	mother	mother
Father's brother	uncle	father	father	father
Father's sister	aunt	mother	... father's sister ...	
Mother's sister's son	cousin	brother	brother	brother
Mother's sister's daughter	cousin	sister	sister	sister
Father's sister's son	cousin	brother	father	nephew
Father's sister's daughter	cousin	sister	father's sister	niece
Mother's brother's daughter	cousin	sister	daughter	mother
Mother's brother's son	cousin	brother	son	mother's brother
Father's brother's son	cousin	brother	brother	brother
Father's brother's daughter	cousin	sister	sister	sister

* English uses the Eskimo terminology

systems and marriage systems (Keesing 1975), there is some evidence to suggest that at least some naming systems correlate with levels of paternity certainty (and hence with inheritance patterns). The Crow and Omaha kinship terminologies have long been recognised as being mirror images of each other: father's sister's children and mother's brother's children (both cousins to us) are classified in the parental generation in one system and the offspring generation in the other (**Table 9.1**).

Using the approach outlined Chapter 3, Hughes (1988) modelled the fitness implications of investing in individuals of different nominal kinship as a function of high (100 per cent) or low (0 per cent) paternity certainty. He was able to show that the Crow system produced the higher fitness returns when paternity confidence is low, while the Omaha system did so when paternity confidence is high. Significantly, the actual Crow and Omaha peoples had matrilineal and patrilineal descent systems, respectively. Alexander (1979) and others (for example, Kurland 1979) have noted that matrilineal descent (and inheritance) systems tend to be common when paternity certainty is low. As we discussed in Chapter 8, where paternity cannot be assured, males prefer to invest in their sisters' offspring rather than their wives' offspring because they can be more certain of genetic relatedness through the female line. This system is known as the *avunculate* and is a common arrangement in Polynesian societies where sexual promiscuity is more common than in the West (though perhaps not as common as anthropologists were originally misled into believing). Interestingly, in medieval European chivalric tales, it is the mother's brother who is invariably asked for help and advice by the young knight before he embarks on the quest that will give him fame and fortune (and, ultimately, the damsel); the father merely appears at the end of the story to validate his son's identity (Sabean 1976). One of the more obvious characteristics of chivalric society, of course, was that it involved a great deal of illicit 'courtly' love: if nothing else, paternity certainty must have been low.

Ellison (1994) has pointed out that the level of polygyny skew that characterises a society may also have implications for how one should calculate relatedness. When the variance in male lifetime reproductive success greatly exceeds that for females (as it may do in highly polygynous societies), individuals will be more closely related through the male line than through the female line. Hence, kinship systems that emphasize patrilinearity may provide a better rule-of-thumb for estimating relatedness than those that emphasize matrilineal or bilateral relatedness. As the reproductive variances of the two sexes approach equality (as they will in exclusively monogamous mating systems), so calculating relatedness through both maternal and paternal lines (bilateral descent) becomes increasingly important. However, bilateral descent involves keeping track of many more lineage relationships (and inter-relationships), so that the task becomes cognitively much more demanding. He suggests that cognitive constraints on information processing may explain why bilateral descent systems (such as those that are now more common in the developed West) have a much shallower generational compass. Whereas it may be fairly easy to keep track of all one's cousins, second cousins and third-cousins-once-removed in a patrilineal descent system, it just becomes impossible to do so in a bilateral descent system. Hence, in our societies we tend to lose track of relationships beyond the scope of first cousins.

All this is not to say that rules regarding marriage, or kinship classification for that matter, are static and unchanging. Fox (1979) has argued quite strongly for the fact that kinship rules can, and do, change in response to changing circumstances. An example of this on the infra-societal scale is provided by Chagnon (1988). He asked 100 Yanomamö to name their relationship to various members of the community (representing a total of about 11 000 individual kinship classifications). (The Yanomamö kinship system is of the Iroquois-Dravidian type.) He found that that adult males classified kin faster than adult females (means of 21.3 and 18.8 individuals per minute, respectively; $P < 0.005$). Males were, however, especially likely to classify in the category 'potential wives' women who should not have been so classified. Women who were redefined as marriageable tended to be younger (and thus more fertile) than those who were not reclassified in this way.

Chagnon argued that males, in particular, are prone to manipulate kinship categories to their own (reproductive) advantage, especially under conditions where there is a shortage of acceptable female kin. Females, presumably, have less to gain by doing so and more to gain by using kinship assignations that more accurately reflect genetic relatedness in ways that are useful for alliances. Genetic relatedness has much less real impact on the willingness to mate than it does on an individual's willingness to form an alliance (see Chapter 3). Even if the principal genetic reason for attending to kinship is to avoid the genetic risks of inbreeding, these risks, as Waser et al. (1986) have shown, are modest in comparison with the risk of not breeding at all, since only one quarter of the offspring of an inbred marriage are likely to be affected. Inbreeding may thus be an acceptable risk under a wide range of circumstances.

However, it is also clear that the pattern observed in the Yanomamö may not apply universally. Salmon and Daly (1996) asked 24 opposite-sex pairs of (English-speaking) Canadians to name as many relatives as they could. The women listed significantly more relatives than their brothers (means of 31.9 vs 27.5; $P < 0.001$), with the women being better on both ascendent and collateral kin, and on patrilateral as well as matrilateral kin. Significantly, they did best on naming matrilateral kin, and particularly maiden (that is, unmarried) matrilateral aunts. A number of studies of social networks in western societies have noted the importance of matrilateral relatives as social resources (emotional support, childcare, economic assistance, and so on) (for example, Young and Wilmott 1957, Stack 1974, Essock-Vitale and Maguire 1985, Dunbar and Spoors 1995). Hence, sex differences in kinship knowledge may reflect differing demands placed on the two sexes in different socio-economic environments.

Kinship also has an important social dimension that we seem to exploit: classifying someone as kin even though they are unrelated seems to have an important psychological influence on how we treat them (a fact long exploited by public orators engaged in political rabble-rousing: Johnson et al. 1987, Connor 1994). This quite unique form of self-deception seems to play an important role in the facilitating social obligations that underpin the coherence of human social groups. By labelling an unrelated individual as kin we draw them into our circle of obligation and, hopefully, influence their behaviour towards us. There has so far been little attempt to explore any of these issues using an evolutionary perspective, despite the fact that it would merit more detailed investigation.

It is important to appreciate here that the biological underpinning to kinship classification does not have to be absolute for an evolutionary explanation to be true. Natural selection is a *statistical* process: so long as the fitness consequences of kinship hold up most of the time, the evolutionary processes will follow inexorably on. This leaves considerable slop in the system for kinship terminology to be exploited for other purposes. Perhaps significantly in these cases, people are rarely confused by the difference between real biological kinship (in so far as they understand this) and classificatory (or fictive) kinship. Politics may enjoin us to pretend that 'brothers in arms' are genuine brothers, but we are rarely fooled into believing that this is really true.

SEX BIASES IN SOCIAL ORGANISATION

We consider just two issues here: the sexual division of labour and sex differences in dispersal. Humans are unusual among mammals in exhibiting a very striking sex difference in the mode of production: men typically engage in hunting or large-scale farming (or, in modern societies, 'go out to work'), while women gather, manage small-scale gardens, run the household and rear children. Irrespective of what the ontogeny of these differences might be, they are sufficiently widespread among human cultures to warrant explanation.

Traditionally, anthropologists have supposed that the division of labour reflects a cooperative arrangement between the sexes to share the costs of rearing, with males taking on the riskier tasks (for example, hunting) in order to spare the females on whom successful reproduction depends (Bird 1999). Aside from the fact that this view carries undercurrents of group selection, we noted in Chapter 4 the possibility that hunting (especially of big game) might actually be a male mate advertising strategy rather than a foraging or parental investment strategy (Hawkes's 'Show-Off' hypothesis). Nonetheless, hunting does provide important nutritional returns for the community, and some hunter-gatherers (for example, the Inuit) may be more or less wholly dependent on hunting at certain times of year.

Belovsky (1987) has used a linear programming model to explore the relationship between hunting and gathering in traditional societies. His model was based on known physiological relationships between energy requirements for an adult foraging for him/herself and a dependent child, the rate at which energy is acquired from different combinations of hunting and gathering and the purely mechanical constraints imposed by stomach capacity and the time available for foraging (**Figure 9.1**). Optimal foraging theory offers two options for constrained foragers, namely the choice between maximising energy intake and minimising foraging time (and hence exposure to predators) (Schoener 1971). Comparison of the predicted ratio of hunting to gathering for energy maximisers and time minimisers generated by the model (see **Figure 9.1**) with the observed values for the !Kung San suggests that the San (at least) are energy maximisers (**Table 9.2**).

In addition, Belovsky (1987) was able to show that the proportional distribution of effort between hunting and gathering depends crucially on the local ecology, with the proportion of the diet derived from hunting declining as habitat productivity increases (**Figure 9.2**). This can be expected to have a significant impact on the sexual division of labour, and beyond this, on the balance of power between the sexes. When women contribute more to the economy, we may expect them to be able to exert more power (though this may not apply once gathering is transmuted into agriculture and men are able to monopolise the resources on which the women depend). We might also expect the use of big game hunting as a 'show-off' device to increase as hunting becomes less important in the diet and the male's contribution to the nutritional costs of rearing diminish. Essentially similar results were reported by Low (1990b) from an analysis of data on 74 traditional societies: she found a positive correlation between the amount of hunting and the coldness of the habitat and a negative correlation with rainfall. Gathering thus became more important in wetter, warmer habitats.

Turning to sex differences in dispersal, it is important to establish at the outset that sex-biased dispersal at reproduction is by no means uncommon among the higher vertebrates (birds and mammals). In most species of vertebrate, one sex disperses away from its natal community to find mates elsewhere. Female dispersal tends to be more

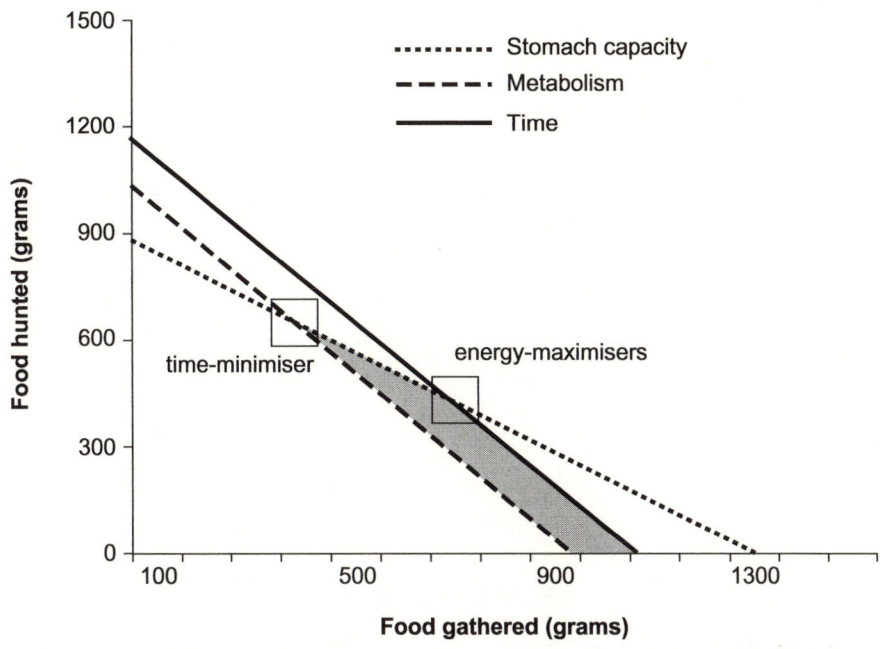

Figure 9.1 Belovsky's linear programming model of optimal diets for hunter-gatherers. The model predicts the dietary intake from different combinations of hunting and gathering when adults forage both for themselves and their offspring. Foragers need to ensure that the balance of hunting versus gathering will yield the minimum amount of energy to meet their daily needs, while not exceeding the amount of time available for foraging or their stomach capacity (since hunter-gatherers cannot store food that they cannot eat), based on empirical data from the !Kung San during the winter season and standard physiological equations. (Belovsky's original model included a line for the daily protein requirement, but this has been omitted since it turns out that any diet that satisfies the energy requirement will also satisfy the protein requirement.) These conditions are met only in the shaded area of the state space, where energy intake lies above the minimum metabolism line but below the two constraint lines. Within the confines of this zone of realisable diets, foragers can choose whether to minimise the time spent foraging (time-minimisers) or maximise energy intake (energy-maximisers). Redrawn with permission from Belovsky (1987).

Table 9.2 Observed and predicted values for diet and foraging variables for !Kung San during winter

	Dietary intake (%) meat	vegetables	Time (min)	Energy (kcal)	Protein (g)
Time minimiser	58	42	349	3080	140
Energy maximiser	32	68	393	3438	147
Observed	31	69	363	3456	147

Note: Time, energy and protein are daily rates.

Source: Belovsky (1987)

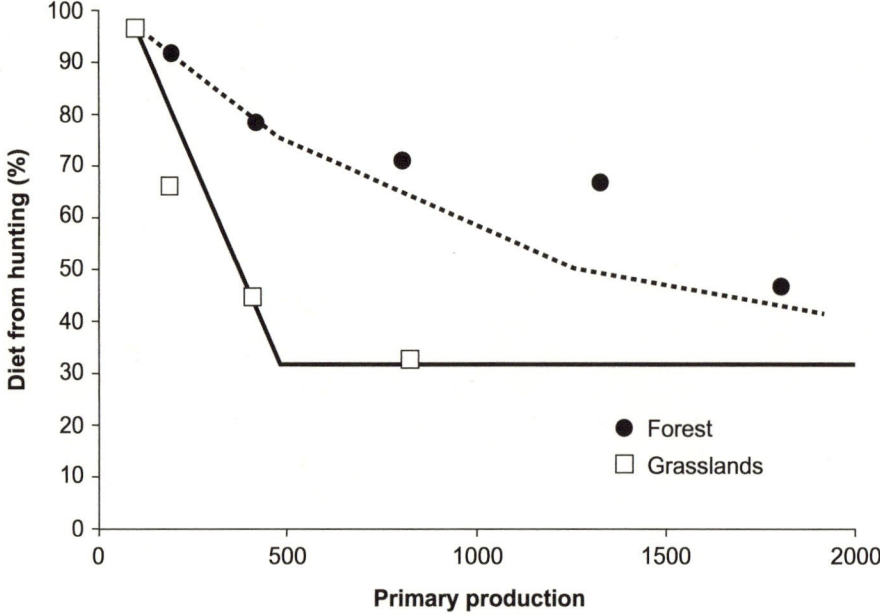

Figure 9.2 Optimal diets predicted by Belovsky's model for hunter-gatherer populations living in forested (broken line) and grassland (solid line) habitats compared to observed diets (circles and squares, respectively) for different tribal groups. In both types of habitat, gathering becomes more important as a source of nutrients (relative to the products of hunting) as habitat primary productivity increases. Redrawn with permission from Belovsky (1987).

common among birds, whereas among mammals it is more typical for males to disperse. It is assumed that this is designed to reduce the risks of inbreeding by ensuring that each sex always mates with individuals that are not its immediate relatives. Among primates, male dispersal is the norm in one important family of Old World monkeys (the baboon-macaque-guenon group), but more recent evidence suggests that both sexes disperse in most other species (although males may disperse further than females) (for reviews, see DiFiore and Rendall 1994; Strier 2000; Cowlishaw and Dunbar 2000). Female dispersal may be more common in chimpanzees and gorillas, however, and this has been interpreted as indicative of female-dispersal being the norm among all great apes (including humans) (Foley and Lee 1989, Rodseth et al. 1991).

Despite the fact that female dispersal (and male philopatry) has been regarded as the norm for modern humans (DiFiore and Rendall 1994, Rodseth et al. 1991), the evidence in fact seems to be more equivocal than we might expect. Although the older anthropological literature insisted that hunter-gatherers lived in patrilocal (or virilocal) bands (for example, Lévi-Strauss 1969, Service 1962), more recent views suggest much greater fluidy, with post-marriage residence often being a matter of convenience and congeniality rather than obligation (though men may be required to do bride service by living with, and hunting for, their in-laws for the first few years) (see Lee 1979, Guenther 1996). (Although pygmies seem universally to follow a patrilocal rule [Hewlett 1996], this may simply reflect their wholesale adoption of the language and culture of the Bantu among whom they live [Cavalli-Sforza 2000]. However, even the pygmies seem to operate a form of matrilocal bride service in which a man is obligated to hunt for his in-laws during the first few years post-marriage: Hewlett and Cavalli-Sforza 1986.) In hunter-gatherer societies, resources are not defendable, so matrilocality (females remaining near their natal group) may be as appropriate as patrilocality (with sons staying on their natal land). Indeed, even among the pygmies, males' ranging areas were much greater than those of females (Cavalli-Sforza 2000), suggesting that males exposed themselves to a much wider range of options in terms of potential brides than women did.

In contrast, patrilocality is almost universal in agriculture-based societies: women move from their natal group to live with their husbands, whose family controls and defends the lineage's farmland or other tangible assets (Lévi-Straus 1969, Otterbein and Otterbein 1965, Rodseth et al. 1991). Seielstad et al. (1998) have shown for a set of 14 (agricultural and pastoral) traditional African tribal groups that Y-chromosome variation is significantly lower than mitochondrial DNA variation, suggesting that women rather than men have traditionally transferred between communities on marriage (at least among these peoples). (The fact that there are no hunter-gatherers in this sample is clearly significant in view of their interpretation that these results apply universally.)

Female-dispersal seems to be typical of some modern European agricultural populations. Among the contemporary 'Nebraska' Amish farmers of Pennsylvania, for example, twice as many married couples settle in the groom's church district as settle in the bride's church district (Hurd 1985). As a result, actual genetic relatedness was significantly higher among the men than among the women of these Amish communities. Similar results were obtained from the upper Parma valley in northern Italy (Cavalli-Sforza and Bodmer 1971). However, the situation can sometimes be more complex: in the Krummhörn during the eighteenth century, for example, women were significantly more likely than men to move between parishes on marriage (40 per cent versus 28 per cent), but, of those that left their natal parishes, men were more likely than women to leave the region altogether (Beise 2001).

These results are in marked contrast, however, with the situation reported for non-agricultural populations. In nineteenth- and early twentieth-century European industrial settings such as the slums in the East End of London, men worked for cash wages and there were no resources that they could garner or defend. Here, women lived significantly closer to their natal homes than to their husbands' natal homes because matrilateral kinship networks (mothers, sisters, aunts, daughters) provided essential support for successful reproduction: when a daughter married, her mother often provided room for the couple in her own home, and later obtained housing for them through contacts with her own rentman, as well as supporting daughters through childbirth, hard times and child-rearing (Young and Willmott 1957, Bott 1971). Similarly, although Koenig (1989) found a weak but significant tendency for females in the USA to disperse further than males after graduating from high school, the difference between the two sexes disappeared once natal location, college location and occupation had been taken into account. In this case, it seemed that women's tendency to disperse further was largely a reflection of career opportunities.

These observations perhaps suggest that the human dispersal system can flip-flop backwards and forwards between the two sexes in response to local economic conditions. When men are able to control resources such as land (or perhaps income-based wealth), female dispersal is more common and married women live in groups where they have no direct kinship support; when men are unable to defend resources, female kinship alliances may become of overwhelming importance and male dispersal is more common. This may in turn affect the character of the relationships between women. When female dispersal is the norm, women may still form alliances with other women, but, since these women will be unrelated, most such alliances will be based on strict reciprocity; when female matrilocality is the norm, kin-based alliances that are less directly reciprocal in character may be more common and non-kin alliances rare by comparison.

Irrespective of how the dispersal system operates, it seems that women are universally much more likely to take on responsibility for maintaining relationships, both with kin and with the extended family as a whole. This seems to be as true of both modern post-industrial societies (Olivieri and Reiss 1987, House et al. 1988, Hogan and Eggebeen 1995, Salmon and Daly 1996, Dunbar and Spoors 1995) as of hunter-gatherers (van den Berghe 1979). The most likely explanation for this is that a kin-bias in willingness to give help, support and advice (see Chapter 3) may be more important for successful reproduction for women (even when they are the dispersing sex) than it is for men, for whom alliance formation is much more a matter of the moment. However, at present, we know too little about the details of this aspect of human behaviour to be able to offer anything more than a speculative answer.

STRUCTURE OF SOCIAL GROUPS

Human social systems are inevitably complex and highly variable. Their form arises, in part, from the demands of marriage and inheritance systems such as those discussed in Chapter 8. However, in addition, there are important ecological and cognitive constraints that limit the sizes of groups that can occur, and these in turn impose constraints on the forms of social organisation that can develop. Hence, the interplay between what suits individuals' interests and what is possible given the ecological circumstances will inevitably be important for any understanding of the evolution and adaptiveness of human social systems.

COGNITIVE CONSTRAINTS ON GROUP SIZE

The most important intrinsic constraint on the size of social groups is likely to be that imposed by the information-processing capacities of the brain itself. Dunbar (1992, 1998a; Joffe and Dunbar 1997) showed that social group size in primates is a direct function of relative neocortex size (the part of brain associated with higher cognitive functions) (see **Box 6.1**). This relationship is assumed to be a consequence of the fact that the size of the neocortex (or some part of it) places an upper limit on the number of individuals that can be held in a given relationship.

Dunbar (1993c) extended this claim to include modern humans (**Figure 9.3**). The relevant group size for modern humans turns out to be in the order of about 150 individuals. This approximates quite closely to the observed sizes of a range of small-scale human groupings (**Table 9.3**). In 'small world' experiments, for example, subjects are given a list of fictitious people in different parts of the world and are asked to send each one a letter by hand, starting with someone they know. Thus, if the target works as a bank clerk in Mexico City, you might hand the letter to your uncle Jack who happens to be a pilot with an airline that runs flights to Mexico; uncle Jack could presumably hand the letter to a colleague on such a flight, who could pass it on to a member of the ground staff in Mexico, who might in turn know someone who works for the relevant bank, who in turn could pass it to someone in the appropriate branch, who could then pass it on to the intended recipient. When

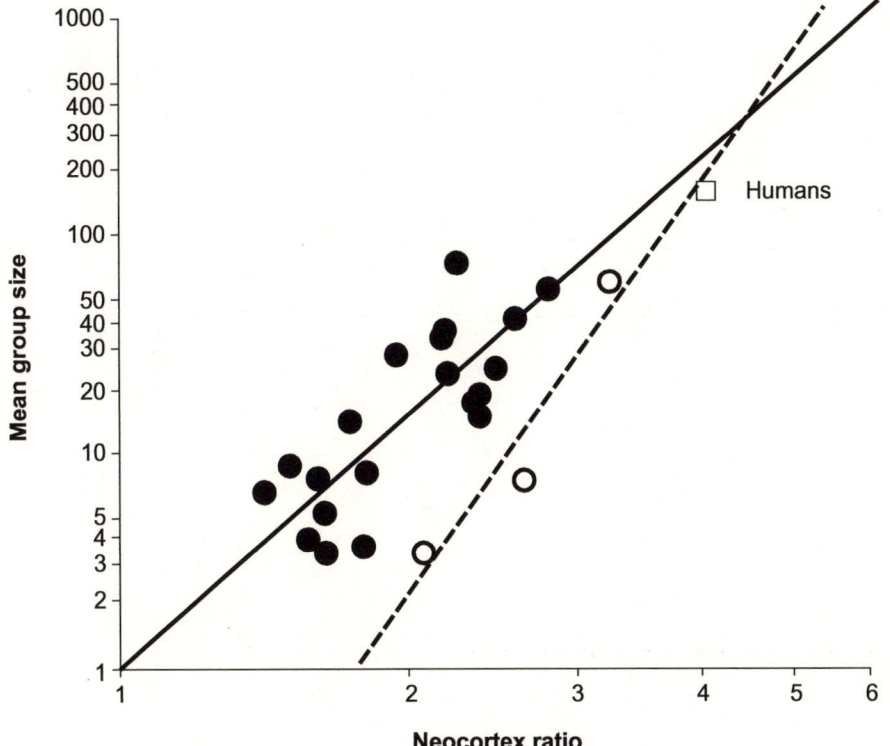

Figure 9.3 Mean group size plotted against neocortex ratio (neocortex volume divided by the volume of the rest of the brain) for New and Old World monkey (solid symbols) and ape (open symbols) genera. The open square marks the observed mean size of clanlike groupings in contemporary and recent hunter-gatherers. Source: Dunbar (1993c).

subjects are given large numbers of these tasks, they tend to run out of new names to ask among their acquaintances after about 125–150 individuals (Killworth et al. 1984). Similar results were obtained from an analysis of UK social networks reflected in whom one sends Christmas cards to. This yielded a mean network size of 153.5 ($N = 43$ respondents), with a significant relationship between perceived emotional closeness and the recency of contact with a given individual within the network as a whole, once other factors (including geographical distance, relatedness, work vs social circles and sex of respondent) are partialled out (Hill and Dunbar, submitted).

Thus, the value of 150 can probably be equated with the number of individuals that one knows well enough to be able to ask a favour of, or perhaps would feel comfortable about joining uninvited if one chanced to come across them in some exotic location (Dunbar 1993b, 1996a). It seems likely that the distinction between those on the inside of this circle and those on the outside reflects the level of intimacy involved. Members of the same group know each other well enough not to need an introduction to establish who they are; they can launch straight into a conversation, catch up on news of mutual acquaintances and so on. This degree of personal knowledge may reflect the level of trust that an individual can place in a given relationship (Dunbar 1996a). It may also involve a sense of obligation, either by virtue of reciprocation for past kindnesses or because of obligations arising out of kinship and other ties. Among the Mae Enga and the Kaluli of New Guinea, for example, these clan groupings of about 150 individuals represent the network of individuals on whom one can call for support during raids or when under threat of attack (Meggitt 1965, Hallpike 1977).

This constraint on group size clearly has significant consequences for the stability of communities once they exceed this limiting size, especially in small-scale tradi-

Table 9.3 Some examples of human social groupings whose size is typically about 150

Grouping	Typical size	Source
Neolithic villages (Middle East, 6500–5500 BC)	150–200	Oates (1977)
Hutterite farming communities (Canada) (mean, $N = 51$)	107	Mange and Mange (1980)
E. Tennessee rural mountain community	197	Bryant (1981)
Research specialities (sciences and humanities) (mode, $N = 13$)	100–200	Becher (1989)
Social network size (mean of two experiments)	134	Killworth et al. (1984)
Social networks (recipients of Christmas cards) (mean, $N = 43$)	154	Hill and Dunbar (submitted)
Church congregations (ideal size)	200	Urban Church Project (1974)
'Nebraska' Amish parishes (mean, $N = 8$)	113	Hurd (1985)
Maniple ('double century') (Roman army: 350–100 BC)	120–130	Montross (1975)
Company (mean and range for 10 World War II armies)	180 (124–223)	MacDonald (1955)
Hunter-gatherer clans (mean and range of 9 tribal societies)	148 (90–222)	Dunbar (1993c)

tional societies that lack formal societal mechanisms for the control of disputes (for example, police forces, law courts) (see **Box 9.1**). Where no such mechanisms exist, group fission may be the only mechanism available to prevent discord breaking out into seriously destructive violence. The Yanomamö of Venezuela, for example, live in approximately 150 villages ranging in size from 25 to 300 people, with warfare within and between villages being all but endemic (Chagnon 1979a). Villages are typically based around related groups of individuals who act together as an alliance, mainly in defence against raiding by neighbours. As villages grow in size, the overall relatedness of individuals in the village decreases. As mean relatedness declines, so too does 'solidarity' such that the dampening effects of kinship on competition are weakened. Under these circumstances, cooperation and reciprocity are not as assured as they might be, and the village begins to fragment into subsets of more closely related individuals. With increasing isolation of clusters and a resulting greater tension between clusters, a critical threshold is reached with fission of the village as the ultimate outcome.

We can be fairly sure that the constraints on social group size at this level are not simply a consequence of memory capacity for faces. The limit on the number of faces that we can put names to appears to be much greater (around 2000 is the often-quoted figure). Rather, the problem seems to lie with the number of relationships that we can juggle with in our mental state space. There are at least two possible ways this cognitive constraint might work. One is that there is an upper limit on the number of relationships we can hold in mind at any one time. The other has more to do with limitations on our ability to manipulate the information in this database when we wish to exploit it in order to engage in complex social strategies. This may have to do with the phenomenon known as 'theory of mind', the ability to imagine what is in the minds of others and use that information in assessing both how they might behave and how they be persuaded to behave. (We discuss this phenomenon in more detail in Chapter 11.)

However, memory might still play a role in another respect. Among humans, mating within the immediate social group is forbidden in most societies; in most of these cases, the restriction on mating is normally confined to the clan-like groupings of 150 that we have identified above. Knowing just who is related to who may thus be crucial for regulating marriages. In an exogamously mating society with reproductive rates comparable to those of a natural fertility population (that is, one that does not use modern contraceptive methods), a group of 150 represents the complete set of living descendents (the three generations currently alive) that originate from a single ancestral pair in the fifth generation (Dunbar 1996c). In other words, the apical ancestors for such a group are the current grandparents' grandparents – or about as far back as any living individual can remember from personal experience who is related to whom.

The group size of 150 that seems to be so characteristic of humans may, in fact, be just one of a series of hierarchically organised cognitively constrained groupings. Another that has been identified in the psychological literature is the so-called *sympathy group*. If one asks people to list all those individuals whose death tomorrow they would be devastated by, the answer seems to be consistently in the region of twelve (Buys and Larsen 1979). Network sizes determined from data on frequency of contact tend to be of similar size (around 12–15) (Hays and Oxley 1986, McCannell 1988, Rands 1988, Dunbar and Spoors 1995, Hill and Dunbar submitted).

In a detailed analysis of social network size (defined as all individuals contacted at least once a month), Dunbar and Spoors (1995) found that, in their UK sample,

BOX 9.1

Self-structuring principles for societies

There is considerable evidence to suggest that a group size in the order of about 150 represents a pivotal point in the nature of relationships between individuals. From an analysis of ethnographic data from some 30 traditional societies, Naroll (1956) found that there is a simple power relationship between the maximum ever-recorded community size for a given society and the number of occupational specialities (or social roles) recorded for it. His graph suggests that there is a critical threshold at about 500 individuals beyond which social coherence breaks down unless some kind of structural organisation is imposed. In effect, police forces become necessary. Unfortunately, Naroll's analyses are given in terms of maximum rather than mean group size; however, since group size distributions are invariably skewed, a maximum of 500 probably suggests a mean value in the region of 100–200.

Similar results have been reported from studies of business organisation. An informal rule suggests that once an organisation exceeds about 150–200 individuals, some kind of line management structure is needed to prevent communication breaking down within the group: below this threshold, information passes freely between individuals, but in larger organisations a combination of rivalries between sub-groups and lack of time/opportunity for contact means that information is not passed on to where it is needed.

Terrien and Mills (1955), for example, showed that the number of control officials needed to ensure an organisation's smooth running was directly correlated with its size. In addition, within the size range 50–500 individuals, job satisfaction correlates negatively, and absenteeism and staff turnover correlate positively, with organisation size (Indik 1965, Porter and Lawlor 1965, Silverman 1970). One reason for these striking findings is likely to be that the number of friends that an individual has within his/her workplace levels off at an organisation size of 90–150 (Coleman 1964). Friendship implies obligation and the willingness to exchange information and do favours.

In both these cases, the difficulty created by lack of familiarity in large groups is overcome by assigning individuals to roles and providing them with easily recognised badges (for instance, uniforms for policemen, 'dog collars' for priests, stars for generals, white coats for doctors). We can then define a generalised relationship for these individuals that prescribes how we should relate to them (at least until we get to know them on a personal level). This allows us to overcome the uncertainties in any first meeting without risking egregious social errors.

One final point concerns the dispersed nature of modern post-industrial societies. Until relatively recently, most people lived in small-scale communities, few of which numbered more than 150–200 individuals in size. Most people's acquaintanceship networks probably overlapped almost completely: quite literally, every one in the village knew everyone else. However, the increased social mobility and migration that have come with industrialisation have resulted in the break-up of traditional communities. We all still have our networks of 150 acquaintances, but they overlap only partially with those of even our most intimate friends. This is partly because our acquaintanceship networks are fragmented into a number of smaller clusters (work friends, sports club friends, church friends, intimate friends and so on). This must have important implications, not least for the intensity with which social groups are bonded: because networks overlap only partially, we will feel less obligation to the wider members of the community than we might have done during previous centuries.

women had significantly larger networks than men (12.4 vs 10.9, respectively) and that women had a significantly larger number of both female friends and kin (defined as related to the respondent by $r \geq 0.125$) in their networks than men did, while men had more male friends. Kin, it seems, were less important for men than for women: the

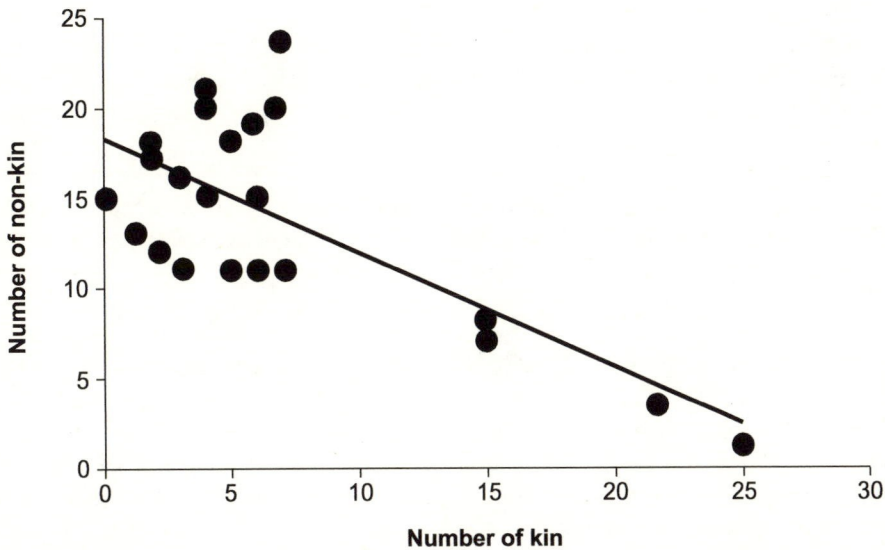

Figure 9.4 Number of kin (defined by $r \geq 0.125$) and non-kin in the social networks of those individuals who had more than ten people in their network (defined as all individuals contacted at least once a month) for a UK sample. In general, the more kin an individual has in his/her social network, the fewer non-kin he/she has, suggesting that kin are given priority in the trade-off between numbers of individuals available and the number of relationships that can be managed. Redrawn with permission from Dunbar and Spoors (1995).

mean number of kin contacted was 3.62 for men and 8.57 for women. In fact, there appears to be a negative relationship between the numbers of kin and non-kin contacted, at least for those who had more than ten individuals in their network (**Figure 9.4**). This appears to be because priority is given to kin, with non-kin only becoming part of the network if there is spare capacity. Individuals who belong to small extended families are likely to have more unallocated slots available in their networks for non-kin; in contrast, those who belong to large extended families may confine most of their social relationships to their kin-group.

In addition to networks of 12–15 individuals (the 'sympathy group'), Dunbar and Spoors also found evidence for a smaller more intimate group (the 'support clique') buried within the network: this consisted of the subset of, on average, 4.7 individuals whom the subject would go to for advice and support in times of crisis. This value turns out to be fairly consistent across cultures: a similar sample from the USA yielded a mean clique size of 3.0 (Marsden 1987), while an analysis of 'hair-care cliques' among San women (that is, women who regularly did each other's hair, a situation providing dedicated time in which the world is often put to rights) yielded a mean size of 3.8 (Sugawara 1984). Support cliques are a subset of each individual's network, and appear to represent a set of relationships that are even more intense than those that create networks, which in turn are more intense than those involved in the larger groupings of 150. In effect, these different grouping types mirror the widely recognised distinction between intimates, friends and acquaintances. One interpretation is that they represent a set of hierarchically inclusive groups that reflect constraints of the level of intimacy that we can have with other individuals (Hill and Dunbar, submitted).

The important issue here is not so much that specific kinds of groups exist in all human societies (they probably don't), but rather that there is an upper limit on the number of individuals that can be held in a particular relationship of trust and intimacy with oneself. The exact composition of that group and its precise function will vary from society to society according to the exigencies of the local ecological, demographic and social contexts. We should also note that this aspect of human social relationships has been hugely under-researched: we know very little about either the normative aspects of relationships or the cognitive factors that underpin them. This is surely a major area for research in the future.

ECOLOGICAL CONSTRAINTS ON SOCIAL ORGANISATION

Ecology must ultimately play a formative part in the evolution of society, if only because the opportunities that the world presents to us inevitably place limits on what living arrangements are possible. Indeed, there is a long and honourable tradition in just this respect within social anthropology (Forde 1949, Steward 1972, Harris 1971, Ellen 1982), although in recent decades this approach has lost favour. We cannot do full justice to this topic here, but in this section we briefly allude to a small number of key points that provide some insights.

The most obvious example of the way ecology influences social organisation is that of hunter-gatherers (see previous section). The constraints imposed by seasonal and yearly variability in the availability of different prey types and by the relatively low density and dispersed distribution of natural food sources means that hunter-gatherers are often forced by ecological necessity to live in small mobile groups. These typically number 30–50 individuals (5–10 nuclear families or the equivalent, depending on local living arrangements). More importantly, the membership of these groups is often unstable: families and individuals may come and go as they please, moving from one camp to another as social or ecological circumstances demand (Steward 1972, Kent 1996). However, individuals do not usually choose whom they stay with at random: those who are willing to share a campsite for any length of time tend to belong to the same higher level grouping (a clan or regional community). Such a higher level community often shares specific interests in common. Among the Australian Aboriginals, clan members may, for example, share a common paternal ancestry. Among the !Kung San bushmen of southern Africa, they share rights of access to the same dry season waterholes (Lee 1979), although these rights themselves may be held by virtue of a shared ancestry. Such systems are now commonly referred to as **fission–fusion social systems**.

FISSION–FUSION SOCIAL SYSTEMS

Fission–fusion social systems are common when one ecological factor demands large groups (for example, access to the one waterhole that won't dry up even in the worst drought) but other ecological factors demand small groups (for example, the resource base can only support small foraging groups). In conditions where ecological competition for limited local resources and the generalised stresses of group-living would naturally tend to cause the break-up and dispersal of large social groups (Dunbar 1988), they offer ecological flexibility without loss of the benefits of large group size. However, it inevitably imposes significant cognitive demands on any species for whom it might be advantageous. The ability to remember that one shares interests and obligations with some individuals but not others among those one does not see every-day is a crucial requirement for fission–fusion systems, and this may explain why apes (and humans) have smaller group sizes for neocortex size than monkeys do (Dunbar

1998a, in press). Without that ability to differentiate friends from strangers, a fission–fusion social system would not be a workable proposition.

An example of the way trade-offs on different ecological dimensions dictate living arrangements is provided by the Turkana pastoralists. The Ngisonyoka Turkana are nomadic pastoralists who live in a dry arid region of northwestern Kenya where rainfall predictability is low, both spatially and temporally, and plant productivity is primarily water-controlled (Dyson-Hudson 1989). On account of low rainfall and poor soil quality, crop cultivation is not possible and the diet of the Ngisonyoka is overwhelmingly animal-based. To support this lifestyle, five different livestock species are kept, including two large-bodied species (cattle and camels) and three smaller-bodied ones (sheep, goats and donkeys). Given that different livestock species have different grazing and maintenance requirements, movement within the region necessarily reflects the way these map onto the different ecological zones that exist in the habitat. Flocks and herds therefore often have to be dispersed in different areas in order to provide the right kinds of forage for the different species. At the same time, however, the Turkana's movements are constrained by the risk of raiding by bandits and other pastoral groups, while cooperation is also required in the digging and maintenance of water holes. Both of these require large social groups to provide sufficient manpower.

Dyson-Hudson (1989) points out that, in order to maximise livestock production and maintain **biomass** stability through time, a dispersion of livestock (and their human caretakers) is necessary, particularly during the dry season when the forage requirements of the five livestock species are not often available in one location. The division of the *camp* (the main residential unit) into satellite camps is a direct response to forage conditions, but this in turn is constrained by the availability of labour. At the same time, however, protection against banditry and large animal predators requires the Turkana to have ready access to other individuals. This is achieved by organising the camps into '*primary neighbourhoods*' that comprise families with particularly close affiliations. These primary neighbourhoods tend to reconstitute themselves even after migration takes place, and so elders of each camp meet regularly to exchange information and to make decisions, including migration plans (Dyson-Hudson 1989). During years when graze is abundant, primary neighbourhoods might cluster into even larger *secondary neighbourhoods*, though these are ultimately limited by the availability of forage and water.

BIOMASS

The point illustrated by the Ngisonyoka is that social organisation reflects the ecological requirements of individuals: in an environment with low resource predictability, residential patterns necessarily need to be flexible, and these will be at least partially determined by evaluations of risk and resource availability that each live-stock-owner makes.

Other ways in which economics can influence behaviour have already been discussed in Chapters 7 and 8. There we saw how the same problem – that of managing lineage survival when large family size generates many competing sons – affected societies which differed widely from each other in terms of their ecological bases. Thus, while the medieval Portuguese nobility introduced primogeniture (eldest son inheritance) in order to sustain the economic viability of their family estates through time and shipped excess sons off to the war, the Krummhörn farmers adopted ultimo-geniture (youngest son inheritance) in order limit the number of wealth transfers made through time and attempted to limit family size so as to have only two male offspring (the 'heir and a spare' strategy). Meanwhile, the Tibetan farmers took an even more extreme course of action, as dictated by their particularly harsh environment,

BOX 9.2

Evolutionary history of society

Both the archaeological and the ethnographic records suggest that human social organisation has changed dramatically over the course of the last dozen or so millennia. From the first appearance of the genus *Homo* some 2 MYA until the agricultural revolution 10 000 years ago, our ancestors lived as nomadic hunter-gatherers.

By the appearance of *Homo sapiens* 0.5 MYA, a fairly modern form of fission–fusion system had probably evolved, with unstable foraging units consisting of a variable number of (probably temporary) male–female pairs, plus the women's dependent offspring. Male–female pairbonding combined with some male provisioning can be expected by this stage because of the costs of rearing imposed by our large brain size (see **Box 5.3**). Judging by modern hunter-gatherers, these foraging groups would have been part of a larger community with a relatively stable membership that occupied a defined home range.

The invention of agriculture, however, was dependent on a settled lifestyle: agriculture, by its very nature, requires a long-term commitment because what is planted now cannot be harvested until half a year or more later (and must be guarded against herbivores in the interim). The archaeological record suggests that the earliest settlements were relatively small (no more than hamlets or very small villages). However, perhaps because settled populations are easier targets for raiding (Johnson and Earle 1987), settlement size grew rapidly until, within a mere 2000 years, recognisable urbanisation had set in. These early 'towns' are represented by Jericho (in the Jordan valley) and Çatal Hüyük (in central Turkey, which covered an area of some 13 ha). The maximum size of settlements has con-

tinued to grow at an exponential rate ever since.

The shift from a hunting-gathering economy to an agricultural one creates opportunities for major shifts in lifestyle and productivity. These in turn open up new possibilities in the social domain (Earle 1994). Thus while hunter-gatherer societies are necessarily small scale (dispersed social groups) and 'democratic' (wealth differentials are minimal, making it difficult for individuals to build economic or political power bases), the sedentary nature of agricultural economies makes monopolisation of reproductively essential resources (principally land) both feasible and profitable.

Competition for control of land (or other economic resources) then leads to the emergence of powerful individuals (often literally 'big men') or family-based alliances. This in turn creates opportunities for raiding that cause families to band together for security in larger villages or towns or to seek protection from locally powerful individuals, thereby enhancing the latter's power base (Johnson and Earle 1987). Johnson and Earle suggest that demography and competition may provide an inexorable force that drives societies into ever-larger units, so long as their economies remain agricultural (or even perhaps industrial). In their view, societies (certainly in the pre-industrial phase) are essentially self-defence alliances that grow in size under upward pressure from an escalating arms' race driven by raiding (the 'producer–scrounger' principle writ large: see Chapter 4). Ultimately, through a process of local domination, 'Big Man' systems give way to kingdoms (often associated with formal religious and political structures), and eventually nation-states (Flannery 1972, see also Friedman and Rowlands 1977).

and not only limited wealth inheritance effectively to one son, but prevented more than one marriage taking place per generation in order both to limit rates of population growth and to provide a committed work force to run a labour-intensive economy. The outcomes of all three alternative solutions were the same (the economic prosperity and long-term survival of the lineage), but the mechanisms involved were very much culture-specific.

The impact of changing economic circumstances on social practices is also evident in the case of the Kenyan Mukogodo. Cronk (1989, 1991) has shown how limited breeding opportunities effectively forced a change from hunting to herding among the Mukogodo (see Chapter 7). Unlike the other groups in their locality, the Mukogodo continued to pursue a hunter-gatherer lifestyle well into the twentieth century; then, between 1925 and 1936, they began to become increasingly reliant on livestock. This shift in economic base turns out to have been driven by the fact that the Mukogodo could not compete in a marital marketplace that was increasingly being dominated by their pastoralist neighbours: livestock came to be the preferred form of bridewealth payment rather than the more traditional beehives that the Mukogodo had formerly exchanged.

Cronk (1989) suggests that the subsistence change reflected a change in mating strategy: livestock acquisition increased intermarriage with the pastoralist neighbours of the Mukogodo, and the introduction of livestock as bridewealth payments meant that, for Mukogodo men to acquire wives, they too had to possess livestock since Mukogodo families now preferred the inflated bridewealth payments offered by livestock. The history of the transition makes this clear. Of the groups living in the Laikipia District, the Mukogodo were the poorest. One immediate consequence of this was inequalities in the amount of bridewealth that could be demanded. Relative to other groups in the area, Mukogodo females were cheaper (in terms of bridewealth payments) than females from other tribes: as a result, Mukogodo wives were attractive to the pastoralists and Mukogodo fathers soon preferred to sell their daughters to pastoralists rather than other Mukogodo unless the latter could match the brideprices offered by the pastoralists (see Chapter 7).

The marked increase in the number of Mukogodo women being married to non-Mukogodo pastoralists inevitably resulted in a net shortage of Mukogodo brides for Mukogodo men. This in turn meant that, if Mukogodo men wanted to marry, they too had to find brides outside of the group – or match the brideprice that could be obtained elsewhere. Since marrying outside the tribe involved bridewealth payments in the currencies used by other social groups (mostly livestock rather than the traditional beehives of the Mukogodo), a vicious circle very quickly ensued: the Mukogodo were forced to accept cattle as bridewealth in order to be able to pay cattle-based brideprices on behalf of their own sons. The alternative explanation for the subsistence shift (that is, that it reflected a change in ecological circumstances, or the need for a regular food source) is not supported by the available data. Cronk (1989) notes that there is no evidence to support the idea that the transition to pastoralism coincided with a decline in food availability from hunting or wild foods.

THE FREERIDER PROBLEM

Social living is founded on cooperation. This requires each individual to compromise on his/her own immediate desires in the interests of group cohesion. The issue here is short- versus long-term interests. Being a member of a group confers benefits that can only be obtained by cooperation, but these benefits only accrue in the long-term; there will therefore always be a temptation to renege on the social contract to steal a short-term advantage. Freeriders thus take the benefits of social cooperation but do not pay the costs. We are all of us familiar with this problem: it is the individuals who never pay their way, who park in the no parking zones, who pay less tax than they should. Anthropologists sometimes refer to this as the **collective action problem**.

COLLECTIVE
ACTION PROBLEM

On the wider scale of what economists refer to as *common pool resources* (resources that no one individual owns), this kind of behaviour generates a 'tragedy of the commons' (see **Box 4.4**). However, the consequences on the small scale can often be just as serious: individuals who relentlessly pursue their own selfish interests inevitably destabilise the group, since others then become unwilling to bear the burden created by the freeriders. As a result, the very foundations of group-living are eroded. In Chapter 2, we discussed the issues involved in terms of the Prisoners' Dilemma and its solutions (see **Box 2.3**).

Hitherto, most studies of the Prisoners' Dilemma have involved simple two-person games and, in this context, cooperation evolves relatively easily. However, a number of recent analyses have suggested that strict reciprocity becomes more difficult to sustain as social group size increases (Joshi 1987, Boyd and Richerson 1988, Enquist and Leimar 1993). Enquist and Leimar (1993), for example, modelled the freerider problem using an ESS approach. (The ESS approach is outlined in **Box 2.5**.) They considered a population of organisms that had to exchange resources in order to be able to reproduce. They were able to show that, as group (or population) size increases, freeriding (accepting the resource offered, but not giving anything away in return on a later occasion) becomes an increasingly successful strategy when the majority of the population are cooperators (**Figure 9.5**). So long as the search time required to find a naïve group is short and/or the investment required to trigger an exchange is small, freeriders will prosper. Freeriders can stay one step ahead of discovery in large populations because there are always new naïve individuals to exploit. This situation is exacerbated when the population is dispersed into small scattered groups. Under these conditions, it is difficult for individuals to exchange information about the misdemeanors of freeriders: if a freerider's strategy is rumbled in one group, he/she can always slip across to the group next door and pick up where they left off.

Freeriding is perhaps the most serious problem faced by social organisms. The size of human groups and their dispersed nature (unavoidable in a typical hunter-gatherer fission–fusion society) must have made this a particularly intrusive problem throughout most of our evolutionary history. Given this, we might expect humans to have evolved counter-strategies designed to keep freeriders in check.

At least five counter-strategies can be identified (Dunbar 1999). These are: (1) confining one's overt generosity to kin (even if kin fail to reciprocate, they at least provide a pay-back in terms of inclusive fitness through the processes of kin selection: see Chapter 3); (2) an increased sensitivity to social cheats (Cosmides 1989, Cosmides and Tooby 1992); (3) mechanisms of information exchange that allow susceptible individuals to identify freeriders more effectively (language is obviously likely to be especially important in this respect: Enquist and Leimar 1993, Dunbar 1996a); (4) the imposition of costly demands on potential cooperators so that freeriders are forced to incur a cost before they can gain the benefit from an exchange (the 'Zahavi handicap' solution); and, finally, (5) the imposition of punishments on freeriders. We deal briefly with each of these here.

The role of kin selection in facilitating altruistic behaviour has been discussed in some detail in Chapter 3. We need note here only the fact that it has been widely recognised that, because the coefficient of relatedness falls away rapidly with increasing kinship distance, kin selection can provide a buffer against the costs of freeriding only in small groups of closely related individuals (although high levels of polygyny skew resulting from small numbers of males siring most offspring in each generation may help to raise levels of relatedness within populations, and so facilitate the role of kin selection over a wider range of group sizes than might otherwise be the case). On

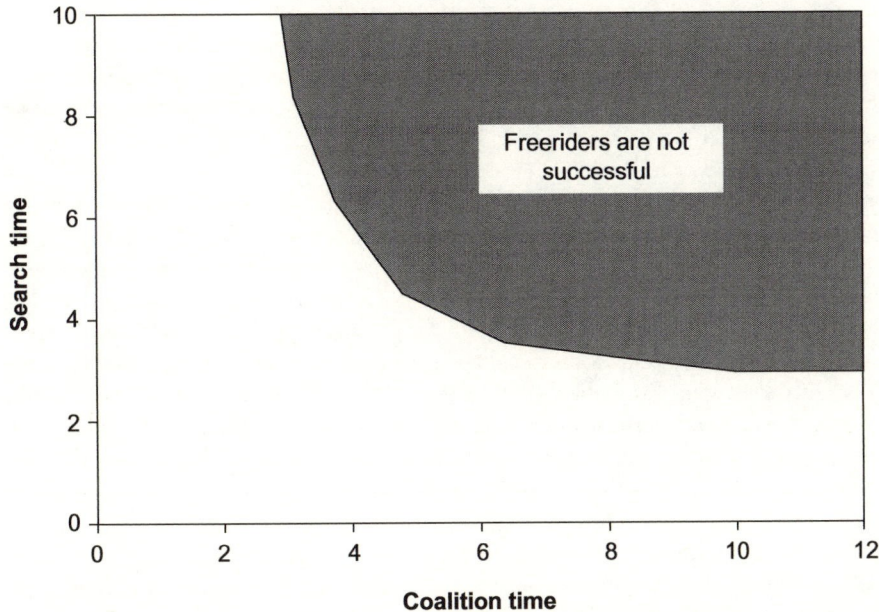

Figure 9.5 Enquist and Leimar's (1993) ESS model of freeriding in populations where cooperation is required for reproduction. In this model, individuals exchange resources that can later be used for reproduction. Freeriders who fail to reciprocate can be successful whenever the coalition time (interaction time required to persuade another individual to offer a resource) or the search time (time required to locate a new naive group in the population) are low. Redrawn with permission from Enquist and Leimar (1993).

balance, then, kin selection per se is unlikely to explain the occurrence of altruistic behaviour on the wide scale that is so characteristic of humans. However, in a recent analysis, Jones (2000) argued that kin selection can become effective over a wider range of degrees of relatedness providing individuals who act in concert to support a mutual relative effectively increase their mean relatedness to each other by doing so. For this to be so, mechanisms to enforce cooperation (such as mutual coercion or even punishment) may be necessary to ensure that the required changes in gene frequency occur within the population. Jones's (2000) models suggest that what he terms *group nepotism* can work effectively to solve collective action problems in groups that are substantially larger (for example, groups of around 500 individuals) than those where individuals act in isolation. He suggests that classificatory kinship (whereby unrelated individuals are included within close kinship categories), aided by a degree of 'genealogical amnesia', may be an important mechanism used to enforce cooperation in large human groups.

The second mechanism invokes specialised cognitive mechanisms that allow us to detect freeriders. Cosmides (1989, Cosmides and Tooby 1992) has argued that humans possess an in-built 'cheat detection' module that is sensitive to recognising infringements of social contracts. (The nature of these kinds of cognitive modules are discussed in more detail in Chapter 10.) Cosmides (1989) showed that people seem to be much better at solving reasoning tasks like the Wason Selection Task (see **Box 10.3**) if the problem is framed as a social contract (or social rule) problem (for example, 'only those over 21 are allowed to drink beer') than if the logically identical task is given in more abstract form (for example, 'every letter H must be associated with the number 3'). These differences in performance led her to conclude that the content of a social contract triggers an automatic 'cheat-detection' algorithm (or module). This algorithm alerts us to the fact that an opportunity for cheating exists and then prompts us to check for potential violators. As the contents of abstract rules do not involve social contracts, they do not trigger this algorithm and performance on these tasks is poorer in consequence.

Cosmides's claims have been criticised on a number of grounds, but we postpone further consideration of this debate to Chapter 11. The only point we want to make here is that there may be plausible cognitive mechanisms that make us particularly sensitive to freeriders. Some evidence in support of this is provided by a study of memory for faces: using a test-retest paradigm (with a week's interval between testing), Mealey et al. (1996) found that subjects were more likely to remember having previously seen someone's photograph if, on first exposure, that person was described as having exploited positions of power for personal gain. Whether this ability is a specific cognitive module or one of the by-products of some more general 'mind-reading' capacity, such as theory-of-mind (see Chapter 11), that can be used to imagine the mental states and intentions of others remains to be seen.

The third mechanism for controlling freeriders is language. Enquist and Leimar (1993) showed that if individuals can exchange information about freeriders, then freeriders find it much harder to invade a population of cooperators (**Figure 9.6**). Even allowing only a small opportunity to share knowledge (the outcomes of 25 per cent of previous encounters are passed on) has a dramatic effect on the ease with which freeriders can operate. Only when the time investment needed to form a coalition is very short does gossip fail to solve the problem.

Language may also be important in another respect. Orstrom et al. (1994) explored the role of face-to-face interaction in common pool resource problems. They used an experimental design in which groups of subjects sitting at separate computer consoles could invest in either of two markets, one of which yielded a fixed rate of return while the other yielded a return that was proportional to the total amount invested by all the subjects in each investment round. A strategy in which all subjects cooperated by investing in the variable-return market provided the best deal for everyone, and this readily became apparent after just a few cycles of the experiment. However, in the variable-return market the dividend rate was an exponential function of investment: low levels of investment produced poor returns and high rates of return were obtained only as the number of investors approached the maximum.

Because almost total cooperation was required for high pay-offs, there was a strong temptation for individuals to renege on any such informal agreement and invest in the alternative market, thereby assuring themselves of a higher pay-off (and a lower return for everyone else). The dilemma was that if even one or two subjects did this, then anyone that held to the cooperative strategy would come out of the experiment with considerably less than they would have gained by reneging. Each subject could see what investments were being made on their linked computer screens, but they could not see which individual was responsible for a particular action. The results showed that, under baseline conditions with no communication allowed, the average payout was only about 20 per cent of what they should have got from the optimal strategy, thanks to the frequency with which subjects adopted the more selfish strategy (**Figure 9.7**).

Allowing the subjects to take a break during which they could talk to each other greatly reduced the frequency of freeriding, even though subjects could not tell who was actually responsible for the infringements: merely being able to exhort everyone

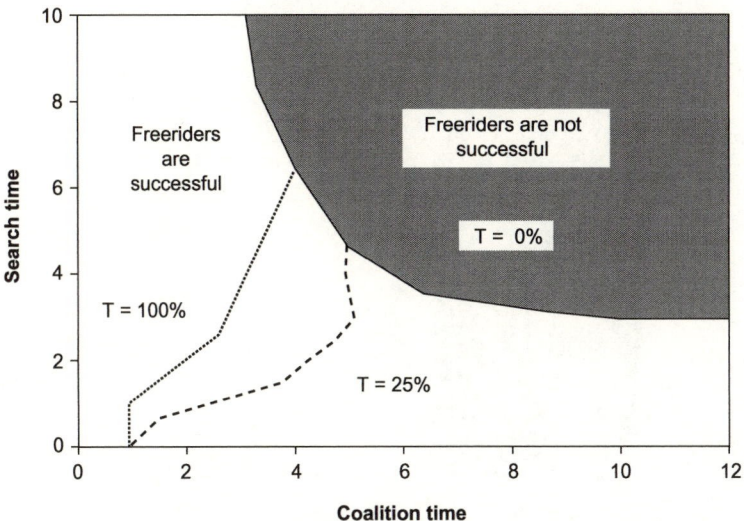

Figure 9.6 When the opportunity to exchange information about the behaviour of freeriders is permitted, the area of the state space where freeriding is possible is significantly reduced. (The zone where freeriding is possible always lies to the left of a given *T*-line.) *T* = proportion of information exchanged. *T* can represent either the proportion of interaction outcomes passed on, the proportion of individuals who share knowledge or the reliability of the information exchanged. Redrawn with permission from Enquist and Leimar (1993).

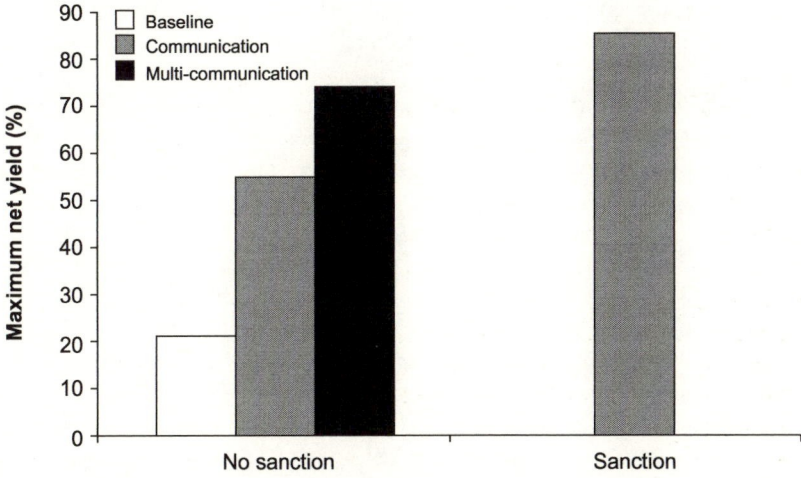

Figure 9.7 When a group of subjects played a computerised common pool resource game that allowed them to invest either cooperatively as a group or selfishly, the temptation to defect and invest for selfish benefit meant that the group as a whole earned significantly less than it would have done had everyone cooperated on the optimal strategy (baseline condition). The opportunity to exhort the group to stick with the optimal strategy or to lambast the defectors (even though their identities were unknown) provided by one or more conversation breaks during the game (communication conditions) significantly reduced the lost income for the group, indicating an important role for face-to-face communication. There was further improvement if conversation breaks were combined with the opportunity to punish (anonymous) defaulters during a game (sanction condition). Reproduced with permission from Dunbar (1999).

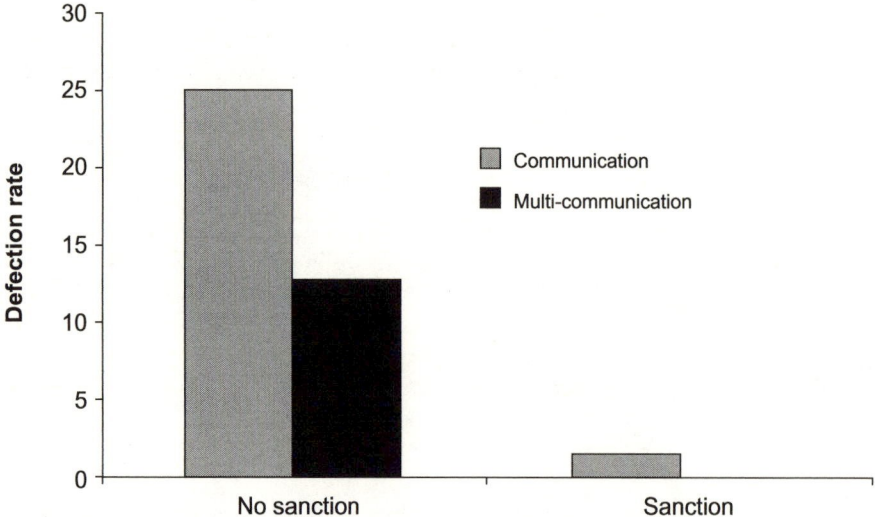

Figure 9.8 In the common pool resource experiment described in Figure 9.7, the frequency of defections (investments against the optimal strategy for the group as a whole) were significantly reduced when subjects had the opportunity to meet face-to-face and talk among themselves, and were virtually eliminated when combined with the opportunity to impose sanctions on the (anonymous) defectors. Reproduced with permission from Dunbar (1999).

to toe the line was usually enough to persuade potential freeriders to conform (**Figure 9.7**). Allowing the group to impose sanctions (financial penalties) on (anonymous) freeriders was better still, though not that much better than simply allowing repeated interludes for conversation. The success of these strategies was reflected in the frequencies of defection: individuals defected from the common strategy less often when exposed to the harangues of their fellows and virtually ceased defecting altogether when subjected to penalties for doing so (even when they themselves remained anonymous) (**Figure 9.8**).

Enquist and Leimar's analysis also suggested that the costs incurred by freeriders might be an important factor influencing their success. Imposing costs on freeriders is often a very successful strategy. It forms the basis for the long drawn-out (and often economically expensive) business of courtship: courtship feeding (in which a female demands a gift of food before she will mate with a male) is particularly common in the animal world (especially among insects: Thornhill 1976). Human females sometimes behave similarly, expecting small (but preferably frequent) tokens of interest, including presents (fur coats, diamond rings), treats (dinner at expensive restaurants, the best seats in the cinema or theatre) and a significant investment of time and attention (failing to pay enough attention is a common complaint). These are all, in effect, tests of the suitor's honesty and reliability, as well as sometimes being tests of his wealth (that is, ability to invest in children).

Nettle and Dunbar (1997) have suggested that language may itself provide a mechanism for imposing costs on prospective social allies. One of the particularly puzzling features of human language is the ease with which languages evolve. There are estimated to be some 6000 distinct languages currently in existence, plus many tens of thousands of dialects (local usages or pronunciation). The number of extinct languages is unknown, since we only know about those for which we have written records. The rate of language change is so rapid, that within a relatively short space of time it is possible for local dialects in neighbouring areas to evolve into mutually incomprehensible languages. Latin, for example, has spawned some eight descendent mutually incomprehensible languages (Italian, French, Spanish, Catalan, Portuguese, Occitan, Rhaetian, Sardinian and Romanian) within the last 1500 years, while the ancestral Proto-Indo-European spoken somewhere in the region of modern Turkey around 6000 BC has spawned the 150-odd languages of the Indo-European family (ranging from English, Russian and the Latinate languages in the West to Persian, Bengali and Urdu in the East, not to overlook a raft of now extinct languages that include Latin, Sanskrit and Hittite), all in less than 8000 years (Crystal 1997, Ruhlen 1994). There has never been any really satisfactory explanation as to why dialects and languages should be so numerous, or why they should be spawned so easily: after all, language is designed to improve communication, so why deliberately make it difficult to communicate?

Nettle and Dunbar (1997) argued that dialects evolve so as to allow us to recognise members of our own small communities, in particular the communities within which we grew up. (Indeed, one of the standard ethnological definitions of a tribe is all the people that speak the same language.) Dialects have the peculiar property that they are hard to learn and are best learned when very young. Adults who learn a new language or move to an area where a different dialect of the mother tongue is spoken rarely learn to speak an accentless version of the new dialect, no matter how hard they try or how long they live there. Thus dialects have two key properties that make them valuable as cues of origin: they are hard to learn (fulfilling the biological dictum that honest cues are costly: **Box 2.7**) and they mark you out unmistakably as having grown up in a particular community. Trudgill (1999), for example, observes that traditional English dialects (that is, those spoken up until as late as the 1970s) allowed a particular speaker to be located to an area as small as 60 km in diameter within England. There is good experimental evidence to suggest that subjects listening to taped speakers regard those with their own regional dialect as having higher integrity and attractiveness (Cheyne 1970, Giles 1971).

The final means of controlling freeriders is to impose significant costs on them. Boyd and Richerson (1992) used an ESS modelling approach to show that cooperation can be evolutionarily stable even when social group size is large, providing failure to cooperate is punished. If the cost to the individual of punishing defaulters is less than the benefit gained from cooperation, then it is possible for a strategy that both cooperates and punishes defaulters and one that only cooperates to coexist. Moreover, their model suggests that a 'moralistic' strategy (punishing those who fail to punish defaulters) can also be evolutionarily stable under a wide range of conditions.

Punishment is broadly successful if both the probability of detection and the penalty are both high enough. Studies of poaching suggest that fines and modest custodial sentences only deter poachers if the risk of apprehension is high (Milner-Gulland and Mace 1998, Cowlishaw and Dunbar 2000). In traditional societies, the penalties for those who defaulted on social obligations or local mores were often severe, and included ostracism and, in the more extreme cases, execution (Lee 1979, Otterbein and Otterbein 1965, Boehm 1986, Mahdi 1986, Knauft 1987). In tribal societies, banishment (a punishment frequently meted out to the more troublesome *berserkers* in Viking communities) carried very serious risks because a combination of adverse environmental conditions and hostile neighbours was often prejudicial to a lone individual's survival.

BOX 9.3

How dialects control freeriders

Nettle and Dunbar (1997) modelled the role of dialects in moderating the success of freeriders using a simple one-dimensional spatial model that consisted of a population of 100 individuals who had to exchange resources in order to reproduce. At the end of each cycle of exchange, the 20 individuals with the greatest wealth were allowed to reproduce with a probability of 0.5 and the 20 poorest died with a probability of 0.5. (This was a device to keep the population size constant, and thus avoid the computational costs escalating out of hand: the details don't affect the outcome of the simulation.)

The simulations showed that freeriders (who accepted resource gifts but did not offer any in return to other individuals) easily invaded the population of cooperators and eliminated them within a few generations. But, once dialects (in this case, six digit numbers attached to each individual) were introduced and individuals would only exchange resources if they shared a similar dialect, only freeriders willing to change their dialect to that of a donor were at all successful (**Figure 9.9**). However, even these could be controlled if the memory for past encounters was long (more than five previous encounters) and the rate of natural dialect change was high. (In the model, dialects were allowed to evolve slowly as a result of the random mutation of one element in each

individual's six-digit dialect at the end of each generation; individuals who exchanged resources on the basis of having five out of six identical dialect characters then adjusted their dialects so that they were identical in all six.) But if the rate of change was increased so that 50 per cent of the dialect changed each generation, then freeriders found it all but impossible to invade (**Figure 9.9**).

The features of successful dialects in this artificial world are very similar to those observed in real life. Natural dialects change rapidly, with each generation developing its own style of speech, words and even pronunciation even within the same language/social community. A rapidly changing dialect is more difficult to mimic, not least because you might mimic the wrong subculture (for example, the middle-aged adults rather than the young teenagers). Being able to identify yourself as a member of the same small natal cohort encourages other members of that cohort to be less demanding of you, to treat you more as a favoured friend than a stranger to be suspicious of. It triggers codes of obligation that are probably underpinned by kin selection. If two people are born into the same small community, the likelihood is that they are more closely related than the average for the population as a whole, at least in traditional societies.

SOCIETY, VIOLENCE AND WARFARE

Warfare has been a part of human social life for as long as history can tell us. So much is evident from the fortified medieval hill towns of central Italy and the Iron Age hillforts of northern Europe: such explicit use of naturally defended sites is otherwise inexplicable. Nonetheless, anthropologists have often been enamoured of the suggestion that there once existed a pristine state of nature when humans lived in harmony with each other and their environment. The ills of modern life (the ravages of epidemics, famine and war) are attributed to the socially destructive nature of modern city life and, in particular, the colonial and capitalist heritage of the industrialised West as reflected in a tendency to favour individualism at the expense of the social group. This has led to something of an attempt to justify claims that hunter-gatherer societies (now or pre-contact with western influence) were relatively peaceful (see for example Knauft 1987, Moore 1990).

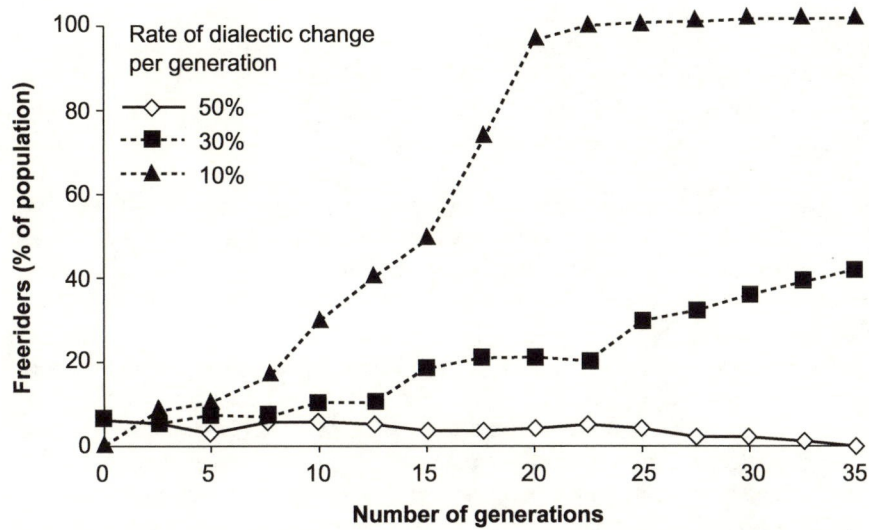

Figure 9.9 In a simple one-dimensional spatial model of cooperation, freeriders who took resources from other members of the population in order to be able to reproduce but declined to reciprocate were soon able to out-reproduce and eventually drive to extinction a population of cooperators. However, if cooperators used a dialect to identify members of the same cooperating community, freeriders were less successful. If individual components in the dialect were allowed to change at a rate of about 50 per cent per generation, freeriders were unable to invade; rates of dialect change as low as 10 per cent per generation were, however, too low to control freeriders. Reproduced with permission from Nettle and Dunbar (1997).

In fact, most of the evidence suggests that hunter-gatherer societies, now or in the past, are neither more nor less peaceful than the most violent of modern cities. Lee (1979) calculated an annual homicide rate of 29.3 per 100 000 people among the !Kung San in the first half of the twentieth century. He contrasted this with rates of between 1.1–11.6 per 100 000 in 11 Ugandan tribes during the colonial period in the late 1940s and 9.2 per 100 000 in mainland USA in 1972 (though individual US cities such as Detroit and Washington DC achieved rates as high as 40 per 100 000 population). The catalogue of murders listed in *Njal's Saga* suggests that Vikings also ran a pretty high risk of dying violently at the hands of their fellows: 30 out of 111 adult males were killed in the disputes and vendettas that dragged on over a 50-year period and left one family lineage completely wiped out and only four out of 14 extended families untouched. We might in fact be pardoned for concluding that, contrary to popular belief, modern societies are much safer places to live, perhaps because a police force imposes some degree of social control.

The issue invariably becomes polarised into one about human nature: are humans naturally violent as a species or not? This, however, is to misunderstand the evolutionary point. Of course individual humans can be violent; and merely by virtue of the fact that individuals have the capacity to be violent, it follows that humans as species are naturally violent. But, as many mathematical game theoretic analyses have shown, uncontrolled violence is an unsustainable evolutionary strategy. Aggression and violence are best used tactically as circumstances demand. Maynard Smith's (1982) ESS models show that mixed populations of Hawks and Doves are typically more stable than all-Hawk or all-Dove populations (see **Box 2.5**).

Violence, or at least the threat of violence, may be a very profitable gambit during a quarrel about resources. Just as two red deer pace up and down and roar at each other in an attempt at intimidation (Clutton-Brock and Albon 1979), so threats of aggression may persuade an opponent to flee without bothering to contest the issue. That violence does, in evolutionary terms pay, is shown by the higher fitness of Viking *berserkers*. Members of a *berserker*'s family were half as likely to be murdered during the events described in *Njal's Saga* as those whose families did not boast a *berserker* (**Figure 9.10**). Despite a high personal risk of mortality (two of the three in *Njal's* community were killed), *berserkers* did seem to have higher fitness (measured in terms of the number of surviving relatives, devalued by their coefficients of relatedness) than non-*berserkers*.

Anthropologists have often noted that people in traditional societies commonly have a deep horror of violence, and have used this as evidence that such peoples are naturally pacific (for example, Knauft 1987). A more plausible interpretation is that they fear violence for the very good reason that they have experienced its consequences more often than they would wish. They do not need to have that experience on a daily basis: the recollection of a single instance of serious violence once in a decade is probably more than enough to frighten most of us.

If violence is a conditional strategy, what precipitates it? To be sure, some individuals simply lose it and become dangerous to have around. Such individuals are

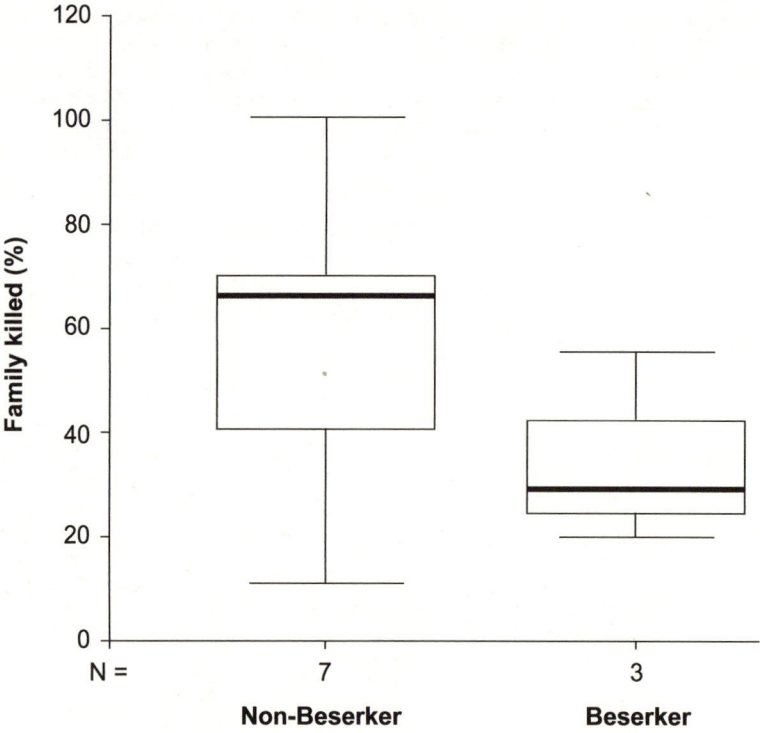

Figure 9.10 Although Viking *berserkers* suffered significantly higher rates of mortality at the hands of their own community, they had a beneficial effect on the survival of the male members of their individual families: the families of the three *berserkers* in the community that forms the context for the Icelandic *Njal's Saga* suffered significantly less mortality than the 7 families that did not contain a recognised *berserker*.

BOX 9.4

How not to do evolutionary analyses

Moore (1990) used the Cheyenne (one of the Dakota family of American Plains Indian tribes) to support the claim that humans were not naturally violent: he claimed that the demographic data for the Cheyenne proved that a 'gene for aggressiveness' would not be selected for. The Cheyenne chiefs divided themselves into two principal types: peace chiefs (who did not fight) and war chiefs (who were responsible for leading the tribe in battle). Moore (1990) showed that the mortality rates on war chiefs (who sometimes staked themselves to the field of battle in order not retreat) were sufficiently high and their reproductive rates sufficiently low (they undertook not to marry, at least until their active fighting days were over) that any 'gene' for violent behaviour would be selected against so heavily that it would soon be eradicated from the population.

Unfortunately, Moore's analysis is simplistic for at least three reasons (Dunbar 1991). First, it assumes that violent individuals always behave in a violent manner, when in fact violence is clearly a conditional strategy: everyone has the 'gene for aggressiveness' but individuals express it to different extents (mainly depending on their circumstances). The proper comparison is, therefore, not one between 'a gene for aggressiveness' and 'a gene for non-aggressiveness' (as Moore assumed) but between the two strategies 'peace chief' and 'war chief' (both of which inherit the gene for aggressiveness). Second, Moore's analysis does not consider the possibility that war chiefs were attractive as illicit lovers by virtue of their displays on the field of battle, and so might have had many offspring conceived by EPCs (see Chapter 5). Third, and perhaps most importantly, it ignores the fact that peace chiefs and war chiefs

had very different choices in life. Peace chiefs held their office by virtue of heredity (it was only possible to be a peace chief if your father had been one); war chiefs, in contrast, tended to be orphans or the children of very low status families. Moore's analysis assumes that all chiefs could choose equally between being war and peace chiefs. While this was, in principle, true for peace chiefs, it was not for war chiefs: their choice was between becoming a war chief (a high-risk strategy with a potentially high pay-off) or remaining as 'braves' at the bottom of the social scale. In either case, their opportunities for marriage were poor. But, for someone with a 'berserker' nature (see **Box 9.5**), becoming a war chief was probably a more promising option because he had a better chance of surviving the risks of battle than the average individual – and doing better than average reproductively if he did survive.

Dunbar (1991) suggested that, in fact, the peace-chief/war-chief complex might have been an ESS. The demographic data provided by Moore (1990) indicate that the reproductive pay-offs to peace chiefs and war chiefs were, respectively, 3.8 and 2.57 offspring (as indexed by the mean number of children per adult male in peace and war bands), giving a pay-off ratio of 1.56:1. The frequency of males aged 20–29 (that is, at the start of their adult and reproductive careers) in peace and war bands in the 1882 census was 29 and 47, respectively, giving a ratio of 1:1.62. Since the pay-off and frequency ratios are almost exact mirror images of each other (as required by ESS theory: see **Box 2.5**), it is likely that these two strategies are held in dynamic equilibrium by their relative costs and benefits.

rarely tolerated in traditional societies. In small-scale societies like those of hunter-gatherers, individuals who became too uncontrollable were much feared: they were likely to become the target of collective action by the rest of the group (see Lee 1969, Knauft 1987). Among the Vikings, the unruly behaviour of *berserkers* back in the community frequently resulted in their being banished or even murdered – usually as a result of concerted action by a group of exasperated members of the community. In *Njal's Saga*, two of the three recognised *berserkers* were killed in this way.

BOX 9.5

When does it pay to go berserk?

When individuals were murdered, relatives of the deceased had two options under Viking law: they could either exact a revenge killing on the original perpetrator or a member of his family, or they could demand 'blood money' in compensation for their loss. While revenge killings were more frequent when the victim was related to the avenger, this was largely dependent on whether or not the murderer was a *berserker*.

Berserkers (whence comes the English word 'berserk') were renowned for their fearlessness in battle (sometimes, but not always, helped on by drinking an infusion of a plant called the bog myrtle), their ability to fight on regardless of any wounds received, and their somewhat impetuous tendency to kill first and ask questions later.

Berserkers were obviously in great demand among raiding crews on their way to rape and pillage in the required Viking fashion, but most people would frankly rather that they didn't live next door (a point amply attested to by the frequency with which acknowledged *berserkers* were banished in the Viking sagas).

The involvement of a *berserker* in a murder increased the costs of a revenge killing considerably, since it was unlikely that a *berserker* could be overcome by an individual acting alone. Consequently, whenever a *berserker* was implicated in murders, the relatives of the deceased were much more likely to settle for blood money than if a non-*berserker* was involved (**Figure 9.11**).

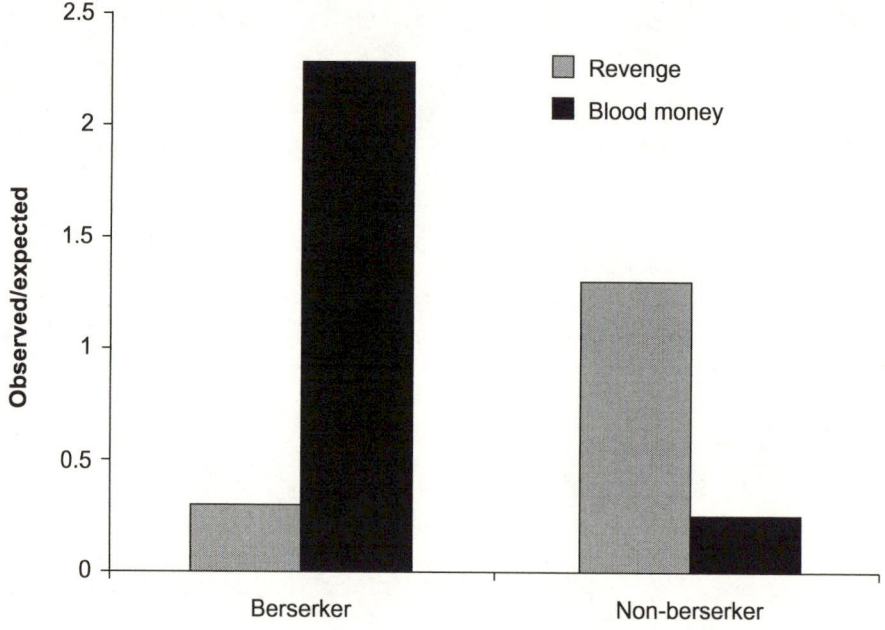

Figure 9.11　When a relative was murdered, Vikings had the choice between a revenge murder or accepting blood money. Because *berserkers* had a fiercesome reputation as dangerous individuals, aggrieved relatives of a murder victim were significantly more likely to accept blood money if the murderer was a *berserker*, but to prefer a revenge killing if the murderer was not a *berserker*. The plotted variable is the ratio of observed murders relative to the number expected on the basis of the proportion of *berserkers* or non-*berserkers* in the population. Source: 34 murders recorded in *Njal's Saga* (data from Dunbar et al. 1995).

Aside from a handful of pathological cases, most instances of everyday violence are concerned with the acquisition (or occasionally the retention) of resources. This may be economic resources or it may be reproductive resources (which usually means women). Daly and Wilson (1988b) reported that an unexpectedly high proportion of homicides are committed by men who have just been deserted (or believe they are about to be deserted) by their partners. In many cases, it is the partner that bears the brunt of this, but occasionally men will also seek vengeance on the offending rival male.

On the larger scale, civil wars are often quite explicitly a consequence of attempts by different tribal, ethnic or social groups to gain economic and political control of a region. Scarcity of economic resources (for example, land) is often the trigger both for migration (for example, the Viking episode and the late medieval Portuguese explorations: see Boone 1988) as well as for warfare on a national scale (Durham 1976, Johnson and Earle 1987, Shaw and Wong 1989, Homer-Dixon et al. 1993). Warfare probably tends to enhance the risks of violence against civilian populations for two reasons. One is that demonstrations of violence are often effective in persuading individuals that challenging the invaders is not a worthwhile strategy (essentially, the *berserker* strategy); the other is that the absence of effective social controls makes it less likely that perpetrators of violence and theft (including rape) will be held accountable for their actions. The fact that the latter is combined with an absence of local social or kinship ties means that invaders lack any of the obligations that might normally dissuade individuals from behaving in this way within their own communities. This is reflected in the high frequencies of rape that typically characterised the Russian army's advance into Germany at the end of the Second World War, as well as the Iraqi invasion of Kuwait, the Serb's Bosnian and Kosovo campaigns and almost every other area where civil authority has broken down.

A much debated issue has been the question of whether there are sex differences in aggression and violence. Conventional wisdom has it that men are more violent than women, and they are certainly responsible for a much higher proportion of homicides than women, both in traditional societies like the !Kung San (Lee 1969) and modern western societies (Daly and Wilson 1988b). This does not, however, necessarily mean that women are less aggressive than men. Campbell (1995) has argued that women (and especially members of female gangs) can be just as aggressive as males, but that they typically express it in rather different ways. Although 73 per cent of Campbell's sample of British girls stated that they had been involved in at least one physical fight with another girl, these fights tended to involve relatively little actual damage to an opponent. Female aggression tends to be verbal rather than physical, and often tends to be provocative (in the sense that women may often stand on the sidelines and incite men to go to war on their behalf). Several of the murders in *Njal's Saga*, for example, were instigated by women.

In her analysis of aggression in teenage girls, Campbell (1995) found that much of the aggression (and occasionally physical violence) exhibited by these girls was related either to defence of the girl's reputation or to competition over individual males (see also Marsh and Paton 1986, Schuster 1983). Accusations of infidelity or promiscuity ranked high among the immediate causes of fights. Cook (1992) reported very similar findings from the island of Margarita off Venezuela. Campbell (1995) argued that female aggression was most likely to erupt into physical violence under conditions where women are free to choose their own mates, male quality is highly variable and there are rather few high quality mates available.

All this is not, of course, to justify or glorify violence in any way. Our concern here is simply to state the facts and to understand what lies behind those facts. Only

BOX 9.6

Evolutionary explanations of rape

Perhaps because the topic is unusually sensitive, attempts to explore rape (and, more generally, coercive sex) have proved particularly controversial – mainly, it seems, because there has been an implicit assumption that evolutionary explanations imply either that such behaviour is inevitable or that it is in some way thereby justifiable. Neither claim is, of course, true. The real issue from an evolutionary point of view hinges around the question as to whether such behaviour is simply the maladaptive tail-end of male (and, occasionally, female) behaviour or is an evolutionarily adaptive mating strategy.

The fact that at least some victims of rape are non-reproductive (children, post-menopausal women and, occasionally, men), combined with the fact that rape victims are occasionally murdered, has been taken to imply that rape is dysfunctional behaviour. This view is given weight by two common themes in the literature on rape, namely that rapists are often unconcerned about victims' interests (Malamuth et al. 1993; Lisak and Ivan 1995) and that they commonly misread women's affective cues (usually by interpreting rejections as invitations: Malamuth and Brown 1994). That at least some instances of rape are the result of behavioural pathology seems uncontroversial. However, it is far from clear that such a label is appropriate for all cases of rape: this is a purely empirical question that requires more detailed analysis.

The evolutionarily more interesting question is whether rape (or, more generally, coercive sex) has the characteristics of a functionally adaptive mating strategy that uses physical aggression as one among many tactics designed to persuade a reluctant partner to mate (Palmer 1991b, Thornhill and Thornhill 1992). There are two possibilities in this respect. One is that rape constitutes the functionally maladapted extreme of normal male (and maybe female) mating behaviour rather in the way murder can be (**Box 3.4**); the other is that rape itself is a facultative response to particular circumstances (in effect, an alternative mating strategy). The use of force

should not distract us, since coercion is often the means whereby an individual achieves particular objectives (see **Box 9.5**).

To claim that the behaviour confers a fitness benefit, we need to show that coercion/rape has consequences that enhance the perpetrator's fitness. Two lines of evidence have been adduced to suggest that it might. One is that rape victims have a significantly higher probability of conception following a single instance of intercourse than is the case in unprotected voluntary copulations (Holmes et al. 1996). One possible explanation for this is that rapists target women who are likely to be fertile. Attacks on children and post-menopausal women notwithstanding, younger women (that is, those who are more fertile) are significantly more likely to be the victims of rape than would be expected given their representation in the population (Greenfield 1997). Although this might equally be explained by their greater social availability (women of this age are more likely to be encountered because they are socially more active), the fact that the majority of forced copulations involve individuals who already know each other argues against this.

Starks and Blackie (2000) showed that, in the USA, the frequency of reported rapes correlates positively (and highly significantly) with the frequency of divorce across years between 1960–95 and, within years, across states. They show from the US demographic data that high frequencies of divorce result in high frequencies of remarriage for men but not women, and argue that this results in what amounts to serial polygamy, one consequence of which is a rising number of males unable to obtain mates. (Similar findings have been reported from Sweden: Forsberg and Tullberg 1995.) Starks and Blackie suggest that rape is one reproductive strategy used by males who have reduced access to sexual partners because of a high polygyny skew.

On the individual scale, the evidence in support of this hypothesis is mixed, with some studies reporting that rapists are sexually frustrated (for example, Kanin 1985) while others report no cor-

BOX 9.6 cont'd

Evolutionary explanations of rape

relation between sexual disadvantage and the use of coercive sexual tactics (for example, a questionnaire-based study of 156 young males: Lalumière et al. 1996b). Instead, the latter study found significant correlations between use of coercive behaviour and both sexual success and desire for wider sexual experiences. This might suggest that rape is not a unitary phenomenon with a single explanation.

Smith et al. (2001) present a simple model to evaluate the costs and benefits of rape as a reproductive strategy. A major component of their model is the social costs of rape (revenge by the victim's relatives and the indirect impact on the rapist's own kin). Although the values they attach to the cost and benefit components in this model are arbitrary, the model does suggest that, in small-scale societies where victim and perpetrator may be known to each other and to other members of the community, the costs of rape can exceed its benefits by an order of magnitude.

This model clearly only applies to small-scale societies where the frequency of within-community rape is often minimised by draconian penalties (Hartung 1992) and a high risk of apprehension. (Note that, in most traditional societies, forced copulations are not considered rape if the victim is a member of another social group.) It will not apply to situations where social control is reduced,

as in the largely anonymous societies of the industrialised world or during (civil) war. The recent history of the Balkans attests to the fact that, once formal social controls are removed, rape can become extremely common.

Most studies of this topic have focused on the rapist, but the victim's situation also invites attention. Somewhat controversially. Thornhill and Thornhill (1990 a, b) have argued that the victim's psychological response to rape may be titrated by the extent to which she has reproductive interests at stake (especially where paternity certainty is involved). From an analysis of interviews with 790 US rape victims, they concluded that the traumatisation experienced by the victim is significantly greater for reproductive age women than pre- or post-reproductive women, for married as opposed to single women of reproductive age and for victims of stranger rape as opposed to rape by family or friend. These differences in psychological response may reflect attempts to minimise the risk that the victim will be blamed for precipitating the attack (or, alternatively, for falsely claiming to have been attacked after consensual intercourse). That this might be a serious consideration is indicated by the fact that, in many traditional cultures, rape victims may be ostracised or even killed for bringing dishonour on the family in much the same way that a woman who engages in premarital sex does.

by doing so can we hope to change human behaviour for the better. Too often, commentators in the social sciences and humanities confuse a biological understanding of violence with the claim that such behaviour is 'natural', inevitable and cannot be changed. The short answer is: of course it can be changed. Changing behaviour, in humans, as much as in animals, depends on altering the balance between the costs and benefits. If violence pays, people will adopt it as a strategy precisely because it is effective. Violence and other forms of anti-social behaviour are much less common in places where there is less to be gained by it (as in most modern societies where everyone is generally relatively wealthy) or where the risk of detection and punishment is high.

*

Having reviewed the various behavioural strategies individuals use to negotiate relationships, rear offspring and live together in mutually beneficial ways, we can go on to investigate the underlying psychological tendencies that help to promote these abilities at the proximate level. In the next chapter, we discuss a number of the cognitive adaptations humans have evolved to facilitate their effective functioning in a complex social world.

Chapter summary

- Kinship classifications are a universal feature of all human societies and serve to differentiate between members of a society who are more and less important to an individual. Not only do they define patterns of obligation, but they may also place constraints on the individuals who can marry.

- Frequent assertions to the contrary notwithstanding, close analysis suggests that social kinship categories often do reflect biological kinship (at least, once allowance has been made for society-specific risks of paternity uncertainty). However, unrelated individuals may sometimes be classified as kin, possibly in order to foster relationships with them.

- The division of labour between the sexes influences the balance of power. In cases where men control the resources that influence reproduction, they also control the activities of women. In circumstances where women contribute more to the economy, they are expected to exert more power. Sex differences in dispersal are also related to the ability of men to defend resources and dictate whether men or women remain in their natal groups on marriage.

- The size of human groups is limited by a cognitive constraint on an individual's ability to remember the details of individual relationships and to use this information in planning and executing social strategies.

- Ecological variables influence social structure by placing limits on the kinds of living arrangements than people make and the size of the groups in which they live.

- Social living is founded on cooperation but, as groups grow larger and more dispersed, opportunities for cheating increase. In order to avoid disintegration, social groups must devise mechanisms to ensure that individuals toe the line on social contracts. Humans may have evolved psychological mechanisms that increase our ability to detect cheats. Some aspects of language may also be effective at controlling group members' behaviour.

- Aggression and violence are not inevitable facets of humans' psychological make-up, but rather represent a strategic and facultative response to particular circumstances. Although men are more overtly violent than women, the latter also use aggressive tactics to obtain resources (usually indirectly in the form of mates). Unlike male aggression, female aggression tends to be verbal rather than physical, and often tends to be provocative (inciting others to violence).

■ A biological basis to human aggression does not automatically lead to the conclusion that this behaviour cannot be changed. Violence is used only when it pays. Under conditions where aggression does not achieve high pay-offs, such anti-social behaviour is rare.

Further reading

Chagnon, N. A. (1974). *Studying the Yanomamö*. New York: Holt, Rhinehart and Winston.

Daly, M. and Wilson, M. 1988b. *Homicide*. New York: Aldine de Gruyter.

Diamond, J. (1997). *Guns, Germs and Steel: a Short History of Everybody for the Last 13 000 Years*. London: Jonathan Cape.

Johnson, A. W. and Earle, T. (1987). *The Evolution of Human Societies: From Foraging Group to Agrarian State*. Stanford: Stanford University Press.

Kent, S. (ed.) (1996). *Cultural Diversity among Twentieth Century Foragers: an African Perspective*. Cambridge: Cambridge University Press.

Knauft, B. M. (1987). 'Reconsidering violence in simple human societies: homicide among the Gebusi of New Guinea'. *Current Anthropology* 28: 457–99.

So far, we have considered the decisions that individuals make about their behaviour and its fitness consequences in some considerable detail. However, we have only briefly considered the psychological processes that mediate these decisions. This chapter attempts to restore the balance somewhat by bringing not only minds and psychology into the discussion but also the underlying 'wetware' of the brain itself. An increased ability to tap into the brain (and hence the mind) has been brought about by great advances in scanning techniques that can measure the activity of normal, living brains as they go about their business. This has allowed a much more accurate mapping of brain structure and function than was previously possible when researchers had to wait until subjects died in order to identify abnormalities in the brains of brain-damaged individuals or those suffering from mental illness.

One issue has come to dominate discussions in developmental cognitive psychology in the past decade, and this is the question as to whether the human mind is **'domain general'** or **'domain specific'**. The issue here is whether the mind is composed of a number of essentially innate modules, each designed to solve problems of a particular kind – the 'Swiss army knife' model of Cosmides and Tooby (1992) – or, alternatively, in effect a single 'domain general' reasoning device that we apply flex-

DOMAIN GENERAL
DOMAIN SPECIFIC

ibly to many different situations. The latter view may still allow for sets of neurones to learn (in the sense that neural nets learn) to handle particular kinds of problems as a result of experience. In one respect, this dichotomy (though intermediate positions do exist) is just another form of the long-running 'nativist' versus 'environmentalist' debate within developmental psychology. The central issue here is whether cognitive abilities (and hence behaviour) are innate (hence, instinctive) or are learned as a result of experience. Perhaps inevitably, this debate has spilled over into evolutionary psychology, where it lies at the heart of the EP view.

NATIVIST
ENVIRONMENTALIST

However, it is important to point out that, in one sense, how this particular issue is resolved is neither here nor there from an evolutionary point of view. The question is undoubtedly interesting, and it may have implications for how fast human behaviour can change (or indeed, whether it can be changed). However, as we were at pains to point out in Chapter 1, Tinbergen's Four Why's mean that functional and ontogenetic questions are quite independent of each other. Hence, in asking functional questions, we need not be troubled by the particular disputes of developmental psychologists. So long as *some* mechanism exists to ensure that the individual can work its way through to a decision, it does not really matter too much *how* that mechanism is created. In other words, from an HBE perspective, the dispute here is really between two factions of cognitive/developmental psychologists, and has little to do with evolution per se.

Nonetheless, we are now in a more privileged position than our immediate predecessors in terms of being able to answer these kinds of ontogenetic questions. The greatly improved brain-mapping techniques that have been developed in the past decade may be able to help us answer at least one of the questions that lie at the roots of evolutionary psychology: is the human brain/mind divided up into a large number of discrete, physically distinct 'domain-specific' modules?

A BRIEF HISTORY OF MODULARITY

PERIPHERAL INPUT SYSTEMS

The term '**modularity** of mind' was first coined in 1983 by the philosopher–linguist, Jerry Fodor. In a short, thought-provoking and highly influential book, Fodor (1983) argued that perceptual processes (for example, vision, hearing, language comprehension and production) were organised into innate (that is, hard-wired, genetically specified) special purpose modules or 'input systems'. According to Fodor (1983), each of these modules functions completely independently of the others, using its own dedicated processes with each one responding to uniquely different inputs from the environment. This 'domain-specificity' means that 'concepts may thus be available for language learning, or face recognition or space perception which are not likewise available for balancing one's checkbook or deciding which omnibus to take to Clapham' (Fodor 1998). Finally, it was also suggested that modules were completely automatic, stimulus-driven, fast and, most importantly, 'mandatory' and 'informationally encapsulated'.

MODULARITY

By mandatory, Fodor meant that the modules would always operate in the same way under all circumstances. Think of listening to a person speaking out loud. Even if you do not speak the language they are using, you can still hear words being spoken; it is very difficult to hear a spoken language as just a stream of pure sound. And if the person is speaking in your native tongue, then it is virtually impossible to do so. We hear words whether we like it or not.

Figure 10.1 In the Müller-Lyer visual illusion, the arrowheads create the illusion that line A is longer. This illusion persists even if subjects are allowed to measure the two lines to confirm that they are equal.

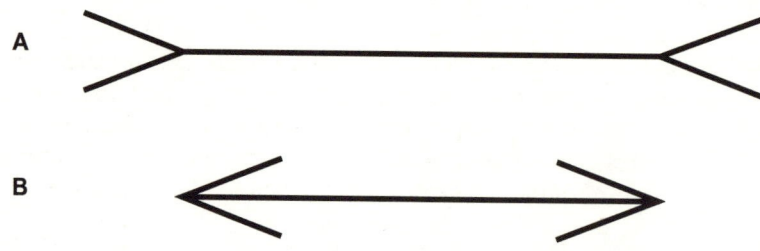

Fodor (1983) used the term 'informationally encapsulated' to suggest that each module was sealed off from all other modules and from higher cognitive centres. That is, the operation of modules could not be influenced by what an individual already knew or believed from past experience. A classic example of this is provided by the Müller-Lyer visual illusion (**Figure 10.1**). The illusion (that the upper line is longer than the lower one) is created by the arrowheads. Even when you know for certain that both lines are the same, the upper one still looks longer. That's what Fodor means by informational encapsulation: the knowledge you possess is completely unable to influence the way in which your visual input systems feed data to your brain.

Fodor suggested that the function of these perceptual modules was to process incoming sensory information swiftly and efficiently, helping to speed up reaction times. For example, imagine coming across a dangerous predator, like a panther. In such cases, Fodor (1983) suggests, it is critical to identify the panther as a panther very quickly. Given that this is the case, it would be unproductive (and highly dangerous) to trawl mentally through a lot of information about panthers that is not relevant to the situation at hand. As Fodor (1983) himself puts it:

> In the rush and scramble of panther identification, there are many things I know about panthers whose bearing on the likely pantherhood of the present stimulus *I do not wish to have to consider*. As, for example, that my grandmother abhors panthers; that every panther bears some distant relation to my siamese cat Jerrold J; that there are no panthers on Mars . . .

Perceptual processes are informationally encapsulated so that they can get on with the work of identifying panthers – or anything else, for that matter – without interference. And if the lack of extra information leads to the occasional mistake, as in the Müller-Lyer illusion, then, as long as the modules err on the side of making false positives (mistaking innocuous objects for panthers) and not vice versa, it doesn't really matter.

Crucially, Fodor argued that 'central cognitive processes' (that is, those involving thought and reasoning) were not organised in a modular fashion. Central processes appear to be unencapsulated (what the system knows influences how it performs) and they are slow, non-mandatory (you can choose not to think about something if you don't want to), much more controlled and geared to achieving global cognitive goals. In essence, Fodor views higher cognitive processes as a 'big holistic lump lacking joints at which to carve' (Sperber 1994) and, as a consequence, Fodor (1983) was pessimistic about our prospect of understanding anything really interesting about the way that they worked.

MASSIVE MODULARITY: THE 'SWISS ARMY KNIFE' MODEL

Evolutionary psychologists, like Cosmides and Tooby (1992), disagree with this notion of domain-general central processing, arguing instead that there would have been strong selection pressures for many problem-solving abilities to be modular – that is, hard-wired and universal among humans. They argue that many of the fundamental cognitive problems of social life, such as detecting cheats, using tools, socially manipulating others, choosing mates, learning language, and so on, would have been selected to operate in a modular fashion in exactly the same way as Fodor's (1983) perceptual modules.

They suggested that cognitive modules were a more reasonable assumption than domain-general processes for several reasons. First, the possession of a single (or small number of) reasoning device(s) would be highly inefficient and error-prone, in much the same way that using a corkscrew to open a tin can is less efficient than using a can-opener specifically designed for the job. Second, they suggested that children would be unable to learn all the things that they do if they relied only on experience and general learning mechanisms. The rate and efficiency with which children acquire skills during development suggests that at least some of these must be hard-wired and present from birth – an argument similar to that put forward by Chomsky (1975) to account for language acquisition. The third reason is essentially the same as that put forward by Fodor (1983) with regard to panther identification. When solving certain cognitive problems, there is a large amount of information that will be entirely irrelevant to the issue at hand. Ensuring that only the relevant information is accessed and processed increases both speed and efficiency. Cognitive modules containing all the information needed to solve the problem and running 'Darwinian algorithms' to work out the correct answer would therefore have been favoured by natural selection over slow, limited general learning mechanisms.

Consequently, Cosmides argued that the mind as a whole, not just its peripheral input systems, can be regarded as a modular 'Swiss Army knife' with a multitude of blades, each one dedicated to a specific task, or, if you prefer a more high-tech analogy, 'a confederation of hundreds or thousands of functionally dedicated computers' (Tooby and Cosmides 2000) – a view which, incidentally, Fodor himself describes as 'modularity gone mad' (Fodor 1987). Cosmides' research programme is therefore dedicated to identifying and seeking evidence for the operation of such higher cognitive modules (see below). Surprisingly, perhaps, although in principle these represent a wide range of possible cognitive modules, almost all their research has focused on a single module, the so-called cheat detection module.

Smith et al. (2001) have criticised this position on the grounds that it leaves no room for flexibility; if there are dedicated, informationally encapsulated, hard-wired modules for every facet of behaviour, how would individuals be able to make the complex series of trade-offs between alternative behaviours that we have highlighted in previous chapters? Are the cues that trigger the modules supposed to be weighted in some way? How do they interact with each other and how exactly are they used in the real world (as opposed to the pencil-and-paper tests of the laboratory)? So far, Tooby and Cosmides have not provided the answers to these questions. Smith et al. (2001) argue that the 'Swiss army knife' would not, in fact, work more efficiently unless there was also an 'intelligent actor [who could] employ it in an adaptive (fitness-enhancing) manner, selecting the right tool for the job at hand, or even improvising if no "built to order" tool is available' (Smith et al. 2001, p. 131).

BOX 10.1

Mental models: error-prone biases or simple rules that make us smart?

An important development in the understanding of human cognition was provided by the so-called 'mental models' view (Johnson-Laird 1982). The essence of this approach was based on the supposition that the human mind (that is, brain) could not be expected to act as an all-purpose empirical processing device, simply because the amount of information present in everyday events was far too great. An organism that tried to learn the defining characteristics of a particular class of events by classic trial-and-error learning would be overwhelmed by irrelevant details long before it was able to extract any valid generalisations.

Instead, the human mind was seen as following a kind of satisficing strategy. It used a number of shortcuts and predispositions to identify key cues. As a result, we have predispositions to attend to or react to certain kinds of phenomena (for instance, fear of snakes) or develop such predispositions through learning as a result of experience. Johnson-Laird (1982) argued that even memory is organised on this basis. Rather than remembering every detail of what happened, we remember a few salient points and then reconstruct the sequence of events that happened by applying a set of general principles.

Some particularly striking examples of these kinds of shortcuts had been investigated earlier by Tversky and Kahneman in a series of studies of human reasoning abilities. Tversky and Kahneman (1983) showed that humans were often misled in their reasoning about probabilistic events (in particular). For example, when asked to express a preference for two medical programmes, one of which will save exactly 600 lives out of every 1000 patients and one which will save everyone with probability 0.6 and no one with probability 0.4, subjects invariably chose the first even though they should have been indifferent and chosen them with equal frequency because the expectations of both programmes are identical (600 lives saved on average) (**Figure 10.2**).

Tversky and Kahneman (1983) suggested that these kinds of errors of reasoning are common because people use rules of thumb to avoid the hard work of having to do all the necessary calculations. Although these rules of thumb work adequately most of the time, their use exposes us to the risk of what have become known as 'framing problems': we are apt to be misled by the way a problem is presented (or framed) to us.

THE MODULARITY OF THOUGHT

Opponents of the Cosmides-Tooby 'super-modularity' view consider it implausible, on the grounds of what Sperber (1994) calls 'commonsense arguments' that thought processes are modular in the same way as perceptual processes. One of the major objections against a 'modularity of thought' is linked to the novelty and cultural diversity of human behaviour. Any individual human has a vast range of conceptual domains, including, among others 'Zen Buddhism, French cuisine, Italian opera, chess playing... and modern science' (Sperber 1994). Most of these domains have appeared very recently in human history and many of them vary quite dramatically from one culture to another. It seems unlikely that these could all be the products of a process of adaptation – if nothing else, it would require that natural selection 'anticipate' new domains arising, which is obviously impossible. Instead, it suggests that the brain can give rise to new functions without necessarily needing to evolve new modules (if indeed there are modules at all). For example, we may reasonably assume that reading and writing developed too recently in human evolution to have

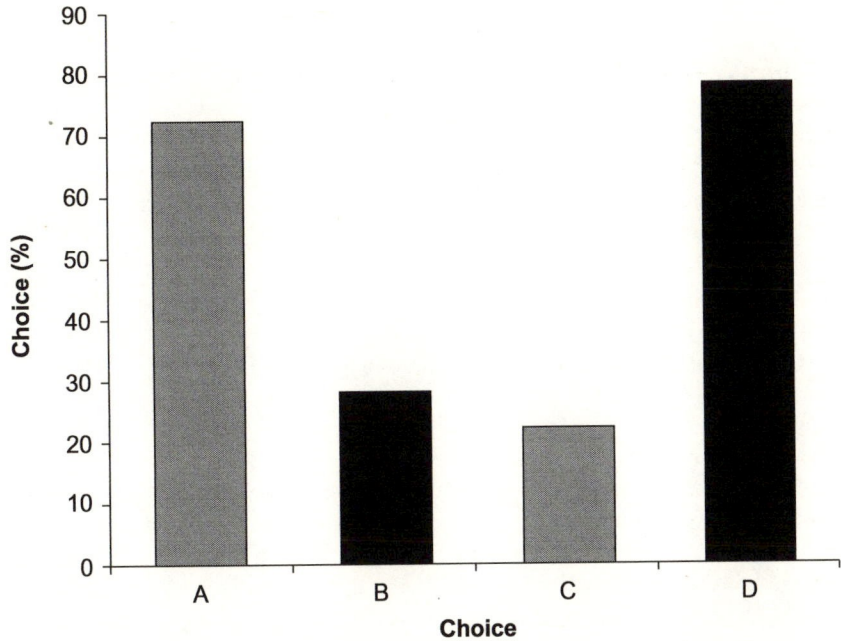

Figure 10.2 The framing problem. Subjects were asked to choose between two medical programmes for a new disease, one of which would cure 200 out of 600 victims (option A), the other of which would cure everyone with a probability of $P = 0.33$ (option B). A second group of subjects was asked to make the same choice, but told that the cure would result in either 400 people dying (option C) or all dying with a probability of $P = 0.67$ (option D). Subjects should have been quite indifferent between the four options, choosing each with equal frequency (50 per cent). In fact, they typically preferred the safe option (certain cure and uncertain death) over the more risky option (uncertain cure and certain death); moreover, which they preferred depended on how the question was phrased. Source: Tversky and Kahneman (1981).

exerted any selection pressures favouring the evolution of dedicated modules. Rather, they must capitalise on and exploit existing modules (or more general computational abilities) whose features can be adapted with a little cognitive effort (that is, learning) to support new functions.

Despite this objection, Sperber (1994) argues that it is, in fact, possible to have a 'modularity of thought' and that taking such a view can account not only for human thought processes, but also for precisely those patterns of cultural innovation and information exchange that have been used as arguments against it. Sperber's argument is built on viewing the mind as an integrated hierarchy of different kinds of modules which, while operating in a Fodorian manner individually, have the emergent property of producing controlled, non-mandatory, flexible thinking.

Sperber (1994) suggests that the mind is partitioned into a number of conceptual domains into which all the sensory information relating to a certain concept is fed. That is, all the sensory information pertaining to the concept 'DOG' – the sight, sound, smell and feel of dogs – is fed into a domain that makes inferences about 'living things'. In humans, it has been suggested that peoples' ordinary understanding of how objects move, the appearance of organisms and the actions of people are based on three such distinctive mental mechanisms: a folk physics, a folk biology and a folk psychology. (In this context, 'folk' means knowledge that is acquired without formal training.) In other words, the input from perceptual 'Fodorian' modules would feed into other modules that organised the information into particular concepts.

Sperber (1994) then suggests that the outputs from these different modules could feed into each other, rather than a single central processing mechanism. In this way, a complex network of conceptual modules could be created: some would receive their input from perceptual modules, some would receive at least some input from other conceptual modules and some could just have other conceptual modules as their input. A network of this kind might be capable of high levels of flexibility even though all the individual modules operated in a Fodorian manner. That is, each of these conceptual modules would be domain-specific, use a 'proprietary database' (all the knowledge that is filed away under that concept), would be triggered only by representations in which the right concept occurs and would be, to some extent, hard-wired. With this level of complexity achieved, modules could then emerge whose function was to handle the problems raised, not externally by the environment, but internally, by the workings of the mind itself.

The final stage in this increase in modular complexity would be the emergence of a module capable of processing information regarding the concept of a concept: Sperber (1994) refers to this as a meta-representational module. For example, when you think about what your friends think of you, and you believe that they believe you are completely fabulous, then you are engaged in meta-representation. In the whole of the animal kingdom, only humans have this ability and, as Sperber (1994) says, it is so peculiar that even Fodor is prepared to consider this to be a modular ability, in the same way that he considers language to be modular (Fodor 1992).

An organism endowed with such a module could form representations of concepts across all potential domains. That is, you could hold the same information in your head in two different modes: you would have a 'first order' representation of 'DOG' in your conceptual module and a representation of that representation of 'DOG' in your second-order meta-representational module. The latter module would 'know' nothing about dogs as such, but it would be able to evaluate the relationship between the dog representation and those from other domains and establish links between them. It could, for example, link the representation of rodents to representations of cartoons and clothing (how else would you come up with Mickey Mouse?). In other words, modular functioning could give rise to imaginative, creative and holistic thought. It would also allow for cultural influences to have an impact on individuals' thought and behaviour: the links between concepts could be dictated largely by the cultural environment to which one is exposed and this could lead to culturally specific and highly diverse behaviours (see Chapter 13). This is quite unlike the Tooby and Cosmides (2000) vision of modularity in which genetically hard-wired, universal 'Swiss army' modules operate in the same way in all humans.

'BEYOND MODULARITY'

Notwithstanding Sperber's (1994) vision of flexible meta-representational modularity, there are still those who continue to object to modularity in principle. Karmiloff-Smith (1996), for example, is a developmental psychologist who objects to modularity as defined by Fodor (1983) because such an approach denies a role for development (in the sense that learning plays a role in development). Indeed, it seems to suggest that there is no such thing as development at all, merely modules that progressively come 'on-line' as the brain matures. Karmiloff-Smith (1996) strongly disputes such a position, arguing that development is a very real phenomenon and that studying the process of development can provide valuable insights into cognitive processes in adult brains. (We explore development in more detail in the next Chapter.)

Karmiloff-Smith (1996) suggests that human infants are born, not with innate modules, but with predispositions and biases corresponding to the major domains of biology, physics and psychology. These biases channel attention toward certain environmental inputs, such as human faces, but they do not act as specialised modules. That is, although infants may prefer to look at face-like stimuli, they can nevertheless direct their attention to many things in their environment. As a consequence of these biases, infants undergo a process of 'modularization' whereby knowledge and skills are channelled into one or other of the three major conceptual domains. As a result, their knowledge of the world becomes increasingly 'domain-specific' (a term which seems to approximate to Fodor's idea of informational encapsulation).

What happens next, however, is that children go 'beyond modularity': in a continual and cyclical process, they redistribute this domain-specific knowledge across other domains, so that it is no longer 'informationally encapsulated' but freely available to all parts of the mind. In other words, they 'demodularize' (as Fodor [1998] describes it). Central to this process is what Karmiloff-Smith calls **Representational Redescription (RR)**, which is the mechanism by which knowledge is redistributed within and across domains. Karmiloff-Smith (1996) envisages that RR takes place in the following manner. Initially, children use the information they obtain from the environment to build up internal mental representations within a single domain. Whenever something new is learned and mastered, it is added to the relevant domain, but no connections are made with other knowledge either across domains or within them: this is the process of 'modularization'.

At this point, children's knowledge is 'implicit': they have knowledge *in* their minds, but that knowledge is not available *to* their minds as conscious thought. Once 'behavioural mastery' is achieved and children can perform a given behaviour correctly and consistently, the process of RR is initiated spontaneously. The implicit knowledge that children possess with regard to particular behaviours is redescribed into a new mental representation that captures just its essential features. This redescription is a more 'abstract' representation of the original knowledge because it has lost many of the perceptual details associated with the original representation. Karmiloff-Smith (1996) gives the example of a zebra. When this is redescribed as a 'striped animal', the representation loses many of its perceptual details, but this makes it easier for the cognitive system to understand the similarity between a real zebra and the road sign for a 'zebra crossing'. Exactly how all this is achieved neurobiologically, however, is not considered: RR is a model of cognitive processes and not brain function per se.

Once RR is underway, a child's representations continue to be redescribed in increasingly abstract forms to both the original domain and to all the other domains present. In each case, reaching behavioural mastery is the trigger for a phase of RR to take place. Karmiloff-Smith (1996) suggests that children have an inbuilt 'drive' to understand their behaviour fully; successful performance alone does not 'satisfy' them and they have to push beyond mere behavioural mastery to acquire a conscious appreciation of the knowledge they possess.

The eventual result of all these ongoing cycles of RR is that the same knowledge comes to be represented in a number of different ways within the child's mind and can be 'mapped across' from one domain to the other (the function which, in Sperber's (1994) model, is attributed to the meta-representational module). In practice, the

REPRESENTATIONAL
REDESCRIPTION

BOX 10.2

'Fast-and-frugal' algorithms

Gigerenzer and Goldstein (1996) believe it is wrong to state that people's performance on decision tasks is error-prone. They argue that this can only be the case if faultless rational reasoning is what individuals are trying to achieve. If they are not, in fact, attempting to find the perfect rational answer, then they cannot be making mistakes. They suggest that individuals merely seek to find solutions that satisfy their particular aspiration levels – they want answers that are good enough to allow them to get by, and not necessarily the best answers overall. To achieve this, they use 'fast and frugal' algorithms that successfully deal with conditions of limited time, knowledge or computational ability (in other words, real life) to provide an answer that works well under the prevailing circumstances. In other words, Gigerenzer and Goldstein (1996, see also Gigerenzer et al. 1999) argue that we need models of **bounded rationality**, and not classical rationality.

Using a variety of computer simulations, Gigerenzer et al. (1999) have shown that fast and frugal algorithms are just as effective at problem-solving as more complicated rational processes. For example, the 'recognition heuristic' (when given a choice of alternatives, pick the one that you recognize) and 'take the best' (when given a choice of alternatives, pick the one that is better/bigger/faster) can both out-perform more complicated algorithms (those that look at all available information, weigh it, and then use detailed calculations to make a choice), even though they consist of a single decision-rule and use only a fraction of the available information (Gigerenzer and Goldstein 1996, Gigerenzer et al. 1999). Fast and frugal algorithms are also much faster than more complicated approaches, giving them the advantage of speed as well as accuracy.

The fact that simple one-dimensional decision-making algorithms can be just as accurate as more complex algorithms but produce results much more swiftly led Gigerenzer to suggest that the cognitive algorithms that people employ in everyday life are likely to be just as fast and frugal, since time is often an important limiting constraint on decision-making processes and choices need to be made quickly. The time advantage alone would probably be enough to give fast-and-frugal algorithms a selective advantage over more complicated processes, but the fact that there is little, if any, loss of accuracy makes it even more likely that this is the route that natural selection would have taken.

---

ability to map across domains means that children can become more creative and imaginative in their thinking because mapping across domains allows them to think about objects and events in a variety of ways. This increase in flexibility and cognitive power is associated with an increasing conscious awareness of knowledge. Eventually, children become aware of their own knowledge to such an extent that they are able to report on their mental processes verbally.

Fodor (1998), however, questions whether mere redescription into another format can explain why information increases in accessibility. He states that it is impossible to regard formats as more or less accessible in and of themselves, since formats can only be accessible *in relation to* something. For example, a statement in English is more accessible to an English speaker and less accessible to a French speaker, but there is nothing about the statement per se that makes it accessible. Similarly, for explicitness. Fodor (1998) argues that every representation a child has must be explicit *about* something; Different formats can be explicit about certain aspects of a representation and others can be implicit about these aspects, but they can't just be explicit or implicit in and of themselves. Fodor (1998) therefore insists that, even if children do change their representational formats over the course of cognitive

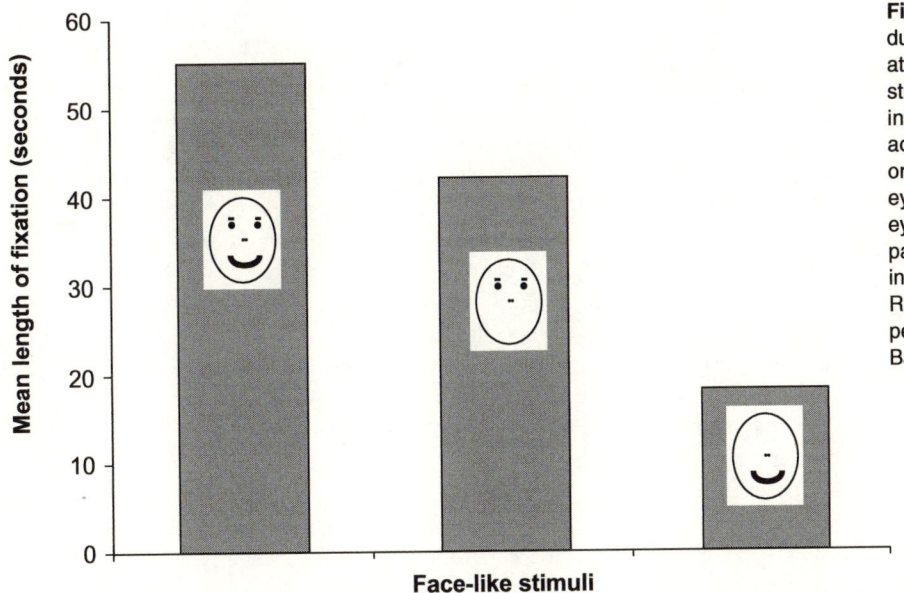

Mean length of fixation (seconds)

Face-like stimuli

Figure 10.3 Mean duration of visual attention to face-like stimuli by very young infants varies according to whether or not the face has eyes, suggesting that eye contact may be particularly informative in social interaction. Redrawn with permission from Baron-Cohen (1995).

development (something which he also disputes), this tells us nothing about how or why the information that is redescribed should become more accessible as a consequence. In fact, it only begs the questions of why the new formats are more accessible than the olds ones, and why changes of format should have any effect at all on the accessibility of representations.

Fodor (1998) suggests that what actually happens as children undergo cognitive development is that the outputs of the modules become more available 'to the mind', but that the information contained in the modules themselves continues to be unavailable (and always remains so). He therefore suggests that children gain greater 'off-line' access to module outputs with the result that they are able to think about and manipulate this information in more complex ways using their central processing mechanisms. This in turn suggests that demodularization would not (and could not), in fact, occur. Since the intramodular information remains encapsulated, there would be nothing for RR to redescribe. Fodor (1998) supports this argument with data showing that, even in adults, certain aspects of linguistic processing are not explicitly available and individuals cannot tell experimenters why they interpret certain utterances in the way they do.

These criticisms notwithstanding, Karmiloff-Smith has suggested that face-processing is a good example of domain-specific modularization, as opposed to innate modularity. Infants respond positively to face-like stimuli from an extremely young age indeed (within about nine minutes of birth: **Figure 10.3**). Such a precocious ability has been taken to suggest that the ability to process faces is innate (that is, genetically specified and present from birth). However, detailed studies of the specialisation and localisation of this ability within the brain show that both of these are actually gradual processes.

Brain imaging experiments on infants, for example, have shown that six-month old babies process faces using several areas of the brain and that processing occurs across both hemispheres (Neville et al. 1993). By the time infants have reached 12 months of age, however, their face-processing circuits have become specialised to particular areas of the visual cortex, mainly the lateral fusiform gyrus and the superior temporal sulcus, and are localised in the right hemisphere alone. By 12 months, infants have had sufficient experience of faces to 'tune up' particular populations of cells to respond only to face-like visual stimuli (Neville et al. 1993, Karmiloff-Smith 2000).

Further support for a process of modularization in face-processing was obtained by Gauthier et al. (2000). In this study, subjects who were experts in recognising birds or cars had their brains scanned while they viewed photographs of their specialist subject. Their scans were compared with those of non-specialists. This revealed that when experts viewed photos of birds or cars, the areas of the brain associated with face-processing were activated, whereas no such response was found in non-experts. Gauthier at al. (1999) concluded from this that 'face-responsive' regions might more accurately be referred to as regions of 'visual expertise' – that they are areas which are actually specialised to respond to any object that the subject perceives as a distinct individual, rather than as a member of a generic category. Since we are all expert individual-face recognisers, these regions are always activated when subjects are shown faces, and hence they have been assumed to be areas for face-recognition alone. However, the fact that these areas are also activated in bird and car experts shows that they can also be used to recognise any kind of object, even those that bear no resemblance to faces whatsoever. In other words, there is no 'face-processing module' as such that is present and active from birth, but rather a population of cells that becomes modularized for faces – or birds or cars or whatever – during the process of development.

As a result of findings like these, Karmiloff-Smith (2000) argues that the mind cannot, by any stretch of the imagination, be viewed as a 'Swiss Army knife' and that this is especially so for the higher cognitive functions. After all, if not all perceptual input systems are hard-wired modules as Fodor (1983) envisaged but are instead the emergent products of modularization, it seems highly unreasonable to expect high-level cognitive processes to be completely modular in the way that Cosmides and Tooby propose. In the next section, we review the literature on a 'module' that has received particular attention in the EP literature: the cheat detection module. Before doing so, however, two other points need to be made.

First, other developmental psychologists have argued for the primacy of certain key primitive functional modules, such as number, causality, an intentional stance and language (for example, Carey and Spelke 1994; Leslie 1982, 1994; Leslie and Keeble 1987; Keil 1994; see also references in Hirschfeld and Gelman 1994). These are seen as providing the foundations on which the later acquisition of more sophisticated modules are developed, mainly through generalising the cognitive skills that underpin them. Some have argued that language plays a crucial role here by enabling different modules to be linked together at a conceptual level in a way that cannot be done without language.

Second, this debate has many of the hallmarks of the old nature/nurture debate in comparative psychology during the 1950s (for a review, see Manning and Dawkins 1992). This particular debate focused on the contrast between instinct and learning in the ontogeny of behaviour. Such debates are particularly difficult to resolve because they essentially rely on demonstrating that a particular behaviour or module exists even before learning has had a chance to come into play. In the nature/nurture debate, the protagonists were often forced into arguing whether or not learning was possible inside the egg or womb. In the end, the debate inevitably became sterile. In the present case, very few of the studies involved in this debate have actually attempted to get to grips with the question as to whether a particular module exists before learning (and especially language) has had a chance to come into the equation. Indeed, all of the studies that focus on 'cheat detection' modules, for example, have been carried out on adults. If neural nets can learn, then studies of adults tell us nothing about the ontogeny of any modules we can discern: they may simply be the product of experience. Because it is difficult to study cognitive phenomena in pre-linguistic children, relatively few studies have done so. However, there have been some attempts to do so, though these have inevitably focused on relatively primitive cognitive abilities. Important examples include the studies of face recognition discussed above (see **Figure 10.3**) and causality (Leslie and Keeble 1987).

SOCIAL EXCHANGE AND CHEAT DETECTION

In Chapter 9, we discussed the idea that humans may be especially sensitive to behaviour indicative of cheating on social contracts. Cheating on social agreements has such devastating consequences for the coherence of social groups, and hence for the survival and successful reproduction of the group's individual members, that some authors have suggested that the selection pressure favouring a dedicated mental module would have been very intense (Cosmides and Tooby 1992, Tooby and Cosmides 1990).

Cosmides (1989, Cosmides and Tooby 1992) hypothesized that cooperative strategies of exchange and reciprocity would require that people became cognitively attuned to detect cheats. Over time this has resulted in the evolution of a specialised psychological mechanism or 'Darwinian algorithm' which is activated by the kind of social exchange problems that would have recurred repeatedly over the course of human evolution. Cosmides has made extensive use of the Wason Selection Task to explore the nature of this mechanism (see **Box 10.3**). The central result seems to be that subjects generally solve easily a conditional reasoning problem (what observations do you need to check to establish the truth of a particular rule or generalisation) when it is presented as a social contract infringement (for example, the drinking age problem), but consistently fail to do so when it is presented as an abstract logical problem and do poorly (though not as badly) when the task is presented as non-social real life problem (for example, a geographical problem about transportation). In other words, the cognitive mechanisms that are activated to solve selection tasks are 'content-specific': they are activated only by social exchange problems and nothing else.

FAMILIARITY AND PERSPECTIVE CHANGE

One potential problem that Cosmides (1989, Cosmides and Tooby 1992) herself recognised and set out to remedy is that people may perform better on social contract problems like the drinking-age problem (**Figure 10.4**) because these tasks are more familiar than those involving a non-social task. Cosmides (1989, Cosmides and Tooby 1992) therefore ran a further series of trials designed to eliminate this possible confound. For this, she used tests in which social contracts were embedded in unfamiliar situations (for instance, if you eat cassava root, then you must have a tattoo on your face') and found exactly the same results: most subjects can solve the problem correctly.

But what if social contracts just help people to reason in a more logical manner? In all the social contract tasks, the right answer for detecting violations was also the logically correct answer (in the format of **Box 10.3**, the P and *not-Q* cards). Embedding rules in a social context may just enable people to think more clearly. Rather than having a specialised Darwinian algorithm for social contracts, it may be that

BOX 10.3

The Wason selection task

Wason (1966, 1983) developed his 'selection task' to test the proposition that people are natural scientists who test hypotheses about the world in a logical way. Subjects are presented with a problem of the following kind. Suppose there is a general rule (perhaps a code for filing documents) stating that a card with a vowel on the front always has an even number on the reverse. This is equivalent to the logical rule: *If P then Q.* Subjects are then given four cards (showing the symbols A, H, 4 and 7) and asked 'which card or cards should you definitely turn over to see if the rule has been violated?'

checked to ensure that the law prohibiting alcohol consumption by minors is being not being broken. Here the answer seems obvious: you need to check that the 16-year-old is not drinking beer and you need to check the age of the beer drinker. The other two don't matter because anyone can drink Coca Cola and 21-year-olds can drink what they like.

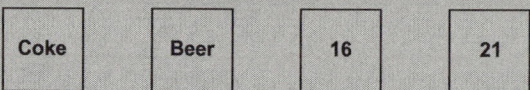

The cards correspond, respectively from left to right, to the logical statements: *P, not-P, Q* and *not-Q.* The logically correct answer is to select the A card (*P*) and the 7 card (*not-Q*) since only this combination can falsify the hypothesis. The rule says nothing about what cards with consonants are supposed to have on the back, so, strictly speaking, it does not matter whether they have odd or even numbers on their obverse. However, this seems to be the logical mistake that most people make, since in any test of this type the majority of people will pick the vowel card alone (*P*), or both the vowel and the even number cards (*P* and *Q*), and only about one quarter of subjects solve this problem correctly (**Figure 10.4**).

When this same problem is presented as a social contract problem, however, about 75 per cent of subjects get the answer right (**Figure 10.4**). One social contract version suggests that the subject is responsible for checking that the licensing laws are being adhered to. On entering a bar, you discover four people sitting around a table: one is drinking a coke, one a beer, one is 16 years old and one is 21, see below. Subjects have to decide which individuals need to be

Another version of the social task assumes that the subject has to make sure that student documents are correctly processed, following a rule that states 'if a student has a D grade, the record has to be marked with a "3". In alternative versions, the subject is told that the previous clerk had incorrectly coded some of the records, either deliberately or through laziness. Faced with the following four cards, which ones would you definitely have to turn over to see if the rule had been broken?

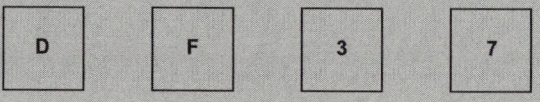

Cosmides' (1989, Cosmides and Tooby 1992) suggests that it is not just the recognition that certain individuals can be disregarded as irrelevant that improves performance on the social contract version, but the specific nature of the rule itself: when rules are presented as 'social contracts' (where if one is to take a benefit, then one must pay the cost), individuals succeed because their 'cheat-detection' module becomes engaged and people then quickly seek out potential cheats, like under-age drinkers. However, Cosmides's interpretation of these results has been subject to considerable criticism.

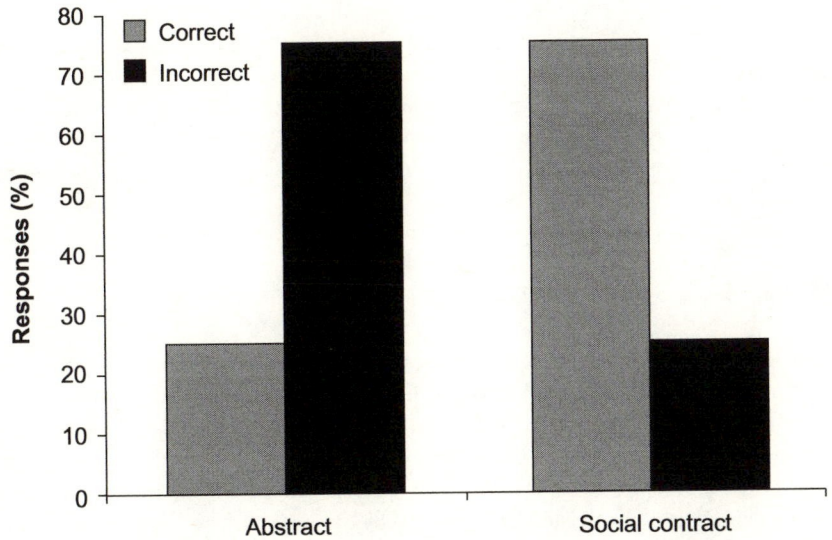

Figure 10.4 The Wason Selection Task asks subjects to identify which of four options need to be checked in order to ensure that a specified rule is not being broken. Subjects invariably perform poorly on the abstract version of the task (a rule involving correspondences between letters and numbers), but do well when the same task is presented as a social contract problem (where a social rule might have been infringed). See **Box 10.3** for details. Reproduced with permission from Dunbar (1999).

social problems allow people to access general reasoning mechanisms more effectively and so reach the logically correct answer. Cosmides (1989, Cosmides and Tooby 1992) investigated this by reversing her social exchange problems ('switched social contracts'). Instead of phrasing the problems in terms of 'if you take the benefit, you must pay the cost', they were phrased as 'if you pay the cost, you may take the benefit'. The socially correct choices for problems phrased in this way are *not-P* and Q, which is the opposite of the logically correct answer. Cosmides argued that if people were sensitive to the social contract content of the problem, then they should choose the socially correct but logically incorrect *not-P* and Q cards as their answer. If the problems were merely activating logical inference mechanisms, then people should still choose the logically correct categories (even though this would result in a foolish choice socially, that is, paying the cost, but not taking the benefit). Cosmides and Tooby (1992) found that subjects were more likely to make the socially correct choice than the one that was logically correct.

Gigerenzer and Hugg (1992) also investigated whether it was specific social contents that produced high-performance, but instead of switching the contracts, they switched the perspective of the subjects while keeping the task structure constant. For example, one problem was framed in terms of 'if an employee gets a pension, then that person must have worked for the firm for at least 10 years', with one set of subjects taking the perspective of an employee, while the others were told to take the view of an employer. From an employer's perspective, if an employee gets a pension but hasn't worked for ten years, then they are cheating. The right answer to this version of the task is thus the logically correct answer: you need to turn the cards marked 'gets a pension' (*P*) and 'worked eight years' (*not-Q*) in order to check that the rule isn't being violated. From an employee's perspective, however, employers would be cheating if you worked for ten years but they didn't give you a pension after all. In this case, the cards that have to be checked are the 'no pension' (*not-P*) and 'worked 10 years' (*Q*). As in Cosmides' switched contracts, this is not a *logically* correct response to the problem, but it is the right *social* answer. If social contracts merely facilitate logical reasoning, then the perspective taken shouldn't matter: subjects should select the logical answer regardless. Gigerenzer and Hugg (1992) found that the response given depended strongly on perspective: 60 per cent of 'employees' selected the *not-P* and *Q* cards, whereas no 'employers' selected these cards.

PERMISSION AND OBLIGATION

While these results might seem pretty conclusive, other authors have questioned Cosmides' (1989, Cosmides and Tooby 1992) findings. Cheng and Holyoak (1989), for example, question whether Cosmides' (1989) results really do show evidence for a domain-specific social exchange module and cheat detection algorithms. They suggest that her results actually show evidence for 'permission and obligation schemas' and not social exchange. Cheng and Holyoak (1989) argue that the 'drinking rule' problem, for example, isn't really an example of social exchange because reaching the age of 21 can't really be viewed as a 'cost' in the wider scheme of things; and, even if it was, to whom is such a cost paid? Despite the fact that the drinking rule is not an example of 'social exchange', it nevertheless produces reliable and replicable successful performance in practically all subjects. If people find it easy to solve despite the fact that it is not a social exchange contract, then there may be no such thing as a specialised social contract module.

Cheng and Holyoak (1989) therefore suggest that all problems phrased in the same terms as the 'drinking rule' problem actually represent permission rules, and that these are solved by a wide-spectrum permission schema. They were also able to show that rules which obliged individuals to take precautions for their own good (for example, being immunised against cholera) also produced high performance. On the basis of this, they argued that a more general processing mechanism must be in operation, one that responds to permissions and obligations and not just social exchanges. They further argued that these permission schemas, although 'domain-specific' in the sense that they were triggered by permission-type problems, were actually the product of an over-arching domain-general process, rather than a specific module.

Cosmides and Tooby (1992) countered these objections by arguing that social exchange can be interpreted as paying costs and benefits to society in general, rather than occurring only between individuals. They also suggested that social exchange rules constitute one specific type of permission, and that another domain-specific module for 'hazard avoidance' or 'precautions' could explain obligation rules like the cholera problem (Cosmides and Tooby 2000).

Cheng and Holyoak (1989) also raised methodological objections about Cosmides' (1989) results, suggesting that cheating and non-cheating versions varied in uncontrolled ways that made it impossible to know for sure whether it was merely the presence or absence of cheats which produced the observed effects. Differences in the way selection tasks are set up are known to influence performance (Yachanin and Tweney 1982) and it could be task differences, rather than the activity of mental modules, that lead to improved performance. For example, phrasing a task in terms of checking for a specific violation (as in cheat detection tasks) leads to better performance than phrasing a task in terms of testing the truth of a hypothesized rule (as in non-cheat detection tasks) (see Lieberman and Klar 1996).

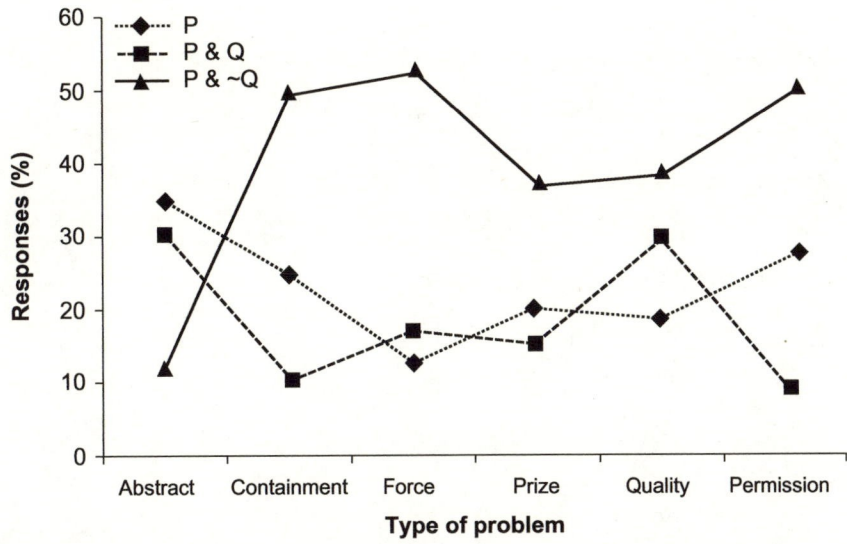

Figure 10.5 When the Wason selection task is presented in different forms that cue the subject in to the possibility that the classifying rule may be asymmetric (as in Cosmides's social contract version: *P* implies *Q*, but *Q* does not imply *P*), they perform nearly as well as subjects on the 'social rule' version, even though they fail on the original (abstract) version. The asymmetric versions are: *containment* ('If a large object is stored, then a large container must be used'), *force* (a physicist's hypothesis: 'if the weak force wins, then the strong force must have been weakened first'); *prize* ('if a product gets a prestigious prize, then it must have distinctive quality'); *quality* ('if the product breaks, then it must have been used under abnormal conditions'); *permission* ('if a letter is sealed, then it must have a 35-cent stamp') represents a social rule infringement. The graphed variables are the percentage of subjects that chose to check only the *P* card, the *P* and *Q* cards or the *P* and *not-Q* cards. Redrawn from Almor and Sloman (1996).

Similarly, Almor and Sloman (1996) have shown that non-cheating conditional rule tasks (such as a scientific hypothesis or a rule about how objects should be stored) can be solved with the same accuracy as social contract tasks if the question is framed in a manner similar to the drinking-age problem; that is, if the nature of the question makes it clear that, while a certain condition must be satisfied if the rule is not to be violated, other conditions are not relevant since there is no logical necessity for them to be true. In other words, just as being 21 years old permits one to drink beer, but doesn't mean that all 21-year-olds *must* drink beer – they can, after all, drink what they like – so the rule that 'if a large object is stored, then a large container must be used' suggests that while a large object must be stored in a large container, it doesn't mean that small objects can't also be placed in large containers as well as small ones.

Selection tasks presented in this way yield results that are significantly better than those obtained for abstract versions of the task (**Figure 10.5**). People perform well because the beliefs they generate about the problem (for example, small objects can go in big or small containers, since it is not explicitly said that they can't) happen to match the logically correct solution. Solving 'real life' versions of the selection task is therefore not dependent on the content of the problem presented, and it does not require the operation of a 'cheat detection' module. It merely requires that the *structure* of the problem should reflect people's intuitive beliefs about dependency relations.

BOX 10.4

Relevance theory and the selection task

Sperber et al. (1995) have suggested that an alternative explanation for Cosmides and Tooby's (1992) results is that people use 'relevance mechanisms' to solve the selection task. Relevance mechanisms are presumed to be domain-specific and are used by individuals to understand intentional communication – to understand what someone actually *means* when they attempt to communicate. Importantly, Sperber et al. (1995) argue that these mechanisms are content-general. That is, any kind of information can be processed as long as it is in the form of a communication from another person (an 'intentional agent'). This is because relevance mechanisms transform intentional communications into general logic statements before any other processing takes place. This conversion means that the specific content of the communication is irrelevant, and it is the logical structure of the communication that is crucial.

When people judge a piece of information to be relevant, it means they have brought it together with other background information to produce a 'cognitive effect'. This process requires 'cognitive effort'. The greater the cognitive effort needed to produce a cognitive effect, the less relevant it is considered to be. Thus, in order to maximise relevance, large cognitive effects should be produced with the minimum amount of cognitive effort. In terms of the selection task, this means that people should be able to solve them, no matter what the content, provided they produce high levels of relevance.

Sperber et al. (1995) showed that this was the case by changing the structure of abstract tasks in ways that would increase their relevance and make their intentions more clear. This led to an increase in performance comparable to that on standard social contract tasks. As with Almor and Sloman (1996) and Lieberman and Klar (1996), Sperber et al. (1995) concluded that the content of a selection task was not important in and of itself – as Cosmides and Tooby (1992) had argued. Instead, they argued that it was the way in which the content of the task influenced the production of general purpose relevance mechanisms that made the difference.

Fiddick et al. (2000) have disputed this, however. They claim it is not possible to explain all the content-specific effects found in selection task experiments with only domain-general relevance mechanisms. Instead, they argue that, in order to produce certain cognitive effects and perform successfully on the selection, it is necessary for people to call on domain-specific algorithms that are '**ecologically-valid**' (that is, appropriate to the particular social situation) rather than valid in a purely logical sense (as relevance theory assumes). They also present data to show that these specialised systems pre-empt more general ones whenever the stimuli fit the conditions of the specialised system (for example, on social contract tasks), but that general relevance mechanisms operate in those instances where no specialised system exists (for example, on abstract tasks). Fiddick et al. (2000) thus argue that Sperber et al. (1995) didn't find these effects because they used only abstract tasks and there was thus no conflict between relevance effects and specialised algorithms.

It is, however, worth noting here that the frequency with which subjects chose the correct cards (that is, the P and *not-Q* cards) on Almor and Sloman's social contract task (their *permissions* tasks) was only about 50 per cent (**Figure 10.5**), or about the same frequency with which subjects chose correctly in Cosmides's original *non-social* tasks (for example, the transport problem), and much lower than the 75 per cent success rate for her social contract task (see **Figure 10.4**). This might suggest that Almor and Sloman's *permissions* task (a rule about when one is permitted to use a stamp on an envelope) is not a true social contract in Cosmides's sense.

TASK UNDERSTANDING

Lieberman and Klar (1996) manipulated both social and non-social tasks in ways that altered task understanding but kept the content the same. They found that by changing the task in this way they could completely reverse the results usually associated with cheat detection tasks, with over 70 per cent scoring correctly on the non-cheating version and only 30 per cent or so scoring correctly on the cheat-detection versions.

Lieberman and Klar (1996) argued that the results could be altered so dramatically because performance on the selection task depends only on how people understand the task they are given and not on social exchange modules or cheat detection algorithms. They identified several aspects of task understanding that can influence performance. One of the most critical is the way in which people interpret the rule they are supposed to be testing. According to formal logic, 'if P, then Q' is deterministic: P necessitates the occurrence of Q. In everyday life, however, things are usually probabilistic: 'if you eat spoiled food, you become ill' is often true, but not always. Consequently, finding 'P and not-Q' instances (eating rotten food and not becoming ill) does not falsify a probabilistic rule. If people assume that a given rule in a selection task is probabilistic in this way, it could affect the cards they choose.

Lieberman and Klar (1996) point out that non-cheating versions of the selection task are often presented as probabilistic, whereas this is less likely for cheating versions. For example, in Gigerenzer and Hugg's (1992) 'overnight' task (their task 3), the cheating version states that hikers have to carry wood up to a cabin if they wish to spend the night there, whereas the non-cheating version offers two possible explanations (that the hikers have to bring wood, but it *might also* be the Swiss Alpine Club members who bring the wood). Thus, in the cheating version, it is clear that a deterministic rule is in operation, but the possibility of an alternative way of getting wood to the cabin in the non-cheating task (via the Swiss Apline Club) increases the likelihood that people will interpret this as a probabilistic rule; consequently, testing whether hikers always bring wood will not falsify the rule.

Another problem with task understanding is that, in formal logic, 'if P, then Q' does not necessarily imply 'if Q, then P'. However, in real life, many people assume that it does. This biconditional interpretation tends to be rare in social rules (as we have pointed out previously 'if you drink beer, then you *must* be over 21' does not imply to most people that 'if you are over 21, you *must* drink beer') but is common in abstract ones ('if a card has a 3 on one side, then it must have a D on the other' is understood to imply that 'if a card has a D, it should have a 3' because that is the way most filing systems work). Again, this can lead to differences in the interpretation of cheating versus non-cheating tasks and, hence, to corresponding differences in performance with people performing better on the social task as they are less likely to make this logical mistake.

Another factor that affects performance is whether the context of the task supplies a clear alternative to the rule suggesting that P and not-Q would be the relevant violation of the rule. Obviously, this is the case in cheat detection tasks, since one is asked to look for a specific kind of cheat. However, non-cheating tasks tend to be more ambiguous. In the non-cheating version of the hiking task mentioned above, for example, it is not clear whether one needs to check on hikers or members of the Swiss Alpine club to test whether the rule is upheld.

The final aspect of task understanding is whether searching for violations of the rule is relevant or useful. In cheat-detection tasks, finding the cheater is the whole point, so searching for someone who breaks the rule is the obvious thing to do.

However, this is not always the case. Sometimes positive evidence showing the rule is in operation can be more informative. For example, if a doctor suspects that a particular drug causes nausea, it is much more natural to examine people who have taken the drug and got sick, than to round up healthy people and ask if they have taken the drug. Lieberman and Klar (1996) show that non-cheating versions are often phrased in a way that makes it unclear whether people should look for cases where the rule is broken rather than find positive supporting evidence. Once again, this could lead to differences in performance on cheater versus non-cheater tasks.

Having identified the manner in which task understanding affects performance, Lieberman and Klar (1996) took Gigerenzer and Hugg's (1992) original problems and reconstructed them so that, while the contents remained the same, the cheating versions had all the characteristics usually found in non-cheating tasks and the non-cheating versions had all the usual characteristics of cheating tasks. With these changes, the content effects usually ascribed to the operation of a social exchange module were no longer apparent, since subjects performed better on tasks where there were no cheats to be detected. Fiddick et al. (2000) take issue with this and argue that in one of the tasks the effect was actually produced by changing a cheat-detection task into a hazard avoidance task, thus activating the hazard/precaution module; however, this argument does not hold for any of the other five task variants that were tested.

Taken together, these results suggest that a dedicated social exchange module is highly unlikely to be operating. Task performance can be predicted by factors other than those included within social exchange theory and the balance of evidence seems to suggest that content-general processing mechanisms are responsible for high performance (although the mechanisms themselves may be domain-specific as Sperber et al. [1995] and Cheng and Holyoak [1989] suggest). Maybe this is what we should expect, given that other aspects of social cognition, from face-processing to mental state reasoning, are clearly not modular in the sense proposed by Fodor (1983). If even perceptual processes like face-processing cannot be considered to be innate genetically-specified mechanisms, there seems plenty of room for scepticism regarding the modularity of higher cognitive functions. Obviously, what we need are better tests of the social exchange hypothesis.

THE ROLE OF EMOTIONS

An ability to represent the emotions of others may be as important as being able to represent their mental states, and deficits in this ability result in the devastating social impairments seen in autism (discussed in more detail in Chapter 11) and in patients with severe damage to the frontal lobes of the brain (see Damasio 1994). Here we explore the topic of emotion in a little more detail to highlight how emotional responses, which most people would not consider to be important in 'rational' decision-making processes, actually function to increase our ability to reason effectively. The literature associated with this topic is huge and we can do no more than give a flavour of current work in this area.

WHAT ARE 'FEELINGS'?

The first point to make is that, in Damasio's (1994) view, it is wrong to consider the workings of the brain and mind as separate from the workings of the body. Rather, the mind should be considered as part and parcel of the body. One important, but generally overlooked, way in which this holds true is in the generation of what Damasio (1994) terms 'background feelings'. This is the sense of your body that you possess at all times – the underlying awareness of the state that your body is in. This arises as a consequence of the various neuronal and hormonal signals arising from the viscera (body organs, including the skin, heart, lungs and so on) that are sent to and processed by the brain, providing a continuous update on the changes that your body state undergoes as you go about your business.

Damasio (1994) claims that it is these background feelings that provide our sense of 'self'. Although it feels as though our 'self' is a product of the mind alone and that this is where 'we' reside, in fact we are using information emanating from our entire body. In this view, you would not be the same person if your brain were somehow transplanted to another body, because your new body would provide different information to your old one; since your brain would respond differently to the way it did before the transplant, you would acquire an entirely different sense of yourself. Trying to separate purely cognitive functioning from the functions of the 'body proper' is therefore an impossible task since they are interrelated in a deeply fundamental way.

The fact that certain neurotransmitters produced in the brain, called **peptides**, are also active in the human immune system and endocrine systems (Pert 1997) adds further weight to this suggestion. Peptides modulate brain processes by influencing the activity of networks of neurons in the brain, and at the same time mobilise cells in the immune and endocrine systems to produce particular bodily reactions. The mind and body are thus inextricably linked. What you think can directly influence your body by changing its biochemical balance. And if you doubt that mind can influence matter in this way, Greenfield (2000) suggests you try this test: think of a time you embarrassed yourself in public. Chances are you'll blush at the memory – a strong bodily reaction produced by a mental representation. Proof of the power of thought!

PEPTIDES

SOMATIC MARKERS

As well as background feelings, we also have much stronger 'feelings' which arise when we experience emotions in response to particular events in our environment. In his developmental theory of adult sexual behaviour, for example, Chisholm (1996) has emphasised the importance of emotions as a mechanism whereby children assess the state of the environment in which they are currently living (see **Box 7.3**). In Damasio's terms, an 'emotion' is the change in body state associated with a stimulus (a racing heart in response to a loud bang), whereas the 'feeling' of an emotion is the experience of this change in body state in juxtaposition to the mental images that initiated the cycle of bodily changes. When we are born, we tend to show only 'primary emotions' such as fear which are innate and pre-organised. As we grow, we develop and make more use of secondary emotions, which are primary emotions tempered by experience; the emotions that we undergo become associated with certain situations and their outcomes, and the level of, say, fear or happiness that we display becomes increasingly subtle and varied.

Damasio (1994) argues that a special category of secondary emotions, termed '**somatic markers**', are used to bias decision-making (often unconsciously) toward certain future outcomes and to avoid others. When a negative somatic marker (for example, a feeling of fear or sadness) is juxtaposed to a particular future outcome, it functions as an alarm bell and warns us off a course of action. When a positive somatic marker is

SOMATIC MARKERS

BOX 10.5

Neurobiology of social reasoning

If hardwired cognitive modules exist, we might expect these to be represented in the brain as specific bits of wetware much in the same way that, for example, Broca's area is associated with speech production. The recent advances in brain scanning techniques offer us a particularly powerful opportunity to put some of these suggestions to the test.

Cosmides and Tooby (2000) have themselves begun to take steps in this direction. They report on selection task performance of individuals with lesions to various brain regions. An individual with damage to the orbito-frontal and superior temporal cortex was able to pass both social contract and hazard precaution problems, as was an individual with damage to the superior temporal cortex alone. However, a third individual with bilateral damage to the amygdala as well as the orbito-frontal and superior temporal region, could pass only hazard precaution problems and not social contracts, suggesting that the amygdala is somehow involved in processing the latter.

Cosmides and Tooby (2000) cite a study by Maljkovic (1987) that they suggest supports this claim. This apparently demonstrated that people suffering from schizophrenia (a mental disorder associated with a disturbance of higher cognitive function) were unimpaired on the selection task, suggesting that social exchange mechanisms are generated in evolutionarily more ancient parts of the brain (such as the amygdala) and support their suggestion that the social exchange module represents an ancient human adaptation. However, it is important to note that just because the amygdala is evolutionarily ancient, it does not mean that it is not involved in higher cognitive processes. The amydgala is associated with several aspects of theory of mind (see Chapter 11) which is one of the most highly advanced cognitive abilities shown by humans. Processes associated with the function of evolutionarily older parts of the brain does not necessarily imply that

these processes are also of an older evolutionary origin than other processes.

In addition, Corcoran and Frith (submitted) found the opposite, with impairment in social contract reasoning by people diagnosed as schizophrenic. An alternative explanation, therefore, is that the social contract problem used in the lesion study was phrased in a way that required some assessment of emotional state. The failure on this task could thus reflect the deficit in emotional processing ability that damage to the amygdala causes, rather than damage to social contract mechanisms *per se*. However, the details of the tasks given to the subject are not provided by Cosmides and Tooby (2000), so it is, at present, impossible to know which is the more likely explanation.

In another study, Adolphs et al. (1998) discovered that people with ventro-medial lesions tended to perform much more poorly on all versions of the test, but were particularly poor at social problems. This was the finding they predicted, since, as we mentioned above, patients with ventro-medial lesions show an inability to function effectively in the social domain even though their general reasoning abilities remain unimpaired.

Damasio et al. (1996) suggest that the reason why such subjects perform so poorly is that, when they consider which cards to turn over, their thoughts are not accompanied by any so-called 'somatic markers' (changes in body state which bias decision-making): they do not feel particularly strongly about any of the cards when they think about them, and thus have no particular motivation to turn any of the cards over. This is in strong contrast to the behaviour of an individual with obsessive-compulsive disorder who inevitably ended up turning all of the cards over because he was worried about 'missing something'. In this case, the somatic markers were working overtime, creating the feeling that there was still something to be done.

experienced, it acts as an added incentive to behave in a particular way. In this way, somatic markers can help control our tendency to 'discount the future' since a negative somatic marker connected with an option that has long-term costs can bias us away from a decision. Similarly, a positive marker associated with an option that has short-term costs but long-term pay-offs can help to increase our 'willpower' to endure current sacrifices (Damasio 1994).

LeDoux (1998) makes a similar argument with regard to certain strong emotions, like fear. He suggests that this information is routed through the brain in two ways. One is a 'quick and dirty' route via the amygdala, the other is via the cortex. The amygdala route is fast and unconscious, allowing a swift 'instinctive' response to danger, while the route through the cortex produces the conscious awareness of the emotion – the 'feeling' of fear. The interconnections between the cortex and the amygdala run both ways, but the amygdala can exert a much stronger influence over the cortex than vice versa. LeDoux (1998) suggests this is why we often let our emotions get the better of us, even when calm rational thought would be a better solution.

Damasio (1994) also suggests that somatic markers can speed up the process of decision-making by ensuring that only the most reasonable options are considered (those that produce positive somatic markers), and others are immediately and unconsciously rejected. Damasio (1994) considers that this facility will be most important in the social domain where decisions often have to made quickly in response to unpredictable stimuli (that is, other people), and that somatic markers may be an integral component of 'theory of mind' (see Chapter 11) by biasing our 'mindreading' abilities toward the most appropriate predictions for others' behaviour and mind states. They may also influence our assessments of the risk associated with certain activities in ways which are not 'rational' in the Tversky and Kahneman (1981) sense (for example, we perceive air travel to be more dangerous than car travel, when the reverse is in fact true), but which may make sense from an emotional perspective: the somatic markers associated with a negative outcome to a plane journey – that is, an air crash – are likely to be very much stronger than those associated with a car crash, since fewer people survive plane crashes than car crashes. In this view then, feelings and emotions are just as 'cognitive' as any other kind of perceptual image we experience, and play a much larger role in decision-making than we realise.

Damasio (1994; Bechara et al. 1993, 1994) was able to demonstrate the value of 'somatic markers' using a 'gambling experiment' to compare the performance of 'normal' individuals with individuals who had suffered lesions to the prefrontal cortex. People with damage to this area of the brain are able to perform highly on most standard psychological tests, but they have enormous trouble functioning in the social domain, and find it extremely difficult to make even the simplest decision, like making a date to have coffee with a friend. Damasio's work (1994, and references therein) suggests that this may be due to a deficit in the functioning of the somatic markers, which means that they are unable to take advantage of rapid, unconscious, visceral cues toward certain options, and instead have to consider every choice consciously, using a laborious cost-benefit analysis, while at the same time being unable to perceive the choice of options as inherently good or bad. Given this, Damasio and his colleagues reasoned that such individuals might be expected to differ markedly on tasks involving 'intuitive' decision-making, like gambling, where gut-feelings often play a large role in determining the choices of normal individuals.

In the gambling experiment, individuals had to select cards from one of four packs, A, B, C and D. They were given $2000 of toy money and told that the aim of the game was to lose as little as possible of the 'loan' and make as much as possible. They had to turn over cards from the four packs and only stop when the experimenter told them. Two of the packs (A and B) gave a high reward of $100 whenever a card was turned over, but certain cards within these packs incurred a 'punishment' which required that quite large amounts of money (up to $1250) were paid to the experimenter. The other two packs (C and D) gave a lower reward ($50) but 'punishments' were also much lower ($100 on average). The rules regarding rewards and punishments were not disclosed to the players and were never changed. There was no way a player could know what would happen until the cards were turned over. While playing the game the players were wired up to a device that measured galvanic skin response (the level of skin conductivity due to sweating) which was used as an indicator of the possible influence of somatic markers on individuals' decisions.

When normal individuals played the game, they tended to start off sampling cards from all four decks, and showed an early preference for A and B (as you might expect). However, once they'd been stung a few times with extortionate pay-outs to the experimenter, they shifted their preference to packs C and D, and avoided the 'dangerous' packs for the rest of the game. Interestingly, as the game continued, there was a shift in skin conductance responses from a peak occurring after a card had been turned from one of the 'bad' decks to one occurring *before* the card was turned. This response also increased in magnitude as the game continued. Over the course of a game, normal people were therefore learning to predict a bad outcome and were signalling the relative badness of the deck unconsciously in a manner that anticipated the players' subsequent conscious choices.

This was in stark contrast to the results from subjects with damage to the prefrontal cortex. After the early period of general sampling, these individuals showed no tendency to avoid the 'dangerous' packs, and continued to select cards from A and B until they bankrupted themselves and had to secure a further loan from the experimenter. This was not a failure of learning as such: the individuals in question knew by the end of the game which packs were 'bad' and which were not. Instead, they appeared to be insensitive to bad future outcomes (or, as Damasio puts it, they showed 'myopia for the future'). While these individuals were sensitive to punishment, in that they tended to avoid a bad pack immediately after being punished from drawing a card from that pack, this effect did not last long, and they soon returned to the bad (but high reward) packs. These individuals seemed to show an extreme tendency to discount the future. The prospect of immediate high reward outweighed the negative long-term consequences of these choices. In line with this, the skin conductance responses of lesion patients showed no anticipatory responses whatsoever. There was no sign at all that their brains were beginning to predict possible negative future outcomes (Bechara et al. 1994).

Damasio (1994) suggests that the skin conductance responses reflect the operation of somatic markers and that, in normal people, these provide a covert, non-conscious estimate of the 'goodness' and 'badness' of a pack based on the ratio of rewards and punishments experienced. This initial sorting occurs before any conscious cognitive process of selection and the subject is then 'guided into a theory about the game' more efficiently. This process doesn't occur in people with temporal lobe damage, because they are no longer sensitive to their somatic markers (or no longer generate them), and their conscious cognitive processes tend to focus on gaining immediate rewards. Put simply, there is a dynamic partnership between so-called cognitive (that is, rational) processes and processes usually called 'emotional' (that is, irrational). Somatic markers therefore appear to be another 'fast and frugal' mechanism for reasoning, allowing individuals to make satisfactory decisions without having to go through a comprehensive and lengthy cost-benefit analysis. 'Mind' alone simply cannot do the job.

*

In this chapter, we have explored in some detail the question of whether the mind is organised into a set of modules. We have left open two key questions. One of these concerns the ontogenetic origins of these modules: are they part of a naturally unfolding **epigenetic** programme or do they develop out of the individual's experiences through learning and use? The other is the question of how these mind modules map onto real brain architecture (the 'wetware'). Answering these two questions may help us to bring the disputes highlighted in this chapter to some kind of conclusion. In the next chapter, then, we focus on these two issues, using the phenomenon known as 'theory of mind' as the principal case study.

EPIGENETIC

Chapter summary

■ Understanding the cognitive mechanisms that underpin behaviour and make individual decision-making possible is ultimately an important component of the evolutionary psychology programme. The main emphasis so far has focused on the question of whether the mind is modular in its structure and, if so, what form these modules take.

■ Fodor argued that certain perceptual and linguistic processes are contained in domain-specific, informationally encapsulated cognitive modules. These allow for fast processing without interference from central processes. Some authors, specifically Tooby and Cosmides, have attempted to extend this idea to include not only perceptual processes, but also high-level cognitive processes.

■ Other authors question the notion of modularity altogether, arguing instead for a process of 'modularization' during development followed by a process of 'demodularization'. Data on the specialisation and localisation of face-processing cells in the brain seems to support parts of this model.

■ Cosmides and Tooby (1992) argue that the ability to detect cheats is a domain-specific modular ability, and use data from the Wason selection task to support their hypothesis. However, data from a number of other studies calls the validity of this into question. Aspects of task understanding seem to be more important in producing good performance than socially relevant contents, suggesting that Cosmides' finding may be an artefact of the experimental design.

■ Emotional processing is as important to social decision-making as high-level cognitive processing. Individuals with damage to brain areas that integrate emotional states with information about mental states perform badly on social tasks, and function poorly in normal life.

Further reading

Damasio, A. (1994). *Descartes' Error: Emotion, Reason and the Human Brain*. London: Papermac.

Damasio, A. (2000). *The Feeling of What Happens*. London: Heinemann.

Fodor, J. (1983). *The Modularity of Mind*. Cambridge, MA: MIT Press.

Fodor, J. (2000). *The Mind Doesn't Work That Way*. Cambridge, MA: MIT Press.

Karmiloff-Smith, A. (1996) *Beyond Modularity*. Cambridge, MA: MIT Press.

LeDoux, J. (1998). *The Emotional Brain*. New York: Simon and Schuster.

Social cognition and its development

Human social skills (and so many of those features like language, culture and politics) are dependent on cognitive mechanisms that appear to be more or less unique to our species, at least in terms of the levels to which they have been developed. Understanding how these mechanisms work and the role they play in decision-making is clearly an important project for evolutionary psychology. Unfortunately, virtually all the studies of social cognition to date have focused either on non-human primates (see for example Tomasello and Call 1997) or on developmental issues relating to very young children (typically the 3-5-year-old age group: Astington 1993, Mitchell 1997). Derivative of the latter has been a major clinically-related interest in individuals who lack these mechanisms (principally autistic individuals: Happé 1994a, Baron-Cohen 1995). Almost nothing is known of how these mechanisms function in normal adults.

Despite this, we consider it important to discuss social cognition at some length for three reasons. First, a number of researchers have identified **theory of mind (ToM)** as a prime candidate for a mental module and evaluation of this claim allows us to develop further insights into the issues raised in Chapter 10. Second, although studies of social cognition have so far had almost no impact on any aspect of evolutionary psychology, we think it important to draw attention to an area that is ultimately

Figure 11.1 Mindreading involves a reflexive series of mental states about the beliefs/desires/intentions (*intentional* states) of oneself and others. From left to right, these three individuals are in first, second and third order levels of intentionality, respectively. Being unaware of the relationship between *husband* and *wife*, *stranger* has a false belief about *wife*'s state of mind because he has mis-read the cues she has been providing; *husband*, in contrast, has made the correct inferences because he is able to interpret *wife*'s cues correctly. In recognising that *stranger* has a false belief about some aspect of the world (in this case, *wife*'s intentions), *husband* is able to correctly pass a false belief task (the benchmark for having *theory of mind*). Theory of mind (ToM) is equivalent to second order intentionality, and develops at around the age of 4–5 years in normal children; it is what makes much of our social and cultural life possible.

certain to be of considerable importance to our understanding of human behaviour and evolutionary psychology. Although Tinbergen's Four Why's entitle us to maintain as much clear water between the functional (HBE) and the mechanisms (conventional EP *sensu stricto*) aspects of evolutionary psychology as we wish, we are convinced that an understanding of the ways in which cognition constrains decision-making will ultimately prove of particular value for the functional approach because it will help us to identify both the limits on the choices that individuals have and the efficiency with which they can choose between those options. Finally, theory of mind appears to be fundamentally important for the evolution of language and thus, ultimately, culture. An understanding of ToM will help us when we come to discuss these topics in Chapters 12 and 13.

THEORY OF MIND

The best known and most intensively studied of the social cognition mechanisms is the phenomenon known as *theory of mind* (Premack and Woodruff 1978). This rather awkward term is meant to suggest that the individual has a belief (a theory) about the content of (another person's) mind – in other words, he/she understands that other people have mental states (thoughts, desires, beliefs) and that these mental states drive their behaviour (see **Figure 11.1**). More importantly, perhaps, the individual can appreciate that, at any one time, the actual content of these mental states can

BOX 11.1

Intentionality

ToM is sometimes referred to by an alternative name (*intentionality*) that derives from philosophy of mind. Intentionality refers explicitly to states of mind that are about beliefs and desires, as reflected in the use of verbs like *believe, think* (that something is the case), *hope, wish, want, suppose, intend.* The significance of this is that it allows ToM to be seen as one step in a reflexive hierarchy of intentionality. Belief–desire psychology effectively becomes first order intentionality: *I believe something to be the case.* ToM is second order intentionality: *I believe that you think that something is the case.*

Extending the logic, we can set up what is in principle an infinite sequence (with the successive orders of intentionality marked in square brackets): *I believe* [1] *that you suppose* [2] *that I want* [3] *you to understand* [4] *that...* and so on. The importance of this lies in the depth of mindreading that successive levels reflect. Double bluff depends on at least third order intentionality. Writing a novel about a triangular relationship depends on fifth order intentionality (the writer has to *intend* that the reader *believes* that character A *thinks* that character B *supposes* that character C *wants* something). Reading the novel requires one less level of intentionality.

differ considerably from our own and from the objective reality of a situation (see **Box 11.3** below).

As we indicated in Chapter 10, Sperber (1994) regards this ability to be modular (his 'meta-representational' module). A number of developmental psychologists (for example, Leslie 1988, 1991, 2000; Baron-Cohen 1991a, b, 1995, 2000a, b) agree with this and view ToM as a specially-dedicated mechanism by which children process information with regard to human social interaction. They use the term 'Theory of Mind' in a very loose sense to refer to a particular kind of cognitive processing system.

Others in the developmental field (for example, Perner 1991, Gopnik 1993, Wellman 1990, Astington 1993), however, take the view that children literally do form a 'theory' of their own mind and those of others in exactly the same way that scientists generate theories to explain the physical world (see **Box 11.2**). They do not view the child's increasing understanding of the mind in modular terms, but see it as an on-going process of revision and reorganisation – something more akin to Karmiloff-Smith's (1996) ideas regarding modularization and RR.

Data from normally developing children and those with specific deficits in mental ability (for instance, autism) have been used to provide evidence for both these views. As yet there is no firm consensus as to which theory provides a more accurate explanation of children's cognitive development.

JOINT ATTENTION AND SOCIAL REFERENCING

In the previous chapter, we noted that face-processing – an important element of social cognition – is acquired extremely early in development. Within minutes of birth, babies are more likely to follow face-like stimuli than scrambled or random patterns (Goren et al. 1975). Tests performed on two-month old babies showed that attention tended to be focused longer on face-like stimuli that showed the eyes compared to stimuli that did not (**Figure 10.3**). By the age of 14 months, this tendency for a young infant to focus its attention preferentially on face-like stimuli in its environment

BOX 11.2

Theory theory

Proponents of the so-called 'theory' theory of mind suggest that children are born with theory-formation mechanisms that allow them to explain the world on the basis of the events that they observe (Gopnik and Astington 1988, Gopnik 1993, Gopnik et al. 2000, Perner 1991). These mechanisms are not modular since they are general-purpose rules that apply across many other domains, including folk physics and folk biology as well as psychology. However, in addition to these general mechanisms, theory theory hypothesizes that babies are born with initial 'starting state' theories about the mind and other domains (like physics and biology). During the course of development, children modify and extend their initial starting state theories using their innate theory-formation mechanisms. Rather like scientists, children develop an understanding of their own minds and others 'by constructing coherent views . . . and then changing these views in light of the evidence they obtain' (Gopnik et al. 2000, see also Harris 1991).

Unlike the modular view of ToM in which children's abilities increase 'passively' as a consequence of maturation of specific modules, the theory theory posits that children play an active role in their own development by making predictions, seeking explanations and considering evidence that is relevant to mind. Theory theory also differs from modularity in that there is no informational encapsulation: theories in one domain can influence theories in other domains and the knowledge encoded in modules can be influenced by what the individual already knows. This means that knowledge can be updated and revised in the light of new information, something that is impossible for a cognitive module.

Children begin with an initial 'starting state' theory of mind because it is impossible to construct a coherent view of the world without one. Scientists, for example, do not discover how the world works by accumulating a large amount of raw, unstructured data and then building a theory from it. Rather, new data are always interpreted and selected in the light of an existing theory, which may then be updated and revised as a consequence (Dunbar 1995a). In human babies, the initial theory of people seems to take the form of a primitive understanding of the link between the babies' own mind and that of others as revealed by their imitative abilities.

According to Gopnik et al. (2000), children are born with an innate capacity to mimic facial expressions (but see Heyes, submitted, for a criticism of this), and this initial psychological engagement between the baby and other individuals means that, later on, further discoveries about people's minds are easily and naturally extended to knowledge of their own minds and vice versa. Brothers (1997) argues in a similar vein that the key to the development of social intelligence is an initial understanding that bodies have 'selves' and that a single 'self' occupies each body. Once this initial link is made, increased social and psychological understanding follow as logical extensions of this initial discovery.

It is important to realise that, according to the theory theorists, we understand our own minds using just the same theories as we use to understand the minds of others (Gopnik 1993). We do not have privileged direct first-person access to what goes on in our own heads, only a theory about what is going on. This may seem very strange, and completely at odds with our everyday experience, but it is possible that our feeling that we have direct access to our own thought processes is just an illusion created by our innate theory-formation mechanisms. Our 'commonsense' idea of what is going on – our folk psychology of our own minds – need not coincide with the way our brain constructs the world (after all, when you look at the horizon, commonsense tells you that the world is flat; but, as we now know, that doesn't mean that it actually is).

This notion that we have a theory of our own minds contrasts sharply with the so-called simulation theory, which suggests that we understand others' minds by using our own minds as a model. This allows us to predict the behaviour of other

BOX 11.2 cont'd

Theory theory

people by simulating what we would do or feel under particular circumstances and then projecting those thoughts and feeling onto them. Some authors suggest this is achieved consciously thanks to a special introspective access to our own mental states (for example, Goldman 1993; Harris 1991), while others state this is an unconscious process (Gordon 1986). Whether simulation theory or theory theory represents the more accurate picture of how people understand their own and others' minds is still open to debate. However, the fact that three-year-olds not only fail to understand other people's false beliefs but also cannot easily describe the contents of their own minds is hard to explain using simulation theory (since these findings imply that three-year-olds do not have privileged access to their own thought processes), and seems to suggest that the theory theory is on the right track (see Gopnik and Astington 1988).

develops into an ability to engage in 'joint attention' (using another individual's gaze direction to focus its own attention on the same object) (Scaife and Bruner 1975). At this age, an infant is also capable of directing another individual's attention to what it is looking at (for example, by pointing), so that the individual's attention is co-ordinated with the baby's.

Baron-Cohen (1991b, 2000b) has suggested that joint attention is mediated by two modular mechanisms – an eye-direction detector (EDD) and an intentionality detector (ID). The ID is a perceptual mechanism that detects and interprets motion in terms of whether or not it has been produced by an 'intentional agent' – an object that has a 'purpose' and shows goal-directed behaviour. As far as infants are concerned, people are the most important intentional agents in the environment, but the ID is not very discriminating and will interpret almost anything that engages in self-propelled motion as an agent with goals and desires (Baron-Cohen 1995). This effect persists into adulthood: when shown a film of geometric shapes (two triangles and a circle) moving around on a white background, subjects interpreted the scene before them as one in which the circle was 'bullying' the small triangle, who was then 'saved' by the large triangle (Heider and Simmel 1944). Children may therefore make mistakes about what exactly an intentional agent is, but it also means that they rarely fail to spot a truly intentional agent when one is around. The ID is thus a 'stupid' module in Fodor's (1983) sense: the module is triggered by certain cues in the environment, even if that results in misidentification.

The eye-direction detector does exactly as the name suggests and responds to eye movement and gaze-direction, inferring from this that if eyes are directed at something then the individual must actually see that thing. In conjunction, the ID and EDD help to produce shared attention by drawing infants' attention to intentional agents and cueing them into the focus of the agents' attention. Baron-Cohen (2000b) considers these early modular components to be essential for the later development of a full-blown meta-representational ToM (**Figure 11.2**). These 'modular' abilities may not be present from birth, however. It's possible that they are acquired by a process of modularization, starting from an initial attentional bias, as proposed by Karmiloff-Smith (1996).

Once children are capable of joint attention, they begin to use 'social referencing' to guide their behaviour (Astington 1993). This is the means by which young babies gauge the feelings of another individual (usually the mother) to an object in the environment and use that information to form their own attitude toward the same

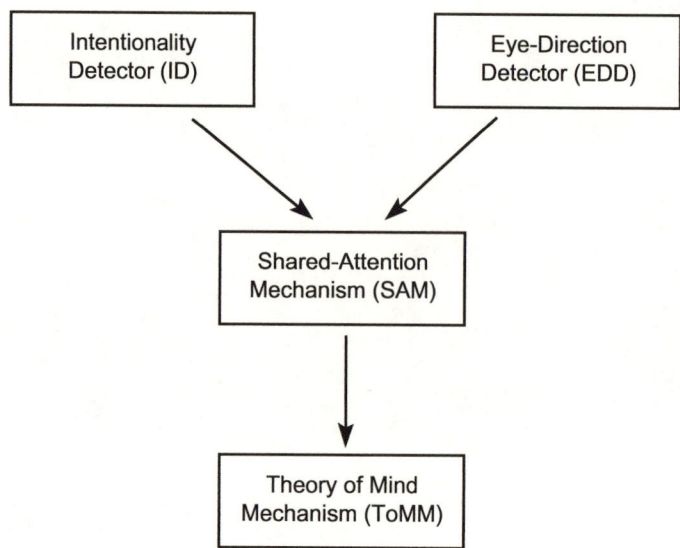

Figure 11.2 Baron-Cohen's conception of the human mind-reading system. The four components represent basic properties of the world and are assumed to come into play during development in the order top to bottom. The four elements target, respectively, volition, perception, shared attention and mindreading. Each layer acts as an essential precursor for the layer below it. The upper pair are Fodorian modules that interact, in combination with the shared-attention module, to give rise to theory of mind as an emergent property. Redrawn with permission from Baron-Cohen (1995).

object. When they are unsure about something (a new toy, a strange person), infants rapidly glance back and forth between the object in question and their mother's face in order to read her expression. One of the most striking experiments demonstrating this is the 'visual cliff'. This is an apparatus used to assess an infant's perception of depth. It has a transparent glass cover, which overlies a patterned surface. This patterned surface drops quite suddenly to a depth below the glass to form a 'cliff'. There is no danger to the infant since the glass cover is continuous and quite level all the way across; there is no way that the baby can fall over the edge.

When a baby is placed on this apparatus, it notices the apparent change in depth and hesitates to crawl across. If the mother, who is on the 'other side' of the cliff shows fear, the baby will not cross. If, on the other hand, the mother smiles and encourages the baby, then it will continue on its way. The infant's behaviour is utterly dependent on the mother's expression. This ability therefore requires, not only an understanding of attention, but also some understanding of emotion, so that babies learn to respond differentially to different expressions (Astington 1993). As with joint attention, this ability does not require a deep understanding of mental states, but it does involve psychological engagement with another individual and is thus an important stepping stone along the way.

DEVELOPMENT OF THEORY OF MIND

In the past two decades since the concept of theory of mind was introduced, approximately a thousand studies have been published on this topic, almost all of which have been concerned with its development. The primary focus of these studies has been how children in the 3–5-year age bracket develop theory of mind. We here provide a very brief summary of this literature. Although the processes of development have their own intrinsic interest, the real value of this analysis for present purposes is that it helps to clarify exactly what ToM is.

PRETEND PLAY AND 'DESIRE PSYCHOLOGY'

At the age of 18–24 months, children begin to engage in pretend play. They are able to use a banana as though it were a telephone, hold doll's tea-parties and if they accidentally spill the 'tea' on one of the dolls, they will dry her off because she's 'wet' (see Harris and Kavanaugh [1993] for experimental evidence of children's understanding of pretence). Young children can also recognise and understand when other people are pretending and will happily join in with these games. Leslie (1991, 2000) regards this as the first sign of an innate ToM module coming into operation. He argues that if a child can recognise when another individual is pretending, and is able to join in with them, then the child must have some understanding of the other individual's mental state. In other words, the child must understand the other individual's intention, and not their behaviour, so that it can link its own behaviour with what the other person is pretending to do (pouring an imaginary cup of tea from an imaginary pot) and not with what they are actually doing (performing an exaggerated motor movement with their arms).

Leslie (1988, 1991) argues that it is the possession of an innate ToM module that allows children to distinguish clearly between pretence and reality. This is essential if children are to learn about the world in a truthful and consistent manner. Children would end up with some very funny ideas if they treated their mother's pretend use of a banana as a telephone as equivalent to her use of a real telephone. According to Leslie (1988), children avoid this problem thanks to the operation of an innate ToM module that prevents any confusion between reality and pretence.

To illustrate how this could work, Leslie (1988) first points out that our perceptual and cognitive systems have evolved to enable us to form true beliefs about the world. (This has to be the case, since a system that resulted in the formation of false beliefs is potentially very dangerous and would be selected out fairly rapidly.) These true beliefs he calls '**primary representations**'. In addition to these primary representations, we also have **secondary representations** that capture our attitudes towards our primary representations (that is, we have beliefs about our beliefs).

PRIMARY REPRESENTATIONS

SECONDARY REPRESENTATIONS

The characteristic feature of these secondary representations is that they are suspended from reality or 'opaque'. That is, they need not correspond directly to reality in the same way that primary representations do. The primary representation 'there is a cream-cake in the fridge' is only true if there is a cake and it is in the fridge. On the other hand, the secondary representation that 'Bob *thinks* there's a cream cake in the fridge' can be true whether or not the cake is actually in the fridge, or even if there is no cake at all (Astington 1993). In other words, when a primary representation is embedded in a secondary representation, it becomes isolated from reality, and no longer has to reflect a 'truth' about the world. According to Leslie (1988) this embedding procedure is what allows pretence to be held separate from reality and this is the job performed by the innate ToM module. In this way, the child's cognitive system ensures that secondary opaque representations like 'I *pretend* the banana is a telephone' or 'Daddy *pretends* to be a bear' are kept separate from true-belief primary representations, thus avoiding any confusion between the properties of bananas, bears, dads and telephones (for a more detailed summary, see Astington 1993).

Leslie (1988) initially called this ability to form secondary representations 'meta-representation' to get across the notion that they are representations of representations. More recently, however, he has adopted the term M-representation after objections by the theory theorists. Perner (1991), for example, argues that, although two-year-olds may understand that representations are mental entities ('thoughts'), they do not understand that representation is also the process by which these mental entities are formed in a person's mind. They understand that the mind contains thoughts but they do not know how they get there. As a result, two-year-olds do not understand misrepresentation – that someone's beliefs can be false – because they do not realise that a person constructs the world in their head, rather than sees it as it really is. Hence, two-year-olds can understand that Bob has a representation *of* his cake but they cannot understand that he represents the cake *as* being in the fridge (that is, that he forms a belief about the location of the cake). Without an understanding of representation as a process, Perner (1991) argues, it is inappropriate to describe two-year-olds as capable of meta-representation as this implies their understanding of representation is more advanced than is actually the case.

Although they have no notion of belief, Wellman (1991, see also Wellman and Woolley 1990) states that two-year-olds have a good understanding of desire. This 'desire psychology' means that they are able to understand that people are motivated to act by internal drives ('desires'), that people engage in goal-directed behaviour as a result of these inner drives, and that they display particular emotions depending on whether or not their goals are achieved (for example, happiness on satisfying a desire). This understanding of desire is also reflected in children's spontaneous language. At two years of age, children begin to make frequent reference to desire (making use of terms like 'want'). However, they make no references to belief terms like 'think' and 'know' until their third birthday (**Figure 11.3**).

This change in their use of language signals a change to a 'belief–desire psychology' in three-year-olds (Wellman 1990). For example, when asked to explain simple human actions like a girl looking for her cat, children will say 'she *thinks* the cat is missing' as well as providing answers like 'she *wants* the cat' (Schult and Wellman 1997). They will also give belief–desire explanations for human 'intentional' actions (for example, a person deciding to stand up) but will give purely physical explanations for a situation in which a person is, say, being blown around by the wind (Schult and Wellman 1997).

Wellman (1991) suggests that this shift occurs, not because of the maturation of an innate ToM module, but because the child's simple desire psychology increasingly fails to account for people's behaviour. For example, two people with the same desires can often behave in completely different ways because they have different beliefs. Such discrepancies prompt children to formulate and test new theories about individual action, resulting in an understanding of belief as an additional internal state that influences behaviour. As a consequence of theory changes, children come to understand that people have beliefs about their desires and that their actions are motivated by their beliefs and not by their desires alone. Interestingly, it is the possession of an initial desire psychology that forces children to reformulate their theory: if children were purely behaviourists with no understanding of internal states, then they would not notice any anomalies in people's behaviour. According to Wellman (1991), it is only because they possess a theory of desire that they are led down the path to more sophisticated understanding.

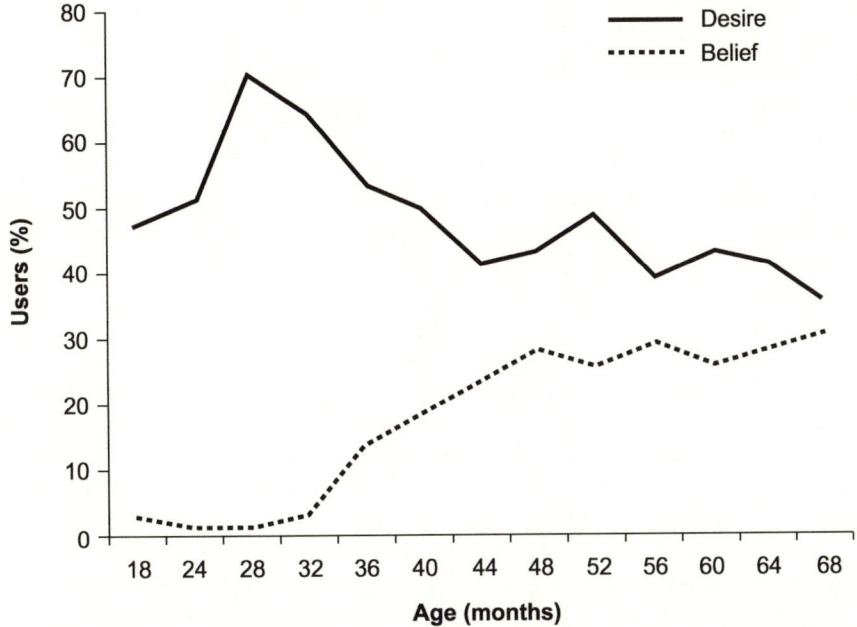

Figure 11.3 The frequency with which children use 'desire' terms in normal conversation is more or less constant across age. In contrast, the frequency with which they use 'belief' words increases sharply after age 3 years, suggesting that an understanding of belief states is acquired at this point. Reproduced with permission from Wellman (1991).

This bears a resemblance to Karmiloff-Smith's (1996) ideas that RR is endogenously driven and allows children to gain a greater understanding of their own and others' minds. As a consequence of this reorganization of their knowledge – or the improved functioning of the ToM module if you prefer that theory – three-year-olds can explain why someone would look in only one of two possible locations for an object (because that's where the person believes the object to be located), but they still fail to understand false beliefs (that someone's beliefs could lead them to look in a location that didn't contain an object at all). Like two-year-olds, they cannot grasp the fact that the process of *representing* the world often leads to a false *representation* of that world.

REPRESENTATIONAL THEORY OF MIND

By the age of 4–5 years, however, this has all changed, and the child has a fully representational theory of mind. There is marked shift in the kind of reasoning of which children are capable and most children are able to pass the false belief task, the benchmark test for ToM (**Figure 11.4**). The false belief task (**Box 11.3**) requires the child to be able to step outside its own personal view of the world so as to be able to see the world from another individual's point of view. Only once it is able to make this differentiation between its own mind and someone else's can it appreciate that another individual can hold a belief that differs from its own (that is, a false belief). Perner (1991) argues that this represents a major reorganisation of the child's knowledge and the development of an understanding of representation as a process.

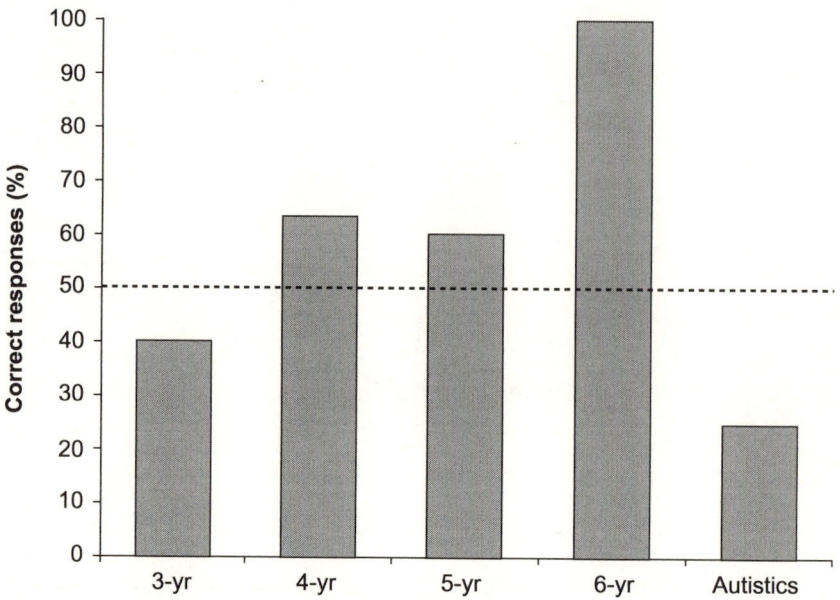

Figure 11.4 Children's success rates on false belief tasks increases rapidly between ages 3 and 5 years, suggesting a critical threshold for the development of theory-of-mind at around age 4. Autistic individuals fail these tasks, even as adults. The task used in this study is the 'Smartie' task. In this task, the child is asked what is inside a tube of Smartie sweets, but is then shown that it actually contains pencils; they are then asked what their best friend would say if asked. Children who cannot distinguish between their own knowledge and someone else's will say 'pencils'; those who understand that someone else can have a false belief will say 'Smarties'. The horizontal line represents chance level of performance. Source: O'Connell and Dunbar (submitted).

By contrast, modular theorists (for example, Leslie 2000) argue for the maturation of processing mechanisms that improve the functioning of the innate ToM module. Whatever view one takes, it is clear that the ability to pass a false belief task heralds the child's new understanding that representation refers to both an entity and a process – true meta-representational thought (Perner 1991).

One reason why false belief tasks are the preferred assay for ToM is that they are the only way to be sure that children really do understand other people's mental states and are not just reporting their own beliefs or assessment of reality (Dennett 1978). True beliefs, by definition, match up to the real state of the world, so that a child who can predict that an individual will act in accord with a true belief cannot be distinguished from a child who merely expects people's behaviour to accord with reality. However, because false beliefs reflect an individual's representation of the world rather than the world as it really is, a child will only be able to predict how people will behave on the basis of false beliefs if it can reason about mental states. There is no other way to get the answer right.

Although passing a false belief test is seen as unequivocal evidence for ToM, and has been taken as indicating a major shift in children's representational abilities (for example, Perner 1991), there are some who argue that it has been accorded a significance that is not really justified. Bloom and German (2000), for example, argue that there is, in fact, more to passing a false belief test than possessing a theory of mind. Although they may sound simple, false belief tasks are actually very difficult in terms of the

BOX 11.3

False-belief task: a benchmark for theory of mind?

False belief tasks are deemed to be a benchmark for theory of mind because they can only be answered correctly if the child can differentiate its own knowledge of the world from the beliefs held by another – beliefs that the child must assume, on the basis of its own knowledge, to be false.

The classic false-belief task is known as the Sally-Ann task. Sally and Ann are two dolls manipulated by the experimenter. Sally places a ball (or other toy) in a basket and then leaves the room. While Sally is away, Ann takes the ball from the basket and hides it in a box. The child is then asked the critical question: 'When Sally comes back, where will she look for her ball?' Children under the age of four fail this task: they say that Sally will look in the box. They are unable to take Sally's perspective and understand that her mental state corresponds to a different version of reality to their own. Children over the age of four, however, almost always get the answer right, and state that Sally will look in the box where she thinks the ball is, rather than where it is actually hidden.

Another frequently used task is the 'Smartie task'. (Smarties are sweets similar to M&Ms that are sold in cardboard tubes.) The child is shown a Smartie tube and asked what is inside. They will usually answer 'Smarties'. The experimenter removes the lid of the tube and shows that it actually contains pencils. The child is then asked: 'If we showed the tube to your best friend X, what do you think they would say is inside?' As with the Sally-Ann task, children under the age of four will answer 'Pencils' because they cannot distinguish between their own knowledge of the world and someone else's, whereas children over four will answer 'Smarties'.

memory and other cognitive demands they make on young children. Bloom and German (2000) suggest that failing a false belief task may have as much to do with the demands made by the task as with a failure to understand mental states. The difference between three- and four-year-olds may not be in their understanding of mental states as such, but in their ability to understand the task they are asked to perform.

Some support for this is provided by experiments using 'false photographs' (Zaitchik 1990, Leslie and Thaiss 1992, see also Leslie 2000). In these experiments, children are presented with a toy room in which, for example, a toy cat is sitting on a chair. A photograph is taken of the room, and then the cat is moved to a new position. The children are then asked: 'where is the cat in the photograph?'. Most three-year-olds fail this task whereas most four-year-olds pass it; yet, this task does not ask about or require any understanding of mental states. In other words, it is the demand that the task makes on the child's cognitive processing capacity that leads to failure and, if this is true of this task, it could also be true of false belief tasks as well (Bloom and German 2000).

Indeed, it is possible to increase three-year-olds performance on false belief tasks by making them easier in various ways. For example, getting the child more actively engaged in the deception increases their performance (Chandler et al. 1989, Sullivan and Winner 1993), as does making the change of location less salient (Carlson et al. 1998), giving a child a memory aid (German and Leslie 2000) or phrasing the critical false belief question more simply and/or more specifically (Lewis and Osbourne 1990, Siegal and Beattie 1991, Surian and Leslie 1999). For example, asking children 'where will Sally look *first* for her ball?' increases three-year-olds' performance on a false belief task substantially, possibly by helping three-year-olds to recognise that they are actually being asked about Sally's beliefs rather than reality (**Figure 11.5**).

Figure 11.5 Three-year-old children's performance on false belief tasks can be significantly improved if their attention is first drawn to salient points in the task ('look first' condition). Redrawn with permission from Leslie (2000).

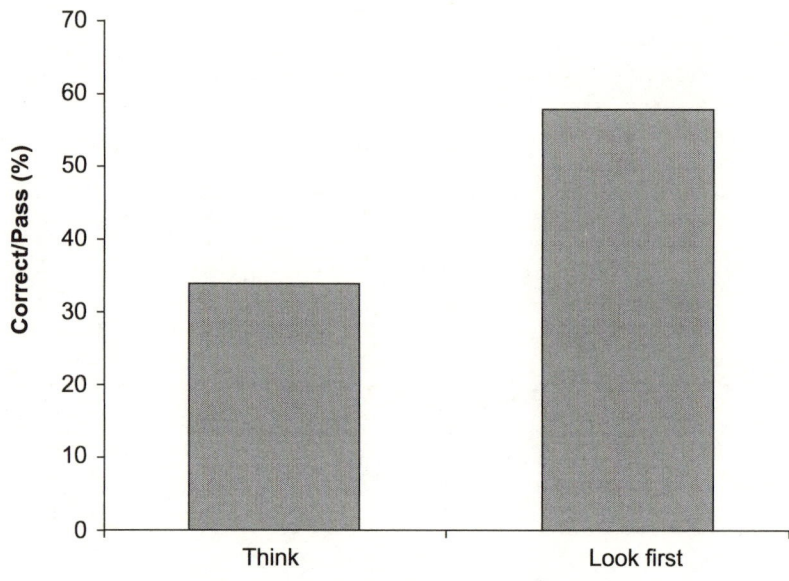

Figure 11.5 Three-year-old children's performance on false belief tasks can be significantly improved if their attention is first drawn to salient points in the task ('look first' condition). Redrawn with permission from Leslie (2000).

These kinds of findings shed a little doubt on Perner's (1991) suggestion that a major reorganisation of cognition takes place at four years of age, although it may just mean that children make the transition to a fully representational mind earlier, before their other capacities have reached an equivalent level of sophistication (possibly as the result of a process like RR: see above).

Leslie and Thaiss (1992), on the other hand, offer a different explanation that builds on Leslie's (1991) argument regarding the modular nature of ToM. They suggest that three-year-olds fail standard false belief tasks because ToM alone is not sufficient for successful performance. They argue that three-year-olds lack an ability they call 'selection processing' (SP) which is essential for the effective operation of the theory of mind mechanism when reasoning about false beliefs. Making false belief tasks easier allows three-year-olds to pass because it reduces the load placed on the child's selection processing capacity. Leslie and Thaiss (1992) argue that the need for selection processing is related to the true-belief 'default setting' of the ToM mechanism (see above). In order to predict a person's behaviour on the basis of a false belief, it is necessary to inhibit this default attribution (true belief) and attribute a non-factual (since it refers to a state that no longer exists) false belief in its place. Leslie (2000) suggests that SP is the mechanism that, operating in conjunction with ToM, allows children to achieve this inhibition of their true beliefs. Once the true beliefs are inhibited, the child can retrieve the representation of the world held at an earlier point in time (before the world changed) and use that to answer the question. Thus, the average four-year-old has an intact ToM mechanism and sufficiently strong SP to pass false belief tasks, whereas the three-year-old's weaker SP leads to failure despite a fully functional ToM mechanism. According to this theory, then, the watershed reached at four years of age represents an increase in processing power and not a major revolution in the understanding of representation as a process.

BOX 11.4

Theory of mind in adults

While we know a great deal about the early development of ToM in children, we have very little idea what happens beyond the age of six years. Indeed, we know very little about the natural history of intentionality among older children or adults, although it is clear that humans seem to be especially predisposed to viewing the world in intentional terms: we sometimes even attribute intentions to inanimate objects (for example, the weather, volcanoes). Just what levels of intentionality can normal individuals aspire to? Are there significant differences between individuals or between the sexes? How do any such differences relate to social skills?

We can glean some glimpses in answer to these questions from a handful of studies that have recently begun to explore some of these issues. Kinderman et al. (1998), for example, undertook the first attempt to explore 'advanced ToM' (that is, intentionality above level two) in normal adults. They presented verbal vignettes to subjects, each telling a story about an episode in about 200 words. Some were accounts about someone being misled by being directed to a particular shop whose location had, unbeknownst to the person giving the directions, changed (were they being deliberately misled or not?), others about someone wanting to persuade their boss to give them a payrise, yet others about someone wanting to make a date with someone else. After each story had been read out, subjects answered questions about the story. Some questions were about the mental states of the people in the story, some were simply factual questions about the circumstances described in the story. The latter questions acted as checks to ensure that any failure to answer mental state questions correctly were not simply due to failure to remember the facts of the story.

As **Figure 11.6** shows, the subjects in this study demonstrated considerable competence in answering mental state questions that involved as many as four orders of intentionality (I *believe* that you *think* that I *intend* to *deceive* you). Indeed, their response rates were no worse than those on purely factual questions, even when these factual questions are multi-element causal chains about physical world events that contain no mental state descriptions at all (in this case, a story detailing the sequence of physical events that occurred when someone accidentally set fire to himself). However, with fifth order intentional questions, success rates fell spectacularly from their previous level of around 90 per cent to around 40 per cent. In contrast, the same subjects experienced no trouble even with sixth order causal sequences (A led to B, which caused C, which resulted in D, which precipitated E, which caused F).

This suggests that the problem lies not with the ability to make inferences about sequences of statements, but rather quite specifically with the ability to keep track of who thinks what about who's mind. More than four levels of mental state reflexivity seems to put enormous stress on our cognitive abilities, and may explain in part why humans need such large brains to support their social groups (Dunbar 1993b, 1996a).

Even less is known about sex differences in competence on these kinds of tasks. However, Swarbrick (2000) reported that women score significantly better on second and third order tasks than men do, suggesting that they may be more socially skilled. These results agree well with other evidence that women are better than men at social tasks, including language fluency, social judgments, empathy, cooperation and pretend play (**Table 11.1**). Indeed, Baron-Cohen and Hammer (1997b) have argued that autism is simply one end of the normal male cognitive spectrum, reflecting the fact that the 'male brain' is less socially competent than is the 'female brain'.

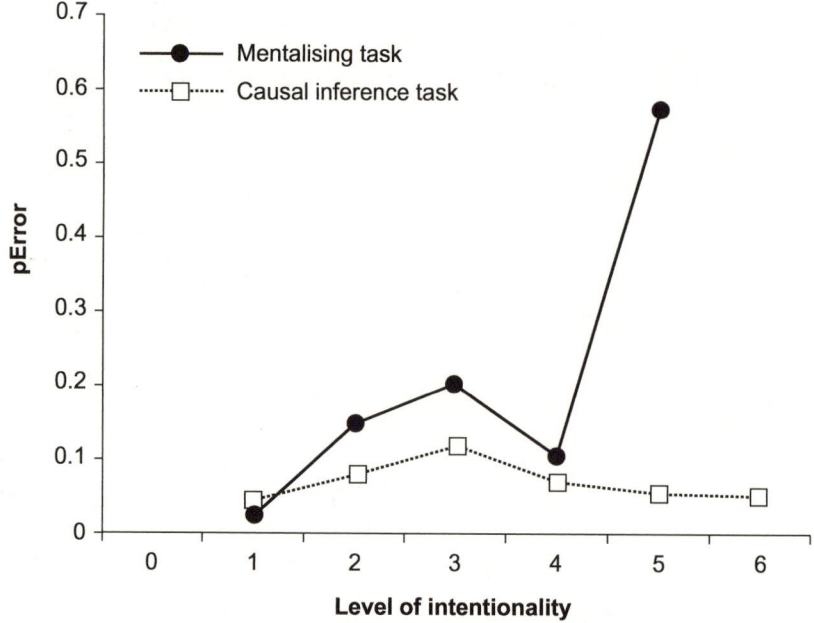

Figure 11.6 Adults perform well (error rates of 5–10 per cent) on simple causal inference tasks of up to six orders of causality, but can match this performance only up to fourth order intentionality with mentalising tasks (advanced theory of mind tasks). This suggests that mindreading is computationally more taxing than conventional causal reasoning about superficial correlations between physical world events. Redrawn with permission from Kinderman et al. (1997).

Table 11.1 Robust sex differences in performance on psychological tasks

Men are better at	Women are better at
mathematical reasoning	language
geometry	pretend play
embedded figures tasks	social judgement tasks
mental rotation	empathy
(some) spatial skills	cooperation
guiding motor skills	matching items
	ideational fluency
	fine motor skills
	arithmetic

Note: of the many hundreds of psychological tasks that have been tested, these are the only ones that have yielded consistent sex differences across studies over time.

Source: Baron-Cohen and Hammer (1997b)

WHEN TOM FAILS

Although there is now some debate as to whether ToM arises suddenly at age four or is already present in younger children as 'implicit' knowledge, there is no doubt that, among normally developing children, the ability to pass a standard false belief test is both rapid and unexpectedly consistent in respect of chronological age (**Figure 11.4**). The remarkable nature of this phenomenon is given both emphasis and poignancy by those cases where children continue to fail on these tasks and never acquire formal ToM, even when not severely disadvantaged in terms of IQ. Such individuals have a condition known as autism. Understanding the nature of these deficits throws considerable light on just what is involved in ToM itself.

AUTISM

Autism affects at least four to five in every 10 000 children and may be genetic in origin (Gillberg and Coleman 1992, Bailey et al. 1995). Children affected by this disorder show severe deficits in social and communication skills (for example, lack of eye contact, poor facial recognition), language (for example, no grasp of metaphor or conversational pragmatics – the non-verbal aspects of conversation that allow speakers to coordinate their utterances) and imagination (for example, little or no pretend play), and these in turn are thought to be a consequence of a severe impairment of ToM abilities (Leslie 1991, 2000; Baron-Cohen 1995, 2000a, b). In addition to these deficits in their social skills, individuals with autism show a number of 'asocial' symptoms such as strong obsessions and fixations, attention to detail, stereotypies, hypersensitivity to sound and an inability to cope with change. 75 per cent of children diagnosed with autism are also mentally retarded, but 25 per cent have IQs that fall within the normal range for their age (so-called 'high-functioning' individuals). These latter individuals still show the characteristic communicative impairments of autism, demonstrating that autism is not just an overall deficit in mental function.

Autistic children show a characteristic failure to pass false belief tests at the age at which normally developing children pass them. Baron-Cohen et al. (1985) found that while 85 per cent of normal four-year-olds could pass a Sally-Ann false belief test, and 86 per cent of Down's syndrome children could do the same, only 20 per cent of children with autism succeeded on the test (**Figure 11.7**). Autistic children also have trouble with true belief tests (Perner et al. 1989), even though these are much easier to pass, and they also show much worse performance than normal three-year olds on these tests (**Figure 11.8**). Some high-functioning autistic children can eventually pass false belief tests, but tend to be much older when they do so (approx. 9–10 years of age). These individuals are also unable to progress beyond this level of belief. They cannot, for example, perform correctly on tasks that require an understanding of a person's beliefs about someone else's beliefs (Baron-Cohen 1989). Happé (1994a) has suggested that these individuals use their innate intelligence to develop general rules of thumb that allow them to get by in social situations; while these rules work well enough most of the time, they fail on those occasions when deep social understanding of other minds is required.

Interestingly, Leslie (2000) showed that autistic children are able to pass the false photograph task at the same age as normally developing children (and indeed show improved performance), demonstrating that their impairment is specific to understanding belief states and is not a failure to understand the task itself (as is the case

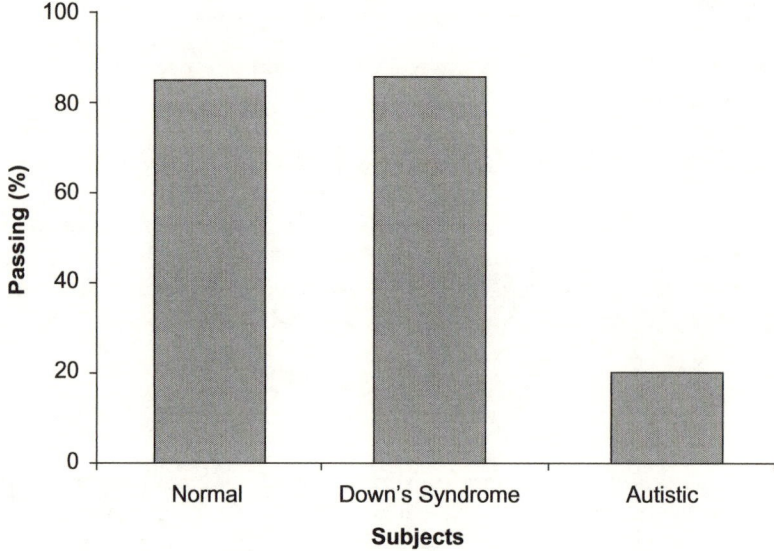

Figure 11.7 Normal 4-year-old and Down's syndrome children are significantly more likely to pass a false belief task than are autistic children of the same age. Since Down's syndrome children suffer from mental handicap (low IQ) while being highly sociable, the results suggest that autistic individuals' failure to pass these tasks must be due to a problem with theory of mind (the ability to understand another individual's mind state) and not to a problem with executive function. Source: Baron-Cohen et al. (1985).

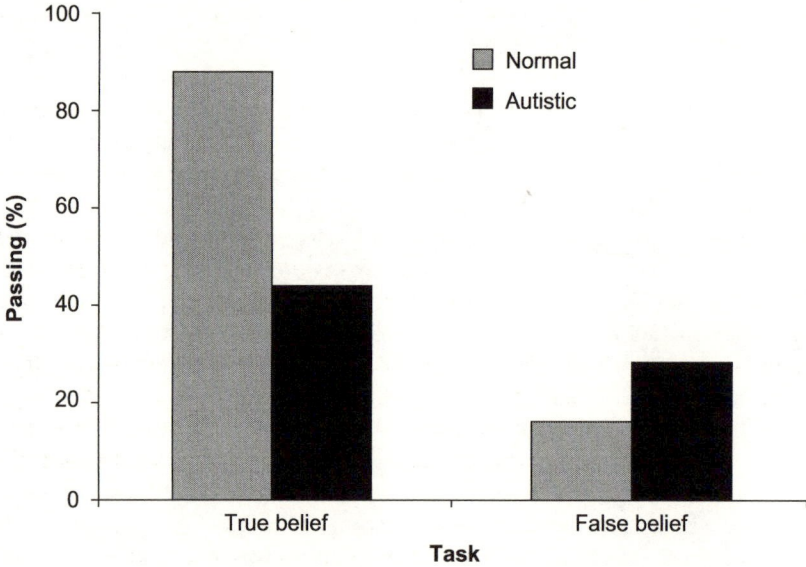

Figure 11.8 While both autistic and normal 3-year-old children find false belief tasks (see Box 11.3) difficult to do, normal children can cope well with true belief tasks. In a true belief task, the subject should be aware that the actor in the story knows that the situation is no longer as it was earlier: for example, in the Sally-Ann task, Sally sees Ann switch the ball, while in the Smartie task the friend has also seen what is in the tube. The fact that autistics do so badly suggests that they are being misled by irrelevant features of the task. Redrawn with permission from Leslie (1991).

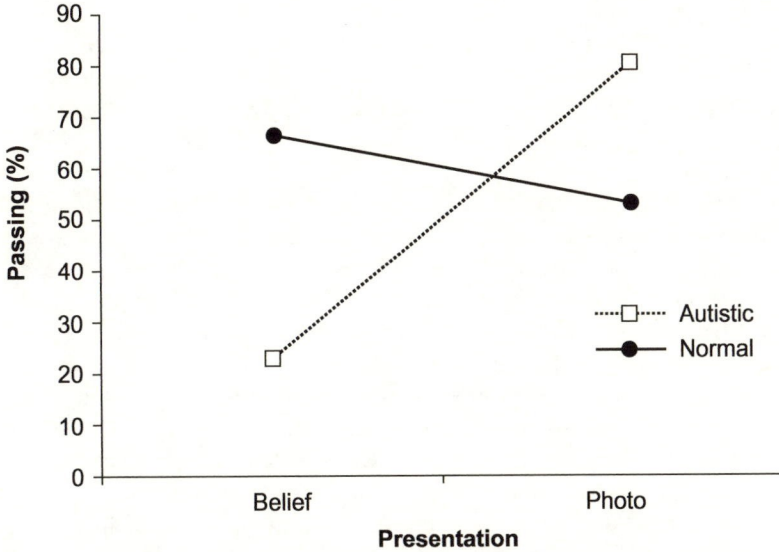

Figure 11.9 Autistic children fail standard false belief tasks, but can pass a pure memory task like the false photograph test at the same age as normal children (4–5 years). In the false photograph task, a photograph is taken of a scene before it is changed and the child is asked what the scene *had* been like. Redrawn with permission from Leslie (2000).

for normal three-year-olds) **(Figure 11.9)**. Autistic children are, in fact, the mirror image of three-year-olds: their selection processing is strong enough to pass the photograph test, but they have impaired ToM and thus perform badly on tasks that require an understanding of beliefs and other mental states. This double dissociation between performance on the false photograph task and false belief tasks across three-year-olds and children with autism has been taken as support for the modular nature of ToM, since it indicates a specific impairment in understanding mental states that is independent from other cognitive functions. The fact that autistic children fail to engage in joint attention as babies (Baron-Cohen 1991b), are unable to lie effectively or engage in pretend play (Leslie 1988) and do not understand desire (for example, they are particularly puzzled by the notion that an unfulfilled desire leads to distress) further points to a specific deficit in the realm of social cognition.

As a result of these findings, Leslie (1991) concluded that autism consists of a single cognitive disorder (the failure of the ToM module) and that all the affective disorders associated with autism (indifference to people, literal interpretation of language) are a secondary consequence of this single impairment. Other authors, however, disagree with this assessment. They argue that the ToM deficit cannot, in fact, explain all the symptoms of autism and so it cannot merely be a deficit in a single module that produces this condition (Happé 2000, Plaisted 2000).

Frith and Happé (1994, see also Happé 2000), for example, argued that autistic individuals show 'weak central coherence' – an inability at the perceptual level to view things as wholes rather than parts – in addition to their ToM impairment. In normal children, they argue, there is a drive to process the holistic properties of a stimulus before looking at its constituent parts (for example, seeing a face as a face, rather than an assortment of individual features like eyes, nose and mouth). If this drive is weakened, as in autism, then a deficit arises in this kind of holistic perception, and focus is instead placed on local features. Consequently, autistic individuals are better than normal or mentally-handicapped controls at making accurate judgements of visual illusions like the Müller-Lyer illusion illustrated in **Figure 10.1** (Happé 1996). They are able to focus on the bits of the illusions without integrating them into the surrounding illusion-inducing context. They are also less affected by upside-down faces in object recognition tasks. Because they recognise faces on the basis of individual features in isolation rather than as a face as whole, inverting the face is much less disruptive for them than it is for normally-developing children (Hobson et al. 1988). Both low- and high-functioning individuals with autism are also better than normal individuals at the Embedded Figures Test (a task where a small shape must be found in a larger design) (Shah and Frith 1983, Jolliffe and Baron-Cohen 1997). It has been suggested that their tendency to focus on local features aids them by reducing the amount of distraction caused by the larger picture. This inability to view the 'big picture' may explain other characteristic features of autism – such as the restricted repertoire of interests, preoccupation with parts of objects, an obsessive desire for sameness and an inability to generalise – that cannot be easily accounted for by a deficit in mental state understanding (Happé 2000).

Frith and Happé thus argue that autism should be reconsidered in terms of these two independent (but interacting) disorders rather than as a failure of a single modular capacity. Happé (2000) further argues that weak central coherence hypothesis should be seen as an explanation of a particular 'cognitive style' rather than a deficit account of autism. That is, weak central coherence can produce patterns of excellence in autistic individuals (the savant abilities like perfect pitch and drawing skills), as well as patterns of poor performance, and so autism should be viewed as the extreme end of a continuum of central coherence, rather than as a disorder per se. She has also suggested that weak central coherence is the most likely contender for the aspect of autism that is transmitted genetically. Parents, especially the fathers, of autistic children show significantly better than average performance on tasks favouring piecemeal local processing. For these fathers, their detail-focused cognitive style can be an asset rather than a deficit, since it may enable them to pursue careers where these skills are at a premium.

Baron-Cohen (2000a, b; Baron-Cohen and Hammer 1997a; Baron-Cohen et al. 1997), for example, has found that fathers of children with autism, as well as grandfathers, tend to be over-represented in occupations such as engineering relative to occupations requiring good social skills, such as social work. He has also shown that students studying in the field of mathematics and physics are also more likely to have a relative with autism than students in the humanities. Baron-Cohen (2000a) has therefore argued that autism reflects an impaired folk psychology accompanied by a superior folk physics (see Chapter 10). These findings – in conjunction with data showing that males with autism outnumber females by four to one and that, in the general population, superior folk physics skills are associated with males rather than females – led Baron-Cohen and Hammer (1997b) to speculate that autism may just be an extreme form of the 'male brain'. Like Frith and Happé's (1994) ideas about central coherence, the male brain should be viewed as a particular cognitive style that varies across individuals (that is, it doesn't mean that all males show signs of autism, or that only males can display these features). Only at the extremes of the continuum are any deficits apparent. For the majority, the elements of their cognitive style are an asset, rather than a handicap.

Exactly why folk physics and folk psychology should trade-off against each other in this way is not clear. There seems to be no reason in principle why one should not be socially as well as mechanically skilled – especially if ToM and central coherence

```
ggggggggggggggggggggg
ggggggggggggggggggggg
ggggggggggggggggggggg
ggggggg
ggggggg
ggggggg
ggggggg
ggggggggggggggggg
ggggggggggggggggg
ggggggggggggggggg
ggggggg
ggggggg
ggggggg
ggggggg
ggggggggggggggggggggg
ggggggggggggggggggggg
ggggggggggggggggggggg
```

Figure 11.10 The Navon task presents a large letter made up by a pattern composed of a different letter. Normal individuals exhibit 'global precedence' (they tend to identify the large letter rather than the small ones), but autistic individuals exhibit 'local precedence' (they tend to recognise the small letters).

are independent of each other. However, it is possible that they interact early on in life in a way that precludes normal development.

Plaisted (2000) offers an alternative interpretation for the asocial symptoms of autism arguing that critical predictions of the central coherence hypothesis were not upheld in tests using the Navon task (Navon 1977). The stimulus used in the Navon task consists of a large letter that is made up of a number of smaller letters (see **Figure 11.10**). Importantly, the large letter is sometimes made up of different letters from that which it depicts. Normal individuals tend to show a 'global precedence' on this kind of task. They are much better at identifying the large 'global' letter than the small letters of which it is made up. If the central coherence hypothesis is correct, then autistic individuals should, by contrast, show a 'local precedence' and be much better at identifying the small letters than the large global letter. Plaisted (2000) reports that this is not, in fact, the case (see also Ozonoff et al. 1994, Mottron and Belleville 1993). As long as autistic individuals are primed before each block of trials so that they know whether they will need to recognise the global or local features of the stimulus (a 'selective attention' task), they are able to perform as well as normal children. This should not happen if autistic children have a specific deficit in global processing abilities. However, it is clear that priming plays a key role: if autistic children are given a divided attention task, in which they have to indicate the presence or absence of a target letter which could be either at a local or global level (that is, there is no priming beforehand), then autistic children do show a local precedence effect (producing global errors) and they are also able to detect local targets more rapidly.

Plaisted (2000) therefore suggests that, rather than a deficit in central coherence mechanisms, autistic children's problem lies in their inability to inhibit irrelevant local information under circumstances where they should be processing information at the global level. Such an argument bears a resemblance to Leslie and Thaiss' (1992) selection processing hypothesis. Following their argument, the above results could be explained as follows: when autistic children are primed so as to increase the salience of the global level, it places less load on SP and they are able to inhibit local information (in the same way that normal children can inhibit their true beliefs) and perform successfully. When any overt (or covert) priming is not forthcoming, on the other hand, the child's selection processing abilities are too weak to enable the inhibition to occur, thus leading to failure.

The results of these and similar tests led Plaisted (2000) to argue that children with autism show a tendency to process features that are held in common between objects rather poorly, but are very good at focusing on features that are unique. Put simply, they tend to see much less similarity between objects than do normal children. This difference in visual processing with regard to shared versus unique properties may therefore explain the impairment in face recognition shown by autistic individuals and, by extension, their overall lack of interest in people.

In order to learn someone's face, it is necessary to recognise those features that remain more or less the same no matter what the face happens to be doing (smiling, laughing, shouting). That is, one has to form a face prototype that covers all possible expressions and orientations. In autism, the tendency to focus on the unique features of particular stimuli, and not on the similarities between them, makes it much harder to form such prototypes. Autistic individuals may be continually drawn to the unique features of a smiling versus a crying face, and not on the overall similarities that exist despite the change in expression. Autistic individuals may find faces unworthy of attention because they contain little meaning for them in terms of individual recognition. This lack of attention to faces may ultimately lead to a deficit in ToM, due to an inability to develop attention and social referencing which are essential precursors of ToM, but it arises out of a deficit in perceptual processes and not one related to social cognition per se.

One final, and important point to emerge from these studies of autism, is that we need to make a clear distinction between the representational mechanism that is necessary *for* ToM and the social understanding that springs *from* ToM (Happé 2000). The latter ability is what allows people to cope in a flexible manner in social situations; realising the significance of certain remarks, gauging changes in the 'emotional temperature' and an awareness of subtle contextual cues are just as important as an ability to understand 'cold' mental states. In other words, central coherence (or selection processing) may well be as important to social understanding as it is to perceptual processes. Happé (2000) suggests that autistic individuals who pass false belief tests are able to do so because the tasks are taken out of a real social context and all the relevant information is made available to them explicitly: the opposite of what occurs in real life. Consequently, although these individuals show an ability to attribute mental states, they still do not function well in normal life because of their inability to view social situations as holistic on-going processes. As Bloom and German (2000) point out, not only is there more to passing a false belief test than theory of mind, there may also be more to theory of mind than merely being able to pass false belief tests.

'HOT' AND 'COLD' COGNITION

Williams syndrome is another rare (1 in 25 000 births) neurodevelopmental disorder with a genetic origin (Bellugi et al. 2000). Children with Williams syndrome have low IQ and are often quite severely mentally handicapped. Unlike children with autism, however, they have exceptionally good language skills (the most impressive

BOX 11.5

Theory of mind and clinical disorders

The socially inappropriate behaviour displayed by individuals suffering from particular psychiatric conditions has been linked to deficits in certain aspects of mental state reasoning. Frith (1992), for example, has explicitly identified schizophrenia as a disorder of meta-representational abilities. In line with this, Frith and Corcoran (1996) produced preliminary evidence that paranoid schizophrenic patients fail second order intentionality tasks more often than normal adults or schizophrenics in remission (**Figure 11.11**). Swarbrick (2000) took this further and found that schizophrenics consistently fail third order tasks that normal adults (and schizophrenics in the remitted phase of their condition) experience no difficulty solving. There is also some preliminary evidence to suggest mindreading abilities are equally impaired during the active phase of bipolar depression (Kerr et al, submitted).

Corcoran and Frith (submitted) have also shown that people diagnosed with schizophrenia show impairments on the Wason selection task. Un-

like normal individuals, people with schizophrenia perform no better on social tasks than they do on abstract tasks. They suggest that this is because people with schizophrenia show a deficit in 'analogical reasoning'. According to Corcoran (2000), normal people infer other people's mental states by first searching their autobiographical memory (memories of specific events that the person has experienced in their own life) to see if any remembered event can help inform them as to what might be going on. Conditional reasoning processes then work upon this memory to produce a solution appropriate to the current situation. Individuals with schizophrenia tend to show impaired autobiographical memory and, as a consequence, are unable to use conditional reasoning to infer others' mental states because they are unable to retrieve relevant social memories. This deficit results in both an inability to attribute mental states and a failure to use appropriate forms of conditional reasoning in social versions of the selection task.

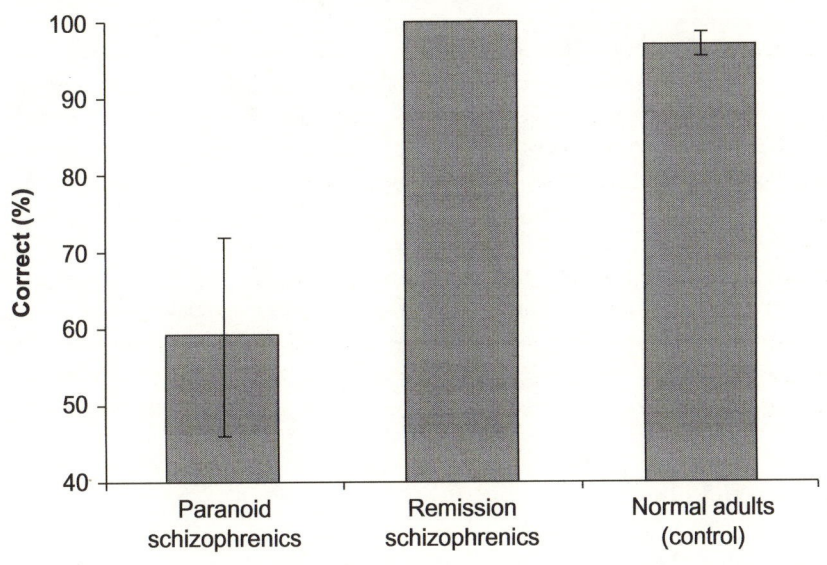

Figure 11.11 In their clinical phase, schizophrenics consistently fail second order intentionality tasks (theory of mind), even though schizophrenics in remission and normal human adults easily pass them. Redrawn with permission from Frith and Corcoran (1996).

of which being an extensive vocabulary) and they are incredibly sociable. They show an intense interest in people and have exceptional face-processing skills. In many ways, they seem like the opposite of autistic individuals. However, despite their sociable tendencies, they show the same level of impairment on false belief tasks as autistic children (Tager-Flusberg and Sullivan 2000).

Using data on children with Williams syndrome, Tager-Flusberg and Sullivan (2000) argue that ToM actually consists of two separate components: a social-cognitive component (concerned with representing the mental states of others) and a social-perceptual component (concerned with representing the emotional states of others). Stone (2000) makes a similar distinction between 'cold' and 'hot' cognition, respectively. Tager-Flusberg and Sullivan (2000) argue that while autistic children show an impairment in both the social-perception and social-cognitive aspects of ToM, those with Williams syndrome show much less impairment on the social-perceptual component, with the result that the latter are able to respond emotionally to other individuals in a way that the former cannot.

This sparing of the social-perceptual component of ToM in combination with a deficit in representational ToM helps to explain the paradoxical nature of Williams syndrome: although individuals are responsive to people, empathetic and friendly, they nevertheless function poorly in the social domain, displaying poor social judgement and having considerable trouble sustaining friendships. Similarly, high-functioning autistic individuals who can solve false belief tasks (possibly by tapping into alternative neurobiological pathways, such as those used for language: Tager-Flusberg 2000) remain socially impaired because they cannot respond empathetically to people or read their emotions. Sophisticated social reasoning of the kind we take for granted is dependent on the integration of these two components.

THE SOCIAL BRAIN

Brain-imaging studies tend to support this notion of a social-perceptual as well as a social-cognitive component to ToM. When normal individuals were asked to read a person's emotional state from a photograph of their eyes alone, the amygdala was activated (Baron-Cohen et al. 2000c). The amygdala forms part of the limbic system, an evolutionarily ancient part of the brain. Studies of monkeys have shown that this area recognizes emotional displays, and is activated in response to facial expressions, especially those depicting fear and threat (Brothers and Ring 1992). The amygdala is also linked to the temporal sulcus, an area of the brain that is critical in determining gaze-direction and judging direct eye-contact (Young et al. 1995). As might be expected, given the nature of their symptoms, there was no activation of the amygdala in autistic individuals during presentation of the 'mind in the eyes' task and their performance on the task was correspondingly poor (Baron-Cohen et al. 2000c). Studies of adults with Williams syndrome, by contrast, show that they are successful at this kind of test (Tager-Flusberg et al. 1998), although no brain-imaging studies have yet been performed to confirm that the amygdala is involved in the same way as in normal individuals.

By contrast, when individuals are presented with tasks requiring an assessment of mental states, regions of the frontal lobes are activated. Unlike the amygdala, which is an evolutionarily ancient part of the brain, the frontal lobes form part of the neocortex – an area which has undergone expansion only in the mammals and, among the mammals, most spectacularly in the anthropoid primates (monkeys, apes and humans). Humans in particular display a remarkable expansion of the neocortex in general and the frontal lobes in particular. It has long been suggested that our exceptional mental abilities are attributable to this selective increase in the size of these brain regions.

Fletcher et al. (1994), Goel et al. (1995) and Schultz et al. (quoted in Klin et al. 2000) all found evidence of activation in the left medial frontal cortex during tasks requiring mental state attributions in normal individuals, whereas this area was not activated in high-functioning autistic individuals (Happé et al. 1996). Individuals with damage to this area of the brain and those with autism have problems both with behaving appropriately in social situations and with understanding certain high-level aspects of language such as metaphor (Stone 2000, Happé 1994b). Metaphor is what gives language its richness, and without it language becomes stilted and dully factual. Metaphor allows very fine nuances of meaning to be hinted at, permitting us to mean something other than what we say; so an inability to understand metaphor can result in quite large social handicaps. The medial frontal cortex is also associated with affective (emotional) processing. One area in particular, the ventro-medial prefrontal cortex, processes emotions differently depending on whether they are positive or negative (that is, whether they will elicit approach [pleasure] or withdrawal [pain] behaviours: see, for example, Damasio 1994).

Another area of the neocortex quite distinct from the medial frontal cortex has also been linked with the generation of particular mental states. This may partly be due to the fact that this area (known as the orbito-frontal region) was studied using a different technique to the studies referred to above (Baron-Cohen et al. 1994). Rather than being read stories and then asked questions about them, subjects in this study were asked to tell the experimenter whether certain words referred to the mind (for example, think, remember, imagine) or to the body (shoulder, teeth). The orbito-frontal cortex was activated only when subjects were processing mental-state terms.

The orbito-frontal cortex is known to be heavily involved in social cognition and damage to this area produces a whole suite of social and emotional changes: those with damage in this area show socially inappropriate humour and behaviour, are self-centred and lack concern for others, and are very bad at social decision-making (Stone 2000, see below). All these symptoms arise from an inability to combine information about mental states with emotional information. Stone et al. (1998a, b), for example, have demonstrated that patients with orbito-frontal damage are capable of passing both first and second-order false belief tests, but fail the 'mind in the eyes' task and 'faux-pas' tasks (where individuals are required to recognise that someone has made a socially embarrassing mistake). Like high-functioning autistic individuals, they are able to understand social problems in a 'cold' purely cognitive manner but cannot integrate this with information on emotional states.

What all these studies indicate is that social reasoning is not a purely cognitive activity, but relies heavily on understanding and representing emotional as well as mental states. Such findings also highlight the likely evolutionary path we have taken from our non-human primate ancestors (Brothers 1990, 1997). The social understanding displayed by monkeys is heavily dependent on emotional or social-perceptual aspects of ToM (see Emery and Perrett 2000, Emery 2000, Emery et al. 1998, Perrett and Emery 1994), but they appear to lack the cognitive meta-representational abilities of humans (Tomasello and Call 1997). The more complex human mentalising abilities are built on top of, and integrated with, a more primitive emotional-perceptual system, supplementing it but not replacing it. Consequently, emotional cognition remains an essential part of human social life.

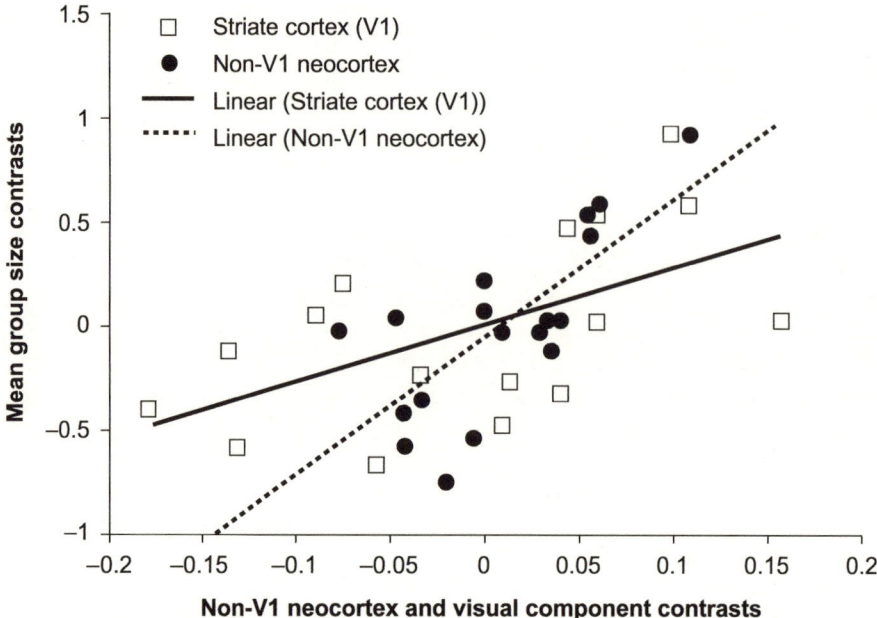

Figure 11.12 When the primary visual cortex (striate cortex, area V1) is distinguished from the rest of the neocortex, a greatly improved relationship is found between mean group size for anthropoid primate genera and the relative volume of the non-striate neocortex. Note that the regression line for the non-striate data is steeper and the variance in the data much lower than for the striate cortex data. The axes are independent contrasts between pairs of taxa (a technique designed to remove the effects of **phylogentic inertia**). Redrawn from Joffe and Dunbar (1997).

It has long been recognised that primates have unusually large brains for body size (Jerison 1973), and that humans have disproportionately large brains even for a primate. The real issue here, however, turns out to be neocortex size, which increases disproportionately across the mammalian and primate orders (Finlay and Darlington 1995). Primates have disproportionately large neocortices compared to other mammals, and it is this that has largely been responsible for brain size increase among the primates. As a result, it is relative neocortex size rather than total brain size that correlates with many social indices in primates (see **Figure 6.1**).

However, the relationship between relative neocortex size and indices of social group size shown in **Figure 9.3** contains within it an unexpected result: the apes (including humans) seem to lie on a separate grade to the right of the simians (monkeys), suggesting that apes require more neocortex volume (that is, computing power) to support a given group size than monkeys do (see also Kudo and Dunbar, in press). This suggests that the apes are doing something that is cognitively more complex.

Joffe and Dunbar (1997) showed that the relationship between group size and neocortex volume could be narrowed down to the areas of the neocortex forwards of the primary visual area (V1) (**Figure 11.12**), an observation that ties in well with the brain-scan evidence that areas of the frontal cortex are crucial for social cognition. Dunbar (in press) has extended this by showing that, at least within the ape–human grade, social group size correlates very closely with the relative size of the frontal lobe. More importantly, Joffe and Dunbar showed that, within primates, there is a tendency for the non-V1 components of the neocortex to increase in volume faster

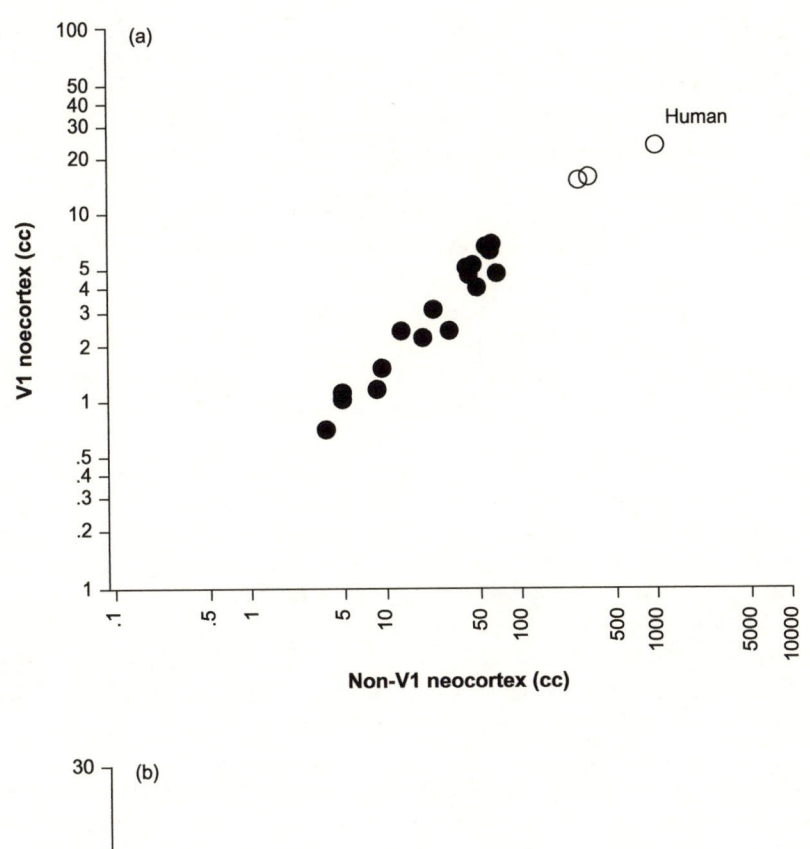

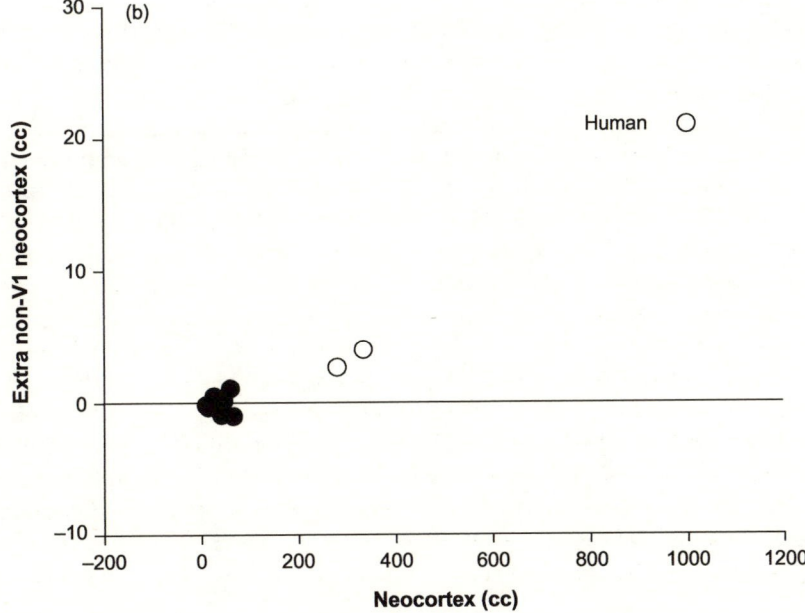

Figure 11.13 (a) A plot of the absolute volume of primary visual cortex (striate cortex, area V1) against the rest of the neocortex for anthropoid primates suggests that the size of striate cortex begins to reach an asymptotic value at the neocortex size of the great apes. (b) If the regression through the simian datapoints in (a) is used to predict the volume of the non-striate neocortex, the residual between the observed and predicted volumes provides a measure of the relative amount of extra neocortex available for executive function. (a) is redrawn from Joffe and Dunbar (1997).

than the primary visual cortex as brain size increases (**Figure 11.13a**). The significance of this finding is that, among the great apes, an increasing quantity of cortical volume becomes available for executive functions, relative to other primates (**Figure 11.13b**). This process reaches its maximum in humans. Recent studies showing that impairments in ToM abilities are independent of those associated with, for example, the loss of general executive function reinforce this suggestion, since they imply that the frontal lobes are necessary for ToM in humans (Stuss et al. 2001, Rowe et al. 2001).

It seems likely that the resulting availability of spare computing capacity in the great apes (in other words, that not required for management of basic somatic processes or the essential processing of sensory input) increases the capacity for (1) social cognition (thus explaining the tentative presence of theory of mind in great apes and, more especially, humans: Dunbar 2000a, in press) and (2) social learning (thus explaining the evidence for culture in these species: Boesch 1996; Whiten et al. 1999).

*

Although theory of mind has been presented as an archetypal example of a mental module, the burden of the evidence and opinions that we have reviewed here tend to reinforce the conclusion that we arrived at in Chapter 10, namely that what often looks like an integrated coherent cognitive function on superficial first view turns out on closer analysis to be better described as the emergent property of a composite of several more fundamental functions. Although we are only just beginning to understand the underlying neurobiology of the brain, what evidence we do now have tends to point in the same direction: functions like language and ToM that have been described as modular seem to have a dispersed construction at the neural level, involving small groups of neurones in quite different locations. Although it can legitimately be argued that functional modularity need not reflect anatomical modularity, nonetheless, the burden of the evidence (and especially that for ToM) increasingly points towards the suggestion that the more complex kinds of social cognition seen in humans may have more to do with the integration of a number of cognitive units during the course of early development than with the existence of unique highly specialised modules. In effect, humans may differ from other primates in their social cognitive abilities more by virtue of simply having 'bigger and better' versions of the same general capacities than by having evolved novel and unique ones.

The cognitive adaptations discussed in this chapter are essentially what makes us human. Another distinguishing feature of our species is the degree to which culture pervades our lives and determines much of what we see, think and do. Is it possible to look at human culture from an evolutionary perspective? Can such a complex multi-layered phenomenon really have biology at its base? In the next two chapters, we tackle these issues by focusing on two aspects of culture: language (Chapter 12) and then culture as the set of social rules that guide human behaviour (Chapter 13). We will show that, while culture may not (and need not) be adaptive, there is no question that it has an evolutionary origin and can be investigated in evolutionary terms.

■ Theory of mind (ToM – the ability to attribute mental states to others) is possibly a uniquely human attribute.

■ Humans are not born with ToM, but rather acquire this ability rather suddenly at around four years of age. The development of ToM appears to be preceded by a phase known as 'belief–desire psychology' during which children learn to understand their own mental states in terms of, first, desires and, then, beliefs.

■ Some authors have viewed ToM as an archetypal modular faculty. Data from children with autism have been adduced in support of this suggestion. Autistic individuals consistently fail false belief tasks (the benchmark for theory of mind) that normal five-year-olds pass with ease. The fact that autism is particularly common in males, combined with some evidence to suggest that it may be inherited, has been interpreted in support of the suggestion that ToM modules are hardwired and genetically determined.

■ However, there are some features of this disorder that cannot easily be explained by the impairment of a single ToM module. This has led other authors to argue that deficits in central coherence, inhibition of local processing or failure of theory formation mechanisms lie at the heart of autism and related conditions.

■ As well as the ability to represent mental states, normally developing individuals also have the ability to represent emotional states. This suggests that ToM may consist of two components: a social-perceptual component and a social-cognitive component. These suggest a route by which human social cognition could have been built up from the emotionally-based social cognition of non-human primates.

■ Recent evidence from brain-scan studies and comparative analyses of primate brain evolution suggest that two areas (the orbito-frontal area of the neocortex and the amygdala) may be especially important in social cognition.

·· **Further reading**

Astington, J. W. (1993). *The Child's Discovery of the Mind*. Cambridge: Cambridge University Press.

Baron-Cohen, S. Tager-Flusberg, H., Cohen, D. J. (ed.) (2000). *Understanding Other Minds: Perspectives from Developmental Neuroscience*. 2nd edition. Oxford: Oxford University Press.

Brothers, L. (1997). *Friday's Footprint: How Society Shapes the Human Mind*. Oxford: Oxford University Press.

Gopnik, A., Mettzoff, A. and Kuhl, P. K. (2000). *How Babies Think*. London: Weidenfeld & Nicholson.

Heyes, C. and Huber, L. (eds) (2000). *The Evolution of Cognition*. Cambridge (MA): MIT Press.

Perner, J. (1991). *Understanding the Representational Mind*. Cambridge (MA): MIT Press.

12 Language

CONTENTS

Language may be the single most important feature that distinguishes humans from other animals. This is not necessarily because language, of itself, influences the way we think and see the world (though many in the humanities and social sciences have so argued) but rather because it allows us to exchange information and opinions, thereby making it possible for one individual to influence how another sees the world. It may be essential in allowing humans to engage in some of the more advanced forms of social cognition. It also inevitably forms the core to human culture, the subject of the following chapter.

Human language is an open-ended system of communication in which grammatical structure allows information of great cognitive complexity to be passed from one individual (the speaker) to another (the listener). Among the 4000 or so species of mammals and 10 000-odd species of birds currently alive, we are the only species that possess true language in this sense. Other species communicate with each other, and their systems of communication may indeed be able to transmit information about the world (see for example Cheney and Seyfarth 1990), but none does so with the degree of sophistication seen in even the simplest human languages (pidgins). That humans are the only species to engage in such a sophisticated a form of communication calls for an explanation.

To suggest, as some have, that language is a mere by-product of having a large brain or some kind of accidental macro-mutation is not good enough. Language, and its principal *modus operandi* (grammar) is simply too complex a phenomenon to have evolved by pure accident. The speed with which children acquire language points to a highly organised faculty: between the ages of 18 months and six years, children acquire a vocabulary of around 15 000 words, which represents about one new word for every 90 minutes of their waking day. By the age of six years, they are more or less fully competent in the grammar of their particular language community. Granted that all human languages have the same (or at least very similar) underlying grammatical structures (and this could in principle be genetically programmed), nonetheless languages differ enormously in the details of how these general grammatical principles are formulated. There are languages with word-ending grammars (Latin, German), languages with prefix grammars (Kiswahili), languages with place (or word order) grammars (English), and so on. A child could not, of course, describe how the grammar it uses works, but it has learned a large number of rules, including, in cases like English, a very large number of irregular constructions – and it does this with surprisingly few mistakes along the way. Indeed, the classic mistakes are often quite logical: *Jimmy has* do-ed *this*, following the general rule that past tenses are created by adding -*ed* to the end of the verb. Children do not learn to read and write with such easy facility.

Many would argue that language deserves an entire book in its own right, and so it may. Our aim here, however, is more modest. We will ask and try to answer just two questions: what function does language serve? and why might it have evolved in the first place? We begin with the more general question of why speech evolved, and then ask what we use this unusual capacity to talk about.

THE EVOLUTION OF LANGUAGE

In this section, we consider three key questions: why, how and when did language and speech evolve? Answering questions that have a strong historical component is inevitably fraught with difficulty. Nonetheless, the questions are crucial and need to be considered, even though at this stage we can offer only a preliminary account.

WHY DID LANGUAGE EVOLVE?

Conventional theories for the origin of language tend to focus on the way language is designed to subserve its primary purpose, namely the communication of information (see for example Pinker and Bloom 1990, Pinker 1994), but they rarely ask what that function might be for. In an authoritative review of language and linguistics, Crystal (1997) devotes just four pages out of 400 to the functions of language, and the functions he identifies focus principally on the communication or storage of information, with some modest additional social functions (cursing, routinised conversational exchanges). In effect, the conventional spotlight has been on how language facilitates the exchange of information rather than on why we have language at all. This inevitably means a focus on grammar and the way grammar aids the coding and comprehension of thoughts. These are of course interesting and worthy projects in their own right, but they fail to address the larger evolutionary question as to why we need to exchange our thoughts in this way.

Conventionally, language can be seen as typified by either of two utterances: *There are bison down by the lake right now* or *This is how you make a handaxe*. Both are inevitably predicated on the view that the most important information-exchange problems faced by our ancestors had something to do with hunting (or even gathering). Language's functions were thus either the exchange of ecological information or instructional.

There are several problems with this view. First, studies of what people actually talk about in both modern industrial societies (Landis and Burtt 1924, Kipers 1987, Bischoping 1993, Dunbar et al. 1997) and traditional societies (Haviland 1991) suggest that most conversations are in fact dominated by social topics ('gossip' see pp. 334–7). We use instructional forms of language only occasionally, and then in rather specialised contexts. Second, hunters commonly prefer to hunt in very small groups and often do so in silence (see, for example, Inuit hunting practices, where modal group sizes for different hunt types vary between one to four individuals: Smith 1991). Third, teaching someone how to make an object (for example, a pot or a handaxe) is best done by demonstration (and a lot of practice) rather than verbal instruction (verbal instructions are usually limited to rather vague utterances like: *Do you see what I'm doing . . .?*). Fourth, there appears to be little or no correlation in the archaeological record between the changes in hominid brain size (on which language ultimately must depend) and the changes in tool complexity (Wynn 1988): if language was intimately related to hunting and tool use, then we might have expected some correlation between these and brain size.

During the last few years, a number of alternative explanations for the evolution of language have been proposed that focus on essentially social functions for language. In this respect, they converge on the broad position long advocated by social linguists (Trudgill 1974, Eakins and Eakins 1978, Coates 1993). These explanations for the evolution of language differ mainly in the social function that they favour as the primary focus. We consider three principal versions of this idea in the next section. First, we examine the evidence in support of the claim that language's principal function is a social one, irrespective of what that particular function may be. For this reason, we make a clear distinction between speech and language and here focus on speech as a general capacity rather than what people actually talk about during conversation (which we explore in the following section).

Dunbar (1993c, 1996a) argued that the principal function of language (and hence speech) is the exchange of social information ('gossip' in a broad sense) and that language evolved to support cohesion within large social groups. The issue revolves around two keys facts. One is that social group size correlates closely with relative neocortex size in primates as a whole (see **Figure 6.1**), with modern humans seeming to fit neatly into this pattern with ('cognitive') group sizes in the order of 150 (**Figure 9.3**). The other is the suggestion that the Old World monkeys and apes, at least, use social grooming as the principal mechanism for bonding their groups. We do not understand exactly how grooming achieves this, though it seems likely that it has something to do with the fact that grooming is particularly good at releasing endogenous opiates (Keverne et al. 1989). The feelings of pleasure and contentment that seem to well over an animal as a result of being groomed (and they clearly do find it very relaxing) may create a sense of trust and contentment in the partner, and this in turn may facilitate alliance formation and the reciprocation of many other social and reproductive benefits.

The problem, however, is that the time devoted to social grooming by Old World monkeys and apes is a more or less linear function of social group size (**Figure 12.1**).

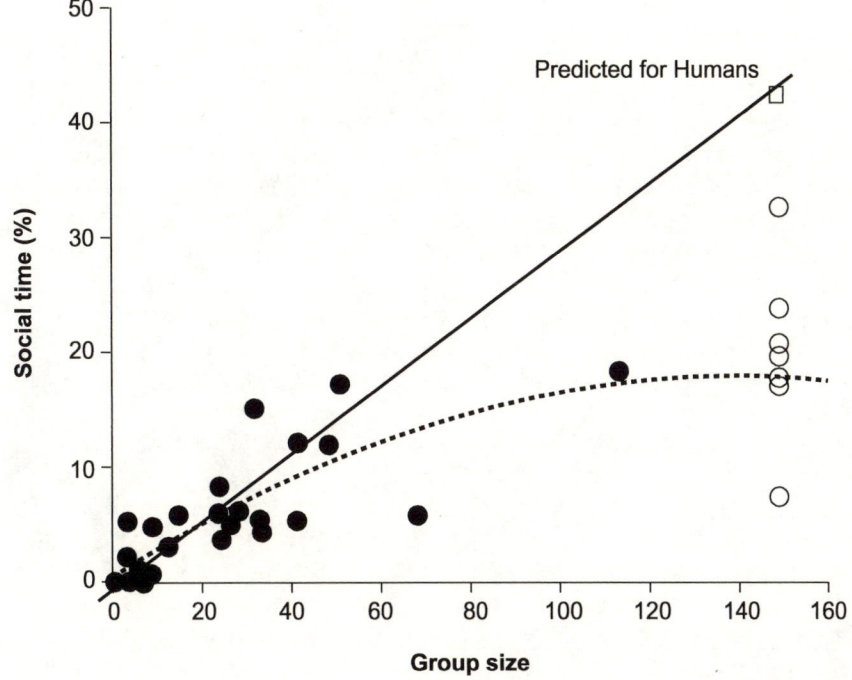

Figure 12.1 Grooming time in Old World monkeys and apes (filled symbols) increases with group size (the plotted data are for individual species). However, the distribution of the datapoints suggests that grooming time may be constrained by time budgeting considerations, with an absolute upper limit at around 20 per cent (broken line). A linear relationship between grooming time and group size (solid line) would predict that modern humans (with social groups of 150) would need to spend 43 per cent of their total day time engaged in social grooming if humans were to bond their groups using the same mechanism as other Old World monkeys and apes (open square). In fact, a sample of seven contemporary societies (open circles) spend a mean of exactly 20 per cent of total day time engaged in social interaction (conversation). The human populations include African pastoralists and horticulturalists, Nepalese hill farmers, New Guinea horticulturalists and contemporary Europeans. Sources: Dunbar (1993c, 1999).

As hominid group size began to creep up above that found in the most social of the primates (and presumably our ape-like ancestors), so the demands on time budgets for social grooming time must have become intense. No species of living nonhuman primate devotes more than 20 per cent of its total daily time budget to social inter-action; this in itself represents a phenomenal amount of time (one fifth of the waking day), considering the fact that the animal has to earn its living from the environment. Ultimately, the biological demands of feeding, travel and resting mean that there will inevitably be an upper limit to the amount of time available to be devoted to social bonding. This in turn will place an upper limit on the size of group that a given population of animals can maintain as a coherent unit. It seems that this upper limit is probably around 70 individuals. Given that modern humans have managed to evolve a stable group size of about 150, how have they managed to do this?

Extrapolating from the graph in Figure 12.1 would suggest that, if humans bonded their groups in exactly the same way that other Old World monkeys and apes do (that is, by social grooming), then they would need to spend around 40–45 per cent of their waking day grooming with each other (Dunbar 1993c). This is more than

BOX 12.1

How many people can you talk to?

Observations of naturally-occurring conversation groups suggest that conversations typically consist of a single speaker at any one moment, even when there are many listeners. Turn-taking is a particularly characteristic feature of human conversations (Sacks et al. 1974, Beattie 1983). There are quite strict rules about who can talk and when, some of which may be formally imposed (the role of the chairman in committees or that of the lecturer at a seminar or sermon). While the number of listeners (that is, those paying explicit attention to a particular speaker) is more variable, the data suggest an asymptotic value at around 2.7 listeners on average (**Figure 12.2**).

This limit on conversational group sizes at about four individuals (one speaker and three listeners) appears to be extremely robust: it can be observed it in almost any social setting. When a fifth person joins a conversation, the others will make every effort to include the new member, but within half a minute or so the group is likely to break up into two separate conversations.

This limit on the size of conversational groups appears to be a consequence of the fact that human auditory acuity (specifically our ability to discriminate speech sounds) is designed to be just efficient enough to include about four people in a conversation at normal background noise levels (Cohen 1971). When more than four people are involved, the distance across the circle formed by the group when separated by socially acceptable distances is such that individuals on opposite sides of the circle cannot easily hear what each other are saying (**Figure 12.3**). In effect, it seems that our speech detection mechanisms are designed to be just efficient enough to allow us to have interaction groups that are sufficiently large to increase total group size by the requisite amount. Since there is no intrinsic limit on the acuity of our speech detection mechanisms (other animals have much more acute hearing than humans), we must assume that natural selection has evolved a mechanism that is adequate for its required purpose, but no more than that.

double the time that ecologically hard-pressed monkeys and apes seem to be able to devote to social interaction. So some mechanism must have been necessary to enable modern humans (and, to a lesser extent presumably, our hominid ancestors) to bond larger groups in the same amount of time, otherwise we simply would not be able to maintain cohesive social groups of the size that we now do. In effect, some process is needed whereby the time available for social interaction can be used more efficiently.

The answer seems to the capacity to speak. As it turns out, the mean amount of time actually spent in social interaction (principally conversation) in a range of modern human societies (including not only populations in the UK but also more traditional horticulturalists and pastoralists) is 20 per cent of waking time – in other words, exactly the upper limit on social time observed in other primates (**Figure 12.1**). If speech allows us to use this time more efficiently than we could by grooming, how is this achieved?

Human language allows us to use social time more effectively to bond larger groups in at least three ways. First, it allows us to interact with more individuals at the same time (that is, it increases the size of the broadcast network). Second, it allows us to acquire (or exchange) information that we would otherwise never find out about (see next section). Third, as we noted in Chapter 9, it allows us to police freeriders (see **Box 9.3**). The latter two of these functions will be discussed in more detail in the following section; here, we concentrate on the more mechanical issue of network size.

Dunbar (1993c) argued that the core problem for language was to raise the size of the bonded group up from the maximum seen in nonhuman primates (about 50

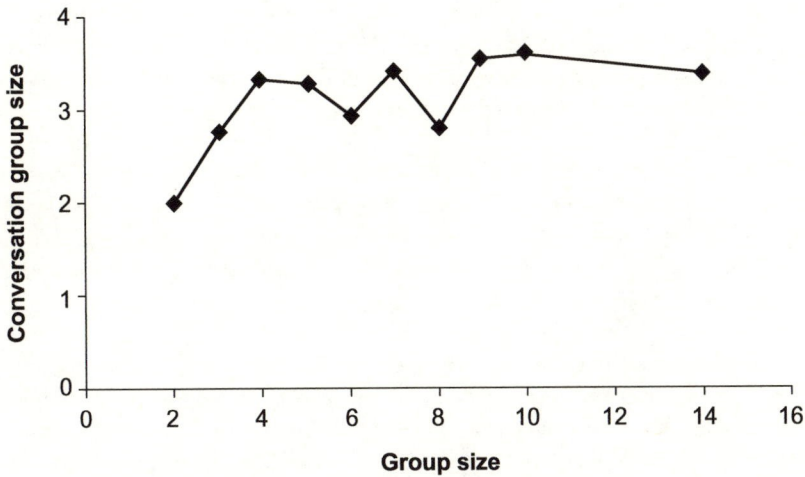

Figure 12.2 As the size of an interacting group increases, it fragments into more independent conversations. The mean number of individuals engaged in conversation (as speaker and listeners) rises to an asymptotic value at just below 4, irrespective of the size of the group within which the conversation is embedded. Reproduced with permission from Dunbar et al. (1995).

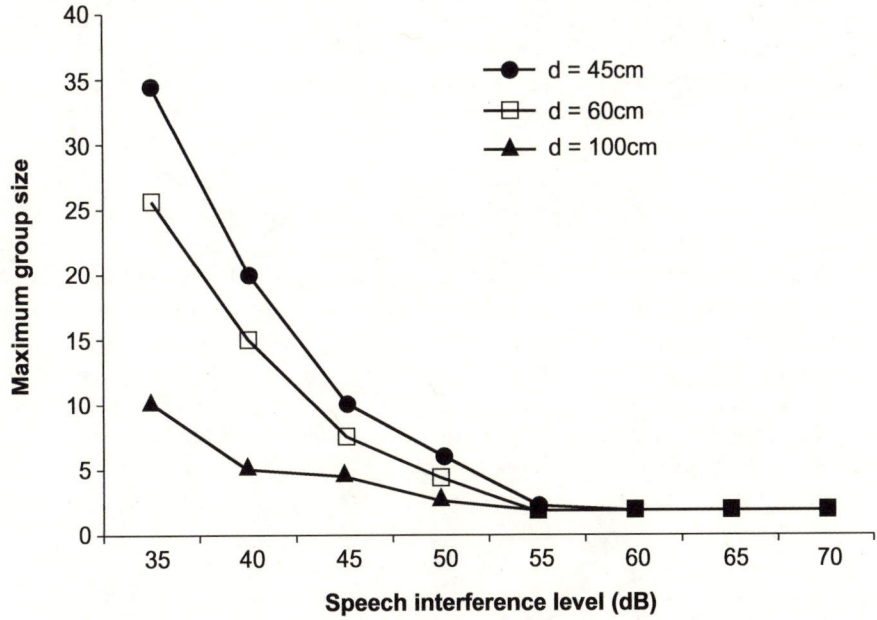

Figure 12.3 The noisy cocktail party phenomenon: speech detectability limits the number of individuals who can be involved in a conversation. The figure plots the maximum number of people that can stand in a circle at three different shoulder-to-shoulder distances (*d* cm) and still hear everything that is said. As background noise levels rise, the distance across the circle between speaker and listener has to be smaller to maintain the same level of speech discrimination; as a result, the number of people that can be in the circle is smaller. The curve for *d* = 60cm is interpolated by extrapolation from the other two curves. Background noise levels in a large office are typically about 50 dB. Source of data: Cohen (1971). Redrawn from Dunbar (1997).

individuals in chimpanzees and baboons) to the 150 reported for modern humans. In effect, this means that language has to be $150/50 \approx 3$ times more efficient than social grooming in terms of its ability to reach members of the social network. Social grooming is essentially a one-on-one activity: animals normally only groom one other individual at a time. Indeed, this is still characteristic even of modern humans: when we pet or fondle each other (the human equivalent of primate grooming), we very seldom do so with more than one individual. Grooming among primates (and fondling among humans!) is a very focused activity in which the groomer avidly concentrates on the task in hand. When a groomer shows less than committed enthusiasm for its task, the recipient often seems to interpret this as lack of interest and may terminate the session by walking off to find a more willing partner.

In effect, then, speech allows us to engage in grooming at a distance, thus making it possible for the 'groomer' to interact simultaneously with several other individuals at the same time. Data on the size of conversational groups in modern humans suggests that there is an upper limit on the size of these groups at four (**Box 12.1**). Conversation groups typically have only one speaker (interrupting or speaking over someone else is regarded as rude and aggressive in all human societies). This implies that the number of listeners has an asymptotic value at three, which just happens to be the number required to increase ape group sizes up to human group sizes.

One final point deserves comment. Although we are often very impressed by language's ability to convey information, it is by no means always a perfect medium for communication, especially when it comes to describing the inner world of our feelings. While we can describe things that happen in the world we live in with consummate ease, we are surprisingly poor at expressing our inner feelings in language. This comes out in two different ways. One is that we rely very heavily on nonverbal communication to interpret the true meaning of a speaker's utterances (see pp. 345–8). Words in themselves are often open to interpretation: *That's very nice* can mean either *I like it* or, spoken with the emphasis and a tonal down-turn on the middle word, *it's completely awful*. We rely on **prosodic cues** (timing, pitch, intonation) and facial and gestural cues to interpret the speaker's intentions.

PROSODIC CUES

Second, language is such a poor medium for conveying our inner feelings that most of us resort to good old-fashioned primate mutual mauling when it comes to building more intimate relationships: in other words, physical contact expressed in hand-holding, stroking, petting, nuzzling, kissing and, in many instances, true grooming of the hair becomes more important than speech as relationships become more intimate. Direct physical contact seems to arouse in us the kinds of feelings of attraction and positive affect that no amount of fine words can achieve. This may be one reason why we are so impressed by those writers who can express in words the inner feelings that we experience.

HOW DID SPEECH EVOLVE?

We can never know what the first words ever spoken were. Indeed, it is doubtful whether we would even recognise them as words even if we did know, although it seems reasonable to suggest that the earliest words were names (either for objects or individuals). A more useful question, however, is: how did hominids get from essentially monkey-like vocalisations to semantically meaningful words?

There are two main suggestions as to how this transition came about. One has been to suggest that, since language-trained apes are better at sign languages than verbal ones, spoken language evolved from a gestural stage (Hewes 1973, Calvin 1983, Corballis 1983, D. Kimura 1993). The alternative suggestion is that language evolved out of conventional primate vocal communication by a simple elaboration of the same kinds of sound patterns (Dunbar 1993c, Burling 1993, MacNeilage 1998).

Evidence adduced in favour of gestural theories of language origin include the following facts: (a) monkeys and apes commonly use gestures as part of their natural communication, (b) a gestural language would have been facilitated in the hominid lineage by the evolution of aimed throwing, (c) deaf-and-dumb children learn sign languages with considerable speed and facility and (d) both aimed throwing and speech tend to be localised on the same side of the brain (left for most of us). Throwing rocks, and later spears, at moving animals during hunting is very demanding computationally and the neural circuits that support this activity are complex and well developed. The fine motor control needed would thus have provided the neural basis for a gestural language, which could later have been transmuted into a vocal one.

However, gestural theories face a number of difficulties that vocal theories do not. First, an extra step is required to transmute gesture into speech, and it is not obvious how or why this should happen. Although we use gestural signs extensively during conversation and can learn symbolically sophisticated sign languages, gesture is not normally used to convey concepts or ideas, but rather simply to add emphasis to speech. We only learn sign languages when a total impairment of the vocal apparatus prevents speech developing at all. Second, the fact that fine motor control is involved in both aimed throwing and speech does not necessarily imply that gesture and *language* have a common origin. Indeed, they do not even exploit the same neural circuits, since the areas of the motor cortex that manage the hands and arms are separate from the areas that manage the tongue, lips and breath control. The (normally left-sided) laterality of motor function for both throwing and vocal control seems to long predate human evolution (MacNeilage 1998, Corballis 1991, Bradshaw and Rogers 1993). Third, as MacNeilage (1998) remarks, it seems inexplicable that gestural language should have been lost, given the claim that it was capable of doing the same kind of job as a vocal language (that is, communicating abstract concepts); and if it was not as efficient as speech in this respect, then how come it evolved for this purpose in the first place? Finally, humans use speech a great deal at night (story-telling around the campfire), and gesture is useless under these conditions.

The more plausible alternative might thus seem to be that speech arose by elaboration from natural primate vocal communication. Perhaps the most obvious source for this is the contact calls found among Old World cercopithecine monkeys (baboons, macaques) and apes: these are often highly elaborate and already capable of conveying considerable prosodic and **semantic** information (Cheney and Seyfarth 1990; Dunbar 1993c, 1996a). Two species (gelada and bonobos) exhibit unusual flamboyance and flexibility in their close calls, and provide a clear model for a first stage (Dunbar 1996a, de Waal 1988). Interestingly, both species are characterised by unusually large social groups for their respective taxa.

SEMANTIC

At this point, two options exist. One is to suggest that these calls acquired semantic meaning: naming has been suggested as the most obvious starting point (Nowak and Krakauer 1999; see below). The other is the suggestion that the process involved an initial phase that had more in common with singing (Dunbar 1996a, Knight 1999). In humans, communal singing appears to be capable of tapping into very ancient emotional processes that facilitate the whipping up and synchronising of several individuals' emotional states. Very similar phenomena also occur among some monkeys

BOX 12.2

When did speech evolve?

There are at least three positions on this: (1) language evolved in early *Homo erectus*, perhaps as early as 1.5–2.0 million years ago (Deacon 1997), (2) language evolved with the appearance of *Homo sapiens* half a million years ago, or perhaps a little later (Falk 1980, Holloway and de la Coste-Lareymondie 1982, Aiello and Dunbar 1993, Worden 1998), (3) language evolved as part of a complex of innovations that mark out the Upper Palaeolithic Revolution that occurred some time around 50 000 years ago (White 1982, Chase and Dibble 1987, Noble and Davidson 1991, 1996, Lieberman 1989).

Although each view can adduce plausible evidence in support of its claims, all this evidence is inevitably indirect. Those who support the latest date, for example, rely on the archaeological evidence of a dramatic change in the form and complexity of tools (including the appearance of art); they argue that these changes reflect important changes in the makers' minds that were only possible with language.

Very early dates can probably be dismissed because it is difficult to see why language should have evolved so early. The design quality of the stone tools made by *Homo erectus* is not so different to those of their predecessors (the australopithecines) as to warrant any greater verbal sophistication. Moreover, it is doubtful whether pairbonding (which Deacon [1997] suggests is the critical factor: see pp. 337–9) evolved this early on. Equally, very late dates seem just as implausible. For example, Lieberman (1989) originally claimed that speech must postdate the split between the Neanderthals and our immediate ancestors (anatomically modern humans) because the Neanderthal hyoid bone was positioned high in the throat like a chimpanzee's, so preventing them producing certain key vowel sounds. However, this was subsequently brought into question by the discovery of a Neanderthal skeleton with its hyoid bone intact – in a low position like ours.

The only attempt so far to approach the problem quantitatively is that of Aiello and Dunbar (1993). They used Dunbar's (1993c) equations

for the relationships between neocortex size, group size and grooming time in primates to predict the date at which group sizes would have been too large to bond using grooming alone. At this point, they reasoned, language must have evolved as a device for supplementing social grooming as a bonding mechanism. Using the fact that relative neocortex volume can be calculated from total brain volume, they estimated neocortex volume from the cranial dimensions of all known hominid fossils, and then used these to predict, first, social group size and, then, grooming time requirement (**Figure 12.4**).

The suggestion is that so long as the grooming time requirement remained less than, say, 25 per cent of total available day time (that is, a little above the limit currently observed in living primates), hominids would have been able to cope quite adequately, especially if they could exploit other forms of vocal communication (such as conventional primate contact calls) to supplement grooming as group size continued to grow. However, the use of contact calling to supplement grooming (as a kind of vocal grooming-at-a-distance) would only bridge the gap over a relatively limited range of group sizes. By the point at which the grooming requirement reached, say, 30 per cent, further growth in group size would only have been possible with the evolution of a more sophisticated form of bonding mechanism (presumably language). **Figure 12.4** suggests that this threshold was breached with the appearance of the earliest members of our species, *Homo sapiens*, some time around half a million years ago. This date seems to agree well with the evidence from neuroanatomy. MacLarnon and Hewitt (1999) have shown that a marked expansion of the spinal cord (and particularly the grey matter) occurs in the thoracic (chest) region in modern humans but not in any other living primates. Nerves in this area enervate the muscles that control breathing, and fine control of these is thought to be crucial for the greater breath control needed for speech. (Speech requires longer periods of slow exhalation at a constant air pressure than is

BOX 12.2 cont'd

When did speech evolve?

typical for vocalisations by monkeys and apes.) MacLarnon and Hewitt found that this expansion can be documented in the earliest *Homo sapiens* fossils (c. 0.5 MYA) but not in older hominid fossils, suggesting that language might have evolved at about this point in hominid evolution. A similar expansion in the size of the hypoglossal canal (the route whereby the nerve enervating the tongue passes through the base of the skull) can also be identified in archaic and modern *Homo sapiens* but not in *Homo erectus* or the great apes (Kay et al. 1998). Since the tongue plays a very important role in pronunciation, this too points to a date around 0.5 MYA for the first appearance of language.

Much of the dispute over the timing of language origins can perhaps be explained by the fact that the various factions hold different views on what language is and how it differs from the commu-nication systems of other primates. Those who favour a late date for language emergence tend to consider language in its fully developed modern symbolic form, whereas those who favour an earlier date typically consider a more primitive form of language (for example, language as vocal grooming). Aiello and Dunbar (1993) suggested that we should view language evolution as involving at least three separate phases: an early contentless form of vocal grooming (during the *Homo erectus* phase), a purely social content phase (early *Homo sapiens*) and a fully modern phase capable of supporting symbolic meaning (after the Upper Palaeolithic revolution). The latter can be viewed as an essentially software rather than hardware change. The former might well have involved more musical forms akin to communal singing.

(for example, gelada contact-calling choruses: Dunbar 1996a) and the chimpanzees (for example, pant hoots: Mitani and Brandt 1994, Arcadi 1996, Ujhelyi 1998). As such, it may have played a particularly important role in the emotional bonding of social groups. In addition, we now know that monkey and ape vocalisations are capable of supporting considerable semantic (in particular referential) content (vervet contact calls: Cheney and Seyfarth 1990). Chorusing may thus have a sufficiently ancient origin to provide the starting point for the process of elaboration that, with the addition of semantic content, eventually gave rise to language as we know it.

Although a gestural origin for language seems to be a more convoluted route for evolution to have taken, recent evidence for so-called 'mirror neurons' in the brain suggest that there may be some aspects of this hypothesis that need to be considered seriously. Mirror neurons were first discovered during experiments on an area of the pre-motor cortex (known as F5) in macaque monkeys. These cells not only fired when the animal performed a motor action (reaching for a piece of food with its hand), but also when the animals observed the experimenters performing the same task. Mirror neurons thus have a role in controlling movement and reading the action of others when the same movements are involved.

Brain imaging studies of humans have revealed similar effects. Watching an experimenter pick up or manipulate objects was found to activate two regions of the brain: the superior temporal sulcus (STS) and Broca's area. The STS is involved in several other aspects of social cognition, such as face recognition and the detection of biological (intentional) motion but Broca's area is associated with a very different aspect of human social interaction: language. Interestingly, area F5 in the macaque brain is found in the region that corresponds to Broca's area in the human brain. Obviously, this area is not associated with speech in the macaque, but is linked mainly with

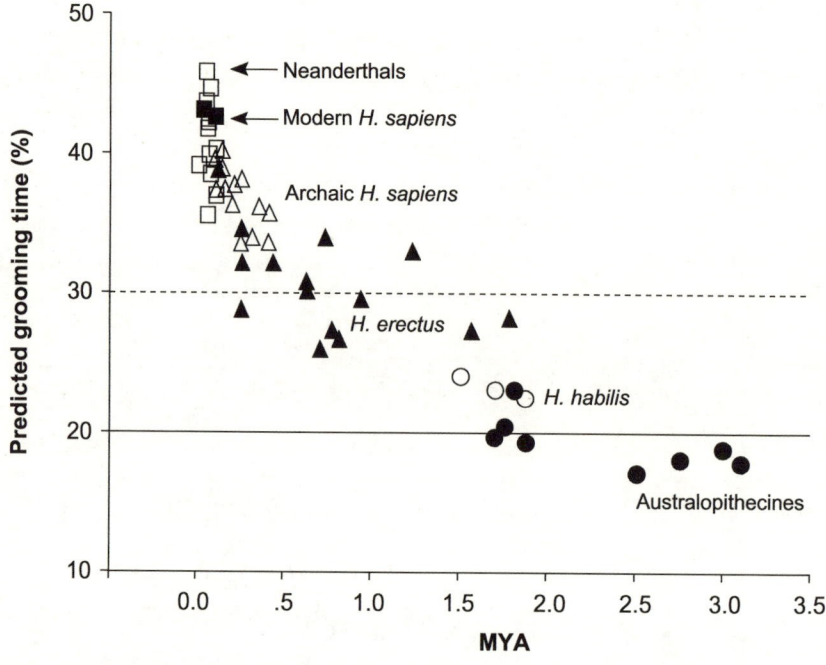

Figure 12.4 Predicted amount of time spent grooming by different fossil hominid populations, plotted against the time period when they lived. Each data point represents a single population (defined as the set of fossils from a particular site over a 50 000 year period). Neocortex ratio was first estimated from cranial volume using an equation based on data from **anthropoid primates**; this was then used to predict social group size using the relationship shown in Figure 6.1; finally, this value was used to predict required grooming time using the relationship for **catarrhine primates** shown in Figure 12.1. The maximum amount of time spent grooming by any primate species (20 per cent of total time budget) is shown as a solid horizontal line; the threshold above which language would have had to evolve (c. 30 per cent, midway between that maximum observed in primates and that required for human groups [see Figure 12.1]) is indicated by the dotted horizontal line. This suggests that language would have evolved with the appearance of the earliest members of our own species, *Homo sapiens*, around 0.5 MYA. Filled circles: Australopithecines; open circles: *Homo habilis*; filled triangles: *Homo erectus*; open triangles: archaic *Homo sapiens*; open squares: Neanderthals; filled squares: modern *Homo sapiens*. Source: Aiello and Dunbar (1993), after Dunbar (1996a).

ANTHROPOID
PRIMATES

CATARRHINE
PRIMATES

hand movements. This led Rizzolatti and Arbib (1998) to speculate that mirror neurons may be the bridge between performing actions and communicating with others. Since mirror neurons in F5 allow individuals to recognise and, to some degree, interpret the actions of others, the presence of these cells may have preadapted the brain for the development of inter-individual communication and, ultimately, speech. Rizzolatti and Arbib (1998) suggest that very slight movements which sometimes occur when we observe the actions of others (caused by a momentary failure of spinal cord inhibition which usually prevents us acting out what we see) may have formed the basis for a primitive form of gestural communication, that built up into a more complex form before being ultimately superseded by speech due to the latter's superiority at a distance or in the dark.

Gallese and Goldman (1998) have also linked mirror neurons with Theory of Mind. They suggest that the action of mirror neurons support the simulation theory of

BOX 12.3

The evolution of languages

The issue as to whether or not the history of language evolution can be traced using similarities in word form among the 5000 or so modern languages has been much debated since the possibility was first mooted in 1786 by Sir William Jones. Jones had recognised that Sanskrit (the now extinct ancient language of northern India) shared many word roots with classical Greek and Latin. Jones's conclusion was later generalised to include an extended family (known as Indo-European) of around 150 European and south Asian languages (including the Celtic, Germanic and Slavic languages of Europe as well as Persian, Urdu, Bengali and a number of other north Indian languages). On the basis of words shared with Semitic languages (of the Afro-Asiatic family) in the Near East, Indo-European is thought to have evolved among farmers living in Anatolia (Turkey) around 6000 BC.

Subsequent analyses suggested the existence of a number of other major language families (or *phyla*), including three in the Americas, around 11 in Asia, north Africa and the Pacific region, one in Australia and four in sub-Saharan Africa (Ruhlen 1994, Cavalli-Sforza 2000). More contentiously, some linguists have claimed, on the basis of similarities in a handful of words (including the words for *finger/one, two, water, mother, father, brother*) that all these families can be traced back to a single ancestor language that was spoken around 40–60 KYA when modern humans first emerged out of Africa (see **Box 1.5**).

Languages evolve rapidly by the invention of new words and the modification of old ones (the same processes that drive dialect evolution [see **Box 9.3**], albeit over a longer timescale). Ancestral Indo-European has given rise to around 150 descendent languages in 8000 years, while Latin has left around eight mutually incomprehensible languages after 1500 years. The processes whereby languages come to occupy particular geographical areas, however, can be attributed to three major causes: (a) the physical replacement of one group of people by another, (b) wholesale language borrowing from a small number of incomers (the colonial effect) and (c) diffusion between neighbouring languages. Comparison of genetic and language trees suggest that wholesale replacement of peoples following major migration events is the most plausible, at least for the major language groups (Cavalli-Sforza et al. 1988, Cavalli-Sforza 2000). Although linguists have vigorously disputed the validity of language classifications for these purposes (Bateman et al. 1990, see also MacEachern 2000), it seems unlikely that the broad patterns reflected in these genetic-linguistic comparisons are wholly wrong.

These patterns have been particularly well studied in Europe (Barbujani and Sokal 1990, Piazza et al. 1995). Here, Basque (a language isolate unrelated to any other modern European languages, but with affinities to the Dene-Caucasian family that includes Chinese, Caucasian and the languages of the northwest American Indians: Ruhlen 1994) may represent the remnants of the language spoken in Europe by the Cro-Magnon hunter-gatherers that replaced the Neanderthals some 40 000 years ago (see **Box 1.5**). This suggestion is born out by the genetic data which suggest that the Basques are also a genetic isolate within Europe.

The ancestor of modern Basque and its speakers appears to have subsequently been replaced wholesale by successive waves of Indo-European-speaking invaders: the first spreading northwestwards from Turkey around 8000 years ago on the back of the agricultural revolution, the second some time after 6000 BC from the northeast by horse-borne nomads from north of the Black Sea (the so-called Kurgan culture) (Ammerman and Cavalli-Sforza 1984; Renfrew 1987; Piazza et al. 1995).

Subsequent invasions within historical times by speakers of Mongolian languages (the Uralic and Altaic families) from the east left pockets of non-Indo-European languages in modern Hungary, Finland, Lappland and Turkey. In the case of Finnish and Hungarian, the migrant populations involved seem to have been quite small (judging by the levels of genetic admixture), with the cultural replacement of languages from more powerful invaders being responsible for the present linguistic patterns in these areas.

ToM (see Chapter 11). This states that we read the minds of others by simulating their behaviour and experiences in our heads; by putting ourselves in their shoes, we come to understand what they must be thinking and feeling. Mirror neurons would seem to back up this theory rather well. If we vicariously experience what others are doing via the actions of mirror neurons, it seems possible that we could also link these experiences with the particular mental states associated with certain actions. The link between mirror neurons and imitation, language and ToM – three key attributes in human cognition and communication – suggests that further work on mirror neurons in both humans and other primates will surely pay dividends in understanding the essential nature of our social adaptations.

THE SOCIAL FUNCTIONS OF LANGUAGE

The suggestion that language is a form of grooming-at-a-distance designed to facilitate the bonding of very large groups, raises the question of why we need syntax at all. If it is just a matter of bonding, then wordless communal singing should be all we need. That we have syntax and that syntax plays a vital role in facilitating the exchange of information during speech calls for an explanation. Can we provide one that sits comfortably within the social bonding hypothesis?

Three specific suggestions as to how what we talk about during conversations might be effective in the bonding of large social groups have been made. Dunbar (1993c, 1996a) argued that the exchange of information about the current and future states of the social network (essentially gossip, in its broadest sense) was a crucial factor in facilitating the cohesion of large social groups (the *Gossip Hypothesis*). Alternatively, Deacon (1997) suggested that language evolved to facilitate the coordination of social contracts (such as marriages) (the *Social Contract Hypothesis*). Finally, Miller (2000) has argued that language evolved by sexual selection in the context of advertising for mates (the *Scheherazade Effect*). These hypotheses differ mainly in respect of the role given to language. We consider each of them briefly in turn.

SOCIAL GOSSIP HYPOTHESIS

Our social world is in a state of constant flux as individuals make and break relationships. To operate in such a world, we need to have an accurate model in our minds of who is currently friends with whom, and this model needs to be constantly updated as relationships come and go. The problem that monkeys and apes face in maintaining this mental database is that they know only what they see. If a monkey does not see two former enemies making up or an ally reneging on its relationship by supporting an opponent, it will never know that this happened. Freeriders can thus have a field day, and this problem is especially intrusive in large dispersed social systems such as those typical of human hunter-gatherers (see **Figure 9.6**).

Language offers humans a unique mechanism for circumventing this problem: we can seek out information on the behaviour of our friends, or others can tell us what happened while we were elsewhere. The exchange of information allows us to keep tabs on the dynamic state of relationships within our social network: it allows us to update our model of the group in response to the changing patterns of friendships. Studies of just what people do talk about in relaxed informal social settings (Landis and Burtt 1924, Kipers 1987, Bischoping 1993, Dunbar et al. 1997) suggest that

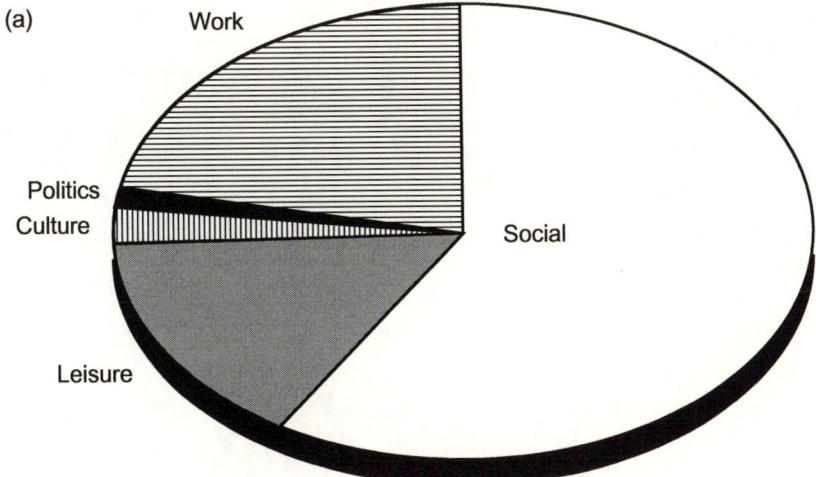

(a)

Work

Politics
Culture

Social

Leisure

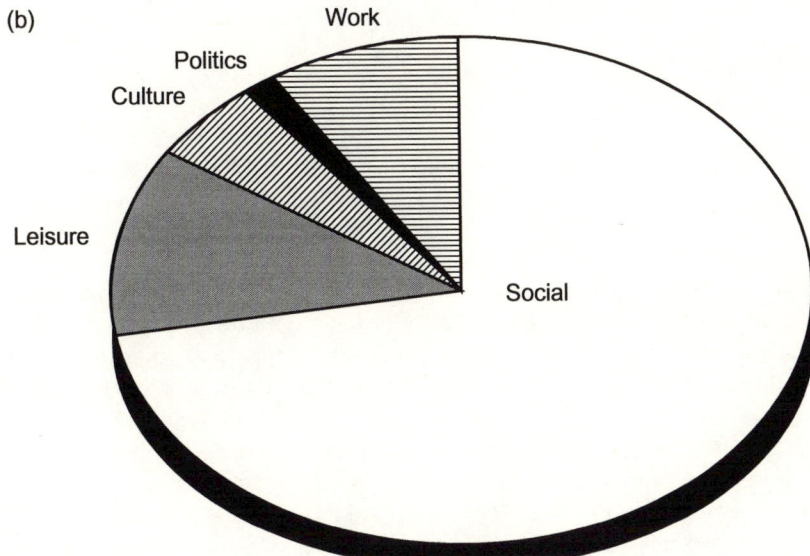

(b)

Work

Politics

Culture

Leisure

Social

Figure 12.5
Proportion of total
conversation time that
(a) males and (b)
females devoted to
particular topics
under natural
conditions (bars,
cafeterias, trains and
other public places).
The only statistically
significant difference
between the sexes is
in work-related (or
academic) topics.
Source: Dunbar et al.
(1997).

around two-thirds of conversation time is devoted to social topics (statements about the personal characteristics or relationships of the speaker, the listener or a third party: **Figure 12.5**). Being able to tap into information networks in this way adds an important dimension to the selective advantages of language: it provides another way in which the time constraint on grooming can be bridged in addition to those discussed in the previous section. Although knowledge via third parties is never as accurate as direct personal knowledge, there is clearly a significant advantage in being able to monitor changes within a social network while you are not present.

In addition, we can envisage other social advantages to language. Enquist and Leimar (1993), for example, argued that gossip has a policing function in large social networks because we can exchange warnings about the misdeeds of freeriders who renege on their obligations under the communal social contracts that are so essential to the success of large social groups (see Chapter 9). Emler (1990, 1994) has emphasized

Table 12.1 Frequencies with which males and females were recorded speaking on different social topics

Topic	% Speaking Time Males	Females
Personal social/emotional	28.6	27.2
Personal experiences	25.9	32.4
Third party social/emotional	29.6	30.7
Critical comments on third parties	4.2	0.6
Asking/giving advice	6.6	0
Hypothetical social situations	5.4	6.1
Percentage of time on social topics	63.1	69.8

Sample: 616 scan samples at 30-second intervals from 16 conversations

Source: Dunbar et al. (1997)

the importance of reputation management in the context of gossip: in addition to denouncing freeriders, social exchanges also allow us to advertise our own qualities (or those of our friends and allies). Reputation management can clearly serve an important function in respect of both mate searching and the management of social networks. In addition, we may also be able to use language to solicit and give advice on how to behave or how to handle a particular situation, especially in those cases where we have never encountered the situation before (Suls 1977). All these various functions can, however, be subsumed under the more general rubric of 'gossip' (in its original broad sense meaning the exchange of social knowledge).

Although policing functions would seem intuitively important, analysis of the content of conversations suggested that, at least in public venues, people talk most about relationships (their own or other peoples'), and surprisingly little about others' social misdemeanours (the policing function) or situations requiring advice (**Table 12.1**). That the policing function of language should be so rare is surprising given the apparent importance of its effect in controlling the activities of freeriders (see **Figure 9.6**). There are three possible reasons for this failure to find evidence for a policing function. One is that freeriding is not as serious a problem as has been suggested. The second is that people do not like to discuss these matters in public places (and all the observations on conversations were carried out in bars and cafeterias where strangers can easily overhear what is being said). The third is that the policing *is* important, but that it is not an everyday occurrence: the ability to criticise or expose freeriders may be absolutely crucial on the handful of occasions when it happens, but these occur only at infrequent intervals.

Given the costs of growing and maintaining brain tissue (see Chapter 6), it is difficult to see how such an expensive faculty could have evolved to support a rare (albeit important) function. Hence, it seems likely that language may not have evolved solely to subserve the function of controlling freeriders; rather, this is probably a function that became available once (a) large groups were formed (and thus a window of opportunity was opened up for freeriding) and (b) freeriders had actually

BOX 12.4

An instinct for gossip?

Wilson et al. (2000) have argued that humans are intrinsically attuned to gossip in that they make a clear distinction between what is appropriate and what is not. They tested the hypothesis that gossip is 'context sensitive' by asking subjects to rate (on a scale that varied continuously between +1 [maximum approval] to −1 [maximum disapproval]) the behaviour of individuals presented as short vignettes. The experimental design was a between-subjects comparison (with each subject being shown only one vignette).

The speaker in a vignette was rated significantly higher (by both male and female subjects) if his/her utterance (a) involved criticism of others who had violated a social norm rather than referring to the same behaviour in a more self-serving manner and (b) involved truthful negative gossip

(*TNG*) than if it involved either false negative gossip (*FNG*) or self-serving gossip (*SSG*) (**Figure 12.6**).

Subjects were also highly sensitive to the quality (that is, reliability) of the gossip they were hearing. When presented with vignettes that reflected either a violation of social norms (a student cheating in an exam) or a caricature of the professor's behaviour (arriving in class with his trousers on inside out), subjects were highly sensitive to the quality of the information in the first case (reports from witnesses were rated higher than hearsay accounts, with hearsay from two witnesses being rated higher than hearsay from a single witness) whereas there were no differences between conditions in the behavioural caricature vignette.

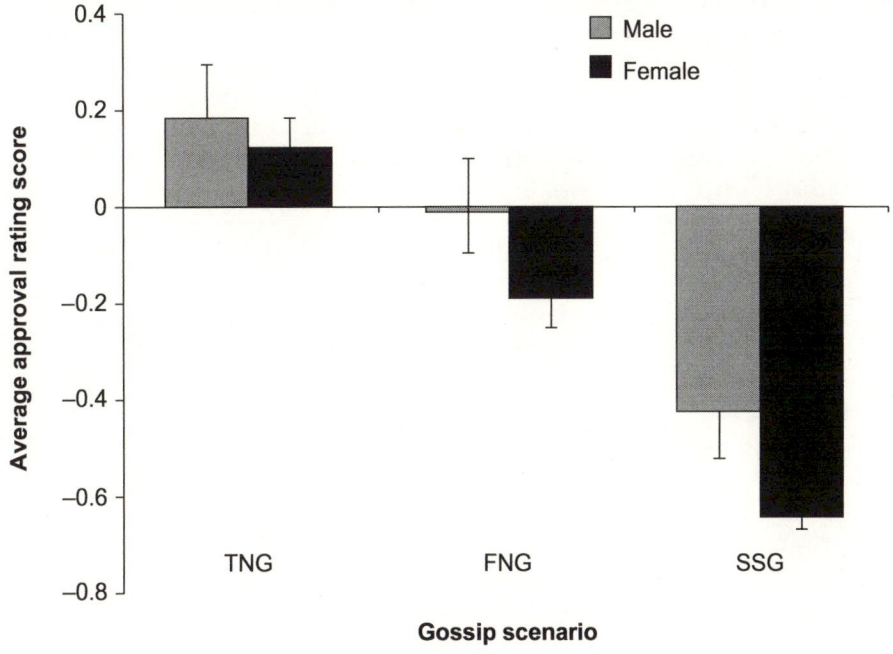

Figure 12.6 Gender differences in approval ratings across different gossip scenarios. Subjects were asked to rate their approval/disapproval of scenarios in which individuals conveyed different kinds of gossip about a norm violation. TNG is truthful negative gossip; FNG is false negative gossip; and SSG is self-serving gossip. Approval rating could range from +1 (maximum approval) to −1 (maximum disapproval). Shaded bars refer to male subjects, filled bars are female subjects. Redrawn with permission from Wilson et al. (2000).

become a serious problem. Long before freeriders became an issue, the bonding of large groups would have been an issue.

Given this, the most parsimonious conclusion is that language (in the form of social gossip) evolved to facilitate the bonding of large groups (themselves required to solve a specific ecological problem); large groups facilitated the spread of freeriders and, as a result of this, the basic functions of language as a device for exchanging information were exploited to control freeriders (see **Box 9.3**).

SOCIAL CONTRACT HYPOTHESIS

Deacon (1997) points out that, in human social systems, marital units (monogamous or polygamous families) live in close association within a larger grouping (the band or village). The problem is that, in the classic hunter-gatherer set-up in which we evolved, males who went hunting to provide meat for their females left their mates at risk of being mated by rivals; equally, their own mates were exposed to the risk that their husbands might themselves mate with other females in neighbouring groups. This, he argues, placed intolerable strains on the relationships involved. Marriage contracts whereby mates declare to each other – and perhaps publicly to the rest of the group – their marital obligations and agree not to mate with other individuals was the solution to this problem.

Male provisioning of female mates in a pairbonding (not necessarily monogamous) society is thus central to Deacon's argument: it was the need for males to be away from their mates hunting (and thus unable to mateguard them) that drove the evolution of verbal contracts. The core to this, he argues, is symbolism. Without the ability to transmute the constellation of events, feelings and intentions that make up these concerns and prohibitions about mating fidelity into abstract symbolic form, these complex living arrangements would not be possible and human social systems would fall apart as unworkable.

Deacon's proposal seems very plausible, and has the merit of bringing into sharp focus some of the more unusual features of human social arrangements. However, it does not provide us with a mechanism for bonding large groups and Deacon takes the existence of these for granted. Since the contractual function of language depends on the existence of large groups, it is difficult to see how the large groups could evolve as coherent social units in the first place. Marriage contracts are not in themselves enough to bond large groups together in the face of the intense pressures that act to disperse individuals in groups (Dunbar 1988).

A second problem is that there are examples of other species solving exactly the same problem without having to resort to language. One example is the little bee-eater, a small brightly coloured bird that lives in large colonies in the African savannah (Emlen and Wrege 1986). The colony consists of a number of small breeding units (essentially, families), each of which has its own breeding burrow in the sandbank occupied by a large colony. Bee-eaters (and especially bee-eater females) face the same problem as humans do, because everyone (males and females alike) heads out to feed during the day at sites far from the colony. During the breeding season, the birds depend on their mates to bring back food for the young, so successful breeding depends on the reliability of the mate and its willingness to forgo opportunities to mate (or form additional pairbonds) with individuals elsewhere in the colony or encountered on the plains while foraging. In addition, the females suffer high levels of harassment from other males whenever they leave or return to the roost on their own. Bee-eaters manage to solve these problems without language, so one wonders why our ancestors were so pressured in this respect. In any case, contemporary hunter-gatherer marital arrangements are notoriously fluid (South American Ache average 11–12 partners in a lifetime: Hill and Kaplan 1988) and it is not entirely clear what the force of the need-to-provision argument actually is.

A third problem is that, as we know only too well from the divorce statistics, verbal contracts are not always that effective at enforcing fidelity: people notoriously promise all manner of things to each other, and then promptly renege on those promises at the first opportunity, even when there are very severe penalties for philandering. Despite threats of stoning in Old Testament Judaism and execution in some of the more strict Islamic codes, people will insist in running off with their lovers. Fidelity is much more effectively engendered by being attentive to one's mate and generally making him/her contented and happy. Only once there is a legal framework and some kind of police force can fidelity be enforced (albeit rather ineffectually) by binding verbal contracts.

If fidelity really was the fundamental problem faced by our ancestors (and it likely was a *very* serious problem), it is not at all obvious that language would have been the best solution. Costly courtship rituals and a more intense form of emotional bonding would seem to be a safer option. And this is surely exactly what we do see in humans: we persistently demand evidence of commitment (lovers constantly test each other in this way) and we engage in that deeply irrational yet irresistible behaviour known as 'falling in love' which makes even the most sensible of people behave in quite absurd ways.

However, Deacon's thesis is not without merit, for we might see the symbolic function that he envisages as a natural development for a language faculty that arose originally to bond large groups. Contractual arrangements would thus be a secondary (and extremely powerful) selection factor reinforcing the rapid evolution of language capacities once these had been kicked off. Also his emphasis on ritual as an important consequence of the symbolic revolution is surely right. We will have more to say about this in the following section.

Deacon (1997) points out that one of the difficulties that theories of language evolution face is that they invariably risk having to assume the very abilities they seek to explain in order to get language off the ground. We cannot assume some kind of half-formed language as a half-way house between no language and full human language, yet still doing the job that fully developed language does. However, by marrying the gossip theory with the social contract theory, we may be able to bridge Deacon's chasm: simple language evolves first to support the servicing of relationships within the group, and once in place provides a window of opportunity for the development of more complex symbolic forms capable of supporting the kinds of functions that Deacon envisages. After that, the rest is history.

SCHEHERAZADE EFFECT

Miller (1999, 2000) offers an alternative solution to Deacon's absent mate problem: his suggestion is that one of the functions of language may be to enable humans to attract and maintain the interest of mates. Like the eponymous Scheherazade in the *Tales of Arabian Nights*, keeping our mates entertained by witty story-telling ensures that they continue to focus their interest on us. If verbal skills are an honest signal of

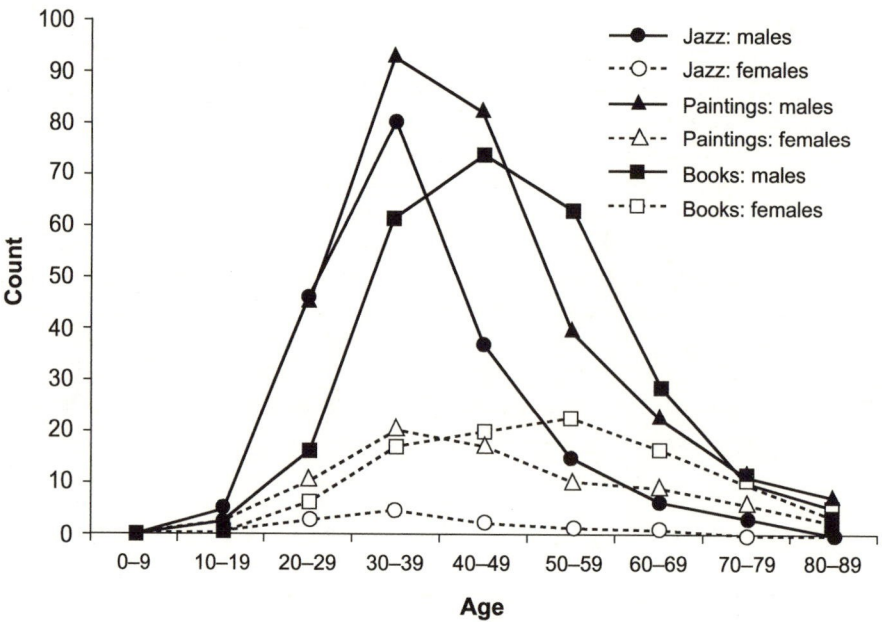

Figure 12.7 Testing Miller's (2000) hypothesis that intellectual creativity is a male mate-advertising strategy: the data plotted here are the outputs of jazz musicians, novelists and classical painters. There are significant differences in the age distributions for the two sexes, as well as in the dimorphism in output (even allowing for reduced female representation in these spheres). Redrawn from Miller (1999).

gene quality, then this has all the hallmarks of the Handicap Principle (the evolution of exaggerated traits that demonstrate superior fitness precisely because they are costly to evolve and maintain: see **Box 2.7**). This version of the social function hypothesis gives language a more exotic role, because it suggests that one of its important functions is to dazzle or entertain rather than merely to facilitate the exchange of information.

Miller's argument stems from his suggestion that the human brain is a sexually selected organ designed to support sophisticated cognitive and verbal skills as part of the mate-searching strategies peculiar to humans. Indeed, he has been able to adduce some indirect evidence in support of this hypothesis (Miller 1999). An analysis of some 16 000 items of cultural output (including the production of jazz and rock records, classical music scores, classical and modern paintings, books, even philosophical tracts) showed (a) that productivity tended to peak at around age 30 (the peak of sexual competition and activity in humans) and (b) that males were up to ten times more productive than women (**Figure 12.7**).

Although, in the social sciences and humanities, such sex differences in productivity have invariably been explained in terms of men's domination of women, Miller (1999) points out that an equally plausible explanation is that it reflects a difference in the two sexes' interests in these forms of public display. Males have an overriding interest in displaying their competences as publicly and widely as possible in order to attract mates, whereas women's interests in this respect tend to be more focused on attracting specific mates (not least in order to minimise harassment by many males). In effect, Miller's hypothesis is that humans are a **lekking** species in which females choose mates from a set of males who display their genetic qualities rather in the way peacocks do through displays that are difficult to cheat precisely because they are demanding and costly.

LEKKING

Miller (1999) has also been able to show that, even though women score better than men on verbal IQ tests and girls develop language earlier and are more fluent in language than boys (Hirsh-Pasek and Golinkoff 1996), men tend to outscore women in both vocabulary size and verbal flamboyance (complex word use). He interprets these as reflecting males' preoccupation with flamboyance in the mating arena.

Like the social contract hypothesis, the Scheherazade Effect also begs questions about how large groups are bonded in the absence of language. Although Miller does not explicitly discuss group size in the context of his hypothesis, there is an implicit assumption that large social groups are already present: the Scheherazade Effect would seem to be most useful when there are large numbers of prospective mates to choose from, and this is only likely when social group size is large. Hence, once again, we face a difficulty in trying to account for the existence of large social groups.

One solution might be to argue, as we did in the case of the social contract hypothesis, that language originally evolved to bond groups, and that this then provided a window of opportunity that sexual selection exploited. However, Miller explicitly wants to present his hypothesis about the evolution of language as an explanation for the evolution of the super-large human brain, and this makes it difficult to splice the two hypotheses together. If we can downplay the relevance of the Scheherazade Effect as an explanation for the initial evolution of large brain size, then it may be possible to integrate the gossip and Scheherazade Effect hypotheses into a single theory (with or without the social contract hypothesis). Indeed, we might still be able to retain the Scheherazade hypothesis as a key factor influencing the *escalation* of brain size in humans once this process has been kicked off by the need to bond large social groups.

IS A SYNTHESIS POSSIBLE?

In summary, we have argued that the most sophisticated (and empirically best supported) explanations for the function of language are social in character. So far, three specific versions of the social function hypothesis have been proposed. Although these could be seen as mutually incompatible, there are good grounds for arguing that they are in fact complementary rather than mutually exclusive. One reason for doing so is that two of them leave unexplained how large social groups are bonded. This may be because neither author is familiar with the extensive primatological literature on group-living (see van Schaik 1983, Dunbar 1988, Janson 2000). The burden of this literature is that groups represent a balance between the selective advantages of group-living (in most cases either protection against predators or the defence of communal resources) and the dispersive forces created by the presence of other individuals (the ecological costs of increased travel and resource competition and the reproductive costs associated with high levels of harassment: Dunbar 1988, 1996b). These costs are all the more intrusive in the super-large groups characteristic of modern humans and some mechanism must exist to neutralise them, otherwise the advantages of group-living may be insufficient to persuade animals to stay together.

There are, however, at least two ways of resolving the apparent conflict implied by the three separate hypotheses. Either the social contract or the Scheherazade Effect version is *the* actual mechanism underpinning the social bonding hypothesis (in which case, presumably, the other must be wrong) or we can see them as later specialisations out of a more primitive social bonding form of language (the gossip version). In the second case, both the social contract and the Scheherazade Effect models are compatible, since they simply represent sequential or parallel developments out of a more general form of social language designed to solve particular problems of group-bonding encountered by our ancestors. The latter solution seems to us the more plausible.

LANGUAGE AND MEANING

Human language is used in a social context, and inevitably it has social meaning. Discourse analysts frequently point out that, when we speak, we invariably do so in an elliptical, often telegraphic, manner (see Gumperz 1982). We use truncated phrases and passing allusions that only members of our language community fully appreciate because only they have the shared knowledge base on which to draw. Examples of this are abundant in much of our literature: European literature is so full of biblical and classical references that only someone deeply versed in the same cultural history can appreciate the often subtle allusions that these conjure up. The language that we use thus helps to place us in our particular cultural heritage, and hence social group.

Dialects offer us another example of the way speech helps to establish membership our particular social groups (Trudgill 1974). Small communities under threat often exaggerate their use of a local dialect in order to set themselves apart from 'foreigners' who have invaded their space. Well known examples include the way Welsh speakers in parts of Wales abruptly shift from English to Welsh when English speakers enter the room, and the way the native inhabitants of Martha's Vineyard exaggerated their ancestral dialect when their retreat on the New England coast began to be invaded by wealthy New Yorkers during the 1960s (Labov 1972). One of the reasons that dialects are effective social badges is that they are hard to learn and often have to be learned very early in life (**Box 9.3**). Dialects are thus an honest signal of group membership that are costly to acquire and difficult for cheats (in other words freeriders) to mimic. A freerider cannot just walk into a naïve social group and begin to speak like other group members.

That dialects have acted as social badges throughout history is evident from allusions to this phenomenon in the Bible. In the Old Testament's Book of Judges, we learn that after the Ephraimites lost a battle with the Gileadites, the latter set about exterminating the Ephraimites as they tried to escape back home. Although the fleeing Ephraimites tried to disguise themselves as innocent civilians, the Gileadites were able to flush them out because, notoriously, Ephraimites could not pronounce the *sh*-sound in words like *shibboleth*. All those who, when challenged, sibilantly pronounced the word as *sibboleth* were slaughtered. On the infamous night of the *Sicilian Vespers* in 1282 when the Sicilians rose up against their French Angevin masters and overthrew them, many Frenchmen lost their lives because they were unable to pronounce the Sicilian word *ceci* (chick pea). The Dutch used a similar tactic to flush out escaping members of the German army fleeing the Allied advance during the later stages of the Second World War. In addition to contrasts in word pronunciation, dialectical differences can also involve differences in word form or meaning (compare the American *faucet* and the English *tap*) or differences in grammatical structure (contrast the southern English rural *she bain't a-coming* with the more urban *she isn't coming*).

Dialects evolve notoriously fast. While local accents may remain broadly the same over time, it is not uncommon for details of pronunciation to change over the timespan of a generation. In the spatial model used by Nettle and Dunbar (1997) to explore the value of dialects as a means of controlling freeriders (**Box 9.3**), the rate at which dialects changed was crucially important: dialects had to change by up to 50 per cent in every generation to keep freeriders under control (**Figure 9.9**). This represents a very high rate of change, but is probably not so far removed from the kinds of dialectical changes that we actually see in practice. Quite marked shifts have been recorded, for example, in the local accent of Liverpool between the heyday of the Beatles in the 1960s and the 1990s. Word usage can change just as fast: the favoured words used by the younger generation are often markedly different to those used by their parents when they were young. This helps to mark out members of a generation even within a regional grouping and so helps to bond the age cohort. To know the local style, you must have been born into the cohort.

Historical examples of dialect change of just this kind can be identified in English. Perhaps the best known is the Great English Vowel Shift that began during the fifteenth century: this resulted in a shortening of all the vowel sounds, so that words like *fine* and *thine* changed from their original pronunciations of *feenay* and *theenay* to their present ones (which explains why English words are seldom pronounced the way they are spelt). A less well-known dialect change had occurred earlier, but can be documented more easily in the written records. During the twelfth century, the habit of dropping the *-en*, *-e* and *-est* present tense verb endings began in the north of England, and gradually moved southwards until after approximately two centuries it had permeated southern England as well (Krygier 1994). This represents a rate of travel of about 44 km per 25-year generation. A distance of about 20 km either side of one's home village is probably about the limit of lifetime travel for most medieval communities, suggesting that the rate at which dialects change approximates the length of a generation (see Chapter 13).

An interesting tactical manipulation of dialect has been documented in working-class British culture. Social linguists have long recognised that working-class girls often speak with a less colloquial (working-class) accent than their brothers – and, indeed, are often actively encouraged in this by their mothers (Trudgill 1974). Although conventional wisdom interprets this as yet another example of sex-biased social attitudes, it looks suspiciously as though the mothers are trying to keep their daughters' future marriage opportunities (in particular, the possibility of hypergamy into the better-off middle classes) as open as possible. Boys, who cannot benefit from hypergamy and are in most cases obliged to remain within their natal communities for life, may find it more advantageous to exaggerate their (dialectical) membership of that community in order to reinforce those peer group bonds that will be crucial to their success later in life.

The use of language to mark social groups in this way relates to another important use of language: ritual. Rituals are a particularly important feature of human societies and serve to reinforce group membership and the collective will. In effect, they act to make people toe the line on the communally agreed social contract, and in this way provide an important psychological mechanism for controlling freeriders. Rituals are, of course, often associated with religious and metaphysical beliefs that provide the community with a common set of beliefs, both about the world and about their own origins. Identifying the common ancestor from which all group members descend is often a central feature (Hughes 1988).

Knight (1998, 1999) has argued that the origins of language lie in just such a need to build and reinforce a common set of beliefs and ritual practices. While an individual can work out a set of beliefs for him/herself, they can only persuade others to adopt these beliefs providing they can explain the beliefs (and the awful consequences of failing to adopt them), and that requires language. Thus, Knight sees language, ritual and, ultimately, culture as being a co-evolved suite designed to reinforce social solidarity (see also Knight 1991, Power 1998). Once again, we can argue that this is either the original social function of language (and, as such, it avoids the trap faced by the social contract and Scheherazade Effect models because it is *all* about bonding large social groups) or a later specialisation out of an earlier social bonding version of language. Either way, it is clear that meaning (and particularly *social* meaning) plays a crucially important part in day-to-day language use as well as in the form and content of the rituals that bind us into groups.

COGNITIVE UNDERPINNINGS

LANGUAGE AND THEORY OF MIND

Speech has typically been seen as a task performed by the speaker, who puts a great deal of effort into trying to ensure that his/her message gets across. This is undoubtedly true, and theory of mind (see Chapter 11) plays an important role in this: speakers often (even if not always!) pay careful attention to whether their message is getting through, and may rephrase the message or try again if they sense that they have not been understood. However, the listener is not merely a passive recipient in this process: listeners devote a great deal of time and effort to trying to fathom out just what it is the speaker is trying to say. Theory of mind (ToM) is crucial in this part of the process (Dunbar 1998b).

Worden (1998) has argued that ToM may be both the most crucial and computationally most expensive component of speech. He points out that the so-called language areas in the brain (Broca's and Wernicke's areas, long assumed to be responsible for the production of speech and meaning, respectively) are very small, especially when compared to the large volume of frontal cortex devoted to higher cognitive function (and in particular, social cognition). Indeed, it is possible that all Broca's area really does is manage the fine control of breathing for speech production. In other words, the big job that needs a great deal of computing power is the mindreading that has to go on in order to give language any of its familiar depth. Without mindreading, conversations would be dull factual exchanges of precisely the *there-are-bison-down-at-the-lake* variety. There would be little literature, no poetry and no religion; and while there would certainly be technology, there would be no science. (In order to be able to ask why the world is as it is, we have to be able to divorce ourselves from the world as we experience it and imagine that it could be otherwise – something that may only be possible with theory of mind.)

The role of ToM in language comprehension is highlighted by the fact that, when we speak, we rarely say exactly what we mean. Metaphor is a major component of word use during normal conversation. Without metaphor, language would lose most of its richness; new uses for old words would not be possible and dialect evolution could not occur. A second feature of natural language use that is significant here is the telegraphic nature of language. As we noted above, most conversational exchanges consist of oblique references and half-completed phrases, often with an attached '. . . *you know what I mean?*' at the end. Without ToM, we would certainly not '. . . know what you mean,' since you rarely tell us. Typically, in conversations, we simply sketch in just enough of the key details for the listener to know what our utterance is all about. We can do this because our listeners are able to use ToM to fill in the gaps.

A particularly graphic example of this is provided by what must be the shortest play in the English language (which we quote from Pinker 1994):

BOX 12.5

Motherese

Although much of the way we speak is learned as local dialects, some aspects seem to be instinctive. A particularly intriguing example of this is the phenomenon known as *motherese*. Motherese is the distinctive style and intonation used (mainly, though not exclusively) by women when talking to very young infants (Fernald 1992, Monnot 1999). Motherese has a number of key features that appear to be employed instinctively by women the world over. These include a higher voice pitch, a softening of the intonation, large pitch contours (*glissandi* rise-fall patterns that can exceed two octaves in pitch) and the use of short repetitive sentences. Young babies seem to find this style of communication attractive (it stimulates smiling) and soothing, whereas they find the deeper voice pitches characteristic of men frightening.

Fernald (1992) argues that motherese instinctively exploits infant's tendencies to react in specific ways to particular sounds. Hence, each type of utterance (comfort sounds, prohibitions, attention-getting sounds, approval sounds) have quite

distinctive pitch contours and prosodic profiles (approval sounds have a sharp rise-then-fall pitch contour, whereas comfort sounds are low-pitched with a tendency to fall away). Infants seem to be especially responsive to these types of sounds in quite appropriate ways (attentive to attention-getting sounds, calmed by comfort sounds) and to be equally responsive (albeit with some exceptions) irrespective of the language in which the words are actually uttered.

That motherese has functional consequences was demonstrated by Monnot (1999) who showed, for a sample of 52 normal fullterm infants, that infant weight gain at 3–4 months of age was significantly correlated ($p < 0.02$ and $p < 0.0001$, respectively) with two measures of the intensity of motherese used by the mothers. These measures were ratings of the prosodic content of utterances (musicality, pitch, rhythm) and their semantic/pragmatic content (indexed as the percentage of speech segments that are infant-centred in the way they are given).

Her: *I'm leaving you!*

Him: *Who is he?*

With those six words, we can read the entire situation: we know exactly what has happened, what is in the process of happening, and have a fair guess as to what is about to happen (Pinker 1994). Moreover, HIM in the dialogue needed only the three words uttered by HER to come to some fairly strong conclusions as to what was going on.

THE ROLE OF NONVERBAL CUES

The great bulk of the literature on language has concentrated on the structural content of language (grammar) and the meanings of words. However, much of what we say is given meaning by the nonverbal behaviour that we wrap around our utterances (McNeill 1992). These nonverbal behaviours, which are clearly both extremely ancient and universal to all living human cultures (Goldin-Meadow and McNeill 1999), include features such as intonation (the rise and fall of voice pitch), gesticulation (pointing, shaking the head, arm-waving) and body stance (leaning towards the speaker, eye contact, and so on). These both control the dynamic flow of conversations (Argyle 1975, Beattie 1983) and provide emphasis to the words, rather in the way we underline words or put them in bold or italic characters or use punctuation marks. Such features allow us to read the real meaning behind words: it is what helps us to

BOX 12.6

Laughter and social bonding

In primates, grooming stimulates the brain to release endogenous opiates (Keverne et al. 1989) and these in turn seem to provide the proximate reinforcer that allows social grooming to act as a bonding mechanism. When language replaced physical grooming as the principal bonding device for humans, we lost the direct physical contact that brought this effect about. Without the stimulus of opiates to reinforce friends' sense of wellbeing and contentment in each others' company, how could language act as a bonding agent?

The short answer is that we do not really know. However, a very plausible *prima facie* case can be made for the suggestion that laughter was coopted to fill this role in humans. Laughter is all but unique to our species, though analogues are clearly recognisable in the play faces of Old World monkeys and apes (van Hooff 1972, Preuschoft 1992, Provine 1997). The frequency and intensity with which humans engage in this curious behaviour is, however, quite unique (Provine 1997). It is also a highly social behaviour: we very seldom laugh or smile when we are alone, and those who do are often labelled as mad.

However, laughter does have one curious and, in this context, very relevant property: it seems to be extremely good at releasing endogenous opiates. This is why we feel rather 'high' and relaxed after a bout of intense laughter. Indeed, there is experimental evidence to suggest that both opiate levels (Berk et al. 1989) and pain thresholds (which are partially determined by opiate titre: Zillman et al. 1993, Weisenberg et al. 1995) are higher after a bout of laughter. One possibility then, is that laughter evolved (or was coopted from its ancestral chimpanzee-like form) to fill the gap as proximate reinforcer left by the removal of physical contact as the principal bonding mechanism (Dunbar 1996b).

Two features of the natural history of laughter seem to be relevant here. One is that it is extremely hard work to keep a conversation going with someone who never laughs or smiles. We rely on the feedback from laughter and smiling to reinforce our belief that the listener is actually interested in us. Women, in particular, seem to use laughter as a means of both expressing interest in and prolonging an interaction with a (male) partner in mixed sex encounters (Grammer 1990, Provine 1997). Lack of response from a listener suggests that they do not regard us as worth spending time with, and we tend to make our excuses as quickly as possible and find someone else to talk to. Second, we spend a great deal of our time while speaking trying to make each other laugh. Jokes are an important part of conversation, as well as good lecturing style. An audience that laughs a lot pays attention to the speaker; an audience that doesn't laugh falls asleep.

realise that apparent praise is actually meant as criticism (the exaggerated 'We-ell done!' when you've just dropped the dinner on the floor).

Some of these general features of speech style may also differ between the sexes (Coates 1994, 1996a, b). Women's conversations tend to involve a great deal more overlap (two speakers speaking simultaneously) than men's; in women's conversations, overlaps tend to involve a reinforcement of the main speaker's utterances (repeating the same words, or saying them together). Women's conversations also involve more positive supportive interjections ('Uh-huh!', 'Oh yes!'). In contrast, men's conversations tend to be more measured (overlaps are rare), but when overlaps do occur they tend to involve speakers talking competitively against each other. Male conversations thus tend to have a more domineering style, while women's tend to appear more supportive and interactive.

In the social linguistics literature (see for example Eakins and Eakins 1978, Coates 1993), these differences have often been interpreted in terms of the relative domi-

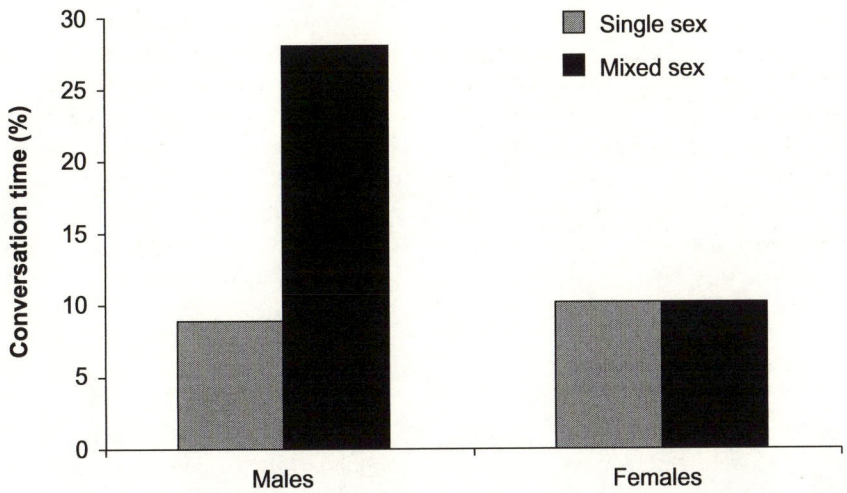

Figure 12.8 The percentage of total conversation time devoted to talking about academic topics among a sample of undergraduates was strongly influenced by the sex of the audience as well as the speaker. Women speakers tended to spend about the same amount of time talking about these topics in single-sex and mixed-sex conversation groups, but male speakers found these topics significantly more interesting to talk about if there were women present. Source: Dunbar et al. (1997).

nance and subordinance of the two sexes. A more plausible interpretation, however, is that they have more to do with the intrinsic social strategies of the two sexes and the role that language plays in facilitating these. Women's conversational styles give the impression of being directed towards bonding, a feature of some considerable concern for women in almost all social systems (and anthropoid primates as a whole: Dunbar 1988). In contrast, men's conversational styles seem to have more to do with mating strategies and a combination of the need both to advertise (to women) and to dominate rivals (other men).

Further insights into sex differences in language use is provided by Dunbar et al. (1997). In their analyses of the content of conversations, they showed that men's and women's conversations did not differ significantly in terms of the amount of time devoted to social gossip, but they did differ in terms of how the sex of the listeners influenced what they talked about in certain key respects. Among students, for example, discussion of academic topics remained at a more or less constant level for women speakers, but men showed striking and significant differences in their interest in such topics depending on whether or not any women were present (**Figure 12.8**). When speaking to other men, yesterday's lecture seems not to have been especially interesting, but when women are present it seems to acquire a significance of such overwhelming proportions as to dominate the conversation. Other technical topics (politics, 'how-to' topics) may be used in the same way (see also Eakins and Eakins 1978, Coates 1993).

One plausible explanation for this phenomenon is that males use technical topics as a kind of vocal lek (the Scheherazade Effect: see above). Leks are arenas that many birds (for example, peacocks, sage grouse) and antelope (for example, hartebeest) use as display venues during courtship: the males gather on what are often traditional sites and display vigorously to passing females, who use these displays to choose males of superior quality with whom to mate (see Krebs and Davies 1997). Conversations seem to fulfil a similar role in human mate choice by allowing males to display their erudition and knowledge, and the superiority of such knowledge compared to potential competitors. Some evidence in support of this is the finding that women tend increasingly to listen rather than speak as the number of males in the conversation group increases (Dunbar et al. 1995). Although conventional wisdom has been to assume that this is a reflection of domineering male behaviour, it can equally be interpreted in terms of lekking behaviour: if women are interested mainly in assessing the respective qualities of the males on offer, taking the floor as the speaker significantly reduces the amount of advertising time available for the males.

ADAPTIVE FEATURES OF LANGUAGE

Following the seminal work of Chomsky (1975), language was widely considered to have no adaptive function, with its structures being largely accidental. Recent analyses that use a more explicitly evolutionary approach have, however, begun to suggest that at least some aspects of speech and grammar are subject to quite explicit selection effects. We give just three brief examples here.

It has long been recognised that different languages have different numbers of vowels: Eskimo and the Australian aboriginal language Aranda have only three, whereas Spanish has five, English twelve and Panjabi twenty (Crystal 1997). Lindblom (1986, 1998) has been able to show that the number of vowels follows very explicit rules that are dictated by our ability to differentiate between sounds. Vowels can be characterised by a three-dimensional state space defined by the first three formants (the fundamental frequency and its harmonics that are created by the way the air resonates in the vocal chords and which give individual speech sounds their particular timbre). It turns out that vowels are distributed within this space in such a way as to enhance the listener's ability to tell them apart and so make the speaker's speech clearer. Precisely which ones a particular language has depends on how many vowels are used in the language, these being selected to ensure that they are spaced as far apart as possible in the formant state space. It is notable that when the Great English Vowel Shift occurred during the fifteenth century, all the English vowels changed rather than just some of them, so repositioning themselves in roughly comparable positions relative to each other. Similar trade-offs have been noted in the structure of words themselves: languages that have many syllables per word or clause tend to have fewer phonemes (the units of sound used to make up words) per syllable (Fenk-Oczlan and Fenk 1997), thereby keeping the temporal unit of analysis at a roughly constant length.

The fact that human languages are so radically different from the communication systems of other animals (and specifically monkeys and apes) has led many linguists to argue that the two cannot be related; instead, language must have evolved by some macro-mutation from some unrelated source (Chomsky 1975, Bickerton 1990). Biologically speaking, this seems implausible, but until recently it has been difficult to show how it might have happened. Recently, a series of rather difficult mathematical models developed by Nowak and colleagues have suggested how language could have evolved by a process of Darwinian selection from a simple primate communication system.

Nowak et al. (1999; Nowak and Krakauer 1999), for example, have used a modelling approach to show how language might have evolved out of naming (a feature that already characterises the communication systems of some monkeys: see Cheney and Seyfarth 1990). Their models assume that the initial function of language is to allow a speaker to name 'objects' in the environment (things or events) to a listener and that both speaker and listener gain a fitness pay-off if the information is successfully communicated between them. The key feature of the model is the risk of misunderstanding (that is, the listener making a mistake about which object or event is being named). The more similar the name (sound), the more likely a mistake is to be made. The language's fitness (indexed as the language's capacity to transfer information) is maximised by having a small number of signals (sounds) to describe a few valuable concepts or objects; increasing the number of signals does not increase the fitness of the language, suggesting that nonhuman communication systems may be error-limited. Nowak et al. (1999) show that this error limit can be overcome by combining sounds (technically, phonemes) into words, since this in effect reduces the number of sounds that can be mistaken for each other. In this case, fitness increases exponentially with the length of the words used. They conclude that word formation (equivalent to the transition from an analogue to a digital communication system) was a crucial step in the evolution of human language.

Nowak et al. (2000) extended this approach to examine the evolution of syntactically structured languages (that is, those that have grammars). They consider a language whose utterances involve *noun + verb* complexes to describe events. They showed that, when the number of relevant events that need to be referred to is very large, syntactical organisation of the stream of words conveys higher fitness. The critical number of events that sets the lower limit for syntax depends on the difficulty of memorising the particular signals and the fraction of all word combinations that describe meaningful events. For example, when a syntactical (grammatically structured) *verb + noun* phrase is twice as difficult to memorise as a non-syntactical utterance (a noun on its own) and one third of all possible *noun/verb* combinations actually refer to possible events, then the minimum object–event combination that is needed to make syntax advantageous is an 18-noun × 18-verb matrix. The dynamics of this system are, however, complex: in smaller systems (say, a 6 × 6 object–event matrix), the number of learning opportunities per individual becomes crucial: the fitness of a syntactical language only exceeds that of a non-syntactical one when this number is below about 400. In other words, when the opportunity to learn (number of exposures) is very large (> 400), a non-syntactical system is more efficient (has higher fitness) when the number of possible events (noun–verb combinations) is small.

What all this rather complex analysis suggests is that so long as there was a modest selective advantage in terms of more effective cooperation between allies, language (in the simple sense of naming objects) and more importantly grammar (the ability to describe events) can evolve quite easily by conventional Darwinian processes. So far from being the result of some all-at-once macro-mutation as Chomsky and other linguists have suggested, language could have evolved piecemeal (albeit rapidly).

*

In this chapter, we have focused our discussion mainly on the functional question of why language evolved and the role it plays in mediating social interaction. However, we have given relatively little attention to the detailed specifics of this role. A particularly important feature of language in this respect is the role it plays in the transmission of culture. Clearly, without language (and specifically speech as the mode of communication), all the richness of mental and social life that we associate with human culture would be impossible. In the next chapter, we consider this all but unique phenomenon in more detail.

Chapter summary

■ Language is unique to humans and, as such, provides a particular challenge for evolutionary psychology to explain. The traditional view is that language evolved to allow the exchange of technical information, but the evidence suggests that it in fact evolved in order to bond large social groups. It may thus have evolved as a substitute for the kind of social grooming that nonhuman primates use to bond their groups (that is, as 'grooming-at-a-distance').

■ Three alternative versions of the social origin of language hypothesis have been offered (the social gossip hypothesis, the social contract hypothesis and the Scheherazade Effect). It is possible to combine all three in a single explanatory framework with the social gossip hypothesis as the original function and the others as benefits that quickly became available once language had evolved.

■ Language appears to have a fairly recent origin within the human lineage. Several lines of evidence point to the first appearance of our own species, *Homo sapiens*, around 500 000 years ago, as the crucial date.

■ Language takes place within a speech community as a social event; hence, much of what we say and how we say it (dialects) mark us out as belonging to particular communities. Dialects evolve very fast in order to maintain that function.

■ Speech is a surprisingly poor medium for the expression of inner feelings, and we use gestures to convey deeper meaning to words. Some aspects of speech (such as 'motherese', the peculiar style of speech that women, in particular, use when talking to babies) exploit these kinds of nonverbal features of communication.

■ Despite what has often been claimed previously, many aspects of language appear to be well adapted to aspects of their function; speech discriminability appears to be a particularly important selection factor in this respect. Theory of mind has been identified as being particularly important in the process of linguistic communication.

Further reading

Deacon, T. (1997). *The Symbolic Species*. Harmondsworth: Penguin.

Dunbar, R. I. M. (1996a). *Grooming, Gossip and the Evolution of Language*. Cambridge (MA): Harvard University Press.

J. R. Hurford, M. Studdert-Kennedy and C. Knight (eds) (1998). *Approaches to the Evolution of Language: Social and Cognitive Bases*, pp. 242–64. Cambridge: Cambridge University Press.

Miller, G. (2000). *The Mating Mind*. London: Heinemann.

Nettle, D. (1999). *Linguistic Diversity*. Oxford: Oxford University Press.

Ruhlen, M. (1994). *The Origin of Language*. New York: Wiley.

Cultural evolution

Culture has long been held to be the one factor that differentiates humans from brute beasts. Cultural transmission of rules of behaviour, as well as of myths and rituals, inevitably enhances the effectiveness with which individuals can reproduce because each generation can learn from its predecessors' discoveries. Although there has been considerable interest in the question of whether or not animals have culture (Mundinger 1980, Nishida 1987, McGrew 1992, Tomasello et al. 1993, Boesch 1996, Boesch and Tomasello 1998), the fact remains that the forms of culture exhibited by non-human species are vastly inferior to that found among humans. It is only humans that have produced religions, literature and the arts, science, technology and the grand architectural constructions with which we are so familiar.

In this chapter, we first consider the thorny issue of how to define culture and then examine some of the ways in which cultural evolution has been studied from an evolutionary point of view. The last few decades have seen the development of two quite independent paradigms in the evolutionarily-informed study of culture. One of these has focused on building mathematical models of the processes of cultural transmission (in effect, studying inheritance mechanisms); the other has had a more empirical focus, being principally concerned with the adaptiveness of culture. There has been relatively little exchange between these two camps, in part because they focus on very different questions but also in part perhaps because mathematical modelling has an intrinsic capacity to frighten off those with a more empirical bent. It seems to us important to try to bring the two approaches closer together.

WHAT IS CULTURE?

We define culture as beliefs or rules of behaviour (or perhaps more generally as knowledge about the world) that are passed on from one individual to another by some form of social learning (including teaching). This is perhaps a narrower definition than some have used. In their classic review, Kroeber and Kluckhohn (1952) concluded that the term *culture* had been used in more than 150 different ways in the anthropological literature. Even when broadly similar definitions are included within the same category, there still remain four radically different classes of definition. These are: (1) rules of behaviour (for example, rituals), (2) ideas in people's minds (for example, myths), (3) material artefacts (the residue left in the archeological record) and (4) literature, art and music. Most of these, however, come down to ideas in people's heads in some form (tools and weapons are, ultimately, the conceptions of their manufacturers), with cultural transmission by imitation or teaching as the key process by which these ideas are acquired. More recently, others (for example, Tooby and Cosmides 1992) have sought to broaden the definition to include both behaviour *and* the mental processes that underpin behaviour. This may be less helpful, however, since it does not allow us to separate out genuine culture from those aspects of animal behaviour that are obviously not cultural (see also Cronk 1995).

Our definition recognises that culture is a peculiarly characteristic feature of the human species (though without excluding its use with nonhuman animals). This raises a number of key questions. (1) What cognitive mechanisms underpin culture? (2) What exactly is transmitted during the course of cultural inheritance? (3) How do the processes of cultural transmission compare to the more conventional processes of genetic transmission?

PSYCHOLOGICAL BASES OF CULTURE

The processes of cultural transmission ultimately depend on a mechanism for social learning (usually either teaching or imitation or both: Tomasello et al. 1993, Tomasello and Call 1997, Flinn 1997, Cronk 1995) and a brain of sufficient size and complexity to understand the point of the behavioural rule to be learned. Psychologists have been particularly insistent that identifying social learning (and, in particular, imitation) as the mechanism is crucial for any claim of cultural behaviour (Tomasello et al. 1993). In this, they at are odds with both socio-cultural anthropologists and those who study cultural behaviour in animals (Boesch 1996, Whiten et al. 1999; Rendell and Whitehead 2001). The reason for this is fairly simple: psychologists want to be certain beyond all possible doubt that any claim for cultural behaviour in an animal really is cultural behaviour rather than being the product of individual learning (see **Box 13.1**). The only satisfactory way of doing this, they argue, is to focus exclusively on the social imitation of those behaviours that have no ecological or other relevance, and thus cannot have been acquired by individual trial-and-error learning. Such

Psychological mechanisms of cultural transmission

Comparative psychologists have been particularly concerned with understanding the learning mechanisms that make the cultural transmission of behaviour possible. Whatever else might be involved, it is clear that cultural transmission must involve social (or observational) learning (that is, acquiring a behaviour by virtue of seeing another individual perform it). From this starting point, three general processes can be identified that might allow one individual to learn a behaviour from another. These are: stimulus enhancement, emulation and imitation.

The simplest of these is stimulus enhancement. This involves an observer's attention being drawn to something in the environment by the behaviour of another individual (the model). The object of attention may be an object (say, a fruit) or a location (in which case, it is sometimes known as 'local enhancement'). Having had its attention drawn to a particular situation, the observer then figures out for itself by trial-and-error how to acquire the resource concerned. This kind of mechanism is probably extremely common among animals, and may account for many aspects of animal behaviour that have formerly been labelled as 'cultural' (for example, sweet potato-washing by Japanese macaques and milk bottle lid-removal by British tits: Tomasello and Call 1997). Some psychologists would not now consider behaviour learned by this mechanism to be culturally transmitted, although there remains some disagreement about this.

Emulation represents a cognitively more sophisticated version of stimulus enhancement. In this case,

the individual's attention is drawn to a particular situation by the behaviour of the model, and it then figures out for itself what the point of this behaviour is. In other words, the observer uses the behaviour of the model to understand the problem by observing how the model's behaviour effects a change in the state of the world. Tomasello and Call (1997) suggest that chimpanzees may learn how to crack open nuts using stones as hammers in this way. They argue that, in this case, the observer has learned something about the environment, but not something about the *behaviour* of the model; consequently, they insist that behaviours learned in this way cannot count as *bona fide* examples of the cultural transmission of behaviour.

The third mechanism is imitation. In this case, the observer mimics or copies exactly what the model is doing, without necessarily understanding why it does it. Tomasello et al. (1993) argue that this is the only form of social learning that can be considered as uncompromising evidence for cultural transmission because it does not involve the observer learning anything for itself. They suggest that only this mechanism can yield the kinds of 'ratchet effect' that are characteristic of cultural transmission in humans (for example, the gradual accumulation over time of trivial modifications to an artefact). They argue that true imitation is found only in humans (being especially characteristic of young children) and perhaps great apes that have been 'enculturated' (raised by humans in a human environment).

---

behaviour must be essentially functionless. Ethologists and anthropologists, in contrast, typically want to consider a broader spectrum of behaviours, but in doing so they risk including behaviours that have in fact been acquired by mechanisms that have no social component to them. Anthropologists also want to consider teaching as the more important mechanism of transmission in humans: humans are, after all, **enculturated** during childhood and adolescence, thus acquiring the whole complex of world-views, self-perceptions, beliefs and rituals that make up a person's identity. For this reason, anthropologists want to see culture as something beyond mere psychological mechanisms like imitation. This, of course, is true enough, but fails to address the more fundamental issue of how we recognise culture (as distinct from knowledge acquired by individual learning) in both ourselves and other species. In the end, the argument about the relative importance of teaching versus imitation is largely irrelevant: teaching is simply guided imitation ('Do it like me!').

ENCULTURATED

Humans (and especially young humans) seem to be especially sensitive to acquiring knowledge and instructions by imitation. Such a mode of acquiring knowledge differs radically from the conventional mechanisms of trial-and-error learning that underpin most animal behaviour. Animals (including humans, of course) learn to solve many problems by working it out for themselves. But individual learning of this kind is a slow process (we make many mistakes on the way) and we end up constantly reinventing the wheel (each generation has to discover the solution to the same problem anew). Culture, it has been suggested, provides a mechanism for overcoming that inertial effect, so allowing each generation to stand on the shoulders of its predecessors.

This intuitively appealing argument does, however, run foul of an unexpected problem. Rogers (1989) pointed out that when the world consists of two kinds of organisms, learners and imitators, the pay-off to the imitators declines as their frequency increases (**Figure 13.1**). The more imitators there are in the population, the fewer people there are who will have worked out the appropriate solution for themselves; hence, imitators will end up copying other imitators whose behaviour may not be optimal, and so perform at no better than chance levels. In other words, if individual learning has a cost (hence the benefit of being an imitator), then social learning (and hence cultural transmission) could not evolve.

However, Boyd and Richerson (1994) were able to show that if imitation makes individual learning cheaper or more efficient, then imitation (and culture) could evolve. One way that might happen is if individual learners are given extra privileges

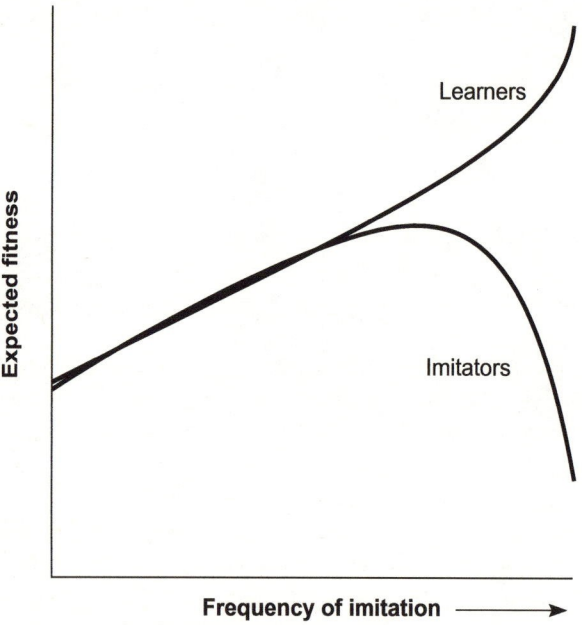

Figure 13.1 The paradox of conformity: in a world that consists of two types of individuals, leaders and imitators, the pay-off to imitators will decline once their frequency increases beyond the point where they start to copy each other rather than copy individuals who have worked out the correct solution for themselves (*learners*). In contrast, the relative fitness of learners will increase disproportionately past this point because they are the only ones able to find the right solution to a problem. Because of this, imitators should only occur at low frequencies in the population. Yet imitating seems to be a particularly common human trait. Redrawn with permission from Boyd and Richerson (1994).

by imitators. Another possibility is that imitators might be selective imitators: they have an intuitive notion of what a good solution to a behavioural problem looks like and are more prone to copy some individuals than others. The first is, of course, something that does seem to be true of humans: we idolise and fête people who do things well (cultural icons) or who are successful innovators.

Whatever the answer to this problem, it raises a more fundamental issue: one of the most puzzling features of humans is their apparently unique willingness (as a species) to conform to the communal will rather than individually striking out on their own. In fact, for two decades or more, a minor industry within social psychology devoted itself to empirical studies of just these kinds of effects (for example, Asch 1951, 1956; Millgram 1974; Latané 1981). Why humans should be so unusually susceptible to these kinds of social pressures remains the deepest and perhaps most intriguing mystery in the whole of human evolution (Henrich and Boyd 1998).

THE UNITS OF CULTURAL TRANSMISSION

Biological reproduction occurs because of the unique chemical properties of a particular molecule: not only does DNA provide the blueprint for the construction of living bodies but it also has the unusual property of being able to copy itself. This ability makes the processes of biological evolution possible because the traits that an individual possesses can be transmitted with a reliable degree of accuracy to its offspring. What, if anything, plays the equivalent role in cultural evolution?

One of the problems with answering this question is that those scholars who have been particularly interested in culture (principally socio-cultural anthropologists) have vigorously resisted the suggestion that culture can be partitioned into units that are in any way equivalent to biological genes. In their view, a society's culture is a single monolithic structure that cannot be subdivided. This view originally derived from the late nineteenth-century French sociologist Émile Durkheim, whose views have been among the most enduring influences on the social sciences during the twentieth century. He insisted that culture was imposed on individual members of society during socialisation. More importantly, because culture was thereby in some sense 'above' biology, cultural facts could not be explained by biology, but only by reference to other cultural phenomena.

One problem with this widely held view is that it misunderstands the relationship between genes and the bodies of which they are a part. A body can no more sensibly be partitioned up into independent units than can a culture. An arm does not have an independent existence apart from the body to which it is attached. Nonetheless, we can usefully partition the body into traits or characters (arms, eye colour, brain volume) in order to discuss their individual properties. We can also relate these characters to the particular units of inheritance (genes) that guide their development. As a result, we can (and biologists have done so for a century and a half) meaningfully speak of the individual components of the body, asking legitimate questions about their development, origins and function. We do not dismiss discussions about the function of eyes (to obtain information about the world 'out there' via the visual spectrum) even though an eye in the absence of the brain to which it is intimately attached is no more useful than a lens without a camera. In short, this common objection to Darwinian discussions of culture is based on a naive understanding of what biologists do rather than any real difference between cultures and bodies.

In fact, the situation may not be quite as bad as the Kroeber and Kluckhohn

BOX 13.2

How memes differ from genes

If a meme is the unit of inheritance (or transmission) in cultural evolution, how does it compare with the equivalent in biological evolution (the gene)? **Table 13.1** summarises the defining characteristics of genes and memes. Note that there are important similarities as well as differences between the two. Both transmit information across generations, but the mechanisms of inheritance differ (biological reproduction versus imitation or teaching) as do the length of a typical generation (long and relatively species-specific in the case of genes, but varying from the very short to the very long in the case of memes) and the mode of transmission (vertical from parents to biological offspring in the case of genes, but by many different routes in the case of memes). Vertical transmission refers to transmission between biological parents and their offspring; cultural transmission can in addition be horizontal (between peers) or oblique (between biologically unrelated members of the parental and offspring generations – for example, teachers and pupils).

Note also that biological generations are linked by acts of reproduction (mating and conception), but cultural generations do not have to be biologically related. What defines a cultural generation is the process of social learning whereby one person (the cultural 'parent') implants an idea into the mind of another (the cultural 'offspring') by teaching or by passive imitation (that is, copying by the offspring).

The value of the processes of cultural transmission is that the generation time can be *very* short, in fact almost instantaneous. The benefit is that some solution to a problem of survival or reproduction that you happen to hit upon, perhaps after a great deal of careful thought and trial-and-error effort, can be passed on to all of us, thereby saving us from having to waste the equivalent amount of time coming to the same conclusion. Cultural transmission thus makes cooperation easier and more effective. Without it, each of us would have to reinvent the wheel, and it may take some of us a very long time to do that: in fact, most of us would probably die before hitting on the solution that might have saved our lives.

One final, and perhaps surprising, difference between memes and genes is that memes often seem to have a higher fidelity of copying than genes do, at least as measured in terms of the correlation between parents and offspring. Cavalli-Sforza et al. (1982) sampled the heritability of a range of cultural traits in North American society. The results suggested that cultural transmission is surprisingly reliable: offspring tend to adopt their parents' social habits and beliefs (**Table 13.2**). Also shown in the table are the equivalent correlations between parents and offspring for a selection of physical traits that are known to have a strong genetic component. Since these correlations reflect both shared genetic and common environmental effects in this case, heritability indices are also given where these are known. (*Heritability* measures the proportion of variation in a character across the population as a whole that can be attributed to genetic as opposed to environmental effects. It is *not* a measure of how much of a character is determined by genes, but rather a measure of how much of the *difference between individuals* is due to genetic effects.) Notice that the reliability of transmission for cultural traits (as measured by the correlation between parents and their offspring) is at least as high as (and sometimes higher than) the equivalent values for morphological traits like stature and length of forearm. Cultural transmission, it seems, is both reliable and surprisingly robust by comparison with genetically transmitted traits.

(1952) review might imply. A more careful examination of their list of definitions suggests that there is an underlying unity to these apparently incompatible definitions that they found in the anthropological literature. They all ultimately involve ideas in people's heads. The differences arise not from what is happening inside people's heads, but in the way these ideas are given physical expression in the world. In some cases, this information is an instruction (how to behave), in others an account (myths about the origins of the universe, a fictional story, a musical score), in yet others a

Table 13.1 Principal defining characteristics of genetic and cultural evolution

	Genetic evolution	Cultural evolution
Unit of selection	gene (DNA)	meme (ideas, beliefs)
Rate of mutation	slow, constant	fast, variable
Mechanism of inheritance	biological reproduction	imitation, teaching
Mode of transmission	vertical	vertical, horizontal, oblique
Heritability	low	low-moderate
Generation length	moderate	variable (instantaneous to millenia)
Rate of evolution	slow	variable (instantaneous to very slow)

plan (how to construct a tool or the artist's conception of what the finished artwork should look like).

In other words, the units of cultural inheritance are rules of behaviour or items of information held in the mind (though they may be temporarily stored in another medium: a book, the decoration on a vase, a computer disk). Examples as diverse as the wearing of baseball caps backwards, genuflecting before the altar in Catholic churches, formal or informal instructions on how to behave towards particular individuals (the Prime Minister or the local police chief), items of gossip about others, the proof of Pythagoras's theorem, the myths and stories peculiar to each society – all these ultimately constitute units of information that can be taught to, or copied by, other minds.

Dawkins (1989) coined the generic term **meme** (derived from the Greek word meaning *imitate*) to refer to such phenomena. Although there has been considerable debate over the value of this term (notably its similarity to the word *gene* and the attendant risk of assuming a similar genetic basis), it is widely considered to be a more useful term than others that have been suggested – (for example, the term *culturgen* proposed by Lumsden and Wilson (1981). Dawkins would define a meme as any unit of information capable of being stored in a brain and transmitted to another organism by social learning. He (and others) have often adopted the colourful metaphor of memes parasitising our minds in just the way that viruses parasitise our bodies, hitch-hiking their way through evolutionary time by jumping from one body/mind to another.

MEME

CULTURE VERSUS BIOLOGY

Anthropologists have long insisted that human culture has nothing to do with the physical world: rather, it is an unfettered creation of the human mind. Indeed, many anthropologists have argued that, so far from being determined by the world in which we live, culture actually determines how we see and interpret that world. Some have even gone so far to suggest that traditional people's cosmologies and understanding of the physical and biological world are simply a reflection of the *social* world in which they live. We view the world as frightening and intimidating because we are frightened and intimidated by the kings that rule over us.

This view reflects an inadequate understanding of biology. First, not everything

Table 13.2 Heritabilities of biologically and culturally transmitted traits

| Cultural transmission[a] | | Biological transmission[b] | | |
Trait	Parent–offspring correlation	Trait	Parent–offspring correlation	Heritability[c]
Religion	0.71	stature	0.51	0.86
Sports	0.22	span (maximum spread of hand)	0.45	–
Politics	0.61	length of forearm	0.42	0.84
Entertainment/interests	0.44	IQ	0.49	–
Habits	0.24	hip circumference	–	0.43
Beliefs	0.49	masculinity/feminity index	–	0.82

a) from Cavalli-Sforza et al. (1982), Table 3.
b) from Cavalli-Sforza and Bodmer (1971), Table 9.9.
c) the proportion of variance due to genetic factors; data from Cavalli-Sforza and Bodmer (1971), Table 9.14.

biological is an adaptation to the physical/biological world. Darwin himself placed great emphasis on sexual selection as a parallel process to natural selection (see Chapter 2). Whereas natural selection reflects adaptation to the biotic world (the environment), sexual selection reflects adaptation to the social world: it favours the evolution of traits that allow individuals to mate more successfully, and in doing so it may result in the evolution of traits (like the peacock's tail) that are actively disadvantageous (by making it difficult for the bird to fly, and thus escape from predators). Second, selection consists of a continuum between the massively beneficial and the massively disadvantageous, with a zone of neutrality around the mid-point where there is neither significant selection for nor significant selection against a trait. In principle, a trait can be hugely advantageous from the sexual selection point of view, but neutral as far as natural selection goes: such a trait allows its bearer to breed more successfully, but does not affect its chances of dying one way or the other. By the same token, of course, a trait might be naturally selected, but sexually neutral, or even neutral on both counts.

When a trait lies in the zone of neutral selection, there is no pressure forcing it one way or another. As a result, a new form of evolutionary process comes into play, generally known as Kimura's **neutral mutation theory** (M. Kimura 1979). In effect, the trait (or at least the genes that produce it) is free to evolve under the dictate of random effects. Chance mutations or accidents of who happens to breed successfully may drive the trait slowly one way or another (and perhaps back again in later generations). The result (known as **genetic drift**) may eventually be the evolution of a novel species. The importance of drift, however, is that unlike natural (and sexual) selection it is not directed, but random. Moreover, because change depends solely on the rate of mutation, evolution will invariably be slow by comparison with that which occurs under selection.

NEUTRAL
MUTATION
THEORY

GENETIC DRIFT

We can draw important parallels for culture here in both respects. First, memes operate in a memic universe that is, at least in principle, orthogonal to the genic universe: genes and memes may be at odds with each other in just the way that natural and sexual selection may be. Second, memes may be under neutral genetic selection, in which case a different dynamic analogous to genetic drift may result. One important implication here is that memes have fitness in just the way that genes do; but, rather than being fitness in the conventional genetic sense, it is mimetic fitness (or fitness in the memic universe).

This raises an important issue: is cultural evolution independent of genetic evolution? The short answer to this question is that it depends on whether or not the cultural processes have any impact on the survival and reproduction of the individuals who exhibit the cultural trait. If they do (an example might be the belief that lions are simply large cuddly domestic cats), then, for obvious reasons, the memes concerned are likely to be tightly constrained by the genes whose brains they parasitise (the so-called 'genetic leash' of Lumsden and Wilson [1981]). As a result, mimetic fitness will be virtually identical to genetic fitness and the evolutionary history of the meme(s) will be determined by the evolutionary history of the genes. However, if the memes do not impact on genetic survival, then mimetic fitness will be independent of genetic fitness and the memes may be able to evolve in accordance with drift or, and this is important, in accordance with selection pressures operating *within the memic universe*. In other words, the fitness of a meme may be determined by its 'fit' with the collection of other memes present in the same brain.

In addition, there is likely also to be an intermediate position at which a meme has positive mimetic fitness but negative genetic fitness (just as genes or traits can be sexually selected but highly disadvantageous in terms of natural selection). One of the central questions in the work on cultural evolution has been whether it is possible for such memes to evolve against the resistance of negative genetic fitness. Most models of gene-culture coevolution suggest that this situation is quite plausible (Cavalli-Sforza and Feldman 1981; Boyd and Richerson 1985). The principal condition favouring meme evolution at the expense of genetic fitness is a high rate of mimetic transmission (or inheritance): providing the frequency of copying is high enough, memes can successfully invade a population even if they drive the genes that build the bodies they exploit to extinction. Of course, this assumes that there is, in effect, an infinite supply of naive bodies (or, rather, minds) for the memes to invade: if the meme is so 'virulent' that it kills off all the minds (bodies), then the meme too must go extinct.

Blackmore (1999) has gone so far as to argue that the superlarge human brain is a direct consequence of selection at the mimetic level, despite the costs of the neural tissue involved. She suggests that the ability to imitate (and thus absorb new memes) has acted as the major force selecting for the rapid increase in brain size within the later stages of hominid evolution by a form of runaway selection at the mimetic level: the more memes an individual learns (and hence the bigger the brain it needs to store them), the more desirable it is to increase the number of memes because this enhances its mimetic fitness (indexed as the frequency with which the individual is imitated by others).

Higgs (2000) has modelled this particular situation mathematically and shown that increased mimetic fitness can drive meme capacity (assumed to be synonymous with brain size) inexorably upwards even when there are *biological* fitness costs to increased brain size. It is significant that the rate of growth is initially very slow but then, after around 5000–10 000 generations, becomes exponential. (As it happens, this is exactly what we see in brain evolution within the hominid lineage over the past two million years: Aiello 1996.) Not surprisingly, the take-off point is later in those cases where the biological costs are higher, and take-off may be prevented altogether if the costs are high enough. Much, however, depends on the shape of the costs of brain growth: costs that scale exponentially with brain size impose stronger constraints on the take-off point than linear costs. Kendal and Laland (2000) have modelled Blackmore's argument using a phenogenotype approach (see p. 372). This model suggests that new habits (memes) could evolve (that is, successfully invade a population characterised by an alternative habit) against the gradient of a genetic cost (for example, large brain size) providing the probability of copying was high and the invading meme has a high fitness advantage in its own right (for example, favouring the use of a highly nutritious plant food). Their results seem to suggest that memes which are selectively neutral in genetic terms may be unlikely to invade a population that already adheres to an alternative meme.

There are two issues that remain unclear, however. One is the fact that Blackmore's argument seems to assume that human brain size is idiosyncratically out of line with brain size in other primate species, but this is in fact only true when brain size is viewed against a biological baseline like body size. If brain size is viewed against the variables that are thought to have driven brain size evolution in primates (namely, social group size), human brain size does not appear to be out of line (**Figure 9.3**). Second, Blackmore's hypothesis is based on the claim that being imitated by others is mimetically advantageous; but being imitated may itself have direct consequences for biological fitness through sexual selection (Miller's Scheherazade effect: see Chapter 12). All that Higgs's (2000) model may be telling us is the well-known fact that sexual selection can overwhelm conventional somatic fitness costs. Indeed, Kendal and Laland's (2000) model suggests that mimetic invasion is unlikely when the fitness consequences of the invading meme are genetically neutral.

IS CULTURE ADAPTIVE?

The extent to which cultural rules help individuals to reproduce more effectively in particular environments has been a particular focus of debate in the study of cultural evolution. This is the issue that has particularly vexed most socio-cultural anthropologists, for example. Nonetheless, the role of culture in facilitating successful reproduction is an important question as far as understanding the evolution of human behaviour is concerned. In this section, we review a small number of studies that argue for the functional adaptiveness of culture. If we can show that at least one cultural rule contributes to an individual's fitness, that is enough to establish the principle that culture *can* be adaptive and may therefore arise through conventional Darwinian processes of selection. It does not, however, necessarily mean that *all* cultural phenomena are adaptive in this sense. Some may be genetically neutral or even disadvantageous: we consider this possibility in the next section.

Malaria has been a major scourge of tropical and sub-tropical human populations for millennia, and many of these populations now exhibit genetic adaptations to the malaria parasite. These include the sickle cell trait found in many west and central African populations and the related condition of thalassemia found among Mediterranean peoples. In both cases, a single genetic mutation results in deformation of the oxygen-carrying haemoglobin cells in the blood (the so-called *sickle cell* trait) and this makes the cell resistant to the malaria parasite. The evolutionary cost incurred by this advantageous trait is that individuals who inherit copies of the sickle cell allele from *both* parents (that is, are homozygous for the mutant allele) suffer from cells that are so badly deformed they cannot carry enough oxygen to sustain a normal life: those afflicted with this condition invariably die during adolescence after a painful and sickly childhood dogged by severe anaemia, painful joints and other debilitating side effects.

Brown (1986) has investigated in some detail the cultural (that is, behavioural) and genetic adaptations to malaria found among Mediterranean peoples. On the island of Sardinia, the peasant community has traditionally engaged in a pattern of transhumance whereby sheep and goat flocks are brought down into the lowlands during the winter and taken up into high mountain pastures during the summer. Brown argued that this pattern was directly related to the level of malarial risk that the people concerned faced. While the highland areas are relatively free of mosquitoes, they offer poorer grazing conditions especially during the cold winters. Since the mosquitoes are active (and breed) mainly during the summers, the shepherds avoid the lowland pastures at this time of year and instead make use of the relatively mosquito-free highlands.

In addition, the communities impose very strict rules on who can go into the more malarious lowlands even during the summer. Women and children are discouraged from leaving the highlands and all permanent villages are placed at these higher altitudes. Even then, further measures are taken to reduce the malaria risk for women (who are especially vulnerable during pregnancy): the scrub and woods around villages are cut back to maintain a vegetation-free zone that mosquitoes cannot cross and the women are discouraged from leaving the confines of the village as far as

BOX 13.3

The evolution of fantasy

A feature of human culture that appears to be quite unique to our species is story-telling. It would be easy to dismiss this as a trivial epiphenomenon of human language, but in fact it has important evolutionary aspects that merit detailed investigation. Although rather little attention has been given to these by evolutionary psychologists, three aspects may be identified as particularly profitable areas for future research.

First, story-telling is only possible with at least second order intentionality (theory of mind) because composer and audience have to be able to imagine that the world could be other than as they find it. The real difference between humans and other animals seems to be that all other species simply accept the world the way it is without asking why this has to be (Dunbar in press). However, if stories are more than just descriptions of fictional worlds, then even more advanced mentalising abilities may be required: we noted in Chapter 11 that the author of a story featuring three individuals involved in a triangular relationship has to be able to achieve fifth order intentionality (a rare ability even among normal adults: see **Figure 11.11**), while the audience has to be able to manage at least fourth order. Thus, the evolution of advanced mentalising abilities has probably played a crucial role in the development of human literary skills. Indeed, it may even be that individuals' literary skills correlate with their mentalising abilities (that is, the number of levels of intentionality that they can regularly manage).

Second, a very high proportion of traditional stories (from the Bible onwards) are concerned with 'origin stories' (how the particular tribe or culture came to be). This suggests that story-telling of a more general kind provides an important device for bonding social groups. This idea has been developed in some detail by Knight (1991), who has argued that a society's myths and legends form part of a complex designed to reinforce and bind the community into a functional coalition. They may also provide relatively safe environments in which dangerous topics can be aired and appropriate warnings passed on: for example, the frequency with which wicked step-mothers appear in European fairy tales is intriguing given the fact that, at least during the period when these stories were composed, most marriages were terminated by the death of the mother during childbirth (Voland 1988), so that any subsequent marriage brought a new unrelated woman into the household.

Finally, we should point out that the stories themselves may be analysed with respect to the fidelity with which they reflect evolutionary principles (Carroll 1995, 1998). Some evidence that they may do so is offered by the analyses of Viking sagas carried out by Dunbar et al. (1995). Whissell (1996) carried out an analysis of 25 modern romantic novels and showed that key features of their story lines matched five key predictions from evolutionary theory. We have ourselves undertaken some preliminary analyses of Jane Austin's novels and shown that the fictional events she describes (marriage choices, inheritance patterns, and so on) do conform remarkably well to evolutionary expectations (see also Whissell 1996). Others have explored similar possibilities in other literary corpuses (for example, Fox 1995, Cooke 1995, Thiessen and Umezawa 1998). So far, however, no detailed statistical analysis of any body of literature has yet been published. One interesting possibility, for example, is that an author's survival (and hence recognition as a 'great author') reflects their intuitive understanding of evolutionary principles and the way these underpin human behaviour.

possible. Anti-malarial strategies in this population thus seem to be a neat example of convergence in both the genetic and behavioural traits in the population aimed at minimising the risks of exposure to malaria-carrying mosquitoes.

Another example documented by Western and Dunne (1979) concerns the rules that the pastoralist Maasai of Eastern Africa use to decide where to locate their homesteads. Like many cattle-keeping peoples, the Maasai occupy temporary villages for varying periods of time but move to a new site (often many kilometres away)

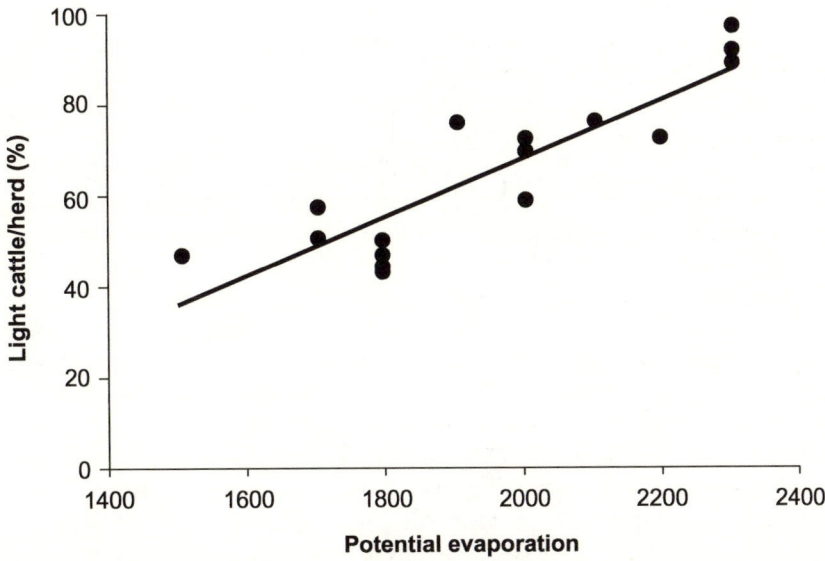

Figure 13.2 The proportion of light-coloured cattle in Maasai herds at different sites in East Africa, plotted against potential evaporation (a correlate of ambient temperature: habitats with high potential evaporation are typically hot and dry). Light-coloured cattle predominate in hotter habitats (low altitudes, nearer the equator) because they can cope better with heat and require less water than animals with dark coats. Maasai selectively cull their herds to ensure that their cattle are as productive as possible in the conditions in which they have to live. Redrawn with permission from Finch and Western (1977).

when either the local grazing becomes too poor or the site becomes disease- or parasite-ridden. Anthropologists have concluded that the decision about where to locate a new homestead is based mainly on social considerations. However, Western and Dunne were able to show that, in fact, the Maasai carry out a very careful assessment of the relative benefits of alternative sites that mainly reflect the consequences for their cattle and the Maasai's own health. An analysis of the physical properties of a large sample of homesteads showed that sites tended to lie uphill from riverbeds (to balance the high risk of malaria for the humans in heavily vegetated habitats near water against access to water for the cattle) on slopes that are not too steep (lest they tire the cattle coming back from water at the end of a long day); the preferred sites have darker coloured soils (because these retain more heat overnight – thermally-stressed cattle produce less milk) and good access to the different kinds of grazing required by the cattle and the small stock (sheep and goats).

Ecologically competent knowledge of this kind is much more widespread among traditional peoples than has typically been assumed in the past (for reviews, see Niamir 1990, Dunbar 1995a). Two more examples of ecologically-accurate social knowledge are worth detailing to reinforce this point. One concerns the preferences the Maasai exhibit for the coat colours of their cattle. Since herds vary enormously in the proportion of dark to light coloured cattle, anthropologists have typically assumed that these simply reflect individual herd owners' colour preferences. However, Finch and Western (1977) showed that, in fact, the proportion of light-coloured cattle in herds correlated very closely with the aridity of the locality (**Figure 13.2**). They were able to show experimentally that light-coloured animals were more resistant to overheating than dark-coloured ones and needed less water, and hence did better in hot dry habitats because they could be grazed further away from water sources. However, in cooler conditions, dark-coloured cattle lost weight less rapidly during lean times of year than did light-coloured cattle, mainly because they wasted less food on thermoregulation. They therefore tended to die less often during famines in these kinds of habitats, although they were at more risk than light-coloured animals in warmer low altitude habitats. Thus so far from being a random choice dictated by local socio-cultural conventions, herders' coat colour preferences are very carefully attuned to local habitat conditions and cattle physiology so as to allow the Maasai to get the best out of their animals.

The second example concerns another East African pastoralist people, the Rendille. Besides tending their camel herds, Rendille specialise in gathering wild honey, which they do by allowing a small bird (the honey-guide of the family Indicatoridae) to direct them to the nest sites the birds have discovered. This symbiotic relationship between bird and human seems to be quite ancient, but it has obviously involved some fairly astute learning on both sides. On locating a bees' nest, the bird seeks out a herdsman and signals to him that it has found a nest by fluttering around the man's head and giving a characteristic call. The herdsman then follows the bird back to the nest, extracts the honeycomb and rewards the bird with a share. Should the herdsman lose sight of the bird while following it, he stops and waits for the bird to find him again. The Rendille claim that they can tell exactly how far away a nest is because the bird's absences and the distance between successive perches are shorter the nearer they are to the nest, while the height at which the bird perches get lower. Isack and Reyer (1989) sampled a large number of instances of honey gathering and found that the Rendille's interpretation of the bird's behaviour was in fact right: the birds really do signal the distance to the nest in this way **(Figure 13.3)**. Once again, the pastoralists' folk knowledge about the world (handed down from one generation to the next) turns out to be essentially correct.

It is not just ecological knowledge that is adaptive in this way. The rules that govern the structure of society itself may also be adapted to local conditions. Nettle (1996) examined the size of language groups in West Africa and showed that the number of speakers of a given language increase with distance from the equator. In other words, the further from the equator a tribe lives, the larger it is likely to be. Nettle argued that this is a response to habitat instability (and indeed, the correlation is better still against measures of seasonality): in unpredictable habitats, it pays to have a wide network of people you can call on for help when things get bad. An alternative (but not necessarily mutually exclusive) explanation is that alliances become increasingly important when population density is high in agriculturally rich habitats; being able to differentiate between reliable and unreliable allies will then be commensurably essential. In a subsequent analysis, Nettle (1999b) confirmed the relationship by showing that the number of languages in a country (adjusted for population size) was directly related to the length of the growing season **(Figure 13.4)**. Mace and Pagel (1995) reported similar results for North American native languages: the number of languages spoken in an area and the geographical range of a given language increased with latitude. More importantly perhaps, when latitude was held constant, there was greater linguistic diversity in areas of greater habitat diversity.

Another example of societal adaptation is provided by the case of Tibetan polyandry that we encountered in Chapter 8. Tibetans are one of the few peoples that habitually practise polyandry. The normal sequence of events is that, at some suitable moment, the parents marry all (or at least most) of their sons off to someone else's daughter and then retire to a small cottage on the periphery of the family farm,

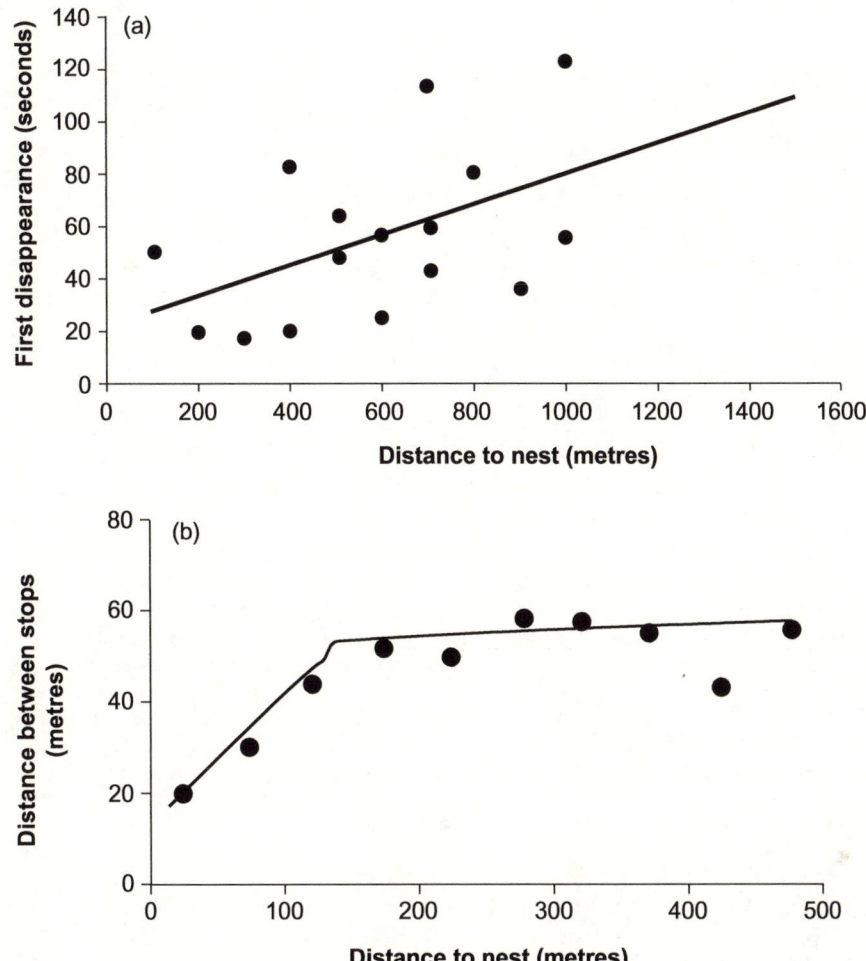

Figure 13.3 Cultural beliefs do reflect reality. Rendille honey gatherers in East Africa believe that the honey-guide bird signals the distance to a wild bees' nest by (a) the length of the bird's absence after first attracting the human's attention and (b) the distance between successive perches. Data from 17 cases of honey-guiding suggest that their beliefs about this natural phenomenon are right. Redrawn from Isack and Reyer (1989).

leaving the newly weds to take over the main farmhouse and run the farm (Crook and Crook 1988). This process (the monomarital principle) has the effect of preventing lineage proliferation: there is only ever one family to inherit the farm (Crook and Crook 1988). Thus the family farm remains intact and does not become fragmented over succeeding generations to the point where it loses its economic viability (see Chapter 8).

Crook (1997) provides us with another particularly intriguing example of the way cultural behaviour may be adapted to resolving some of the psychological costs incurred by this practice. In Tibet, demonic possession associated with psychological disturbance (and mediated by the intervention of skilled shamans) appears to be used as a means of allowing individuals to adjust their sense of identity in order to cope with pressures within the set of marital relationships (particularly in the case of the women, who often have to balance the competing demands of as many as three

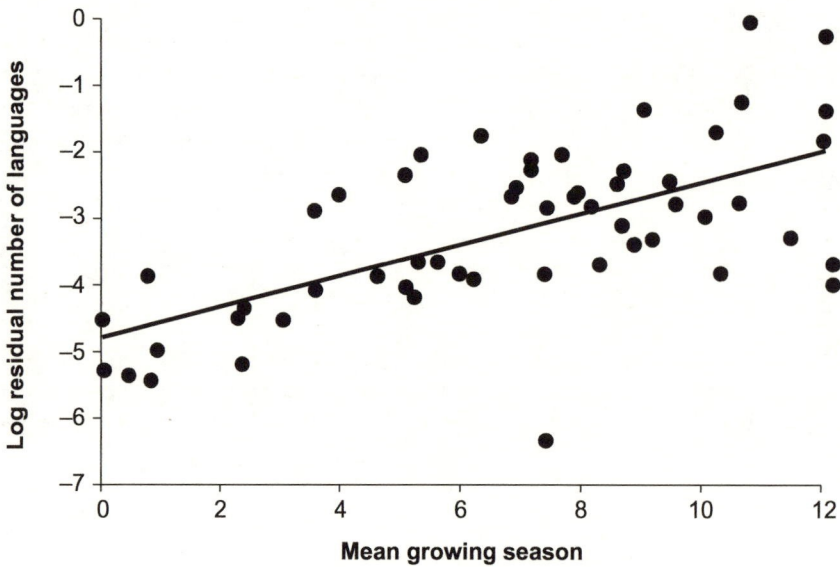

Figure 13.4 The number of languages spoken in a country (controlling for size of country) is a linear function of the length of the growing season (a measure of ecological predictability). Data are for all countries in Africa and South America. The longer the growing season (that is, the more productive the habitat, and the higher the population density as a result), the more languages are spoken per unit area, and hence the smaller the language community (number of people speaking any given language) in each case. It seems that as population density (and thus ecological competition) increases, so tribal groupings become more differentiated in order to allow individuals to identify natural allies more easily. Redrawn with permission from Nettle (1999b).

husbands simultaneously). In this way, social discord is limited and held in control, an important function in a society where collaboration within families is crucial to survival and reproductive success.

Finally, in Chapter 6, we examined the issue of how wealth and social status relate to reproductive success in some detail. The burden of the evidence we reviewed was that these purely social variables did have direct consequences for an individual's lifetime reproductive output. More importantly, Rogers (1990) was able to show that, under appropriate circumstances, fitness was maximised if individuals consciously opted to maximise wealth at the expense of personal reproduction. In other words, there are circumstances under which pursuit of what appear to be purely cultural goals will maximise genetic fitness more successfully than attempting to maximise reproductive goals directly.

Languages provide us with a rich source of information on cultural change, not least because one of the main ways in which language evolution occurs is through the demands of trade. In order to be able to communicate (and so trade), one group of people often abandon their language in favour of another's (usually that of the politically more powerful group); in some cases, a new language may effectively be imposed by an immigrant elite (either as a *lingua franca* or because of a desire to emulate the culture of the elite). English itself provides a clear example of this, since its present form represents the outcome of an amalgam of the language of the original Anglo-Saxon peasants with that of the Vikings (mainly through trading) and French (mainly through conquest by the Normans). It is difficult to see this as anything but a form of cultural adaptation intended to ensure successful survival in a new environment.

A comparison of languages and genes provides us with some instructive examples (see **Box 12.3**). The dominance of Indo-European languages across northern India (those like Urdu, Gujarati, Panjabi and Bengali that derive from Sanskrit) results from the arrival of a relatively small group of well-organised warrior-pastoralists (the Aryans) about 3500 years ago; by imposing themselves as the ruling class (*arya* means nobleman in Sanskrit), they were able to disseminate their own language at the expense of the Dravidian languages that had been spoken there prior to their arrival. Similarly, of course, the English and Spanish settlers of the New World imposed their languages despite being in a numerical minority, while the Romans managed to impose Latin on much of southern Europe (where it is now represented by its daughter languages, including French, Italian, Spanish, Portuguese, Catalan and Romanian). Sometimes, the process works the other way around, as when relatively unsophisticated invaders adopt the language of the country they have conquered, presumably because the latter is seen as being more elite. In the aftermath of the Roman Empire's collapse, for example, the Franks (a Germanic tribe) invaded and settled northern France, while another Germanic-speaking group (the Lombards, originally from Scandinavia) conquered northern Italy; both eventually abandoned their native languages in favour of those of their adoptive countrymen.

In sum, at least in some cases, cultural systems do seem to be dynamically fluid and adapt sensitively to changing local conditions in such as way as to facilitate the successful survival and reproduction of the individuals concerned. However, by no means all cultural traits exhibit such an effect: many appear to be under genetically neutral selection pressure, and thus to have an evolutionary life of their own. We now turn to a discussion of these.

CULTURAL EVOLUTION UNDER NEUTRAL SELECTION

When memes are under neutral selection in genetic terms, cultural evolution may proceed independently, subject either to selection within a memic universe or under the memic equivalent of drift. In this respect, an important class of memes will be those that provide a metaphysical explanation for real world events. An example is provided by Hughes's (1988) analyses of pedigrees.

Hughes (1988) showed that when individuals choose to associate, they do so on the basis of shared genetic relatedness to members of the pubertal cohort (the cohort with the highest reproductive value) (see Chapter 3). However, this cohort changes every generation, making it difficult to provide any constancy and coherence to the kinship group. To solve this problem, the pedigree is hung off some remote ancestor who provides a stable point from which the real focus of grouping can be sprung. Hughes showed mathematically that it did not matter who the anchor point actually was once you were more than about three generations back in time. Consequently, he could incorporate the sun and the moon as mythical ancestors into his pedigrees without affecting in any way the patterns of kinship relatedness within the group of living individuals. In other words, any person or thing could be substituted at this point in the pedigree, so introducing a degree of coherence and stability to the pedigree. Exactly who was chosen could vary from one group to another as a result of mimetic drift (or even mimetic selection so as to fit with the rest of the group's particular cultural universe).

BOX 13.4

The evolution of the teddy bear

Cultural change may be driven by adaptation to an existing genetic predisposition. One example of this is provided by a phenomenon to which we have all probably contributed: the evolution of the teddy bear. When bears were first invented during the early 1900s, they were very bearlike with pronounced snouts and low foreheads. However, Hinde and Barden (1985) have shown that, as the century progressed, their design became increasingly babylike, with foreshortened snouts and higher foreheads (**Figure 13.5**).

The concept of a teddy bear is clearly not a genetically inherited trait, so we can be reasonably sure that what we are looking at here is a cultural trait. However, it is a cultural trait that seems to be under the guidance of another genuinely

genetic trait, namely the cues that attract us to babies (high foreheads and small faces). A likely explanation is that the designers of teddy bears respond to their customers' past buying patterns when designing the next season's bears. The designs that sell better are copied and adapted more often the following year than the more bearlike ones that remain unsold on the shelves.

Morris et al. (1995) used a forced-choice paradigm to assess 4–8-year-old children's preferences for bears with contrasting features. They found that, while there was an increasing tendency for more babylike bears to be preferred with age, younger children did not exhibit a preference for babylike features. They therefore concluded that it must be adults' preferences that are reflected in **Figure 13.5**.

The point here is that the myths (or memes) provide a framework or story that gives coherence to some feature of real life experience that would otherwise lack a clear and easy-to-remember justification, perhaps because the real underlying biological or physical processes are too deeply buried to be understood, given the current technical knowledge of the people concerned.

An example of this is provided by the anti-malarial strategies of the Sardinian pastoralists mentioned earlier. Recall that Brown (1986) found that women were discouraged from leaving the immediate confines of the village and that this helped to reduce their exposure to malaria-carrying mosquitoes by keeping them away from the scrub habitats where mosquitoes lived. The Sardinians justified these restrictions by asserting that women who strayed too far away from the village would be affected by the Evil Eye, which would bring them bad luck and ill health, especially if they were pregnant. In reality, it was malaria that they risked contracting. Traditional medical knowledge lacked a clear understanding of the *cause* of the problem even though it had a clear grasp of the *fact* of the problem; something was needed to ensure that the women would conform to the behavioural rule designed to minimise their exposure and, in the superstitious culture of the Sardinian peasantry, the risk of Evil Eye fulfilled that purpose. A rule that fits well with other local cultural preconceptions will clearly work better than one that does not. Boyd and Richerson (1994) review examples in which the uptake of new technologies was directly influenced by how well the particular technology meshed with existing beliefs and understandings (see below; also Rogers and Shoemaker 1971).

Dialects provide us with another example. Nettle and Dunbar (1997) showed that dialects that changed rapidly over time were an effective means for controlling freeriders within cooperating social groups (**Box 9.3**). Dialects, in effect, provide a marker for kinship because they are learned at a very early age. Provided the dialects changed by drift at a rate in excess of 50 per cent per generation, freeriders (even those capable of mimicking) were unable to keep track of the changes in local dialects (**Figure 9.9**). Exactly what the dialect should be in any given case was simply an

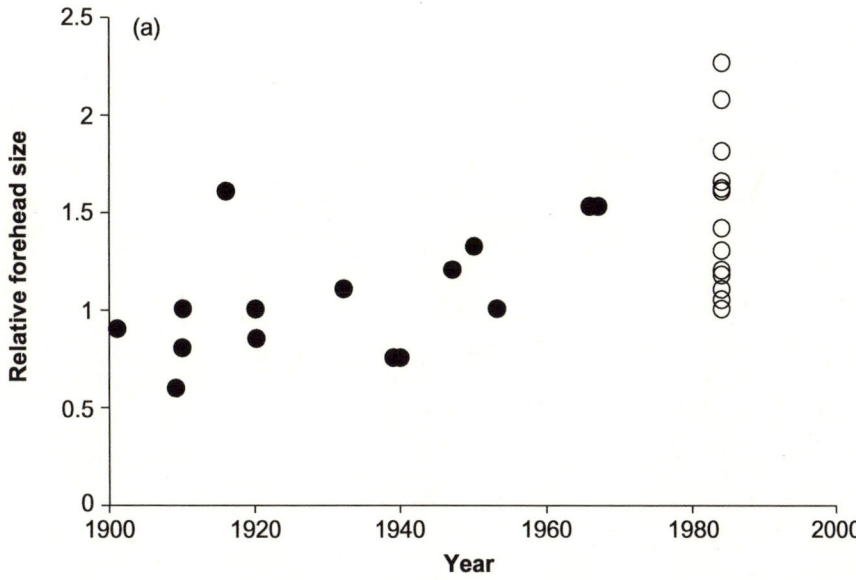

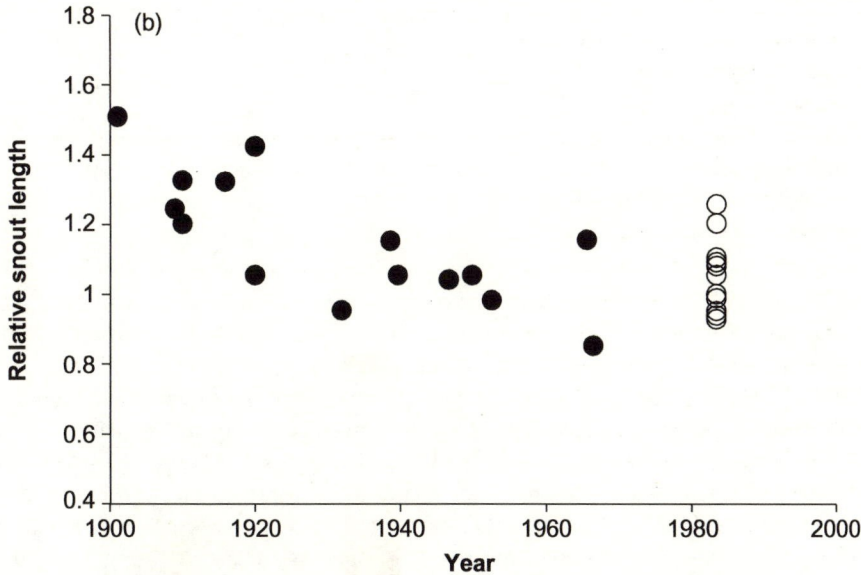

Figure 13.5 The evolution of the teddy bear. The shape of the teddy bear's face has become more babylike over time since the first ones were produced in the 1900s. This is reflected in (a) an increase in the size of the forehead (indexed as the ratio of eye-crown height to eye-chin height) and (b) a decrease in the length of the snout (indexed as the ratio of snout length to total height of head). Filled circles are individual bears at the Cambridge Folk Museum (UK) plotted against date of manufacture; open circles are data for a sample of bears on sale in Cambridge (UK) shops. Redrawn with permission from Hinde and Barden (1985).

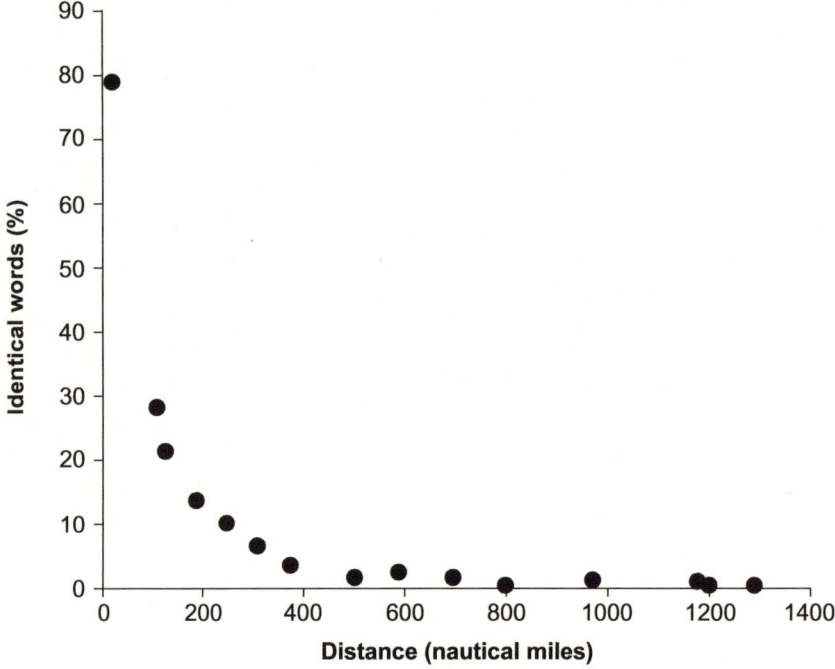

Figure 13.6 The proportion of cognate (i.e. identical) words in Micronesian languages declines as an exponential function of the distance apart of the islands on which they are spoken in just the same way that genetic similarity declines, implying that words 'mutate' at a relatively constant rate with distance (and perhaps time) of separation. Redrawn from Cavalli-Sforza and Feldman (1981).

accident of mimetic drift. Anything would do, providing it was different from those used by other groups. Thus, once again, the phenomenon has a clear functional purpose and is under intense selection, but the specific instantiation of the phenomenon does not and is thus subject to mimetic drift.

We can see the outcome of this in spatial terms when we examine similarities in word form in different languages of the same language group. Just as genes diverge with geographic distance as a result of drift, so do dialects. A nice example is provided by the dialects of the Micronesian islands in the western Pacific (**Figure 13.6**). Cavalli-Sforza and Wang (1986) found that the similarity in cognate words (that is, those having essentially the same meaning) declined as a negative exponential function of the distance the islands were apart, suggesting that the rate of 'mutation' on words is roughly constant (as it is with genes). The shape of the distribution is more bowed than might be expected of genetic data, possibly because words do not normally mutate back to an original form (as genes sometimes do).

PROCESSES OF CULTURAL EVOLUTION

Cultural evolution has been mainly explored through mathematical models designed to explain its underlying mechanisms. So far, three general kinds of model have been developed. Though broadly similar, these approaches differ in their conception of how closely memes are tied to genetic fitness, as well as in the way they conceive the reproductive process involved (that is, the mechanism of inheritance). One approach assumes that memes are ultimately constrained by genes, and this has led to the development of gene–culture co-evolution models that see genetic fitness as the crucial criterion. The second approach assumes that genes and memes interact in such a way as to maximise their joint fitnesses. The third approach, dual inheritance theory, assumes that memes operate in a more or less genetically neutral environment. In reality, each focuses on a slightly different part of the same continuum of biological reality: all three models may be true, but each applies only to a certain limited range of circumstances. We consider each in turn.

Since all the models involve rather daunting mathematics, we confine our remarks to the essential points and ignore technicalities. It should be noted that all these approaches are heavy on theory and light on data. This is not because data do not exist, but because it is often difficult to obtain the right kinds of data to test the assumptions and predictions of the models. Like many similar developments in the history of science, there has been a lot of early theorising which will in due course stimulate a strong empirical phase. Rather than dismissing modelling of this kind as difficult and irrelevant to real world issues, we should seek to profit from it by designing our empirical studies in the light of what the models have to tell us. There have already been a number of quite successful attempts to apply meme theory to song dialects in birds, for example (Mundinger 1980, Lynch et al. 1989).

EXTENDED PHENOTYPE MODELS

This approach focuses on those cases where genetic and memic fitnesses are closely correlated. The first model to be developed was that by Lumsden and Wilson (1981). In this model, the cultural elements (termed *culturgens*) are conceived as being designed to increase the effectiveness of specific genetic components. In this sense, they are close to Dawkins's (1979) idea of an *extended phenotype* in which some behavioural product is designed to reinforce or extend the organism's genetic (or bodily) components. An example of this is the way a nest enhances a bird's ability to reproduce successfully. Although memes may have some degree of independence, ultimately they are held in check by the strings that 'attach' them to the genes on whose bodies they depend. One reason for taking this view is that if a meme encourages a behaviour that results in the death of the body (and hence the extinction of the genes that create that body), then the meme too must go extinct.

Lumsden and Wilson were principally interested in the way cultural elements and the neuropsychological bases on which they depend co-evolve. They modelled these processes on the kinds of co-evolution with which we are familiar in the natural world, namely host–parasite co-evolution. In these cases, the two components – host and parasite, for example – gradually become more adapted to each other's characteristics: this enables parasites to survive without killing the host on which they depend, while hosts become more resistant so that they are less damaged by the parasite's activities. Lumsden and Wilson envisaged culture and the neuropsychological mechanisms as converging in the same way and imagined that neurological processes become sensitive to just those kinds of cues that are appropriate for particular cultural behaviours, while the cultural element itself becomes more suited to the kinds of neurological processes available to it. Deacon (1997) has argued that language and the human brain co-evolved in exactly this way.

The Lumsden-Wilson model is based on the mathematics of host–parasite co-evolution.

BOX 13.5

Two examples of gene–culture co-evolution

Lactose tolerance among the cattle-keeping peoples of Europe and the Horn of Africa and the relationship between the sickle cell condition and yam cultivation in West Africa provide two of the clearest examples of gene–culture co-evolution. Both explicitly involve behavioural effects that have directly influence the genetic evolution of the peoples concerned.

Although all humans consume milk as babies, the ability to digest milk (through the action of the enzyme lactase) is lost shortly after weaning in most humans. As a result, the consumption of milk and many milk products results in diarrhoea and eventually weight loss and death. The only exceptions to this are a relatively small set of cattle-keeping peoples (including all modern Europeans) whose common ancestors evolved lactose tolerance as adults, thanks to a point mutation in the gene responsible for producing lactase.

Durham (1991) argued that this came about because of co-evolution between a behavioural adaptation and a genetic adaptation that allowed these individuals to exploit milk under conditions of nutritional deficit. Shortages of calcium and vitamin D (the latter may be especially important in northern latitudes where its synthesis in the skin using UV sunlight is more difficult) have been suggested as the likely triggers. Both nutrients are readily available from the milk of cattle and sheep/goats. Thus the genetic mutation that promotes lactase production would have been selectively reinforced by the cultural predisposition to consume dairy products,

and vice versa. Feldman and Cavalli-Sforza (1989) have shown that a co-evolutionary model with a single genetic allele (lactose tolerance versus intolerance) and a two-state cultural phenotype (milk users versus non-users) can explain the evolution of lactose tolerance providing there is cultural inheritance of milk use down the generations.

Sickle cell anaemia is a condition that afflicts a significant proportion of the population in the malarial areas of West Africa. This distressing (and lethal) condition is now known to be the result of acquiring copies of a mutant haemoglobin gene from both parents. When a copy is inherited from just one parent (and the normal gene from the other parent), the sickle gene confers significant resistance to malaria. Durham (1991) has suggested that the sickle trait co-evolved with the practice of yam cultivation, and showed that the distribution of the sickle allele among the Kwa-speakers of Sierra Leone corresponds very closely to the distribution of surface water surplus (the latter being a measure of the availability of standing pools of water that the malaria-carrying mosquito requires for breeding). He argued that the practice of yam cultivation opened up the natural forest/bush to create farmland where surface water might accumulate in areas of high rainfall and low evaporation, so facilitating the spread of the mosquito. Because populations with low frequencies of the sickle trait were unable to survive in areas of high malaria risk, this in turn facilitated the spread of the sickle trait.

Their book is not an easy read, and doubts have been raised about its mathematical validity. As a result, it has perhaps received less attention than the other two models discussed below. Nonetheless, its emphasis on gene–culture co-evolution provides a very different perspective to that given by other analyses, and for this reason it continues to attract discussion at a more general level.

PHENOGENOTYPE CO-EVOLUTIONARY MODELS

PHENOGENOTYPE

This class of models consider the case where genes and memes are semi-independent, and calculates the conditions that in effect maximise the conjoint fitness of these two components (the **phenogenotype**). These models are conventional population genetic models but their mathematics is complicated by the fact that they depend on both the parents' genotypes *and* phenotypes as well as the offspring's genotype (Cavalli-Sforza and Feldman 1981, Feldman and Zhivotovsky 1992: for more readable introductions, see Laland et al. 1995a; Laland 1993; Feldman and Laland 1996).

Most of these models focus on genotype-dependent cultural transmission (in most cases, vertical transmission from parent to child), and thus share some conceptual basis with the Lumsden and Wilson model considered above. However, their use of models based on **diploid** sexual genetics combined with the maximisation of phenogenotype fitness places them in a rather different league from the Lumsden-Wilson approach. Phenogenotype co-evolution models have been applied to a number of different phenomena, including the evolution of vocal and sign languages (Aoki and Feldman 1987, 1989, 1994), altruism (Feldman et al. 1985), handedness (Laland et al. 1995b), mating preferences (Laland 1994) and sex ratios (Kumm et al. 1994; Laland et al. 1995a).

DIPLOID

Aoki and Feldman (1987), for example, considered the case of language evolution in which (a) the genetic capacities for signal transmission (speaking) and reception (hearing) co-evolve and (b) these in turn co-evolve with culturally transmitted information (meanings). They show that a genotypic polymorphism can evolve to a stable equilibrium (a simple example would be a population consisting of both speakers and non-speakers) and that the genetic polymorphism must be due to the influence of cultural transmission. (In the absence of cultural transmission, the stable equilibrium is a monomorphic population.) In a further development of the model, Aoki and Feldman (1989) emphasised the importance of vertical transmission of cultural traits, arguing that the empirical evidence favours this form of parent–child cultural transmission over horizontal (sib–sib or peer–peer) transmission.

Laland (1994) used this approach to model the evolution of a sexually selected trait. He considered a genetically transmitted trait with two alleles that is expressed in one sex (say, males) who mate with females who exhibit one of two culturally inherited mating preferences (those that mate indiscriminately with the two male genotypes and those that prefer to mate with just one type of male). Laland showed that, when a culturally transmitted (female) mating preference achieves a significant frequency through cultural drift or social learning, it can drag a genetically inherited (male) trait to fixation even if that trait has negative fitness consequences for the males concerned. When the inheritance bias in favour of the culturally inherited trait is strong, fixation of the genetically inherited trait can occur in 20–50 generations. Even when the bias is weak, it may require fewer than 100 generations to achieve fixation. These represent extremely high rates of genetic change.

One example of this might be the way male-biased natal sex ratios seem to have evolved within some Indian populations: Laland's model demonstrates that this could arise because of a social preference for sons that imposed strong selection pressure in favour of individuals (presumably in either or both sexes) who produced a disproportionate number of male foetuses. That could occur either because the males produce more Y-bearing sperm or because females implant more Y-bearing concepta. This provides us with a particularly convincing example of how cultural practices might influence a population's genetic characteristics.

▌ **THE DUAL INHERITANCE MODEL**

Boyd and Richerson (1985) consider the opposite extreme to Lumsden and Wilson, where genes and memes are functionally independent – in other words, the case where cultural elements (memes) have neutral genetic fitness. They term this approach *dual inheritance theory*. The focus of their model lies in the actual mechanisms of learning and Boyd and Richerson identify a number of different possibilities. The simplest of these is *guided variation*: this is the familiar kind of trial-and-error learning. It results when an individual who acquires some idea of what it should be doing from another individual and then uses the successes and failures of experience to adjust its behaviour. In *direct bias*, the individual takes its choice of behaviours from the individuals around it, but then uses trial-and-error to decide which is the best one to adopt; direct bias differs from guided variation only insofar as the range of behavioural options available to the individual is limited by those provided by other individuals in the local population. In *indirect bias*, the individual looks to see who is the most successful (or attractive?) of the models available to it, and then copies that model's behaviour. An alternative form of indirect bias is *frequency-dependent bias*: in this case, the individual looks around to see which is the commonest form of a given behaviour, and adopts that (on the assumption, presumably, that the most common will also be the most successful – even if only in purely social terms – within that particular cultural environment). In these models, vertical and horizontal transmission have similar properties (aside from differences in generation time).

Boyd and Richerson (1985) showed that even weak biases too small to detect empirically were sufficient to increase the frequencies of cultural variants favoured by the bias. Cultural transmission is favoured whenever environmental conditions have moderately high autocorrelation over time (that is, are relatively predictable, even if only on a cyclic basis), yet year-to-year environmental variability is high: under these conditions, cultural evolution allows more rapid behavioural adaptation to current conditions than responses acquired either by individual trial-and-error learning or by genetic inheritance. In contrast, genetic inheritance is favoured when autocorrelation is very high (the future is *very* predictable) and learning is costly and inaccurate. More importantly, perhaps, it is even possible for cultural variants that are genetically maladaptive to prosper at the expense of variants that are genetically more advantageous under some conditions. Conversely, in rapidly changing environments, conventional trial-and-error learning is favoured.

In reviewing a range of studies on the diffusion of innovations, Boyd and Richerson (1985) concluded that there is considerable evidence to support the suggestion that direct bias and guided variation (that is, learning) are important. Farmers observe neighbours using a new pesticide and then try it themselves; whether they subsequently adopt it or not depends on their own experiences with it (Rogers and Shoemaker 1971). Boyd and Richerson suggest that cultural transmission by indirect bias and frequency-dependent bias have evolved because they obviate the individual's need to work out the solution for him- or herself. They thus offer considerable benefits in terms of reduced costs. Even if the mechanism of transmission (the learning or copying process) incurs a cost (evolving or maintaining a large brain, or the time/energy costs of acquiring information or for practice), the gains from cultural transmission far outweigh the losses.

However, once cultural transmission is in place, it may generate costs of its own: people will be reluctant to adopt new practices if they conflict with other aspects of their culture or are costly to test. Boyd and Richerson noted that an attempt to introduce the habit of boiling water to a rural Peruvian community failed despite its health benefits because the peasants' theory of disease conflicted with that of the outreach workers, the effort required to boil water was too great and many people disliked the taste.

Werren and Pulliam (1981) have used this general approach to consider the more specific question of the evolution of altruism (helping behaviour) as a purely cultural phenomenon. They suggested that a social rule that favours helping individuals who are 'culturally similar' (that is, possess the same social rule) may be able to overwhelm the intrinsic genetic drag created by individuals' natural selfishness. Even if the behaviour in question is costly to the actor, the beneficiaries are individuals who have adopted the same rule, and this will increase the likelihood that the rule will be transmitted to the next generation (thereby enlarging the cohort of altruistic individuals). Allison (1992) extended this argument to show that, when transmission is horizontal (peer–peer) or oblique (teacher–pupil), the level of (cultural) relatedness within a group (that is, the extent to which individuals share cultural rules by inheritance from a common source) depends critically on the extent to which individuals acquire their cultural models from exogenous sources (that is, from individuals who have a different cultural rule). When there are several equally effective exogenous cultural parents (that is, models), the equilibrium level of cultural 'relatedness' is an inverse function of group size (the bigger the social group, the lower the mean level of cultural relatedness). However, when there is a single external model (a single trend-setter or fashion leader), then group size has little effect on the equilibrium level of relatedness.

HOW FAST DOES CULTURE CHANGE?

Although there has been considerable interest in how innovations diffuse through communities (for example, Hagerstrand 1967, Rogers and Shoemaker 1971, Brown 1981), most studies have been concerned with geographical and socio-economic processes rather than with cultural diffusion in the sense considered here. Brown's (1981) extensive review emphasises the importance of economic pressures in promoting the adoption of innovations, especially when this is managed by organisations that seek to promote or control the spread of the innovation. Nonetheless, we might expect the speed of change to be a function of how intrusive these effects are in terms of their influence on individual fitness.

Voland et al. (1997) provide us with a case study of the spread of a behavioural rule, namely those child-rearing practices that influence the preference for (and hence the relative mortality of) sons versus daughters. They found that, within populations over time, shifts in the preference for sons versus daughters occurred in response to changes in demographic parameters (specifically population growth rate) in the previous generation. For two separate populations, the correlation between the two variables was highest when there was a lag of 30 years between the precipitating stimulus (a change in local population growth rates) and the response (preference for sons), and lowest for lags of 0 and 60 years (**Figure 13.7**). One interpretation of these data is that people assess the rules of behaviour (in this case, informal rules on how to treat sons and daughters) handed down by their own parents in the light of how well these rules worked for their parents. If there appears to be a mismatch between the rule and the environmental conditions that the rule was designed to cope with, then the offspring generation adopts an alternative rule that seems a better fit.

Another example is provided by Borgerhoff Mulder (1995). There was a striking tendency for the bridewealth extracted from potential Kipsigis grooms who had received at least secondary education to fall during the 1980s as brides' parents began to view educated husbands as advantageous for their daughters (and thus worth offering

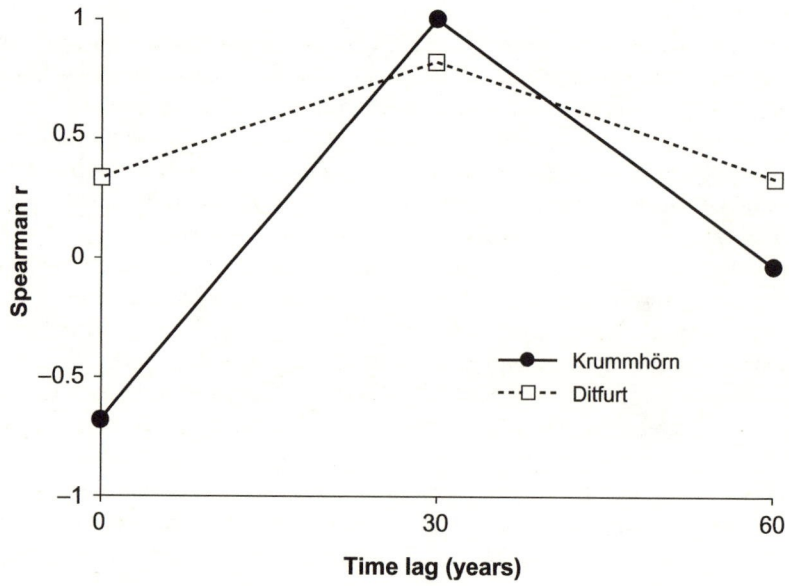

Figure 13.7 Spearman correlation between an index of male-biased investment in offspring among farmers and population growth rates at different time lags in the past for two north German historical populations (Ditfurt and Krummhörn). Voland et al. (1997) showed that when population growth rates are high, farmers invest more heavily in their sons because they can use them to acquire more land to establish new farms, but when populations are stagnant (because they are unable to expand into new land), they invest more heavily in daughters (Fig. 7.16). The results shown here suggest that there is typically a generation's lag between a change in the ecological driving variable (population growth rate) and the farmers' response (a switch in preference from one sex to the other). Redrawn with permission from Voland et al. (1997).

especially good bargains to as inducements to marry their daughters). Since significant numbers of young men with secondary education only began to become available during the 1960s, this suggests a lag effect in the order of about two decades.

The literature on the diffusion of innovations provides some examples of the natural spread of innovations in relatively uncontrolled environments (that is, in the absence of external pressures or controls). Carter and Williams (1957), for example, noted that there was a 40-year time lag between the first successful use of the tunnel oven in the English potteries and its widespread adoption by pottery firms. Brown (1981) found a 40-year lag in the spread of commercial dairying in a rural part of Mexico during the middle years of the twentieth century and a 30-year lag in the adoption of coffee by native farmers in the Kisii district of Kenya. In the latter study, Brown reviewed data on the adoption of six agricultural crops (coffee, tea, pyrethrum, hybrid maize, passion fruit and grade cattle), but only one (coffee) exhibited the classic S-shaped curve suggestive of saturation. The others (whose adoption periods ranged from 10-20 years) all had steeply accelerating curves, suggesting that they had yet to reach saturation (the point where everyone has adopted the habit).

The case of teddy bear evolution provides us with another example (**Box 13.4**). Here, we can actually see the sequence of changes that have taken place as the design has evolved (**Figure 13.5**). The snout length data show a particularly clear relationship in which the index follows a negative exponential curve as it approaches its new lower equilibrium value at a ratio of about 1.1 from its initial starting point

BOX 13.6

What happens when cultural change is too slow?

The Vikings maintained a small but viable colony in the southern tip of Greenland for about 400 years from the last decade of the tenth century AD. At its height, the colony numbered some 3000 souls scattered among 280 farmsteads along the south-western coastal fringe; by the twelfth century, it even boasted its own bishop and parliament. But some time after 1408 (when the last ship left Greenland to return to Europe), the colony died out. The archaeological evidence suggests that the colony finally succumbed to starvation as the Little Ice Age set in during the middle decades of the fifteenth century (McGovern 1981). Skeletal remains from the cemeteries indicate increasing nutritional stress as the century progressed.

It seems that the Greenland Vikings were un-able to let go of their Scandinavian farming practices in order to adopt the more successful hunting lifestyle of the Inuit from the north of Greenland with whom they were just beginning to come into contact. Despite the fact that their European farming practices were becoming increasingly less viable due to the deteriorating climatic conditions, the Vikings seem to have dismissed the pagan Inuit and their culture as beneath consideration for a (by then) devoutly Christian society. It seems likely that the pace of climate change hit the unfortunate Vikings at a rate too fast for them to be able to recognise and adapt to; as a result, they paid the inevitable price. Meanwhile, the Inuit, with their Arctic-adapted culture, continued to maintain viable populations in Greenland right through the next six centuries to the present day.

The case of the neuro-degenerative disease *kuru* among the Fore people of New Guinea provides another possible example of the same phenomenon (Durham 1991). The Fore appear to have acquired the custom of mortuary feasts (in which the flesh of the dead are consumed by the mourners) from neighbouring tribes some time in the nineteenth century. Unfortunately, it seems that one member of their community became infected with a scrapie-like disease and passed it on to his relatives when they consumed him at his funeral some time in the 1920s. Since the brain seemed to be the main site of the infection and the deceased's brain was consumed primarily by women at mortuary feasts, women were the main victims. However, the latency of onset of the disease turns out to be unusually long (up to 23 years), so that the Fore never quite made the connection between cannibalism and the disease. Instead, they blamed sorcery among their enemies and in due course began to take steps to reduce the risk of sorcery (some of which had the incidental effect of reducing exposure to infected brains, thereby confirming their own assessment of the problem). It was only when the government prohibition of cannibalism took effect in the 1970s that the incidence of *kuru* finally declined, thereby saving the Fore from certain extinction.

at 1.5. The transition appears to have been completed by about 1940, 35 years after the first teddy bears went on sale.

Linguistic phenomena are likely sources of data on cultural change, but few studies have been carried out within communities. Studies of dialect patterns within New England communities, for example, suggest that time periods in the order of several decades may be involved (Labov 1972). There are broad studies of largescale changes such as the English verb changes during the late medieval period where changes in pronunciation or verb form took upwards of two centuries to spread throughout England as a whole (Krygier 1994). However, this represents the time taken for the change in pronunciation to spread from its point of inception in the north of England across the whole country, a north–south distance of around 350 km (equivalent to a rate of spread of about 44 km per 25-year generation). How that translates into a rate of adoption *within* a community remains uncertain. However, since this value presumably represents the diameter of a circle centred on one's home village, a radius of 20–25 km coincides surprisingly closely with the typical distances between the birthplaces of husbands and wives in traditional communities (the distance individuals are prepared to go in search of a spouse). Cavalli-Sforza (2000), for example, gives values of 30–40 km for tropical hunter-gatherers and 10–20 km for low population density African farmers, while Trudgill (1999) notes that traditional English dialects can be assigned to areas with a radius no larger than about 30 km. It also coincides surprisingly well with the estimated rate of spread of agriculture through Europe during the Neolithic period (approximately 1 km per year, or 25 km per generation), although the evidence in this case supports population migration rather than simple cultural diffusion as the mechanism of advance (Ammerman and Cavalli-Sforza 1984).

Taken together, these examples suggest that shifts in cultural pattern occur on a timescale of a single human generation (25–30 years). In most of these cases, these shifts in behavioural rule are responses to changing economic conditions and are thus likely to be under significant levels of selection. These rates of change appear slow by comparison with everyday experience, but are in fact extremely fast by the standards of genetic evolution: the single generation required for a cultural mutation to spread right the way through a population contrasts starkly with the general rule of thumb that it takes about 1000 generations for a new mutant gene to achieve fixation under conventional levels of selection.

All this notwithstanding, one point that does need to be born in mind is that the rate of cultural change will depend heavily on the frequencies with which individuals are exposed to new cultural memes. All of the examples we have considered above have been deliberately chosen to reflect the situation in traditional societies, where contacts between individuals from different communities are few and far between. In other words, they were chosen to reflect the conditions that have applied throughout most of our evolutionary history. That situation has been radically changed in the past century by the advent of mass global communication. The significance of this is that new memes can spread rapidly over very large distance in a very short space of time; in addition, individuals may be repeatedly bombarded with a new idea through the media. The effect of this is likely to be a dramatically increased rate of dispersal, such that memes that once took a generation to spread through a population can now do so in a matter of weeks.

More generally, of course, the situation will depend on whether or not a meme is under any form of selection. When cultural rules are subject to neutral selection, the situation may be very different. In these cases, drift alone is responsible for evolutionary change and this is likely to be slower. In addition, the rate of change will depend crucially on a number of factors, including the mode of transmission, the probability of a successful copying on any given exposure and the degree of geographical or social substructuring (R. I. M. Dunbar, unpublished). In addition, the age at which cultural traits are acquired may be important. Hewlett and Cavalli-Sforza (1986) and Guglielmino et al. (1995) have examined the processes of cultural inheritance for a large number of cultural traits in a wide range of African cultures and have shown that the most common mode of inheritance is vertically from parents. When trait acquisition occurs very early in life (as when individuals grow up in a particular family environment), then these may be quite robust in their resistance to change later in life: as a result, particular social styles may persist across many generations. Examples include family structures and marriage patterns (monogamy vs polygamy, marital residence), kinship terminology and inheritance patterns. Le Bras and Todd (1981) noted a similar tendency for family style to be highly localised in modern

France (and, perhaps more remarkably, to correlate with the medieval origins of the populations concerned). As we have already noted, Cavalli-Sforza et al. (1982) found surprisingly high correlations between the beliefs and habits of adult offspring and their parents in the USA (**Table 13.2**).

One important class of cultural rules that we need to consider concerns instances when cultural rules are actively detrimental. Aunger (1994) provides what is probably the only fully analysed case of this kind. It concerns food avoidance practices among tribal peoples inhabiting the Ituri Forest, Congo. Aunger calculated that, for two forager societies and two horticultural societies, food avoidance practices resulted in a shortfall of about 1 per cent in the annual nutritional intake compared to what it would have been without adherence to specific avoidances. This is a small effect, and comes close to neutral selection that would allow avoidance rules to persist and evolve by drift. However, in one of the subpopulations, dietary rules did have a significant impact on fitness: women of two Sudenaic tribal groups in the sample (the Mamve and Lese) incurred a 5 per cent loss of lifetime fertility by adhering to their dietary practices. Aunger points out that what differentiates this subgroup from all the others in the sample is the fact that all its members practise virilocality: women leave their natal homes on marriage (often at a very early age) and thereby lack access to the key sources of information on dietary patterns (namely, their mothers and older female kin) that everyone else seems to rely on. They thus find themselves in an alien environment with only their childhood practices to rely on and therefore tend to perpetuate practices learned in childhood despite the fact that these may no longer be relevant in their marital locations (and may hence have a significant negative impact on their fitness). What we do not know in this case, however, is whether the observed effect is stable in the long-term. Have the Lese women only recently found themselves in this predicament or have they always behaved like this? If this is a novel situation for them, will they change their behaviour in time as they come to appreciate its adverse consequences?

An alternative possibility is that the Lese and Mamve may be examples of cultural practices that are inexorably leading to the eventual extinction of both these tribes. The models of cultural evolution by Cavalli-Sforza and Feldman (1981) and Boyd and Richerson (1985) both show that, under the appropriate conditions, memes can in fact drive genes to extinction. We can never tell whether this is this case until the process is well advanced (or, more accurately, until after the fact of their extinction).

CULTURE CHANGE AND GROUP SELECTION

One final point to consider is whether group selection could provide an explanation for cultural change in human populations. Group selection refers to the process of natural selection in which the group (or species) is the unit of selection: evolution occurs when entire groups are selected for or against because of the particular traits they possess or practise. Group selection was widely accepted as the principal form of natural selection prior to the 1960s, but current thinking emphasises the overriding importance of selection at the level of the individual (or, more properly, the gene). Although group selection is biologically possible, the conditions under which it can occur are very restrictive (Maynard Smith 1976). These conditions include fragmented populations with low inter-group migration rates (so preventing individuals introducing new genes into a group from outside), but high rates of both group fission and group extinction (which ensures that intense selection is imposed on entire lineages of genes that are less well adapted). These conditions are likely to be rare in the natural world. Wilson (1997, Wilson et al. 2000) has argued a particularly strong case for a role for group selection in human cultural evolution. There has been a long tradition within anthropology (known as *functionalism*) that has (in effect) endorsed this view.

Soltis et al. (1995) tested this proposition by asking whether the rates of group extinction from inter-tribal warfare in New Guinea were high enough to make cultural group selection a feasible proposition. Murders in the New Guinea highlands commonly precipitate long-running vendettas that often result in one or more communities being wiped out (or at least so reduced in numbers that the remnant population seeks shelter with, and is ultimately absorbed into, a neighbouring tribe). Among the Mae Enga, for example, there were 29 inter-community conflicts over a 50-year period, five of which resulted in the extinction of an entire community (Meggitt 1977). These yielded in community extinction rates in five tribal populations in New Guinea of 6.4–100 per cent (median 41 per cent) per century.

Cultural group selection would work in this context providing two conditions were fulfilled: (a) there was some mechanism to preserve between-group cultural isolation (so preventing leakage of cultural elements from one group to another – the equivalent of low migration rates in conventional genetic group selection models [see above]) and (b) the mechanism of group selection was sufficiently rapid to explain the observed rate of cultural change. Since community fission is the main way that new communities are created, with little between-community migration, the first condition is satisfied. Using this and the observed rates of community extinction, Soltis et al. (1995) calculated that it would take 40–80 generations for a rare cultural trait to successfully invade an established population. This represents a time span of 500–1000 years. Thus cultural group selection could be an explanation for changes that occurred over longer time intervals, but not those that occurred over shorter intervals (that is, less than 500 years). Soltis et al. concluded that cultural group selection is too slow to account for many of the documented cases of culture change. Earlier we noted that timescales in the order of a single generation seem to be common. Other examples suggesting timescales much shorter than 500 years for the spread of cultural innovations include the impact that the introduction of sweet potatoes had on New Guinea societies: many important cultural and social changes swept through New Guinea's diverse tribes in a matter of a generation or two following the sweet potato's introduction in the eighteenth century. Similarly, the introduction of horses into North America in the sixteenth century gave rise to the Plains Indian cultural complex in less than three centuries after horses first arrived (and probably a great deal less time after the Indians themselves first began to use horses).

Soltis et al. (1995) concluded that most cultural change takes place as a result of individual decisions about adopting new cultural practices or belief-systems, and that models like those of Boyd and Richerson (1985) based on biased inheritance provided a more appropriate explanation in most (even if not all) cases. In addition, they note that the results of this analysis imply that claims that cultural practices are in some way group-beneficial are empirically unsustainable.

*

It would be all but impossible to sum up in a few words everything that we have reviewed in this book. We have deliberately set out to explore the full richness of human experience and the role that evolutionary ideas have to play in elucidating that experience. But perhaps one important lesson can be drawn from all that we have argued and explored, and this is the inevitable complexity of human social behaviour. The evolutionary approach has often been seen by its critics as reductionist and narrow in its focus. We hope that we have dispelled that fundamental misunderstanding. The strategic decision-making approach demanded by the Darwinian perspective obliges us to ask about contextual variables, about the costs and benefits that individuals have to evaluate when deciding what to do. Precisely because of this we are forced to take a very broad view, drawing on the findings of many other disciplines from both the social and the biological sciences. The mistake, perhaps, has been to assume that Darwinian evolutionary theory necessarily implies genetic determinism. There is a very real sense in which Darwinian evolutionary theory provides the unifying framework that makes sense of all these other disciplines, giving a degree of robustness and integration that has hitherto never been present without necessarily detracting from the value and purpose of those other disciplines.

This raises a further issue. Just as with other animals, human behaviour is a complex phenomenon. This does not mean that we cannot explain it. The complexity arises because of two features that characterise all advanced biological systems, namely the large number of contextual variables that impact on the costs and benefits that dictate an individual's decisions and the constraints that time and energy budgets impose on absolute freedom of choice. At the same time, we contend in the strongest possible terms that the rules that guide human behaviour are relatively simple and – more importantly – are universal. The complexity of human behaviour derives not from these rules but from the wide variety of demographic, ecological and social environments in which individual humans seek to apply these rules.

In more conventional animal behaviour, the interaction between the contextual variables and the internal constraints (principally time and energy) commonly mean that multiple optima can be identified. Although in a perfect world, an individual animal might be able to find the perfect solution to the problems of survival and reproduction (and so maximise its fitness), in practice the world is far from perfect and constraints oblige animals to trade preferences on one dimension for preferences on the other dimensions that also have to be taken into account in its decision-making. An individual that devotes all its time to searching for mates would spend no time feeding or watching for predators, and (in most natural environments) its reproductive efforts would be in vain.

The fact that fitness is maximised by finding the best set of compromises to the individual problems of life and reproduction in the situation the individual finds itself in provides more than ample opportunity for alternative strategies to emerge (Dunbar 1983). The result is often a confusing mass of alternative strategies being pursued by different animals (sometimes within the same social group). That confusion, however, fades away once we appreciate that the real world forces individuals (and, of course, species as collections of individuals) to compromise on ideal behavioural strategies in order to reach the best outcome. We continue to find it surprising that protagonists in the debates over the Darwinian approach seem content to accept that phenotypic flexibility is characteristic of animals in general, but promptly ignore this possibility when they turn to discuss humans, surely the most phenotypically flexible of all species.

The issue of phenotypic flexibility should remind us that humans bring with them an additional layer of complexity that is not present in other species, namely the cultural transmission of rules of behaviour. The issue here is not so much that culture of itself changes the criteria by which human behaviour needs to be judged.

Taken on its own merits, culture merely reflects the addition of social learning to conventional genetic inheritance as the mechanism for the trans-generational transmission of information. Were that all there was to it, genetic fitness would remain the sole guiding principle by which humans (and the observing scientists) would seek to judge the fitness of behavioural decisions. The real sea change is introduced by the fact that social transmission (the mechanism underpinning culture) opens up the possibility that behaviour itself may become involved in its own non-genetic fitness race. In other words, the fact that culture may (*but need not!*) have its own fitness criterion (memic fitness) raises the possibility that a particular human behaviour may represent a case of memic fitness-maximising rather than being part of the more conventional genic fitness game.

We tend to ignore this inconvenient complexity when studying human behaviour, perhaps for good heuristic reasons (we *ought* to establish that there are no genetic fitness considerations involved before concluding that a particular phenomena represents a case of memic fitness being maximised). Perhaps our understanding of the role of more conventional genetic fitness in human decisions is now sufficiently good that we can begin to take more serious cognisance of its cultural counterpart.

Chapter summary

- Culture can be defined as rules of behaviour and ideas that are transmitted from one individual to another by social learning (including teaching). The extent to which this phenomenon is developed in humans is unique among animals.

- Cultural behaviour can be studied from a Darwinian perspective, although few such studies have in fact been carried out to date. The units of selection and transmission for cultural phenomena are often referred to as *memes*; their properties share some similarities with conventional biological genes, but also have a number of differences (including the fact that transmission can occur horizontally and obliquely as well as vertically).

- Cultural rules of behaviour are adaptive in that they can be shown to be designed to solve ecological and reproductive problems that individual social groups face. Because such rules are learned, cultural traits can change faster in response to changing environmental conditions than can genetic traits. Nonetheless, rates of cultural change appear to be on the order of a generation.

- When memes are under neutral genetic selection, their evolutionary history is subject to change as a result of either drift or selection within a memic universe.

- Three general models of cultural evolution have been developed (gene–culture co-evolution, phenogenotype co-evolution and the dual inheritance model). These differ in the assumptions they make about the interaction between genetic and memetic fitness and about the process of inheritance (or transmission from one generation to the next).

- Although human culture has been suggested as a likely example of group selection, the only direct test of this so far has produced negative results. Cultural phenomena change too fast to evolve by conventional group selection processes alone.

Further reading

Aunger, R. (ed.) (2000). *Darwinising Culture*. Oxford: Oxford University Press.

Boyd, R. and Richerson, P. J. (1985). *Evolution and Human Culture*. Chicago: University of Chicago Press.

Durham, W. H. (1991). *Coevolution: Genes, Culture and Human Diversity*. Stanford University Press: Stanford.

Sober, E. and Wilson, D. S. (1998). *Do Unto Others: the Evolution and Psychology of Unselfish Behaviour*. Cambridge (MA): Harvard University Press.

Sperber, D. (1996). *Explaining Culture*. Oxford: Blackwell.

Tomasello, M. (2001). *The Cultural Origins of Human Cognition*. Cambridge (MA): Harvard University Press.

Glossary

Adaptively relevant environment A term coined by Irons (1998) to describe the key environmental variables that selected for human behavioural and psychological traits and whose presence is required for these traits to be expressed adaptively.

Affinal kin Individuals related through marriage, not by blood.

Age-specific fertility The average number of female children produced in a population by women of a given age. If both sexes of children are included, it should be termed the *birth rate*.

Allele One of the alternative forms of a gene. An everyday example is the gene for eye colour, which has two forms (or alleles), blue and brown.

Altricial Condition in which young are underdeveloped and helpless at birth.

Amino acids Small organic molecules which form the building blocks of proteins.

Anisogamy Condition in which sex cells (gametes) differ from each other. Eggs are typically much larger than sperm.

Anthropoid primates The sub-order of primates that includes the monkeys and apes.

Australopithecines A sub-family of extinct hominids (the lineage to which modern humans belong) that lived between 4.5–1.0 million years ago. The fossils are characterised by adaptations for bipedal locomotion (walking on two legs), a small ape-like brain and robust jaws and teeth.

Baldwin effect Process by which phenotypic behavioural change leads to genetic changes in a population. The evolution of lactose tolerance in certain human populations is an example of this effect.

Bateman's principle This states that the variance in male reproductive success is much greater than that of females, so that males have greater reproductive potential than females. Male reproductive success increases with the number of matings achieved, whereas female reproductive success does not increase after the first mating.

Biological Market Theory A theory which states that interactions between cooperating animals can be likened to an economic market, with the value of the commodity exchanged (food, sex, protection) varying according to its supply and demand.

Biomass The weight of all the organisms forming a given population.

Body mass index (BMI) A measure of weight scaled for height. It is calculated as $weight/height^2$, in kilograms and metres respectively.

Bounded rationality Assumes that individuals do not make decisions based on a complete (or 'perfect') knowledge of all possible variables affecting the decision (as assumed in classical rationality) but that they use limited information to make their choices.

Bridewealth Money and/or goods transferred from the groom's family to the bride's family upon marriage.

Catarrhine primates The monkeys and apes of the Old World (Africa and Asia). From the Latin meaning 'narrow nosed'.

Collective action problem When goods are public and cannot be monopolised or denied to individuals even when they don't contribute to producing the goods in the first place, the economically rational choice for an individual is not to contribute to the pool of resources. However, if too many individuals adopt this strategy then the goods will no longer be available to any individuals (see also **Tragedy of the Commons**).

Common pool resources Public goods that cannot be monopolised by or denied to individuals, even when they don't contribute to producing the goods in the first place (see also **Collective action problem, Tragedy of the Commons**).

Demographic transition The decline in mortality and fertility over the past 150 years in Europe, resulting in completed family sizes that are typically at or below replacement rate (two offspring per couple) and a zero (or even negative) population growth rate.

Diploid The condition of having two sets of chromosomes present within the nucleus of a cell, one inherited from the female parent and one from the male parent.

DNA An abbreviation of deoxyribonucleic acid, a complex molecule which encodes genetic information and is the basis of genetic inheritance in all known organisms.

Domain-general processes Psychological mechanisms, such as category formation, memory and learning, that are independent of the content of what is to be learned or remembered.

Domain-specific processes Psychological mechanisms that function only in response to highly specific environmental stimuli and events.

Dual inheritance theory (DIT) A theory which states that culture and biology are separate but interacting processes affecting human behavioural response. The interaction between the two can produce results that are predicted by neither process on its own.

Ecological validity Ensuring that psychological experiments and tests are sufficiently similar to conditions that pertain in the real world, such that individuals' responses are the result of the effect that the experiment is designed to investigate, and are not distorted by the artificial nature of the set-up.

Enculturation The process by which humans (and some animals, such as chimpanzees) come to understand and adopt the social conventions of the society in which they live.

Environment of Evolutionary Adaptedness (EEA) The conglomeration of selection pressures that have operated on humans over the course of evolution.

Environmentalist In the context of child development, this refers to the view that children obtain all their knowledge of the world from the outside influences of the environment. At its most extreme, it assumes that children's minds are *tabula rasa* (blank slates) and that they are born with no knowledge whatsoever (see also **Nativism**).

Epigenetics The way in which an organism develops as the product of the interaction between genes and environment. May also refer to the notion that organisms are born with a set of instructions that are translated within a particular environment, rather than being performed.

Evolutionarily Stable Strategy (ESS) A strategy which, if adopted by all individuals within a population, cannot be invaded by any other strategy.

Extra-pair copulations Matings that take place outside of a monogamous relationship.

Fission-fusion social system A fluid arrangement whereby social groups break up and rejoin according to the prevailing environmental conditions.

Fitness The number of copies of a particular gene contributed to future generations.

Fluctuating asymmetry Small, but measurable, deviations in the bilateral symmetry of body parts. Across individuals, the direction of asymmetry (which side is larger or smaller) varies, hence 'fluctuating'.

Frequency dependent selection A process by which the evolutionary strategy adopted by an organism is dependent on the strategy adopted by all other members of the population.

Gene A length of DNA coding for a particular protein. Can also refer to phenotypic traits.

Genetic drift A process by which genetic change takes place within a population as a result of who happens to reproduce, when this is not influenced by the force of natural selection and so is essentially random. Genetic drift is more prominent in small populations than large ones.

Genomic imprinting Genes that are expressed or inactivated depending on which parent they are inherited from. Genes that are expressed only when inherited from the father are said to be maternally imprinted, while those that are expressed when inherited from the mother are paternally imprinted.

Genotype The genetic composition of an organism.

Hamilton's Rule As part of kin selection theory, a formula which demonstrates that individuals should perform beneficial acts for their kin, provided that the benefits exceed the costs when devalued by the degree of relatedness between the actor and the recipient.

Handicap principle As part of sexual selection theory, this states that exaggerated morphological and behavioural traits are displayed only by those individuals of sufficient quality to bear the survivorship costs of such a trait.

Heritability The proportion of phenotypic variation in a population that can be attributed to genetic differences between individuals.

Hominid All species that belong to the family Hominidae, which includes *Australopithecus* and *Homo*.

Hypergyny When women marry men who are above them in a socio-economic scale.

Inclusive fitness A measure that combines an individual organism's direct genetic contribution to future generations through its own offspring, with the contribution made by aiding the reproduction of relatives (who share the same genes by common descent), minus the contribution made by relatives to the organisms own reproductive output.

Independent contrasts An analytical technique which allows comparative tests to be made across species in order to investigate evolutionary patterns. It controls for the confounding effect that species may show similar traits as a consequence of common descent, rather than as a consequence of a similar response to a specific selection pressure.

Inter-birth interval The period of time between successive live births.

Intersexual selection A form of sexual selection that is driven by female choice of suitable mating partners.

Intrasexual selection A form of sexual selection driven by competition within the same sex for suitable mating partners. In most cases, intrasexual selection acts more strongly on males than females.

Kin selection A theory stating that altruistic acts between relatives will be favoured by selection if the benefit to the recipient exceeds the cost to the donor, devalued by the degree of relatedness between them. (see also **Hamilton's rule**).

Lekking A mating pattern whereby males gather in large concentrations in particular areas of habitat where they display to, and mate with, females.

Lifetime reproductive success (LRS) The total number of living offspring contributed by an individual to the next generation.

Local resource competition (LRC) A process by which offspring compete for limited resources with one or both parents and have a negative effect on the parents' direct reproductive success. This effect may be sex-biased with only one sex affecting parental output.

Local resource enhancement (LRE) A process by which offspring help to increase one or both parent's direct reproductive success. The effect may be sex-biased with only one sex affecting parental output.

Machiavellian intelligence hypothesis A hypothesis that states that the increase in relative brain size seen within the primate order can be attributed to the selection pressures exerted by living in permanent social groups.

Major Histocompatibility Complex (MHC) Part of the genome involved in the generation of immune responses.

Marginal value theorem A theory which states that animals should leave a food resource when the amount of energy gained from feeding on the resource is lower than the amount of energy that would be gained by finding a new undepleted resource. The marginal value is the level of resource gain achieved per unit of effort put into acquiring the resource.

Matrilocal A social system whereby women remain in their natal group upon marriage and men leave their natal group to join their wives.

Meme A unit of cultural information that replicates itself in human minds and is transmitted by imitation and teaching.

Mendelian genes Genes (or, strictly speaking, characters) that operate according to Mendel's principles of inheritance. Mendel was an Austrian monk who conducted plant breeding experiments and demonstrated that each the observed traits of organisms are determined by the inheritance of two genes, and that each of these genes is equally likely to be transmitted when sex cells are formed in an organism.

Modularity A term coined by Fodor (1983) which views certain perceptual processes as functionally independent, automatic systems that operate unconsciously to channel relevant information to higher cognitive processes. The notion of modularity has been extended by Cosmides and Tooby (1992) and Sperber (1994) to include the higher cognitive processes themselves (much to Fodor's dismay).

Modules See **Modularity**

Molecular clock The use of genetic mutations to estimate the amount of time elapsed since two species diverged from their common ancestor. This assumes that the rate at which genetic mutations occur remains roughly constant over time, so that the difference in the number of mutations between any two species is also a measure of time.

Monomarital principle In polyandrous social systems, the practice of allowing only one marriage to take place in a family per generation (see **Polyandry**).

Multi-regional hypothesis The notion that modern humans arose from a number of different, geographically isolated *Homo erectus* populations that left Africa around 1.0–1.5 million years ago and distributed themselves across the globe (see **Out of Africa hypothesis**).

Nativism In child development, the notion that babies are born with a fundamental knowledge of the world, such as an understanding of gravity and motion. Also, argues that certain human attributes, such as language, are hard-wired in the brain from birth.

Neoteny The retention of juvenile features into adulthood. In humans, neotenous facial features are considered highly attractive in females but not in males.

Neutral mutation theory A theory that describes how genetic change occurs by the process of random genetic drift alone. It also assumes that many mutations have no physical effects or consequences (positive or negative) for the organisms that carry them and that the accumulation of these mutations over time can help pinpoint the time at which speciation occurred (see **Genetic drift, Molecular clock**).

Ontogeny The process of development, covering all the events that occur between the fertilisation of the egg and the achievement of final adult form.

Opportunity cost The cost that an organism incurs by taking part in one activity at the expense of another.

Optimality models Analytical models that predict how organisms behave based on the assumption that they are attempting to maximise their inclusive fitness, or some other measure that is closely related to fitness (for example, rate of energy gain per unit time in foraging models).

Out of Africa hypothesis The notion that all modern humans evolved from a single population of archaic *Homo sapiens* that moved out of the African continent around 100 000–200 000 years ago (see also **Multi-regional hypothesis**).

Parental investment Any investment by a parent in an individual offspring that increases the offspring's chance of surviving but imposes a cost in terms of the parent's ability to invest in other offspring.

Patrilocal A social system whereby women leave their natal group upon marriage to live in their husband's natal group (see also **Matrilocal**).

Peptides Molecules composed of two or more amino acids. They play a role in the immune, endocrine (hormonal) and nervous systems, and provide a link between all three.

Phenogenotype The expressed traits of an organism arising from the combined effects of biological, environmental and cultural influences.

Phenotype The morphological and behavioural traits displayed by an organism as a result of the interaction between its genotype and the environment.

Phenotypic gambit The tactic used by scientists to test hypotheses about the function of behaviour. This assumes that behaviour is adaptive and serves to increase an organism's inclusive fitness. It allows researchers to ignore the effects of processes such as random drift, and the underlying psychological mechanisms that produce behaviour so as to concentrate on their reproductive outcomes.

Phenotypic plasticity The ability of a single genotype to manifest itself in different morphological forms or display different behaviours in response to variation in environmental conditions.

Phylogenetic inertia The inheritance of features by common descent from an ancestor that are retained but do not necessarily represent an adaptive response by the organism to its environment in the present day.

Pleistocene A geological period that began around two million years ago and ended 10 000 years ago.

Polyandry A mating system in which a single female mates with several males.

Polygynandry A mating system in which males and females mate with each other promiscuously.

Polygyny A mating system in which a single male mates with several females.

Polygyny threshold model A model which shows that when males are able to monopolise the resource that females need for reproduction, it pays females (in terms of fitness) to mate polygynously than to accept a monogamous mate of lower quality if the differences between males in their access to resources is great enough.

Primary representations An individual's true beliefs about the world.

Primogeniture A system of wealth inheritance whereby the first-born son inherits the entire family estate.

Prosody The non-linguistic melodic components of speech, such as pitch and tone, that help to convey meaning and emotional content.

Public goods problem see **Collective action problem**.

Reciprocal altruism The exchange of beneficial acts between individuals over time. The benefit of the acts to the recipients must be greater than the costs to the actor.

Representational Redescription (RR) A hypothetical process by which information originally contained within cognitive modules becomes redistributed across all modules.

Second to fourth digit ratio (2D:4D) The ratio of the length of the index finger to the ring finger. Low ratios characterize high levels of pre-natal testosterone.

Secondary representations The attitudes an individual adopts towards his or her primary representations. That is, the beliefs they hold about their own beliefs (see also **Primary representations**).

Semantic The meaning that words convey.

Sensory Bias hypothesis This states that many of the displays/signals used in animal mating systems are the result of the exploitation of existing biases in the receiver's sensory system.

Show-off hypothesis The hypothesis proposed by Kristen Hawkes suggesting that, within hunter-gatherer societies, men's foraging goals may be strongly linked to their mating strategies.

Social brain hypothesis see **Machiavellian intelligence hypothesis**.

Somatic markers The 'gut feelings' and sensations produced by the body in response to an individual's experiences. These become juxtaposed with certain mental states and the combination, when triggered at a later date, can be used to guide decision making, particularly in the social domain.

Stochastic dynamic programming A technique used to model decision-making whereby both the unpredictability of the environment and the organism's physical condition are able to vary and influence the decisions made.

Theory of Mind (ToM) The ability to attribute mental states to other individuals. To have a theory about the underlying mental processes (thoughts, beliefs, desires) that drive other individuals' behaviour.

Tolerated Theft A phenomenon whereby an individual will give up a valuable resource to another because the costs of defence of the resource outweigh the benefits of continuing to monopolise it.

Tragedy of the Commons A term coined by Hardin (1968) to describe the problems inherent in a system where public goods cannot be monopolised by or denied to individuals, even when they don't contribute to producing the goods in the first place (see also **Collective action problem, Common pool resources**).

Trivers-Willard effect When parents vary in their quality and offspring cost different amounts to rear effectively, a parent will maximise its fitness by biasing investment towards the sex of offspring that provides the highest return (measured in terms of the offspring's own reproductive success or fitness) on the level of investment that the parent is capable of providing.

Ultimogeniture A system of wealth inheritance whereby the last-born son inherits the entire family estate (see also **Primogeniture**).

Ultra-sociality A term used to describe the large-scale co-operation (often between strangers) that characterises human societies.

Waist-to-hip ratio (WHR) A measure of the circumference of the waist relative to that of the hips; male ratios tend to be higher than those for females. Values of 0.7 in women are rated as most attractive (compared to values above and below this) by both males and females.

References

Abbott, D. H., Keverene, E. B., Moore, G. F. and Yodyinguad, U. (1986). 'Social suppression of reproduction in subordinate talapoin monkeys, *Miopithecus talapoin*', in: J. Else and P. C. Lee (eds) *Primate Ontogeny, Cognition and Behaviour*, pp. 329–41. Cambridge: Cambridge University Press.

Adolphs, R., Tranel, D. and Damasio, A. R. (1998). 'The human amygdala in social judgment'. *Nature*, 393: 470–4.

Aiello, L. C. (1993). 'The fossil evidence for modern human origins in Africa: a revised view. *American Anthropologist* 95: 73–96.

Aiello, L. C. (1996). 'Terrestriality, bipedalism and the origin of language', in: G. Runciman, J. Maynard Smith and R. I. M. Dunbar (eds) *Evolution of Social Behaviour Patterns in Primates and Man*, pp. 269–90. Oxford: Oxford University Press.

Aiello, L. C. and Dean, C. (1990). *An Introduction to Human Evolutionary Anatomy*. London: Academic Press.

Aiello, L. C. and Dunbar, R. I. M. (1993). 'Neocortex size, group size and the evolution of language'. *Current Anthropology* 34: 184–93.

Aiello, L. C. and Wheeler, P. (1995). 'The expensive tissue hypothesis: the brain and the digestive system in human evolution'. *Current Anthropology* 36: 199–221.

Alexander, R. D. (1974). 'The evolution of social behaviour'. *Annual Review of Ecology and Systematics* 5: 325–83.

Alexander, R. D. (1979). *Darwinism and Human Affairs*. Seattle: University of Washington Press.

Alexander, R. D. (1987). *The Biology of Moral Systems*. Hawthorne: Aldine de Gruyter.

Alexander, R. D. and Noonan, K. M. (1979). 'Concealment of ovulation, parental care, and human social evolution', in: N. Chagnon and W. Irons (eds) *Evolutionary Biology and Human Social Behaviour: an Anthropological Perspective*, pp. 436–53. North Scituate (MA): Duxbury.

Alexander, R. D., Hoodland, J. L., Howard, R. D., Noonan, K. M. and Sherman, P. W. (1979). 'Sexual dimorphisms and breeding systems in pinnipeds, ungulates, primates, and humans', in: N. A. Chagnon and W. Irons (eds) *Evolutionary Biology and Human Social Behavior: an Anthropological Perspective*, pp. 402–35. North Scituate (MA).: Duxbury Press.

Allison, P. D. (1992). 'Cultural relatedness under oblique and horizontal transmission'. *Ethology and Sociobiology* 13: 153–69.

Almor, A. and Sloman, S. (1996). 'Is deontic reasoning special?' *Psychological Review* 103: 374–80.

Altmann, J. (1980). *Baboon Mothers and Infants*. Cambridge (MA): Harvard University Press.

Ammerman, A. J. and Cavalli-Sforza, L. L. (1984). *The Neolithic Transition and the Genetics of Populations in Europe*. Princeton: Princeton University Press.

Anderies, J. M. (1996). 'An adaptive model for predicting !Kung San reproductive performance: a stochastic dynamic modelling approach'. *Ethology and Sociobiology* 17: 221–45.

Anderson, J. L., Crawford, C. B., Nadeau, J. and Lindberg, T. (1992). 'Was the Duchess of Windsor right? a cross-cultural study of the socioecology of ideals of female body shape'. *Ethology and Sociobiology* 13: 197–227.

Anderson, K. G., Kaplan, H. and Lancaster, J. (1999a). 'Paternal care by genetic fathers and step-fathers: I. reports from Albuquerque men'. *Evolution and Human Behaviour* 20: 405–31.

Anderson, K. G., Kaplan, H., Lam, D. and Lancaster, J. (1999b). 'Paternal care by genetic fathers and stepfathers. II. Reports by Xhosa high school students'. *Evolution and Human Behaviour* 20: 4433–51.

Andersson, M. (1982). 'Female choice selects for extreme tail length in a widowbird'. *Nature, London*, 299: 818–820.

Andersson, M. (1994) *Sexual Selection*. Princeton (NJ): Princeton University Press.

Aoki, K. and Feldman, M. W. (1987). 'Toward a theory for the evolution of cultural communication: coevolution of signal transmission and reception'. *Proceedings of the National Academy of Scences, USA*, 84: 7164–8.

Aoki, K. and Feldman, M. W. (1989). 'Pleiotropy and preadaptation in the evolution of human language capacity'. *Theoretetical Population Biology* 35: 181–94.

Aoki, K. and Feldman, M. W. (1994). 'Cultural transmission of a sign language when deafness is caused by recessive alleles at two independent loci'. *Theoretical Population Biology* 45: 101–20.

Arcadi, A. C. (1996). 'Phrase structure of wild chimpanzee pant hoots: patterns of production and interpopulation variability'. *American Journal of Primatology* 39: 159–78.

Argyle, M. (1975). *Bodily Communication*. London: Methuen.

Arnold, K. E. (2000). 'Kin recognition in rainbow fish (*Melanotaenia eachamensis*): sex, sibs and shoaling'. *Behavioural Ecology and Sociobiology* 48: 385–91.

Asch, S. E. (1951). 'Effects of group pressure upon the modification and distortion of judgements', in: H. Guetzkow (ed.) *Groups, Leadership and Men*, pp. 177–90. Pittsburgh: Carbegie.

Asch, S. E. (1956). 'Studies of independence and conformity: a minority of one against a unanimous majority'. Psychological Monographs 70, No. 416.

Asfaw, B., White, T., Lovejoy, O., Latimer, B., Simpson, S. and Suwa, G. (1999). '*Australopithecus garhi*: a new species of early hominid from Ethiopia'. *Science* 284: 629–35.

Astington, J. W. (1993). *The Child's Discovery of the Mind*. Cambridge (MA): Cambridge University Press.

Aunger, R. (1994). 'Sources of variation in ethnographic interview data: food avoidances in the Ituri forest, Zaire'. *Ethnology* 33: 65–99.

Aunger, R. (ed.) (2000). *Darwinising Culture*. Oxford: Oxford University Press.

Axelrod, R. (1984). '*The Evolution of Cooperation*'. New York: Basic Books.

Axelrod, R. and Hamilton, W. D. (1981). 'The evolution of cooperation'. *Science* 221: 1390–96.

Bailey, A., LeCouteur, A., Gottesman, I., Bolton, P., Simonhoff, E., Yuzda, E. and Rutter, M. (1995). 'Autism as a strongly genetic disorder: evidence from a British twin study'. *Psychological Medicine* 25: 63–77.

Baize, H. R. and Schroeder, J. E. (1995). 'Personality and mate selection in personal ads: evolutionary preferences in a public mate selection process'. *Journal of Social Behavior and Personality* 10: 517–36.

Baker, R. (2000). *Sperm Wars*, 2nd edition. London: Pan.

Baker, R. R. and Bellis, M. A. (1989). Number of sperm in human ejaculates varies in accordance with sperm competition theory. *Animal Behaviour* 37: 867–869.

Baker, R. R. and Bellis, M. A. (1995). *Human Sperm Competition: Copulation, Masturbation and Infidelity*. London: Chapman and Hall.

Ball, H. L. and Hill, C. M. (1996). 'Reevaluating "twin infanticide"'. *Current Anthropology* 37: 856–63.

Barber, N. (1995). 'The evolutionary psychology of physical attractiveness: sexual selection and human morphology'. *Ethology and Sociobiology* 16: 395–424.

Barbujani, G. and Sokal, R. R. (1990). 'Zones of sharp genetic change in Europe are also linguistic boundaries'. *Proceedings of the National Academy of Sciences, USA*, 87: 1816–19.

Barker, D. J. P., Osmond, C. and Golding, J. (1990). 'Height and mortality in the counties of England and Wales'. *Annals of Human Biology* 17: 1–6.

Barkow, J. H. (1990). 'Beyond the DP/DSS controversy'. *Ethology and Sociobiology* 11: 341–51.

Barkow, J. H., Cosmides, L. and Tooby, J. (1992). *The Adapted Mind: Evolutionary Psychology and the Generation of Culture*. Oxford: Oxford University Press.

Barlow, D. (1995). 'Genomic imprinting in mammals'. *Science* 270: 1610–13.

Barnard, A. (1994). 'Rules and prohibitions: the form and content of human kinship', in: T. Ingold (ed.) *Companion Encyclopedia of Anthropology*, pp. 783–812. London: Routledge.

Baron-Cohen, S. (1989). 'The autistic child's theory of mind: a case of specific developmental delay'. *Journal of Child Psychology and Psychiatry* 30: 285–97.

Baron-Cohen, S. (1991a). 'The theory of mind deficit in autism: how specific is it?' *British Journal of Developmental Psychology* 9: 301–14.

Baron-Cohen, S. (1991b). 'Precursors to a Theory of Mind: understanding attention in others', in: A. Whiten (ed.) *Natural Theories of Mind*, pp. 233–52. Oxford: Blackwell.

Baron-Cohen, S. (1995). *Mindblindness: an Essay on Autism and Theory of Mind.* Cambridge (MA): MIT Press.

Baron-Cohen, S. (2000a). 'Autism: deficits in folk psychology exist alongside superiority in folk physics', in: S. Baron-Cohen, H. Tager-Flusberg and D. J. Cohen (eds) *Understanding Other Minds: Perspectives From Developmental Neuroscience*, pp. 73–82. Oxford: Oxford University Press.

Baron-Cohen, S. (2000b). 'The cognitive neuroscience of autism: evolutionary approaches', in: M. Gazzaniga (ed.) *The New Cognitive Neurosciences*, pp. 1249–58. Cambridge, (MA): MIT Press.

Baron-Cohen, S. and Hammer, J. (1997a). 'Parents of children with Asperger syndrome: what is the cognitive phenotype?' *Journal of Cognitive Neuroscience* 9: 548–54.

Baron-Cohen, S. and Hammer, J. (1997b). 'Is autism an extreme form of the male brain?' *Advances in Infancy Research* 11: 193–217.

Baron-Cohen, S., Leslie, A. M. and Frith, U. (1985). 'Does the autistic child have a theory of mind?' *Cognition* 21: 37–46.

Baron-Cohen, S., Ring, H. A, Moriarty, J., Schmitz, P. and Costa, D. P. (1994). 'Recognition of mental state terms: clinical findings in children with autism and a functional neuroimaging study of normal adults'. *British Journal of Psychiatry* 165: 640–9.

Baron-Cohen, S., Ring, H. A., Bullmore, E. T., Wheelwright, S., Ashwin, C., and Williams. S. C. R. (2000c). 'The amygdala theory of autism'. *Neuroscience and Biobehavioural Reviews* 24: 335–64.

Baron-Cohen, S., Wheelwright, S., Stott, C., Bolton, P. and Goodyer, I. (1997). 'Is there a link between engineering and autism?' *Autism* 1: 101–9.

Barrett, L. and Dunbar, R. I. M. (1994). 'Not now, dear, I'm busy'. *New Scientist* 142: 30–4.

Barrett, L. and Henzi, S. P. (2000). 'Are baboon infants Sir Phillip Sydney's offspring?' *Ethology* 106: 645–58.

Barrett, L., Dunbar, R. I. M. and Dunbar, P. (1995). 'Mother-infant contact as contingent behaviour in gelada baboons'. *Animal Behaviour* 49: 805–10.

Barrett, L., Henzi, S. P., Weingrill, T., Lycett, J. E. and Hill, R. A. (1999). 'Market forces predict grooming reciprocity in female baboons'. *Proceedings of the Royal Society, London, B*, 266: 665–70.

Barton, R. A. (1996). 'Neocortex size and behavioural ecology in primates'. *Proceedings of the Royal Society, London, B*, 263: 173–77.

Barton, R. A. and Dunbar, R. I. M. (1997). 'Evolution of the social brain', in: A. Whiten and R. W. Byrne (eds) *Machiavellian Intelligence II*, pp. 240–63. Cambridge: Cambridge University Press.

Bateman, A. J. (1948) 'Intra-sexual selection in Drosophila'. *Heredity* 2: 349–68.

Bateman, R., Goddard, I., O'Grady, R., Funk, V. A., Mooi, R., Kress, W. J. and Cannell, P. (1990). 'Speaking with forked tongue: the feasibility of recording human phylogeny and the history of language'. *Current Anthropology* 31: 1–24.

Bateson, P. P. G. (1994). 'The dynamics of parent offspring conflict relationships in mammals'. *Trends in Evolution and Ecology* 9: 399–403.

Bateson, P. P. G., Mendl, M. and Feaver, J. (1990). 'Play in the domestic cat is enhanced by rationing of the mother during lactation'. *Animal Behaviour* 40: 514–25.

Beattie, G. (1983). *Talk: an Analysis of Speech and Non-Verbal Behaviour in Conversation.* Milton Keynes: Open University Press.

Bechara, A., Damasio, A. R., Damasio, H. and Anderson, S. (1994). 'Insensitivity to future consequences following damage to human prefrontal cortex'. *Cognition* 50: 7–12.

Bechara, A., Tranel, H., Damasio, H. and Damasio, A. R. (1993). 'Failure to respond autonomically in anticipation of future outcomes following damage to human prefrontal cortex'. *Society for Neuroscience* 19: 791.

Becher, T. (1989). *Academic Tribes and Territories.* Milton Keynes: Open University Press.

Beise, J. (2001). *Verhaltensökogie menschlichen Abwanderungsverhaltens – am Beispiel* der historischen Bevolkerung der Krummhörn (Ostfriesland, 18 und 19 Jahrhundert. PhD thesis, University of Giessen (Germany).

Bellugi, U., Lichtenberger, L., Jones, W., Lai, Z. and St. George, M. (2000). 'The neurocognitive profile of Williams' syndrome: a complex pattern of strengths and weakness'. *Journal of Cognitive Neuroscience* 12 (supplement): 7–29.

Belovsky, G. E. (1987). 'Hunter-gatherer foraging: a linear programming approach'. *Journal of Anthopological Archaeology* 6: 29–76.

Belsky, J., Steinberg, L. and Draper, P. (1991). 'Childhood experience, interpersonal development and reproductive strategy: an evolutionary theory of socialisation'. *Child Development* 62: 647–70.

Bereczkei, T. (1998). 'Kinship network, direct childcare, and fertility among Hungarians and Gypsies'. *Evolution and Human Behavior* 19: 283–98.

Bereczkei, T. and Csanaky, A. (1996). 'Evolutionary pathway of child development: lifestyles of adolescents and adults from father-absent families'. *Human Nature* 7: 257–80.

Bereczkei, T. and Dunbar, R. I. M. (1997). 'Female-biased reproductive strategies in a Hungarian Gypsy population'. *Proceedings of the Royal Society, London, B*, 264: 17–22.

Bereczkei, T. and Dunbar, R. I. M. (in press). 'Helping-at-the-nest and sex-biased parental investment in a Hungarian gypsy population'. *Current Anthropology*.

Berk, L. S., Tan, S. A., Fry, W. F., Napier, B. J., Lee, J. W., Hubbard, R. W., Lewis, J. E. and Eby, W. C. (1989). 'Neuroendocrine and stress hormone changes during mirthful laughter'. *American Journal of Medical Science* 298: 390–6.

Berkman, L. F. (1984). 'Assessing the physical health effects of social networks and social support'. *Annual Review of Public Health* 5: 413–32.

Berté, N. A. (1988). 'K'ekchi' horticultural labor exchange: productive and reproductive implications', in: L. Betzig, M. Borgerhoff Mulder and P. Turke (eds) *Human Reproductive Behaviour: a Darwinian Perspective*, pp. 83–96. Cambridge: Cambridge University Press.

Bertram, B. C. R. (1982). 'Problems with altruism', in: Bertram, B. C. R., Clutton-Brock, T., Dunbar, R. I. M. and Wrangham, R. W. (eds) *Current Problems in Sociobiology*, pp. 251–68. Cambridge: Cambridge University Press.

Betzig, L. (1982). 'Despotism and differential reproduction'. *Ethology and Sociobiology* 3: 209–22.

Betzig, L. (1986). *Despotism and Differential Reproduction: a Darwinian View of History*. New York: Aldine de Gruyter.

Betzig, L. (1989a). 'Causes of conjugal dissolution: a cross-cultural study'. *Current Anthropology* 30: 654–76.

Betzig, L. (1989b). 'Rethinking human ethology: a response to some recent critiques'. *Ethology and Sociobiology* 10: 315–24.

Betzig, L. (1991). 'A little more mortar for a firm foundation'. *Behavioral and Brain Sciences* 14: 264.

Betzig, L. and Turke, P. (1986). 'Food sharing on Ifaluk'. *Current Anthropology* 27: 397–400.

Betzig, L., Borgerhoff-Mulder, M. and Turke, P. (eds) (1988). *Human Reproductive Behaviour: a Darwinian Perspective*. Cambridge: Cambridge University Press.

Bevc, I. and Silverman, I. (2000). 'Early separation and sibling incest: a test of the revised Westermarck theory'. *Evolution and Human Behavior* 21: 151–62.

Bickerton, D. (1990). *Species and Language*. Chicago: University of Chicago Press.

Bickerton, D. (1998). 'Catastrophic evolution: the case for a single step from protolanguage to full human language', in: J. R. Hurford, M. Studdert-Kennedy and C. Knight (eds) *Approaches to the Evolution of Language: Social and Cognitive Bases*, pp. 341–58. Cambridge: Cambridge University Press.

Binmore, K. (1994). *Game Theory and the Social Contract*. vol. 1: *Playing Fair*. Cambridge (MA): MIT Press.

Bird, R. (1999). 'Cooperation and conflict: the behavioural ecology of the sexual division of labour'. *Evolutionary Anthropology* 8: 65–75.

Birkhead, T. (2000). *Promiscuity: an Evolutionary History of Sperm Competition and Sexual Conflict*. London: Faber and Faber.

Bischof, N. (1972). 'The biological foundations of the incest taboo'. *Social Science Information* 11: 7–36.

Bischoping, K. (1993). 'Gender differences in conversation topics 1922–1990'. *Sex Roles* 28: 1–17.

Björntorp, P. (1991). 'Adipose tissue distribution and function'. *International Journal of Obesity* 15: 67–81.

Blackmore, S. (1999). *The Meme Machine*. Oxford: Oxford University Press.

Blanchard, R. and Bogaert, A. F. (1996). 'Homosexuality in men and number of older brothers'. *American Journal of Psychiatry* 153: 27–31.

Bliege Bird, R. and Bird, D. W. (1997). 'Delayed reciprocity and tolerated theft: the behavioural ecology of food-sharing strategies'. *Current Anthropology* 38: 49–78.

Bloom, P. and German, T. P. (2000). 'Two reasons to abandon the false belief task as a test of theory of mind'. *Cognition* 77: B25–B31.

Blurton Jones, N. (1984). 'A selfish origin for human food sharing: tolerated theft'. *Ethology and Sociobiology*. 5: 1–3.

Blurton Jones, N. (1986). 'Bushman birth spacing: a test for optimal interbirth intervals'. *Ethology and Sociobiology* 7: 91–105.

Blurton Jones, N. (1987). 'Bushman birth spacing: direct tests of some simple predictions'. *Ethology and Sociobiology* 8: 183–203.

Blurton Jones, N. (1990). 'Three sensible paradigms for the research on evolution and human behaviour'. *Ethology and Sociobiology* 11: 353–59.

Blurton Jones, N. (1994). 'A reply to Harpending'. *American Journal of Physical Anthropology* 93: 391–7.

Blurton Jones, N. and Sibly R. M. (1978). 'Testing adaptiveness of culturally determined behaviour: do bushman women maximise their reproductive success by spacing births widely and foraging seldom?' in: N. Blurton Jones and V. Reynolds (eds) *Human Behaviour and Adaptation*. London: Taylor and Francis.

Blurton Jones, N., Hawkes, K. and Draper, P. (1994). 'Foraging returns of !Kung adults and children: why didn't !Kung children forage?' *Journal of Anthropological Research*, 50: 217–48.

Blurton Jones, N., Hawkes, K. and O'Connell, J. F. (1989). 'Modeling and measuring costs of children in two foraging societies', in: V. Standen and R. A. Foley (eds) *Comparative Socioecology*, pp. 367–90. Oxford: Blackwell.

Blurton Jones, N., Hawkes, K. and O'Connell, J. F. (1999). 'Some current idea about the evolution of the human life history', in: P. C. Lee (ed.) *Comparative Primate Socioecology*, pp. 140–66. Cambridge: Cambridge University Press.

Blurton Jones, N., Smith, L., O'Connell, J. F., Hawkes, K. and Kamuzora, C. L. (1992). 'Demography of the Hadza, an increasing and high density population of savanna foragers'. *American Journal of Physical Anthropology* 89: 311–18.

Boehm, C. (1986). 'Capital punishment in tribal Montenegro: implications for law, biology and theory of social control'. *Ethology and Sociobiology* 7: 305–20.

Boesch, C. (1996). 'The emergence of culture among wild chimpanzees', in: W. Runciman, J. Maynard Smith and R. I. M. Dunbar (eds) *Evolution of Social Behaviour Patterns in Primates and Man*, pp. 251–68. Oxford: Oxford University Press.

Boesch, C. and Tomasello, M. (1998). 'Chimpanzee and human cultures'. *Current Anthropology* 39: 591–614.

Boone, J. L. (1988). 'Parental investment, social subordination and population processes among the 15th- and 16th-century Portuguese nobility', in: L. Betzig, M. Borgerhoff Mulder and P. Turke (eds) *Human Reproductive Behaviour: a Darwinian Perspective*, pp. 83–96. Cambridge: Cambridge University Press.

Borgerhoff Mulder, M. (1987). 'Cultural and reproductive success: Kipsigis evidence'. *American Anthropologist* 89: 617–34.

Borgerhoff Mulder, M. (1988a). 'Early maturing Kipsigis women have higher reproductive success than later maturing women and cost more to marry'. *Behavioral Ecology and Sociobiology* 24: 145–53.

Borgerhoff Mulder, M. (1988b). 'Kipsigis bridewealth payments', in: L. Betzig, M. Borgerhoff Mulder and P. Turke (eds) *Human Reproductive Behaviour: A Darwinian Perspective*, pp. 65–82. Cambridge: Cambridge University Press.

Borgerhoff Mulder, M. (1988c). 'Reproductive success in three Kipsigis cohorts', in: T. H. Clutton-Brock (ed.) *Reproductive Success*, pp. 419–38. Chicago: University of Chicago Press.

Borgerhoff Mulder, M. (1989). 'Reproductive consequences of sex-biased inheritance in Kipsigis', in: V. Standen and R. A. Foley (eds) *Comparative Socioecology*, pp. 405–27. Oxford: Blackwell.

Borgerhoff Mulder, M. (1990). 'Kipsigis women's preference for wealthy men'. *Behavioral Ecology and Sociobiology* 27: 255–64.

Borgerhoff Mulder, M. (1991). 'Human Behavioural Ecology', in: J. R. Krebs and N. B. Davies (eds) *Behavioural Ecology: An Evolutionary Approach*, 3rd ed. pp. 69–103. Oxford: Blackwell.

Borgerhoff Mulder, M. (1992). 'Reproductive decisions', in: E. A. Smith and B. Winterhalder (eds) *Evolutionary Ecology and Human Behaviour*, pp. 339–74. Hawthorne: Aldine de Gruyter.

Borgerhoff Mulder, M. (1994). 'Commentary on Smith and Smith'. *Current Anthropology* 35: 615–16.

Borgerhoff Mulder, M. (1995). 'Bridewealth and its correlates: quantifying changes over time'. *Current Anthropology* 36: 573–604.

Borgerhoff Mulder, M. (1998). 'The demographic transition: are we any closer to an evolutionary explanation?' *Trends in Ecology and Evolution* 13: 266–70.

Borgerhoff Mulder, M. (2000). 'Optimising offspring: the quantity-quality tradeoff in agropastoral Kipsigis'. *Evolution and Human Behaviour* 21: 391–410.

Boswell, J. (1988). *The Kindness of Strangers: the Abandonment of Children in Western Europe from Late Antiquity to the Renaissance*. New York: Random House.

Bott, E. (1971). *Family and Social Network*. London: Tavistock Publications.

Bourdieu, P. and Passeron, J. (1977). *Reproduction: Education, Society and Culture*. London: Sage.

Bowlby, J. (1969). *Attachment and Loss. Vol. 1: Attachment*. New York: Basic Books.

Bowlby, J. (1973). *Attachment and Loss, Vol. 2: Separation, Anxiety and Anger*. New York: Basic Books.

Bowman, L. A., Dilley, S. R. and Keverne, E. B. (1978). 'Suppression of oestrogen-induced LH surges by social subordination in talapoin monkeys'. *Nature, London*, 275: 56–8.

Boyd, R. (1995). 'Is the Prisoner's Dilemma a good model of reciprocal altruism'? *Ethology and Sociobiology* 9: 211–22.

Boyd, R. and Richerson, P. J. (1985). *Culture and the Evolutionary Process*. Chicago: University of Chicago Press.

Boyd, R. and Richerson, P. (1988). 'The evolution of reciprocity in sizeable groups'. *Journal of Theoretical Biology* 132: 337–56.

Boyd, R. and Richerson, P. (1992). 'Punishment allows the evolution of cooperation (and anything else) in sizable groups'. *Ethology and Sociobiology* 13: 171–95.

Boyd, R. and Richerson, P. J. (1994). 'Why does culture increase human adaptability?' *Ethology and Sociobiology* 16: 125–43.

Bradshaw, J. and Rogers, L. (1993). *The Evolution of Lateral Asymmetries, Language, Tool Use and Intellect*. New York: Academic Press.

Brothers, L. (1990). 'The social brain: a project for integrating primate behaviour and neurophysiology in a new domain'. *Concepts in Neuroscience* 1: 27–51.

Brothers, L. (1997). *Friday's Footprint: How Society Shapes the Human Mind*. Oxford: Oxford University Press.

Brothers, L. and Ring, H. (1992). 'A neuroethological framework for the representation of minds'. *Journal of Cognitive Neuroscience* 42: 107–8.

Brotherton, P. N. and Manser, M. B. (1997). 'Female dispersion and the evolution of monogamy in the dik-dik'. *Animal Behaviour* 54: 1413–24.

Brown, C. R., Brown, M. B. and Schaffer, M. L. (1991). 'Food-sharing signals among socially foraging cliff swallows'. *Animal Behaviour* 42: 551–64.

Brown, L. A. (1981). *Innovation Diffusion*. London: Methuen.

Brown, P. J. (1986). 'Cultural and genetic adaptations to malaria: problems of comparison'. *Human Ecology* 14: 311–32.

Brown, P. J. and Konner, M. (1987). 'An anthropological perspective on obesity'. *Annals of the New York Academy of Science* 499: 29–46.

Bruce, A. C. and Johnson, J. E. V. (1994). 'Male and female betting behaviour: new perspectives'. *Journal of Gambling Studies* 10: 183–98.

Bryant, F. C. (1981). *We're All Kin: a Cultural Study of a Mountain Neighbourhood*. Knoxville: University of Tennessee Press.

Buckle, L., Gallup, G. G. and Rodd, Z. A. (1996). 'Marriage as a reproductive contract: patterns of marriage, divorce and remarriage'. *Ethology and Sociobiology* 17: 363–77.

Bugos, P. and McCarthy, L. (1984). 'Ayoreo infanticide: a case study', in: G. Hausfater and S. B. Hrdy (eds) *Infanticide*, pp. 503–20. Hawthorne: Aldine de Gruyter.

Burley, N. (1979). 'The evolution of concealed ovulation'. *American Naturalist* 114: 835–58.

Burling, R. (1993). 'Primate calls, human language and nonverbal communication'. *Current Anthropology* 34: 25–54.

Burton, L. M. (1990). 'Teenage childbearing as an alternative life-course strategy in multigeneration black families'. *Human Nature* 1: 123–43.

Buss, D. M. (1985). 'Human mate selection'. *American Scientist* 73: 47–51.

Buss, D. M. (1987). 'Sex differences in human mate selection criteria: an evolutionary perspective', in: C. Crawford, D. Krebs and M. Smith (eds) *Sociobiology and Psychology: Ideas, Issues and Applications*, pp. 335–52. Hillsdale (NJ): Erlbaum.

Buss, D. M. (1989). 'Sex differences in human mate preferences'. *Behavioral and Brain Sciences* 12: 1–49.

Buss, D. M. (1994). *The Evolution of Desire: Strategies of Human Mating*. New York: Basic Books.

Buss, D. M. (1995). 'Evolutionary psychology: a new paradigm for psychological science'. *Psychological Inquiry* 6: 1–30.

Buss, D. (1999). *Evolutionary Psychology*. London: Allyn & Bacon.

Buss, D. M. and Barnes, M. F. (1986). 'Preferences in human mate selection'. *Journal of Personality and Social Psychology* 50: 559–70.

Buys, C. J. and Larsen, K. L. (1979). 'Human sympathy groups'. *Psychological Report* 45: 547–53.

Byrne, R. W. (1995). *The Thinking Ape*. Oxford: Oxford University Press.

Byrne, R. W. and Whiten, A. (eds) (1988). *Machiavellian Intelligence: Social Expertise and the Evolution of Intellect in Monkeys, Apes, and Humans*. Oxford: Oxford University Press.

Calvin, W. H. (1983). 'A stone's throw and its launch window: timing, precision and its implications for language and the hominid brain'. *Journal of Theoretical Biology* 104: 121–35.

Campbell, A. (1995). 'A few good men: evolutionary psychology and female adolescent aggression'. *Ethology and Sociobiology* 16: 99–123.

Cann, R. L., Stoneking, M. and Wilson, A. C. (1987). Mitochondrial DNA and human evolution. *Nature, London*, 325: 31–6.

Caporael, L., Dawes, R. M., Orbell, J. M. and van de Kragt, A. J. C. (1989). 'Selfishness examined'. *Behavioral and Brain Sciences* 12: 683–739.

Carey, S. and Spelke, E. (1994). 'Domain-specific knowledge and conceptual change', in: L. A. Hirschfeld and S. A. Gelman (eds) *Mapping the Mind: Domain Specificity in Cognition and Culture*, pp. 201–33. Cambridge: Cambridge University Press.

Carlson, S. M., Moses, L. J. and Hix, H. R. (1998). 'The role of inhibitory processes in young children's difficulties with perception and false belief'. *Child Development* 69: 672–91.

Caro, T. and Borgerhoff Mulder, M. (1987). 'Problem of adaptation in the study of human behaviour'. *Ethology and Sociobiology* 8: 61–72.

Carroll, J. (1995). 'Evolution and literary theory'. *Human Nature* 6: 119–34.

Carroll, J. (1998). 'Literary study and evolutionary theory: a review essay'. *Human Nature* 9: 273–92.

Carter, C. F. and Williams, B. R. (1957). *Industry and Technical Progress: Factors Governing the Speed of Application of Science*. Oxford: Oxford University Press.

Cashdan, E. (1993). 'Attracting mates: effects of paternal investment on mate attraction strategies'. *Ethology and Sociobiology* 14: 1–24.

Cashdan, E. (1996). 'Women's mating strategies'. *Evolutionary Anthropology* 5: 134–43.

Cashdan, E. (ed.) (1990). *Risk and Uncertainty in Tribal and Peasant Economies*. Boulder: Westview Press.

Casimir, M. J. and Rao, A. (1995). 'Prestige, possessions and progeny: cultural goals and reproductive success among the Bakkarwal'. *Human Nature* 6: 241–72.

Cavalli-Sforza, L. L. (2000). *Genes, People and Languages*. London: Allen Lane.

Cavalli-Sforza, L. L. and Bodmer, W. (1971). *The Genetics of Human Populations*. San Francisco: Freeman.

Cavalli-Sforza, L. L. and Feldman, M. (1981). *Cultural Transmission and Evolution: a Quantitative Approach*. Princeton: Princeton University Press.

Cavalli-Sforza, L. L. and Wang, W. S-Y. (1986). 'Spatial distance and lexical replacement'. *Language* 62: 38–55.

Cavalli-Sforza, L. L., Feldman, M. W., Chen, K. H. and Dornbush, S. M. (1982). 'Theory and observation in cultural transmission'. *Science* 218: 19–27.

Cavalli-Sforza, L. L., Menozzi, P. and Piazza, A. (1994). *History and Geography of Human Genes*. Princeton (NJ): Princeton University Press.

Cavalli-Sforza, L. L., Piazza, A., Menozzi, P. and Mountain, J. (1988). 'Reconstruction of human evolution: bringing together genetic, archaeological and linguistic data'. *Proceedings of the National Academy of Sciences, USA*, 85: 6002–6.

Chagnon, N. A. (1974). *Studying the Yanomamö*. New York: Holt Rhinehart and Winston.

Chagnon, N. A. (1975). 'Genealogy, solidarity and relatedness: limits to local group size and patterns of fissioning in an expanding population'. *Yearbook of Physical Anthropology* 19: 95–110.

Chagnon, N. A. (1979a). 'Mate competition, favoring close kin, and village fissioning among the Yanomamö Indians', in: N. A. Chagnon and W. Irons (eds) *Evolutionary Biology and Human Social Behavior: an Anthropological Perspective*, pp. 86–131. North Scituate (MA): Duxbury Press.

Chagnon, N. A. (1979b). 'Is reproductive success equal in egalitarian societies?', in: N. A. Chagnon and W. Irons (eds) *Evolutionary Biology and Human Social Behavior: an Anthropological Perspective*, pp. 374–401. North Scituate (MA): Duxbury Press.

Chagnon, N. A. (1988). 'Life histories, blood revenge, and warfare in a tribal population'. *Science* 239: 985–92.

Chagnon, N. A. and Bugos, P. (1979). 'Kin selection and conflict: an analysis of a Yanomamö axe fight', in: N. A. Chagnon and W. Irons (eds) *Evolutionary Biology and Human Social Behaviour: an Anthropological Perspective*, pp. 213–37. North Scituate (MA): Duxbury Press.

Chandler, M., Fritz, A. S. and Hala, S. (1989). 'Small-scale deceit: deception as a marker of 2-, 3- and 4-year olds' early theories of mind'. *Child Development* 60: 1263–77.

Charnov, E. (1976). 'Optimal foraging: the marginal value theorem'. *Theoretical Population Biology* 9: 129–36.

Charnov, E. L. (1993). *Life History Invariants*. Oxford: Oxford University Press.

Chase, P. G. and Dibble, H. (1987). 'Middle Palaeolithic symbolism: a review of current evidence and interpretations'. *Journal of Anthropological Archaeology* 6: 263–96.

Cheney, D. L. and Seyfarth, R. M. (1990). *How Monkeys See the World*. Chicago: University of Chicago Press.

Cheng, P. W. and Holyoak, K. J. (1989). 'On the natural selection of reasoning theories'. *Cognition* 33: 285–313.

Cheyne, W. M. (1970). 'Stereotyped reactions to speakers with Scottish and English regional accents'. *British Journal of Social and Clinical Psychology* 9: 77–9.

Child, I. L., Storm, T. and Veroff, J. (1958). 'Achievement themes in folk tales related to socialization practice', in: J. W. Atkinson (ed.) *Motives in Fantasy, Action and Society*, pp. 000–000. New York: van Nostrand.

Chisholm, J. S. (1993). 'Death, hope and sex: life-history theory and the development of reproductive strategies'. *Current Anthropology* 34: 1–24.

Chisholm, J. S. (1996). 'The evolutionary ecology of attachment organization'. *Human Nature* 7: 1–38.

Chisholm, J. S. (1999). 'Attachment and time preference: relations between early stress and sexual behaviour in a sample of American university women'. *Human Nature* 10: 51–84.

Chisholm, J. S. and Burbank, V. K. (1991). 'Monogamy and polygyny in Southeast Arnhem Land: male coercion and female choice'. *Ethology and Sociobiology* 12: 291–313.

Chomsky, N. (1975). *Reflections on Language*. New York: Pantheon Press.

Clarke, A. B. (1978). 'Sex ratio and local resource competition in a Prosimian primate'. *Science* 201: 163–5.

Clift, S. M., Wilkins, J. C. and Davidson, E. A. (1993). 'Impulsiveness, venturesomeness and sexual risk-taking among heterosexual GUM clinic attenders'. *Personality and Individual Differences* 15: 403–10.

Clutton-Brock, T. H. (1991). *The Evolution of Parental Care*. Princeton: Princeton University Press.

Clutton-Brock, T. H. and Albon, S. D. (1979). 'The roaring of red deer and the evolution of honest advertisement'. *Animal Behaviour* 69: 145–70.

Clutton-Brock, T. H. and Harvey, P. H. (1980). 'Primates, brains and ecology'. *Journal of Zoology* 207: 151–69.

Clutton-Brock, T. H., Albon S. D. and Guinness, F. E. (1988). 'Reproductive success in male and female red deer', in: T. H. Clutton-Brock (ed.) *Reproductive Success: Studies of Individual Variation in Contrasting Breeding Systems*, pp. 325–43. Chicago: University of Chicago Press.

Coale, A. J. and Treadway, R. (1986). 'A summary of the changing distribution of overall fertility, marital fertility and the proportion married in the provinces of Europe', in: A. J. Coale and S. C. Watkins (eds) *The Decline of Fertility in Europe*, pp. 31–181. Princeton: Princeton University Press.

Coates, J. (1993). *Women, Men and Language*. Harlow: Longman.

Coates, J. (1994). 'No gaps, lots of overlaps: turn-taking patterns in the talk of women friends', in: D. Gradol, J. Maybin and B. Strier (eds) *Researching Language and Literacy in Social Contexts*, pp. 177–92. London: Multilingual Matters.

Coates, J. (1996a). 'One-at-a-time: the organisation of men's talk', in: S. Johnson and L. Meinhoff (eds) *Discourses of Masculinity*, pp. 107–29. Oxford: Blackwell.

Coates, J. (1996b). *Women Talk*. Oxford: Blackwell.

Cohen, J. E. (1971). *Casual Groups of Monkeys, Apes and Men*. Cambridge (MA): Harvard University Press.

Coleman, J. S. (1964). *'Introduction to Mathematical Sociology'*. London: Collier-Macmillan.

Collins, S. (2000). 'Men's voices and women's choices'. *Animal Behaviour* 60: 773–80.

Connor, R. (1995). 'Altruism among non-relatives: alternatives to the Prisoner's dilemma'. *Trends in Ecology and Evolution* 10: 84–6.

Connor, R. and Curry, R. (1995). 'Helping non-relatives: a role for deceit?' *Animal Behaviour* 49: 389–93.

Connor, W. (1994). *Ethnonationalism: the Quest for Understanding*. Princeton: Princeton University Press.

Cook, H. B. K. (1992). 'Matrifocality and aggression in Margariteño society', in: K. Björquist and P. Niemelä (eds) *Of Mice and Women: Aspects of Female Aggression*, pp. 149–62. New York: Academic Press.

Cooke, B. (1995). 'Microplots: the case of *Swan Lake*'. *Human Nature* 6: 183–96.

Cooper, J. P. (1976). 'Patterns of inheritance and settlement by great landowners from the fifteenth to the eighteenth centuries', in: J. Goody, J. Thirsk and E. P. Thompson (eds) *Family and Inheritance: Rural Society in Western Europe 1200–1800*, pp. 192–327. Cambridge: Cambridge University Press.

Corballis, M. C. (1983). *Human Laterality*. New York: Academic Press.

Corballis, M. C. (1991). *The Lop-Sided Ape*. Oxford: Oxford University Press.

Corby, C. (1997). *Use of Deception as a Human Mating Strategy*. MSc thesis, University of Liverpool.

Corcoran, R. (2000). 'Theory of mind in other clinical conditions: is there a selective "theory of mind" deficit exclusive to autism?', in: S. Baron-Cohen, H. Tager-Flusberg and D. J. Cohen (eds) *Understanding Other Minds: Perspectives from Developmental Neuroscience*, pp. 73–82. Oxford: Oxford University Press.

Corcoran, R. and Frith, C. (submitted). 'Thematic reasoning and theory of mind: accounting for social inference differences in schizophrenia'.

Cosmides, L. (1989). 'The logic of social exchange: has natural selection shaped how humans reason? Studies with the Wason selection task'. *Cognition* 31: 187–276.

Cosmides, L. and Tooby, J. (1987). 'From evolution to behavior: Evolutionary psychology as the missing link', in: J. Dupré (ed.) *The Latest on the Best: Essays on Evolution and Optimality*, pp. 277–306. Cambridge (MA): MIT Press.

Cosmides, L. and Tooby, J. H. (1992). 'Cognitive adaptations for social exchange', in: J. H. Barkow, L. Cosmides and J. H. Tooby (eds) *The Adapted Mind*, pp. 163–228. Oxford: Oxford University Press.

Cosmides, L. and Tooby, J. (2000). 'The cognitive neuroscience of social reasoning', in: M. Gazzaniga (ed.) *The New Cognitive Neurosciences*, pp. 1259–70. Cambridge (MA): MIT Press.

Cowlishaw, G. and Dunbar, R. I. M. (1991). 'Dominance rank and mating success in male primates'. *Animal Behaviour* 41: 1045–56.

Cowlishaw, G. and Dunbar, R. I. M. (2000). *Primate Conservation Biology*. Chicago: University of Chicago Press.

Cowlishaw, G. and Mace, R. (1996). 'Cross-cultural patterns of marriage and inheritance: a phylogenetic approach'. *Ethology and Sociobiology* 17: 87–97.

Crawford, C. B. (1993). 'The future of sociobiology'. *Trends in Evolution and Ecology* 8: 183–6.

Crawford, C. B., Salter, B. E. and Jang, K. L. (1989). 'Human grief: is its intensity related to the reproductive value of the deceased?' *Ethology and Sociobiology* 10: 297–307.

Cronk, L. (1989). 'Low socioeconomic status and female-biased parental investment: the Mukogodo example'. *American Anthropologist* 91: 414–29.

Cronk, L. (1991). 'Wealth, status, and reproductive success among the Mukogodo of Kenya'. *American Anthropologist* 93: 345–60.

Cronk, L. (1995). 'Is there a role for culture in human behavioral ecology?' *Ethology and Sociobiology* 16: 181–205.

Cronk, L. (1999). *The Whole Complex: Culture and the Evolution of Human Behaviour.* Boulder (Co): Westview Press.

Crook, J. H. (1980). *The Evolution of Human Consciousness.* Oxford: Oxford University Press.

Crook, J. H. (1997). 'The indigenous psychiatry of Ladakh. Part I. Practice theory approaches to trance possession in the Himalayas'. *Anthropological Medicine* 4: 289–307.

Crook, J. H. and Crook, S. J. (1988). 'Tibetan polyandry: problems of adaptation and fitness', in: L. Betzig, M. Borgerhoff-Mulder and P. Turke (eds) *Human Reproductive Behaviour,* pp. 97–114. Cambridge: Cambridge University Press.

Crook, J. and Osmaston, H. D. (eds) (1994) *Himalayan Buddhist Villages.* Bristol, University of Bristol Press.

Crystal, D. (1997). *The Cambridge Encyclopedia of Language.* 2nd edition. Cambridge: Cambridge University Press.

Cunningham, M. R., Barbee, A. P. and Pike, C. L. (1990). 'What do women want? facialmetric assessment of multiple motives in the perception of male facial physical attractiveness'. *Journal of Personality and Social Psychology* 59: 61–72.

Daly, M. and Wilson, M. (1981). 'Abuse and neglect of children in evolutionary perspective', in: R. D. Alexander and D. W. Tinkle (eds) *Natural selection and Social Behaviour,* pp. 405–16. New York: Chiron Press.

Daly, M. and Wilson, M. (1982). 'Whom are newborn babies said to resemble?' *Ethology and Sociobiology* 3: 69–78.

Daly, M. and Wilson, M. (1983). *Sex, Evolution and Behavior,* 2nd edition. Belmont (CA): Wadsworth.

Daly, M. and Wilson, M. (1984). 'A sociobiological analysis of human infanticide', in: G. Hausfater and S. B. Hrdy (eds) *Infanticide: Comparative and Evolutionary Perspectives,* pp. 487–502. New York: Aldine de Gruyter.

Daly, M. and Wilson, M. (1985). 'Child abuse and other risks of not living with both parents'. *Ethology and Sociobiology* 6: 197–210.

Daly, M. and Wilson, M. (1988a). 'Evolutionary psychology and family homicide'. *Science* 242: 519–24.

Daly, M. and Wilson, M. (1988b). *Homicide.* New York: Aldine de Gruyter.

Daly, M. and Wilson, M. (1996). 'The evolutionary psychology of homicide'. *Demos* (8 December): 39–45.

Daly, M. and Wilson, M. (1999). 'Human evolutionary psychology and human behaviour'. *Animal Behaviour* 57: 509–19.

Daly, M. and Wilson, M. I. (2000). 'A reply to Smith et al'. *Animal Behaviour* 60: F27–F29.

Daly, M., Wilson, M. and Weghorst, S. (1982). 'Male sexual jealousy'. *Ethology and Sociobiology* 3: 11–27.

Damasio, A. (1994). *Descartes' Error: Emotion, Reason and the Human Brain.* London: Papermac.

Damasio, A. (2000). *The Feeling of What Happens.* London: Heinemann.

Dardes, D. K. and Koski, P. R. (1988). 'Using the personal ads: a deviant strategy?' *Deviant Behaviour* 9: 383–400.

Darwin, C. (1859). *The Origin of Species.* London: Murray.

Darwin, C. (1871). *The Descent of Man and Selection in Relation to Sex.* London: Murray.

Davis, J. N. (1997). 'Birth order, sibship size, and status in modern Canada'. *Human Nature* 8: 205–30.

Davis, J. N. and Daly, M. (1997). 'Evolutionary theory and the human family'. *Quarterly Review of Biology* 72: 407–35.

Dawkins, R. (1979). *The Extended Phenotype.* Oxford: Oxford University Press.

Dawkins, R. (1982). 'Replicators and vehicles', in: B. C. R. Bertram, T. H. Clutton-Brock, R. I. M. Dunbar, D. I. Rubenstein and R. W. Wrangham (eds) *Current Problems in Sociobiology,* pp. 45–64. Cambridge: Cambridge University Press.

Dawkins, R. (1983). 'Universal Darwinism', in: D. S. Bendall (ed.) *Evolution from Molecules to Man,* pp. 403–25. Cambridge: Cambridge University Press.

Dawkins, R. (1986). *The Blind Watchmaker*. Harlow: Longman.

Dawkins, R. (1989). *The Selfish Gene*, 2nd edition. Oxford: Oxford University Press.

de Waal, F. (1988). 'The communicative repertoire of captive bonobos (*Pan paniscus*) compared to that of chimpanzees'. *Behaviour* 106: 183–251.

Deacon, T. (1997). *The Symbolic Species: the Coevolution of Language and the Human Brain*. Harmondsworth: Allen Lane.

Dennett, D. (1978). 'Beliefs about beliefs'. *Behavioral and Brain Sciences* 1: 568–70.

Dennett, D. (1995). *Darwin's Dangerous Idea: Evolution and the Meanings of Life*. New York: Simon and Schuster.

Diamond, J. (1997). *Guns, Germs and Steel: A Short History of Everybody for the Last 13,000 Years*. London: Jonathon Cape.

Dickemann, M. (1979). 'Female infanticide, reproductive strategies, and social stratification: a preliminary model', in: N. A. Chagnon and W. Irons (eds) *Evolutionary Biology and Human Social Behaviour*, pp. 321–67. North Scituate (MA): Duxbury Press.

Dickemann, M. (1981). 'Paternal confidence and dowry competition: a biocultural analysis of purdah', in: R. Alexander and D. Tinkle (eds) *Natural Selection and Social Behaviour*, pp. 417–38. New York: Chiron Press.

DiFiore, A. and Rendall, D. (1994). 'Evolution of social organization: a reappraisal for primates by using phylogenetic methods'. *Proceedings of the National Academy of Science, USA* 91: 9941–5.

Divale, W. and Harris, M. (1976). 'Population, warfare and the male supremacist complex'. *American Anthropologist* 78: 521–38.

Dixson, A. F. (1998). *Primate Sexuality*. Oxford: Oxford University Press.

Domb, L. G. and Pagel, M. (2001). 'Sexual swellings advertise female quality in wild baboons'. *Nature, London* 410: 204–6.

Donald, M. (1991). *Origins of the Modern Mind: Three Stages in the Evolution of Culture and Cognition*. Cambridge (MA): Harvard University Press.

Douglas, K. (2001). 'Playing fair'. *New Scientist* 169 (10 March): 38–42.

Draper, P. (1989). 'African marriage systems: perspectives from evolutionary ecology'. *Ethology and Sociobiology* 10: 145–69.

Draper, P. and Harpending, H. (1982). 'Father absence and reproductive strategy: an evolutionary perspective'. *Journal of Anthropological Research* 38: 255–73.

Duby, G. (1977). *The Chivalrous Society*. London: Arnold.

Duby, G. (1997). *Women of the Twelfth Century. Vol 2: Remembering the Dead*. Chicago: University of Chicago Press.

Duck, S. and Miell, D. (1982). 'Mate choice in humans as an interpersonal process', in: P. Bateson (ed.) *Mate Choice*, pp. 377–88. Cambridge: Cambridge University Press.

Dunbar, R. I. M. (1982). 'Adaptation, fitness and the evolutionary tautology', in: B. C. R. Bertram, T. H. Clutton-Brock, R. I. M. Dunbar, D. I. Rubenstein and R. W. Wrangham (eds) *Current Problems in Sociobiology*, pp. 9–28. Cambridge: Cambridge University Press.

Dunbar, R. I. M. (1983). 'Lifehistory tactics and alternative strategies of reproduction', in: P. P. G. Bateson (ed.). *Mate Choice*, pp. 423–33. Cambridge: Cambridge University Press.

Dunbar, R. I. M. (1985). 'Stress is a good contraceptive'. *New Scientist* 105: 16–18.

Dunbar, R. I. M. (1987). 'Sociobiological explanations and the evolution of ethnocentrism', in: V. Reynolds, V. Falger and I. Vine (eds) *The Sociobiology of Ethnocentrism: Evolutionary Dimensions of Xenophobia, Discrimination, Racism and Nationalism*, pp. 48–59. London: Croom Helm.

Dunbar, R. I. M. (1988). *Primate Social Systems*. London: Chapman and Hall.

Dunbar, R. I. M. (1989). 'Social systems as optimal strategy sets: the costs and benefits of sociality', in: V. Standen and R. A. Foley (eds) *Comparative Socioecology: the Behavioural Ecology of Humans and Other Mammals*, pp. 131–50. Oxford: Blackwell.

Dunbar, R. I. M. (1991). 'On sociobiological theory and the Cheyenne case'. *Current Anthropology* 32: 169–73.

Dunbar, R. I. M. (1992). 'Neocortex size as a constraint on group size in primates'. *Journal of Human Evolution* 20: 469–93.

Dunbar, R. I. M. (1993a). 'Behavioural adaptation', in: G. A. Harrison (ed.) *Human Adaptation*, pp. 73–98. Oxford: Oxford University Press.

Dunbar, R. I. M. (1993b). 'On the evolution of alternative reproductive strategies'. *Behavioral and Brain Sciences* 16: 291.

Dunbar, R. I. M. (1993c). 'The co-evolution of neocortical size, group size and language in humans'. *Behavioral and Brain Sciences* 16: 681–735.

Dunbar, R. I. M. (1995a). *The Trouble with Science*. Cambridge (MA): Harvard University Press.

Dunbar, R. I. M. (1995b). 'The mating system of Callitrichid primates. I. Conditions for the coevolution of pairbonding and twinning'. *Animal Behaviour* 50: 1057–70.

Dunbar, R. I. M. (1996a). *Grooming, Gossip, and the Evolution of Language*. London: Faber and Faber.

Dunbar, R. I. M. (1996b). 'Determinants of group size in primates: a general model', in: G. Runciman, J. Maynard Smith and R. I. M. Dunbar (eds) *Evolution of Social Behaviour Patterns in Primates and Man*, pp. 33–58. Oxford: Oxford University Press.

Dunbar, R. I. M. (1996c). 'On the evolution of language and kinship', in: J. Steele and S. Shennan (eds) *The Archaeology of Human Ancestry: Power, Sex and Tradition*, pp. 380–96. London: Routledge.

Dunbar, R. I. M. (1997). 'Groups, gossip and the evolution of language', in: A. Schmitt, K. Atzwanger, K. Grammer and K. Schafer (eds) *New Aspects of Human Ethology*, pp. 77–90. New York: Plenum Press.

Dunbar, R. I. M. (1998a). 'The social brain hypothesis'. *Evolutionary Anthropology* 6: 178–90.

Dunbar, R. I. M. (1998b). 'Theory of mind and the evolution of language', in: J. Hurford, M. Studdert-Kennedy and C. Knight (eds) *Approaches to the Evolution of Language*, pp. 92–110. Cambridge: Cambridge University Press.

Dunbar, R. I. M. (1999). 'Culture, honesty and the freerider problem', in: R. I. M. Dunbar, C. Knight and C. Power (eds) *The Evolution of Culture*, pp. 194–213. Edinburgh: Edinburgh University Press.

Dunbar, R. I. M. (2000a). 'Causal reasoning, mental rehearsal and the evolution of primate cognition', in: C. Heyes and L. Huber (eds) *Evolution of Cognition*, pp. 205–21. Cambridge (MA): MIT Press.

Dunbar, R. I. M. (2000b). 'On the origin of the human mind', in: P. Carruthers and A. Chamberlain (eds) *Evolution and the Human Mind: Modularity, Language and Meta-Cognition*. pp. 238–53. Cambridge: Cambridge University Press.

Dunbar, R. I. M. (in press). 'Why are apes so smart?', in: P. M. Kapeller and M. Perriera (eds) *Primate Life Histories*. Cambridge (MA): MIT Press.

Dunbar, R. I. M. and Spoors, M. (1995). 'Social networks, support cliques, and kinship'. *Human Nature* 6: 273–90.

Dunbar, R. I. M., Clark, A. and Hurst, N. L. (1995a). 'Conflict and cooperation among the Vikings: contingent behavioural decisions'. *Ethology and Sociobiology* 16: 233–46.

Dunbar, R. I. M., Duncan, N. and Marriott, A. (1997). 'Human conversational behaviour'. *Human Nature* 8: 231–46.

Dunbar, R. I. M., Duncan, N. D. C. and Nettle, D. (1995b). 'Size and structure of freely forming conversational groups'. *Human Nature* 6: 67–78.

Durham, W. (1976). 'Resource competition and human aggression. Part I: a review of primitive war'. *Quarterly Review of Biology* 51: 385–415.

Durham, W. H. (1991). *Coevolution: Genes, Culture and Human Diversity*. Stanford: Stanford University Press.

Dwyer, P. D. and Minnegal, M. (1993). 'Are Kubo hunters show-offs?' *Ethology and Sociobiology* 14: 53–70.

Dyson-Hudson, R. (1989). 'Ecological influences on systems of food production and social organisation of South Turkana pastoralists', in: V. Standen and R. Foley (eds) *Comparative Socioecology: the Behavioural Ecology of Humans and Other Mammals*, pp. 165–94. Oxford: Blackwell.

Eakins, B. W. and Eakins, R. G. (1978). *Sex Differences in Human Communication*. Boston: Houghton Mifflin.

Earle, T. (1994). 'Political domination and social evolution', in: T. Ingold (ed.) *Companion Encyclopedia of Anthropology*, pp. 940–61. London: Routledge.

Eaton, S. B., Konner, M. and Shostak, M. (1988). 'Stone agers in the fast lane: chronic degenerative diseases in evolutionary perspective'. *American Journal of Medicine* 84: 739–49.

Ellen, R. (1982). *Environment, Subsistence and System: the Ecology of Small-Scale Social Formations*. Cambridge: Cambridge University Press.

Ellison, P. T. (1994). 'Extinction and descent'. *Human Nature* 5: 155–6.

Emery, N. J. (2000). 'The eyes have it: the neuroethology, function and evolution of social gaze'. *Neuroscience and Biobehaviour Reviews* 24: 581–604.

Emery, N. J. and Perrett, D. I. (2000). 'How can studies of the monkey brain help us understand "theory of mind" in autism in humans?', in: S. Baron-Cohen, H. Tager-Flusberg and D. J. Cohen (eds) *Understanding Other Minds: Perspectives from Developmental Neuroscience*, pp. 274–305. Oxford: Oxford University Press.

Emery, N. J., Machado, C. J., Capitanio, J. P., Mendoza, S. P., Mason, W. A. and Amaral, D. G. (1998). 'The role of the amygdala in dyadic social interaction and the stress response in monkeys'. *Society for Neuroscience Abstracts* 24: 312–14.

Emlen, S. T. and Oring, L. W. (1977). 'Ecology, sexual selection, and the evolution of mating systems'. *Science* 197: 215–23.

Emlen, S. T. and Wrege, P. H. (1986). 'Forced copulations and intra-specific parasitism: two costs of social living in the white-fronted bee-eater'. *Ethology* 71: 2–29.

Emlen, S. T., Emlen, J. M. and Levin, S. A. (1986). 'Sex-ratio selection in species with helpers-at-the-nest'. *American Naturalist* 127: 1–8.

Emler, N. (1990). 'A social psychology of reputation'. *European Review of Social Psychology* 1: 171–93.

Emler, N. (1994). 'Gossip, reputation and social adaptation', in: R. Goodman and A. Ben-Ze'ev (eds) *Good Gossip*, pp. 119–40. Lawrence: University of Kansas Press.

Enquist, M. and Leimar, O. (1993). 'The evolution of cooperation in mobile organisms'. *Animal Behaviour* 45: 747–57.

Essock-Vitale, S. M. and McGuire, M. T. (1985). 'Women's lives viewed from an evolutionary perspective. II. Patterns of helping'. *Ethology and Sociobiology* 6: 155–73.

Essock-Vitale, S. M. and McGuire, M. T. (1988). 'What 70 million years hath wrought: sexual histories and reproductive success of a random sample of American women', in: L. Betzig, M. Borgerhoff Mulder and P. Turke (eds) *Human Reproductive Behaviour*, pp. 221–35. Cambridge: Cambridge University Press.

Euler, H. A. and Weitzel, B. (1996). 'Discriminative grandparental solicitude as a reproductive strategy'. *Human Nature* 7: 39–50.

Falk, D. (1980). 'Hominid brain evolution: the approach from paleoneurology'. *Yearbook of Physical Anthropology* 23: 93–107.

Feldman, M. W. and Cavalli-Sforza, L. L. (1989). 'On the theory of evolution under genetic and cultural transmission with application to the lactose absorption problem', in: M. W. Feldman (ed.) *Mathematical Evolutionary Theory*, pp. 145–73. Princeton (NJ): Princeton University Press.

Feldman, M. W. and Laland, K. N. (1996). 'Gene-culture coevolutionary theory'. *Trends in Ecology and Evolution* 11: 453–57.

Feldman, M. W. and Zhivotovsky, L. A. (1992). 'Gene-culture coevolution: toward a general theory of vertical transmission'. *Proceedings of the National Academy of Sciences, USA*, 89: 11935–8.

Feldman, M. W., Cavalli-Sforza, L. L., Peck, J. R. (1985). 'Gene-culture co-evolution: models of the evolution of altruism with cultural transmission'. *Proceedings of the National Academy of Sciences, USA*, 82: 5814–18.

Fenk-Oczlan, G. and Fenk, A. (1997). 'Co-evolution of cognitive functions and natural language', in: A. Schmitt, K. Atzwanger, K. Grammer and K. Schäfer (eds) *New Aspects of Human Ethology*, p. 198. New York: Plenum Press.

Fernald, A. (1992). 'Human maternal vocalisations to infants as biologically relevant signals: an evolutionary perspective', in: J. H. Barkow, L. Cosmides and J. Tooby (eds) *The Adapted Mind*, pp. 391–428. Oxford: Oxford University Press.

Fiddick, L., Cosmides, L., Tooby, J. (2000). 'No interpretation without representation: the role of domain-specific representations and inferences in the Wason selection task'. *Cognition* 77: 1–79.

Fieldman, G., Plotkin, H., Dunbar, R. I. M., Robertson, J-M. and McFarland, D. J. (submitted). 'Blood is thicker than water'.

Finch, V. and Western, D. (1977). 'Cattle colors in pastoral herds: natural selection or social preference?' *Ecology* 58: 1384–92.

Finlay, B. L. and Darlington, R. B. (1995). 'Linked regularities in the development and evolution of mammalian brains'. *Science* 268: 1578–84.

Fisher, R. A. (1930). *The Genetical Theory of Natural Selection*. Oxford: Clarendon Press.

Flannery, K. (1972). 'The cultural evolution of civilizations'. *Annual Review of Ecology and Systematics* 3: 339–426.

Flaxman, S. M. and Sherman, P. W. (2000). 'Morning sickness: a mechanism for protecting mother and embryo'. *Quarterly Review of Biology* 75: 113–48.

Fleagle, J. G. (1999). *Primate Adaptation and Evolution*. London: Academic Press.

Fletcher, P. C., Happé, F., Frith, U., Baker, S. C, Dolan, R. J., Fracowiak, R. S. J. and Frith, C. (1994). 'Other minds in the brain: a functional imaging study of "theory of mind" in story comprehension'. *Cognition* 57: 109–28.

Flinn, M. V. (1988). 'Step- and genetic parent/offspring relationships in a Caribbean village'. *Ethology and Sociobiology* 9: 335–69.

Flinn, M. V. (1989). 'Household composition and female strategies in a Trinidadian village', in: A. E. Rasa, C. Vogel and E. Voland (eds) *The Sociobiology of Sexual and Reproductive Strategies*, pp. 206–33. New York: Chapman and Hall.

Flinn, M. V. (1997). 'Culture and the evolution of social learning'. *Ethology and Sociobiology* 17: 23–67.

Flinn, M. V. and Alexander, R. D. (1982). 'Culture theory: the developing synthesis from biology'. *Human Ecology* 10: 383–400.

Flinn, M. V. and England, B. (1995). 'Childhood stress and family environment'. *Current Anthropology* 36: 854–66.

Flinn, M. V. and Low, B. S. (1986). 'Resource distribution, social competition and mating patterns in human societies', in: R. W. Wrangham and D. I. Rubenstein (eds) *Ecological Aspects of Social Evolution*, pp. 217–43. Princeton: Princeton University Press.

Flisher, A. J., Ziervogel, C. F., Charlton, D. O., Leger, P. H. and Roberston, B. A. (1993). 'Risk-taking behaviour of Cape Penninsula high-school students. 6. Road related behaviour'. *South African Medical Journal* 83: 486–90.

Fodor, J. (1983). *The Modularity of Mind*. Cambridge (MA): MIT Press.

Fodor, J. (1987). 'Modules, frames, fridgeons, sleeping dogs and the music of the spheres', in: J. Garfield (ed.) *Modularity in Knowledge Representation and Natural-Language Understanding*, pp. 26–36. Cambridge (MA): MIT Press.

Fodor, J. (1992). 'A theory of the child's theory of mind'. *Cognition* 44: 283–96.

Fodor, J. (1998). 'There and back again: a review of Annette Karmiloff-Smith's *Beyond Modularity*', in: J. Fodor (ed) *In Critical Condition*. pp. 127–42. Cambridge (MA): MIT Press.

Fodor, J. (2000). *The Mind Doesn't Work That Way*. Cambridge (MA): MIT Press.

Foley, R. (1995a). 'The adaptive legacy of human evolution: a search for the environment of evolutionary adaptedness'. *Evolutionary Anthropology* 4: 192–203.

Foley, R. A. (1995b). *Humans Before Humanity*. Oxford: Blackwell.

Foley, R. A. and Lee, P. C. (1989). 'Finite social space, evolutionary pathways and reconstructing hominid behavior'. *Science* 243: 901–6.

Foley, R. A. and Lee, P. C. (1991). 'Ecology and energetics of encephalisation in hominid evolution'. *Philosophical Transactions of the Royal Society, London, B*, 334: 223–32.

Ford, C. S. and Beach, F. A. (1951). *Patterns of Sexual Behavior*. New York: Harper.

Forde, D. (1949). *Habitat, Economy and Society*. London: Methuen.

Forsberg, A. J. L. and Tullberg, B. S. (1995). 'The relationship between cumulative number of cohabiting partners and number of children for men and women in modern Sweden'. *Ethology and Sociobiology* 16: 221–32.

Fox, R. (1979). 'Kinship categories as natural categories', in: N. A. Chagnon and W. Irons (eds) *Evolutionary Biology and Human Social Behavior: an Anthropological Perspective*, pp. 132–44. North Scituate (MA): Duxbury Press.

Fox, R. (1995). 'Sexual conflict in the epics'. *Human Nature* 6: 135–44.

Friedman, J. and Rowlands, M. R. (eds) (1977). *The Evolution of Social Systems*. London: Duckworth.

Frisch, R. E. (1978). 'Population, food intake and fertility'. *Science* 199: 22–30.

Frisch, R. E. (1985). 'Fatness, menarche, and female fertility'. *Perspectives in Biological Medicine* 28: 611–33.

Frith, C. D. (1992). *The Cognitive Neuropsychology of Schizophrenia*. Hove: Lawrence Erlbaum Associates.

Frith, C. D. and Corcoran, R. (1996). 'Exploring theory of mind in people with schizophrenia'. *Psychological Medicine* 26: 521–30.

Frith, U. and Happé, F. (1994). 'Autism: beyond "theory of mind"'. *Cognition* 50: 115–32.

Furnham, A., Tan, T. and McManus, C. (1997). 'Waist-to-hip ratio and preferences for body shape: a replication and extension'. *Personality and Individual Differences* 22: 539–49.

Fuster, V. (1984). 'Extramarital reproduction and infant mortality in rural Galicia (Spain)'. *Journal of Human Evolution* 13: 457–63.

Gabler, S. and Voland, E. (1994). 'The fitness of twinning'. *Human Biology* 66: 699–713.

Galdikas, B. and Wood, J. (1990). 'Birth spacing patterns in humans and apes'. *American Journal of Physical Anthropology* 63: 185–91.

Gallese, V. and Goldman, A. (1998). 'Mirror neurons and the simulation theory of mind'. *Trends in Cognitive Science* 2: 493.

Gangestad, S. W. and Buss, D. (1993). 'Pathogen prevalences and human mate preferences'. *Ethology and Sociobiology* 14: 89–96.

Gangestad, S. W., Thornhill, R. and Yeo, R. A. (1994). 'Facial attractiveness, developmental stability, and fluctuating asymmetry'. *Ethology and Sociobiology* 15: 73–85.

Gaulin, S. and Robbins, C. (1991). 'Trivers–Willard effect in contemporary North American society'. *American Journal of Physical Anthropology* 85: 61–9.

Gauthier, I., Skudlarski, P., Gore, J. C. and Anderson, A. W. (2000). 'Expertise for cars and birds recruits brain areas involved in face recognition'. *Nature Neuroscience* 3(2): 191–7.

Geary, D. C., Rumsey, M., BowThomas, C. C. and Hoard, M. K. (1995). 'Sexual jealousy as a facultative trait: evidence from patterns of sex differences in adults from China and the US'. *Ethology and Sociobiology* 16: 355–83.

German, T. P. and Leslie, A. M. (2000). 'Attending to and learning about mental states', in: P. Mitchell and K. Riggs (eds) Children's Reasoning and the Mind. Hove: Psychology Press.

Geronimus, A. T. (1987). 'On teenage childbearing and neonatal mortality in the United States'. *Population and Development Review* 13: 245–79.

Geronimus, A. T. (1994). 'The health of African-American women and infants: implications for reproductive strategies and policy analysis', in: G. Sen and R. C. Snow (eds) *Power and Decision: the Social Control of Reproduction*, pp. 77–100. Cambridge (MA): Harvard University Press.

Geronimus, A. T. (1996). 'What teen mothers know'. *Human Nature* 7: 323–52.

Gigerenzer, G. and Goldstein, D. G. (1996). 'Reasoning the fast and frugal way: models of bounded rationality'. *Psychological Review* 103: 650–69.

Gigerenzer, G. and Hugg, K. (1992). 'Domain-specific reasoning: social contracts, cheating and perspective change'. *Cognition* 43: 127–71.

Gigerenzer, G., Todd, P. M. and the ABC Research Group (1999). *Simple Heuristics that Make us Smart*. Oxford: Oxford University Press.

Giles, H. (1971). 'Patterns of evaluation in reactions to RP, South Welsh and Somerset accented speech'. *British Journal of Social and Clinical Psychology* 10: 280–81.

Gillberg, C. and Coleman, M. (1992). *The Biology of the Autistic Syndromes*. 2nd edition. New York: Cambridge University Press.

Goel, V., Grafman, J., Sadato, N. and Hallett, M. (1995). 'Modeling other minds'. *Neuroreport* 6: 1741–6.

Goldberg, T. L. (1995). 'Altruism towards panhandlers: who gives?' *Human Nature* 6: 79–90.

Goldin-Meadow, S. and McNeill, D. (1999). 'The role of gesture and mimetic representation in making language the province of speech', in: M. C. Corballis and S. E. A. Lea (eds) *The Descent of Mind: Psychological Perspectives on Hominid Evolution*, pp. 155–72. Oxford: Oxford University Press.

Goldman, A. (1993). 'The psychology of folk psychology'. *Behavioral and Brain Sciences* 16: 15–28.

Goldstein, M. (1976). 'Fraternal polyandry and fertility in a high Himalayan valley in N. W. Nepal'. *Human Ecology* 4: 223–33.

Goody, J. (1976). *Production and Reproduction: a Comparative Study of Domestic Domains*. Cambridge: Cambridge University Press.

Goody, J., Thirsk, J. and Thompson, E. P. (eds) (1976). *Family and Inheritance: Rural Society in Western Europe 1200–1800*. Cambridge: Cambridge University Press.

Gopnik, A. (1993). 'How we know our own minds: the illusion of first-person knowledge of intentionality'. *Behavioral and Brain Sciences* 16: 29–113.

Gopnik, A. and Astington, J. W. (1988). 'Children's understanding of representational change and its relation to the understanding of false belief and appearance-reality distinction'. *Child Development* 59: 26–37.

Gopnik, A., Capps, L. and Meltzoff, A. N. (2000). 'Early theories of mind: what the theory tells us about autism', in: S. Baron-Cohen, H. Tager-Flusberg and D. J. Cohen (eds) *Understanding Other Minds: Perspectives from Developmental Neuroscience*, pp. 50–72. Oxford: Oxford University Press.

Gopnik, A., Mettzoff, A. and Kuhl, P. K. (2000). *How Babies Think*. London: Weidenfeld & Nicholson.

Gordon, R. M. (1986). 'Folk psychology as simulation'. *Mind and Language* 1: 158–71.

Goren, C., Sarty, M. and Wu, P. (1975). 'Visual following and pattern-discrimination of face-like stimuli by new-born infants'. *Pediatrics* 56: 545.

Gould, S. J. and Vrba, E. (1983). 'Exaptation – a missing term in the science of form'. *Palaeobiology* 8: 4–15.

Grafen, A. (1984) 'Natural selection, kin selection and group selection', in: J. R. Krebs, N. B. Davies (eds) *Behavioural Ecology: an Evolutionary Approach*, 2nd edition, pp. 62–84. Oxford: Blackwell.

Grafen, A. (1990). 'Biological signals as handicaps'. *Journal of Theoretical Biology* 144: 517–46.

Grammer, K. (1989). 'Human courtship behaviour: biological basis and cognitive processing', in: A. E. Rasa, C. Vogel and E. Voland (eds) *The Sociobiology of Sexual and Reproductive Strategies*, pp. 147–69. London: Chapman and Hall.

Grammer, K. (1990). 'Strangers meet: laughter and non-verbal signs of interest in opposite sex encounters'. *Journal of Non-Verbal Behavior* 14: 209–36.

Grammer, K. and Thornhill, R. (1994). 'Human (*Homo sapiens*) facial attractiveness and sexual selection: the role of symmetry and averageness'. *Journal of Comparative Psychology* 108: 233–42.

Grammer, K., Fieder, M. and Filova, V. (1997). 'The communication paradox and possible solutions', in: A. Schmitt, K. Atzwanger, K. Grammer and K. Schäfer (eds) *New Aspects of Human Ethology*, pp. 91–120. New York: Plenum.

Grammer, K., Kruck, K. and Magnusson, M. (1998). 'The courtship dance: patterns of non-verbal synchronisation in opposite sex encounters'. *Journal of Non-verbal Behavior* 22: 3–29.

Grammer, K., Kruck, K., Juette, A. and Fink, B. (2000). 'Non-verbal behaviour as courtship signals: the role of control and choice in selecting partners'. *Evolution and Human Behaviour* 21: 371–90.

Grant, B. R. and Grant, P. R. (1993). 'Evolution of Darwin's finches caused by a rare climatic event'. *Proceedings of the Royal Society, London, B.* 251: 111–17.

Grayson, D. K. (1993). 'Differential mortality and the Donner Party disaster'. *Evolutionary Anthropology* 2: 151–9.

Greenfield, L. (1997). 'Sex offences and offenders: an analysis of data on rape and sexual assault'. Bureau of Justice Statistics, US Department of Justice.

Greenfield, S. (2000). *The Private Life of the Brain*. London: Allen Lane.

Greenlees, I. A. and McGrew, W. C. (1994). 'Sex and age differences in preferences and tactics of mate attraction: analysis of published advertisements'. *Ethology and Sociobiology* 15: 59–72.

Guenther, M. (1996). 'Diversity and flexibility: the case of the Bushmen of southern Africa', in: S. Kent (ed.) *Cultural Diversity among Twentieth Century Foragers: an African Perspective*, pp. 65–86. Cambridge: Cambridge University Press.

Guglielmino, C. R., Viganotti, C., Hewlett, B., and Cavalli-Sforza, L. L. (1995). 'Cultural variation in Africa: role of mechanisms of transmission and adaptation'. *Proceedings of the National Academy of Sciences, USA*, 92: 7585–9.

Guilford, T. and Dawkins, M. S. (1992). 'Receiver psychology and the evolution of animal signals'. *Animal Behaviour* 42: 1–14.

Gumperz, J. J. (1982). *Discourse Strategies*. Cambridge: Cambridge University Press.

Haddix, K. A. (2001). 'Leaving your wife and your brothers: when polyandrous marriages fall apart'. *Evolution and Human Behaviour*, 22(1): 47–60.

Hagerstrand, T. (1967). *Innovation Diffusion as a Spatial Process*. Chicago: University of Chicago Press.

Haig, D. (1993). 'Genetic conflicts in human pregnancy'. *The Quarterly Review of Biology* 68: 495–532.

Haig, D. (1998). 'Genetic conflicts of pregnancy and childhood', in: S. C. Stearns (ed.) *Evolution in Health and Disease*, pp. 77–90. Oxford: Oxford University Press.

Haig, D. and Graham, C. (1991). 'Genomic imprinting and the strange case of insulin-like growth factor II receptor'. *Cell* 64: 1045–6.

Hajnal, J. (1965). 'European marriage patterns in perspective', in: D. V. Glass and D. E. C. Eversley (eds) *Population in History*, pp. 101–43. London: Arnold.

Haldane, J. B. S. (1935). 'Population genetics'. *New Biology* 18: 34–51.

Hallpike, C. R. (1977). *Bloodshed and Vengeance in the Papuan Mountains*. Oxford: Oxford University Press.

Hames, R. (1979). 'Garden labour exchange among the Ye'kwana'. *Ethology and Sociobiology* 8: 259–84.

Hames, R. (1988). 'The allocation of parental care among the Ye'kwana', in: L. Betzig, M. Borgerhoff Mulder and P. Turke (eds) *Human Reproductive Behaviour: a Darwinian Perspective*, pp. 237–52. Cambridge: Cambridge University Press.

Hames, R. (1990). 'Sharing among the Yanomamö: part 1, the effects of risk', in: E. Cashdan (ed.) *Risk and Uncertainty in Tribal and Peasant Economies*, pp. 89–106. Boulder: University of Colorado Press.

Hames, R. (1996). 'Costs and benefits of monogamy and polygyny for Yanomamö women'. *Ethology and Sociobiology* 17: 181–99.

Hamilton, W. D. (1964). 'The genetical evolution of social behaviour'. I, II. *Journal of Theoretical Biology* 7: 1–52.

Hamilton, W. D. (1996) *Narrow Roads of Gene Land*. Oxford: W.H. Freeman.

Hamilton, W. D. and Zuk, M. (1982). 'Heritable true fitness and bright birds: a role for parasites?' *Science* 218: 384–7.

Hammer, D. H., Hu, S., Magnuson, V. L. et al. (1993). 'A linkage between DNA markers on the X chromosome and male sexual orientation'. *Science* 261: 321–7.

Hammer, M. F. and Zegura, S. L. (1996). 'The role of the Y chromosome in human evolutionary studies'. *Evolutionary Anthropology* 5: 116–33.

Happé, F. (1994a). *Autism: an Introduction to Psychological Theory*. London: University College London Press.

Happé, F. (1994b). 'Communicative competence and theory of mind in autism: a test of relevance theory'. *Cognition* 48: 101–19.

Happé, F. (1996). 'Studying weak central coherence and theory of mind in autism: children with autism do not succumb to visual illusions. A research note'. *Journal of Child Psychology and Psychiatry* 37: 873–7.

Happé, F. (2000). 'Parts and wholes, meaning and minds: central coherence and its relation to theory of mind', in: S. Baron-Cohen, H. Tager-Flusberg and D. J. Cohen (eds) *Understanding Other Minds: Perspectives From Developmental Neuroscience*, pp. 203–21. Oxford: Oxford University Press.

Happé, F., Ehlers, S., Fletcher, P. C., Frith, U., Johansson, M., Gillberg, C., Dolan, R., Fracowiak, R. S. J. and Frith, C. (1996). '"Theory of mind" in the brain: evidence from a PET scan study of Asperger syndrome'. *Neuroreport* 8: 197–201.

Harcourt, A. H., Harvey, P. H., Larson, S., and Short, R. V. (1981). 'Testis weight, body weight, and breeding systems in primates'. *Nature (London)* 293: 55–7.

Hardin, G. (1968). 'The tragedy of the commons'. *Science* 162: 1243–8.

Harpending, H. C. (1994). 'Infertility and forager demography'. *American Journal of Physical Anthropology* 93: 385–90.

Harpending, H. C. and Rogers, A. (1990). 'Fitness in stratified societies'. *Ethology and Sociobiology* 11: 497–509.

Harre, N., Field, J., and Kirkwood, B. (1996). 'Gender differences and areas of common concern in the driving behaviors and attitudes of adolescents'. *Journal of Safety Research* 27: 163–73.

Harris, J. R. (1999). *The Nurture Assumption: Why Children Turn Out the Way They Do*. New York: Free Press.

Harris, M. (1971). *Culture, Man and Nature*. New York: Crowell.

Harris, P. L. (1991). 'The work of the imagination', in:. A. Whiten (ed.) *Natural Theories of Mind*, pp. 283–304. Oxford: Blackwell.

Harris, P. L. and Kavanaugh, R. D. (1993) 'Young children's understanding of pretence'. *Monographs of the Society for Research in Child Development*. 58, number 1.

Hartung, J. (1982). 'Polygyny and the inheritance of wealth'. *Current Anthropology* 23: 1–12.

Hartung, J. (1985). 'Matrilineal inheritance. New theory and analysis'. *Behavioural and Brain Sciences* 8: 661–88.

Hartung, J. (1992). 'Getting real about rape'. *Behavioral and Brain Sciences* 15: 390–2.

Harvey, P. M. and Pagel, M. (1991). *The Comparative Method in Evolutionary Biology*. Oxford: Oxford University Press.

Haviland, J. B. (1991). *Gossip, Reputation and Knowledge in Zinacantan*. Chicago: University of Chicago Press.

Hawkes, K. (1990). 'Why do men hunt? Benefits for risky choices', in: E. Cashdan (ed.) *Risk and Uncertainty in Tribal and Peasant Economies*, pp. 145–66. Boulder (Co): Westview Press.

Hawkes, K. (1991). 'Showing off: tests of another hypothesis about men's foraging goals'. *Ethology and Sociobiology* 11: 29–54.

Hawkes, K. (1993). 'Why hunter-gatherers work: an ancient version of the problem of public goods'. *Current Anthropology* 34: 341–61.

Hawkes, K., O'Connell, J. F., and Blurton Jones, N. (1989). 'Hardworking Hazda grandmothers', in: V. Standen and R. A. Foley (eds) *Comparative Socioecology: the Behavioural Ecology of Humans and Other Mammals*, pp. 341–66. Oxford: Blackwell.

Hawkes, K., O'Connell, J. F., and Blurton Jones, N. (1991). 'Hunting income patterns among the Hazda: big game, common foods, foraging goals and the evolution of the human diet'. *Philosophical Transactions of the Royal Society, London, B*, 334: 243–51.

Hawkes, K., O'Connell, J. F., and Blurton Jones, N. (1997). 'Hazda women's time allocation, offspring provisioning, and the evolution of long postmenopausal life spans'. *Current Anthropology* 38: 551–77.

Hays, R. B. and Oxley, D. (1986). 'Social network development and functioning during a lifetime transition'. *Journal of Personality and Social Psychology* 50: 305–13.

Heider, F. and Simmel, M. (1944). 'An experimental study of apparent behaviour'. *American Journal of Psychology* 57: 243–59.

Henrich, J. and Boyd, R. (1998). 'The evolution of conformist transmission and the emergence of between-group differences'. *Evolution and Human Behaviour* 19: 215–42.

Henss, R. (1995). 'Waist-to-hip ratio and attractiveness. a replication and extension'. *Personality and Individual Differences* 19: 479–88.

Herbert P. R., Rich-Edwards, J. W., Manson, J. E., Ridker, P. M., Cook, N. R., O'Connor, G. T., Buring, J. E. and Hennekens, C. H. (1993). 'Height and incidence of cardiovascular disease in male physicians'. *Circulation* 88: 1437–43.

Herz, R. S. and Cahill, E. D. (1997). 'Differential use of sensory information in sexual behaviour as a function of gender'. *Human Nature* 8: 275–86.

Hewes, G. W. (1973). 'Primate communication and the gestural origin of language'. *Current Anthropology* 14: 5–24.

Hewlett, B. S. (1988) 'Sexual selection and paternal investment among Aka pygmies', in: L. Betzig, M. Borgerhoff-Mulder and P. Turke (eds) *Human Reproductive Behaviour*, pp. 263–75. Cambridge: Cambridge University Press.

Hewlett, B. S. (1991). 'Demography and childcare in preindustrial societies'. *Journal of Anthropological Research* 2: 78–88.

Hewlett, B. S. (1996). 'Cultural diversity among African pygmies', in: S. Kent (ed.) *Cultural Diversity among Twentieth Century Foragers: an African Perspective,* pp. 215–44. Cambridge: Cambridge University Press.

Hewlett, B. S. and Cavalli-Sforza, L. L. (1986). 'Cultural transmission among Aka pygmies'. *American Anthropologist* 88: 922–34.

Hewlett, B., Lamb, M. E., Leyendecker, B. and Scholmerlich, A. (2000). 'Parental investment strategies among Aka foragers, Ngandu farmers and Eur-American urban industrialists', in: L. Cronk, N. A. Chagnon and W. Irons (eds) *Adaptation and Human Behaviour: an Anthropological Perspective*, pp. 155–78. New York, Aldine de Gruyter.

Heyes, C. (2000). 'Evolutionary psychology in the round', in: C. Heyes and L. Huber (eds) *The Evolution of Cognition*, pp. 3–22. Cambridge (MA): MIT Press.

Heyes, C. (submitted). Four routes of cognitive evolution.

Higgs, P. G. (2000). 'The mimetic transition: a simulation study of the evolution of learning by imitation'. *Proceedings of the Royal Society London, B* 267: 1355–61.

Hilberman, E. and Munson, K. (1978). 'Sixty battered women'. *Victimology* 2: 460–70.

Hill, C. M. and Ball, H. L. (1996). 'Abnormal births and other "ill omens": the adaptive case of infanticide'. *Human Nature* 7: 381–401.

Hill, E. (1999). 'Lineage interests and nonreproductive strategies: an evolutionary approach to medieval religious women'. *Human Nature* 10: 109–34.

Hill, K. (1988). 'Macronutrient modifications of optimal foraging theory: an approach using indifference curves applied to some modern foragers'. *Human Ecology* 16: 157–97.

Hill, K. and Hawkes, K. (1983). 'Neotropical hunting among the Ache of eastern Paraguay', in: R. Hames and W. Vickers (eds) *Adaptive Responses of Native Amazonia*, pp. 223–67. New York: Academic Press.

Hill, K. and Hurtado, A. M. (1991). 'The evolution of premature reproductive senescence and menopause in human females: an evaluation of the "grandmother" hypothesis'. *Human Nature*, 2: 313–50.

Hill, K. and Hurtado, A. M. (1996). *Ache Life History: the Ecology and Demography of a Foraging People*. New York: Aldine de Gruyter.

Hill, K. and Hurtado, A. M. (1997a). 'The evolution of premature reproductive senescence and menopause in human females: an evaluation of the "grandmother" hypothesis', in: L. Betzig (ed.) *Human Nature: a Critical Reader*, pp. 118–40. Oxford: Oxford University Press.

Hill, K. and Hurtado, A. M. (1997b). 'How much does grandmother help?' in: L. Betzig (ed.) *Human Nature: a Critical Reader*, pp. 140–43. Oxford: Oxford University Press.

Hill, K. and Kaplan, H. (1988). 'Trade-offs in male and female reproductive strategies among the Ache', in: L. Betzig, M. Borgerhof Mulder and P. Turke (eds) *Human Reproductive Behaviour*, pp. 277–305. Cambridge: Cambridge University Press.

Hill, R. A. and Dunbar, R. I. M. (submitted). 'Social network size in humans'.

Hinde, R. A. and Barden, L. A. (1985). 'The evolution of the teddy bear'. *Animal Behaviour* 33: 1371–3.

Hirschfeld, L. A. and S. A. Gelman (eds) (1994). *Mapping the Mind: Domain Specificity in Cognition and Culture*. Cambridge: Cambridge University Press.

Hirsh-Pasek, K. and Golinkoff, R. M. (1996). *The Origins of Grammar: Evidence from Early Language Comprehension*. Cambridge (MA): MIT Press.

Hobson, R. P., Ouston, J. and Lee, T. (1988). 'What's in a face? The case of autism'. *British Journal of Psychology* 79: 441–53.

Hogan, D. P. and Eggebeen, D. J. (1995). 'Sources of energy help and routine assistance in old age'. *Social Forces* 73: 917–36.

Holloway, R. and de la Coste-Lareymondie, C. (1982). 'Brain endocast asymmetry in pongids and hominids: some preliminary findings on the paleontology of cerebral dominance'. *American Journal of Physical Anthropology* 58: 101–10.

Holmes, M. M., Resnick, H. S., Kilpatrick, D. G. and Best, C. L. (1996). 'Rape-related pregnancy: estimates and descriptive characteristics from a national sample of women'. *American Journal of Obstetrics and Gynecology* 175: 320–4.

Homer-Dixon, T. F., Boutwell, J. H. and Rathjens, G. W. (1993). 'Environmental change and violent conflict'. *Scientific American* 268: 16–23.

Hoogland, J. L. (1983). 'Nepotism and alarm calling in the black-tailed prairie dog, *Cynomys ludovicianus*'. *Animal Behaviour* 31: 472–79.

House, J. S., Umberson, D. and Landis, K. R. (1988). 'Structure and processes of social support'. *Annual Review of Sociology* 14: 293–318.

Houston, A. I., Clarke, C. W. and McNamara, J. M. (1988). 'Dynamic models in behavioural and evolutionary ecology'. *Nature, London*, 332: 29–34.

Howard, R. D. (1978a). 'Factors influencing early embryo mortality in bullfrogs'. *Ecology* 59: 789–98.

Howard, R. D. (1978b). 'The evolution of mating strategies in bullfrogs, *Rana catesbiana*'. *Evolution* 32: 850–71.

Howell, C. (1976). 'Peasant inheritance customs in the Midlands, 1280–1700', in: J. Goody, J. Thirsk and E. P. Thompson (eds) *Family and Inheritance: Rural Society in Western Europe 1200–1800*, pp. 112–55. Cambridge: Cambridge University Press.

Howell, N. (1979). *Demography of the Dobe !Kung*. New York: Academic.

Howland, J., Hingson, R., Mangione, T. W. and Bell, N. (1996). 'Why are most drowning victims men? Sex differences in aquatic skills behaviors'. *American Journal of Public Health* 86: 93–6.

Hrdy, S. B. (1981). *The Woman That Never Evolved*. Cambridge (MA): Harvard University Press.

Hrdy, S. B. (1992). 'Fitness trade-offs in the history and evolution of delegated mothering with special references to wet-nursing, abandonment, and infanticide'. *Ethology and Sociobiology* 13: 409–42.

Hrdy, S. B. (2000). *Mother Nature*. London: Chatto and Windus.

Hughes, A. (1988). *Evolution and Human Kinship*. Oxford: Oxford University Press.

Hurd, J. P. (1985). 'Sex differences in mate choice among the "Nebraska" Amish of central Pennsylvania'. *Ethology and Sociobiology* 6: 49–57.

Huxley, R. R. (2000). 'Nausea and vomiting in early pregnancy: its role in placental development'. *Obstetrics and Gynecology* 95: 779–82.

Indik, B. P. (1965). 'Organisation size and member participation: some empirical tests of alternative hypotheses'. *Human Relations* 18: 339–50.

Ingman, M., Kaessmann, H., Pääbo, S. and Gyllensten, U. (2000). 'Mitochondrial genome variation and the origin of modern humans'. *Nature, London*, 408: 708–13.

Irons, W. (1979). 'Natural selection, adaptation and human social behaviour', in: N. A. Chagnon and W. Irons (eds) *Evolutionary Biology and Human Social Behaviour: An Anthropological Perspective*, pp. 213–37. North Scituate (MA): Duxbury Press.

Irons, W. (1998). 'Adaptively relevant environments versus the environment of evolutionary adaptedness'. *Evolutionary Anthropology* 6: 194–204.

Irwin, C. (1989). 'The sociocultural biology of Netsilingmut female infanticide', in: A. Rasa, C. Vogel and E. Voland (eds) *The Sociobiology of Sexual and Reproductive Strategies*, pp. 234–64. London: Chapman and Hall.

Isack, H. A. and Reyer, H. U. (1989). 'Honeyguides and honey gatherers: interspecific communication in a symbiotic relationsip'. *Science* 243: 1343–6.

Ivey, P. (2000). 'Cooperative reproduction in Ituri Forest hunter-gatherers: who cares for Efe infants?' *Current Anthropology* 41: 856–66.

Jackson, L. A. (1992). *Physical Appearance and Gender: Sociobiological and Sociocultural Perspectives*. Albany: State University of New York Press.

Janson, C. (2000). 'Primate socio-ecology: the end of a golden age'. *Evolutionary Anthropology* 9: 73–86.

Jasienska, G. and Ellison, P. (1998). 'Physical work causes suppression of ovarian function in women'. *Proceedings of the Royal Society London, B*, 265: 1847–51.

Jerison, H. J. (1973). *Evolution of the Brain and Intelligence*. New York: Academic Press.

Joffe, T. H. (1997). 'Social pressures have selected for an extended juvenile period in primates'. *Journal of Human Evolution* 32: 593–605.

Joffe, T. H. and Dunbar, R. I. M. (1997). 'Visual and socio-cognitive information processing in primate brain evolution'. *Proceedings of the Royal Society London, B*, 264: 1303–7.

Johnson, A. W. and Earle, T. (1987). *The Evolution of Human Societies: From Foraging Group to Agrarian State*. Stanford: Stanford University Press.

Johnson, G. R., Ratwick, S. H. and Sawyer, T. J. (1987). 'The evocative significance of kin terms in patriotic speech', in: V. Reynolds, V. Falger and I. Vine (eds) *The Sociobiology of Ethnocentrism*, pp. 157–73. Beckenham: Croom Helm.

Johnson, R. C. (1996). 'Attributes of Carnegie medallists performing acts of heroism and of the recipients of these acts'. *Ethology and Sociobiology* 17: 355–62.

Johnson, S. B and Johnson, R. C. (1991). 'Support and conflict of kinsmen in Norse Earldoms, Icelandic families and the English Royalty'. *Ethology and Sociobiology* 12: 211–20.

Johnson-Laird, P. (1982). *Mental Models*. Cambridge: Cambridge University Press.

Jolliffe, T. and Baron-Cohen, S. (1997). 'Are people with autism and Asperger's syndrome faster than normal on the embedded figures test?' *Journal of Child Psychology and Psychiatry* 38: 527–34.

Jones, D. (1995). 'Sexual selection, physical attractiveness, and facial neoteny'. *Current Anthropology* 35: 723–48.

Jones, D. (1996). 'An evolutionary perspective on physical attractiveness'. *Evolutionary Anthropology* 5: 97–110.

Jones, D. (2000). 'Group nepotism and human kinship'. *Current Anthropology* 41: 779–810.

Josephson, S. C. (1993). 'Status, reproductive success and marrying polygynously'. *Ethology and Sociobiology* 14: 391–6.

Joshi, N. V. (1987). 'Evolution of cooperation by reciprocation within structured demes'. *Journal of Genetics* 1: 69–84.

Judge, D. S. (1995). 'American legacies and the variable life histories of women and men'. *Human Nature* 6: 291–324.

Kaberry, P. M. (1971). 'Political organisation among the northern Abelam', in: R. M. Berndt and P. Lawrence (eds) *Politics in New Guinea*. Nedlands: University of Western Australia Press.

Kacelnik, A. and Krebs, J. R. (1997). 'Yanamamö dreams and starling payloads: the logic of optimality', in: L. Betzig (ed.) *Human Nature*, pp 21–35. Oxford: Oxford University Press.

Kanin, E. J. (1985). 'Date rapists: differential sexual socialisation and relative deprivation'. *Archives of Sexual Behaviour* 14: 219–31.

Kaplan, H. and Hill, K. (1985a). 'Food sharing among Ache foragers: tests of explanatory hypotheses'. *Current Anthropology* 26: 233–45.

Kaplan, H. and Hill, K. (1985b). 'Hunting ability and foraging success among Ache foragers'. *Current Anthropology* 26: 131–3.

Kaplan, H. and Hill, K. (1992). 'The evolutionary ecology of food aquisition', in: E. A. Smith and B. Winterhalder (eds) *Evolutionary Ecology and Human Behaviour*, pp. 167–202. New York: Aldine de Gruyter.

Kaplan, H., Hill, K. and Hurtado, A. M. (1990). 'Risk, foraging and food sharing among the Ache', in: E. Cashdan (ed.) *Risk and Uncertainty in Tribal and Peasant Economies*, pp. 107–44. Boulder: Westview Press.

Kaplan, H., Hill, K., Hawkes, K. and Hurtado, A. M. (1987). 'Foraging decisions among the Ache hunter-gatherers: new data and implications for optimal foraging models'. *Ethology and Sociobiology* 8: 1–36.

Kaplan, H., Lancaster, J., Bock, J. A. and Johnson, S. (1995). 'Fertility and fitness among Albuquerque men: a competitive labour market theory', in: R. I. M. Dunbar (ed.) *Human Reproductive Decisions: Biological and Social Perspectives*, pp. 96–136. London: Macmillan.

Kaplan, R. H. and Toshima, M. T. (1990). 'The functional effects of social relationships on chronic illness and disability', in: B. R. Sarason (ed.) *Social Support: An Interactional View*. New York: Wiley.

Karmiloff-Smith, A. (1996). *Beyond Modularity*. Cambridge (MA): MIT Press.

Karmiloff-Smith, A. (2000). 'Why babies brains are not like Swiss Army knives', in: H. Rose and S. Rose (eds) *Alas, Poor Darwin: Arguments Against Evolutionary Psychology*. London: Jonathan Cape.

Kay, R. F., Cartmill, M., and Balow, M (1998). 'The hypoglossal canal and the origin of human vocal behaviour'. *Proceedings of the National Academy of Sciences, USA*, 95: 5417–19.

Kee, F., Nicaud,V. and Tiret, L. (1997). 'Short stature and heart disease: nature or nurture?' *International Journal of Epidemiology* 26: 748–56.

Keesing, R. M (1975). *Kinship Groups and Social Structure*. New York: Holt Rhinehart and Winston.

Keil, F. C. (1994). 'The birth and nurturance of concepts by domains: the origins of concepts of living things', in: Hirschfeld, L. A. and S. A. Gelman (eds) *Mapping the Mind: Domain Specificity in Cognition and Culture*, pp. 234–54.

Kelly, S. and Dunbar, R. I. M. (2001). 'Who dares, wins: heroism versus altruism in women's mate choice'. *Human Nature* 12: 89–105.

Kendal, J. R. and Laland, K. N. (2000). 'Mathematical models for mimetics'. *Journal of Mimetics* 4: 100–8.

Kenrick, D. T. and Keefe, R. C. (1992). 'Age preferences in mates reflect sex differences in human reproductive strategies'. *Behavioral and Brain Sciences* 15: 75–133.

Kent, S. (ed.) (1996). *Cultural Diversity among Twentieth Century Foragers: an African Perspective*. Cambridge: Cambridge University Press.

Kerr, N., Dunbar, R. I. M. and Bentall, R. P. (submitted). 'Theory of mind deficits in bipolar affective disorder'.

Keverne, E. B., Martel, F. L., Nevison, C. M. (1996a). 'Primate brain evolution: genetic and functional considerations'. *Proceedings of the Royal Society, London, B*, 262: 689–96.

Keverne, E. B., Martel, F. L. and Nevison, C. M. (1996b). 'Genomic imprinting and the differential roles of parental genomes in brain development'. *Developmental Brain Research* 92: 91–100.

Keverne, E. B., Martensz, N. and Tuite, B. (1989). 'Beta-endorphin concentrations in cerebrospinal fluid of monkeys are influenced by grooming relationships'. *Psychoneuroendocrinology* 14: 155–61.

Key, C. A. (2000). 'The evolution of human life history'. *World Archaeology* 31: 329–50.

Key, C. A. and Aiello, L. C. (1999). 'The evolution of social organisation', in: R. I. M. Dunbar, C. Knight and C. Power (eds) *The Evolution of Culture*, pp. 15–33. Edinburgh: Edinburgh University Press.

Key, C. A. and Aiello, L. C. (2000). 'A prisoner's dilemma model of the evolution of paternal care'. *Folia Primatologica* 71: 77–92.

Keyfitz, N. (1977). *Applied Mathematical Demography*. New York: Wiley.

Kidwell, J. S. (1982). 'The neglected birth order: middleborns'. *Journal of Marriage and the Family* 44: 225–35.

Killworth, P. D., Bernard, H. P. and McCarty, C. (1984). 'Measuring patterns of acquaintanceship'. *Current Anthropology* 25: 385–97.

Kimura, D. (1993). *Neuromotor Mechanisms in Human Communication*. Oxford: Oxford University Press.

Kimura, M. (1979). 'The neutral theory of molecular evolution'. *Scientific American* 241: 98–126.

Kinderman, P., Dunbar, R. I. M. and Bentall, R. P. (1998). 'Theory-of-mind deficits and causal attributions'. *British Journal of Psychology* 89: 191–204.

Kipers, P. (1987). 'Gender and topic'. *Language and Society* 16: 543–57.

Kirkpatrick, R. C. (2000). 'The evolution of human homosexual behaviour'. *Current Anthropology* 41: 385–414.

Kirkwood, T. (2000). *Time of Our Lives*. Oxford: Oxford University Press.

Kleiman, D. (1977). 'Monogamy in mammals'. *Quarterly Review of Biology* 52: 39–69.

Klein, R. (1999) *The Human Career*. 2nd edition. Chicago: University of Chicago Press.

Klin, A., Schultz, R. and Cohen, D. J. (2000). 'Theory of mind in action: developmental perspectives on social neuroscience', in: S. Baron-Cohen, H. Tager-Flusberg and D. J. Cohen (eds.) *Understanding Other Minds: Perspectives from Developmental Cognitive Neuroscience*, pp. 357–90. Oxford: Oxford University Press.

Klindworth, H. and Voland, E. (1995). 'How did the Krummhörn elite males achieve above-average reproductive success'. *Human Nature* 6: 221–40.

Knauft, B. M. (1987). 'Reconsidering violence in simple human societies: homicide among the Gebusi of New Guinea'. *Current Anthropology* 28: 457–99.

Knight, C. D. (1991). *Blood Relations: Menstruation and the Origins of Culture*. New Haven (Conn): Yale University Press.

Knight, C. D. (1998). 'Ritual/speech coevolution: a solution to the problem of deception', in: J. R. Hurford, M. Studdert-Kennedy and C. Knight (eds). *Approaches to the Evolution of Language: Social and Cognitive Bases*, pp. 68–91. Cambridge: Cambridge University Press.

Knight, C. D. (1999). 'Sex and language as pretend play', in: R. Dunbar, C. Knight and C. Power (eds.) *The Evolution of Culture*, pp. 228–47. Edinburgh: Edinburgh University Press.

Knodel, J. and Lynch, K. A. (1984) 'The decline in remarriage: evidence from German village population in the eighteenth and nineteenth centuries'. *Research Report 84–57 of the Population Studies Center*. Ann Arbor: University of Michigan.

Koenig, W. D. (1989). 'Sex-biased dispersal in the contemporary United States'. *Ethology and Sociobiology* 10: 263–78.

Krebs, J. and Davies, N. (1993). *An Introduction to Behavioural Ecology*, 3rd edition. Oxford: Blackwell.

Krebs, J. R. and Davies, N. B. (eds.) (1997). *Behavioural Ecology: An Evolutionary Approach*, 4th edition. Oxford: Blackwell.

Krebs, J. R. and Kacelnik, A. (1991). 'Decision-making', in: J. R. Krebs and N. B. Davies (eds) *Behavioural Ecology: An Evolutionary Approach*, 3rd edition, pp. 105–36. Oxford: Blackwell.

Krings, M., Stone, A., Schmitz, R. W., Krainitzki, H., Stoneking, M. and Pääbo, S. (1997). 'Neandertal DNA sequences and the origin of modern humans'. *Cell* 90: 19–30.

Kroeber, A. L. and Kluckhohn, C. (1952). *Culture*. New York: Vantage.

Krygier, M. (1994). *The Disintegration of the English Strong Verb System*. Frankfurt-am-Main: Peter Lang.

Kudo, H. and Dunbar, R. I. M. (in press). 'Neocortex size and social network size in primates'. *Animal Behaviour*.

Kuester, J., Paul, A., and Arnemann, J. (1994). 'Kinship, familiarity and mating avoidance in Barbary macaques, *Macaca sylvanus*'. *Animal Behaviour* 48: 1183–94.

Kumm, J., Laland, K. N. and Feldman, M. W. (1994). 'Gene-culture coevolution and sex ratios: the effects of infanticide, sex-selective abortion, sex selection and sex-biased parental investment on the evolution of sex ratios'. *Theoretical Population Biology* 46: 249–78.

Kurland, J. A. (1979). 'Paternity, mother's brother, and human sociality', in: N. A. Chagnon and W. Irons (eds) *Evolutionary Biology and Human Social Behavior: an Anthropological Perspective*, pp. 145–80. North Scituate: Duxbury Press.

Labov, W. (1972). *Sociolinguistic Patterns*. Philadelphia: University of Pennsylvania Press.

Ladurie, E. le R. (1976). 'Family structures and inheritance customs in sixteenth-century France', in: J. Goody, J. Thirsk and E. P. Thompson (eds) *Family and Inheritance: Rural Society in Western Europe 1200–1800*, pp. 37–70. Cambridge: Cambridge University Press.

Lahr, M. M. and Foley, R. (1994). 'Multiple dispersals and modern human origins'. *Evolutionary Anthropology* 3: 48–60.

Laland, K. N. (1993). 'The mathematical modelling of human culture and its implications for psychology and the human sciences'. *British Journal of Psychology* 84: 145–69.

Laland, K. N. (1994). 'Sexual selection with a culturally transmitted mating preference'. *Theoretical Population Biology* 45: 1–15.

Laland, K. N., Kumm, J. and Feldman, M. W. (1995a). 'Gene-culture coevolutionary theory: a test case'. *Current Anthropology* 36: 131–56.

Laland, K. N., Kumm, J., van Horn, J. D. and Feldman, M. W. (1995b). 'A gene-culture model of human handedness'. *Behavioural Genetics* 25: 433–45.

Lalumière, M. L., Quinsey, V. L. and Craig, W. M. (1996a) 'Why are children from the same family so different from one another: a Darwinian note'. *Human Nature* 7: 281–90.

Lalumière, M. L., Chalmers, L. J., Quinsey, V. L. and Seto, M. C. (1996b). 'A test of the mate deprivation hypothesis of sexual coercion'. *Ethology and Sociobiology* 17: 299–318.

Lancaster, J. B. and King, B. J. (1985). 'An evolutionary perspective on menopause', in: J. K. Brown and V. Kerns (eds) *In Her Prime*, pp. 13–20. Boston (MA): Bergin and Garvey.

Landis, M. H. and Burtt, H. E. (1924). 'A study of conversations'. *Journal of Comparative Psychology* 4: 81–9.

Latané, B. (1981). 'The psychology of social impact'. *American Psychologist* 36: 343–65.

Le Bras, H. and Todd, E. (1981). *L'Invention de la France*. Paris: Hachette.

LeDoux, J. (1998). *The Emotional Brain*. New York: Simon and Schuster.

Lee, R. B. (1979). *The !Kung San: Men, Women, and Work in a Foraging Society*. Cambridge: Cambridge University Press.

Leimar, O. and Hammerstein, P. (2001). 'Evolution of cooperation through indirect reciprocity'. *Proceedings of the Royal Society, London, B*, 268: 745–53.

Leslie, A. M. (1982). 'The perception of causality in infants'. *Perception* 11: 173–86.

Leslie, A. M. (1988) 'Some implications of pretense for mechanisms underlying the child's theory of mind', in: J. W. Astington, P. L. Harris and D. R. Olson (eds) *Developing Theories of Mind*, pp. 19–46. New York: Cambridge University Press.

Leslie, A. M. (1991). 'The theory of mind impairment in autism: evidence for a modular mechanism of development?', in: A. Whiten (ed.) *Natural Theories of Mind*, pp. 63–78. Oxford: Blackwell.

Leslie, A. M. (1994). 'ToMM, ToBY, and Agency: core architecture and domain specificity', in: L. A. Hirschfeld and S. A. Gelman (eds) *Mapping the Mind: Domain Specificity in Cognition and Culture*, pp. 119–48. Cambridge: Cambridge University Press.

Leslie, A. M. (2000) '"Theory of Mind" as a mechanism of selective attention', in: M. Gazzaniga (ed.) *The New Cognitive Neurosciences*, pp. 1235–47. Cambridge (MA): MIT Press.

Leslie, A. M. and Keeble, S. (1987). 'Do six-month-old infants perceive causality?' *Cognition* 25: 265–88.

Leslie, A. M. and Thaiss, L. (1992) 'Domain specificity in conceptual development: neuropsychological evidence from autism'. *Cognition* 43: 225–51.

Lessells, C. M. (1986). 'Brood size in Canada geese – a manipulation experiment. *Journal of Animal Ecology* 55: 669–89.

LeVay, S. (1991). 'A difference in hypothalamic structure between heterosexual-homosexual men'. *Science* 253: 1034–6.

Levine, N. E and Silk, J. B. (1997). 'Why polyandry fails: sources of instability in polyandrous marriages'. *Current Anthropology* 38: 375–98.

Lévi-Strauss, C. (1969). *The Elementary Structures of Kinship*. London: Eyre and Spottiswoode.

Lewis, C. and Osbourne, A. (1990). 'Three-year-olds' problem with false belief: conceptual deficit or linguistic artifact?' *Child Development* 61: 1514–19.

Lewis, J. S. (1986). *In the Family Way: Childbearing in the British Aristocracy 1760–1860*. New Brunswick: Rutgers University Press.

Lewis, K. (2001). 'A comparative study of primate play behaviour: implications for the study of cognition'. *Folia Primatologica* 71: 417–21.

Lieberman, N. and Klar, Y. (1996). 'Hypothesis testing in Wason's selection task: social exchange cheating detection or task understanding?' *Cognition* 58: 127–56.

Lieberman, P. (1989). 'The origins of some aspects of human language and cognition', in: P. Mellars and C. B. Stringer (eds.) *The Human Revolution*, pp. 391–414. Edinburgh: Edinburgh University Press.

Lindblom, B. (1986). 'Phonetic universals in vowel systems', in: J. J. Ohala and J. J. Jaeger (eds) *Experimental Phonology*, pp. 13–44. New York: Academic Press.

Lindblom, B. (1998). 'Systemic constraints and adaptive change in the formation of sound structure', in: J. R. Hurford, M. Studdert-Kennedy and C. Knight (eds). *Approaches to the*

Evolution of Language: Social and Cognitive Bases, pp. 242–64. Cambridge: Cambridge University Press.

Lindert, P. H. (1986). 'Unequal English wealth since 1670'. *Journal of Politics and Economics* 94: 1127–62.

Lisak, D. and Ivan, C. (1995). 'Deficits in intimacy and empathy in sexually aggressive men'. *Journal of Interpersonal Violence* 10: 296–308.

Lovejoy, C. O. (1981). 'The origin of man'. *Science* 211: 341–50.

Low, B. S. (1979). 'Sexual selection and human ornamentation', in: N. Chagnon and W. Irons (eds) *Evolutionary Biology and Human Social Behaviour*, pp. 462–86. North Scituate: Duxbury Press.

Low, B. S. (1988). 'Pathogen stress and polygyny in humans', in: L. Betzig, M. Borgerhoff Mulder and P. Turke (eds) *Human Reproductive Behaviour*, pp. 115–28. Cambridge: Cambridge University Press.

Low, B. S. (1989). 'Cross-cultural patterns in the training of children: an evolutionary perspective'. *Journal of Comparative Psychology* 103: 311–19.

Low, B. S. (1990a). 'Marriage systems and pathogen stress in humans'. *American Zoologist* 30: 325–39.

Low, B. S. (1990b). 'Human responses to environmental extremeness and uncertainty: a cross-cultural perspective', in: E. Cashdan (ed.) *Risk and Uncertainty in Tribal and Peasant Economies*, pp. 229–55. Boulder: Westview Press.

Low, B. S. (1991). 'Reproductive life in 19th-century Sweden: an evolutionary perspective on demographic processes'. *Ethology and Sociobiology* 12: 411–48.

Low, B. S. (2000). *Why Sex Matters: a Darwinian Look at Human Behaviour*. Princeton: Princeton University Press.

Low, B. S. and Clarke, A. L. (1991). 'Resources and life course: patterns in the demographic transition'. *Ethology and Sociobiology* 13: 463–94.

Lumsden, C. J. and Wilson, E. O. (1981). *Genes, Mind and Culture*. Cambridge (MA): Harvard University Press.

Luttbeg, B., Borgerhoff Mulder, M. and Mangel, M. (2000). 'To marry again or not: a dynamic model of the demographic transition', in: L. Cronk, W. Irons and N. Chagnon (eds) *Human Behaviour and Adaptation: an Anthropological Perspective*, pp. 345–68.

Lycett, J. E. and Dunbar, R. I. M. (1999). 'Abortion rates reflect optimisation of parental investment strategies'. *Proceedings of the Royal Society, London, B*, 266: 2355–8.

Lycett, J. E. and Dunbar, R. I. M. (2000). 'Mobile phones as lekking devices among human males'. *Human Nature* 11: 93–104.

Lycett, J. E., Dunbar, R. I. M. and Voland, E. (2000). 'Longevity and the costs of reproduction in a historical human population'. *Proceedings of the Royal Society, London, B*, 267: 31–5.

Lynch, A., Plunkett, G. M., Baker, A. J. and Jenkins, P. F. (1989). 'A model of cultural evolution of chaffinch song derived from the meme concept'. *American Naturalist* 133: 634–53.

MacDonald, C. B. (1955). 'Company'. *Encyclopedia Britannica*, 14th edition. London: Encyclopedia Britannica Ltd.

Mace, R. (1993). 'Nomadic pastoralists adopt strategies that maximise long-term household survival'. *Behavioural Ecology and Sociobiology* 33: 329–34.

Mace, R. (1996a). 'When to have another baby'. *Ethology and Sociobiology* 17: 263–73.

Mace, R. (1996b). 'Biased parental investment and reproductive success in Gabbra pastoralists'. *Behavioral Ecology and Sociobiology* 38: 75–81.

Mace, R. (1997). 'Commentary on Levine and Silk'. *Current Anthropology* 38: 396.

Mace, R. (1998). 'The coevolution of human fertility and wealth inheritance strategies'. *Philosophical Transactions of the Royal Society London, B*, 353: 389–97.

Mace, R. and Holden, C. (1999). 'Evolutionary ecology and cross-cultural comparison: the case of matrilineal descent in sub-Saharan Africa', in: P. C. Lee (ed.) *Comparative Primate Socioecology*, pp. 387–405. Cambridge: Cambridge University Press.

Mace, R. and Pagel, M. (1994). 'The comparative method in anthropology: a phylogenetic approach'. *Current Anthropology* 35: 549–64.

Mace, R. and Pagel, M. (1995). 'A latitudinal gradient in the density of human languages in North America'. *Proceedings of the Royal Society, London, B*, 261: 117–21.

Mace, R. and Sear, R. (1997). 'The birth interval and the sex of children in a traditional African population: an evolutionary analysis'. *Journal of Biosocial Science* 29: 499–507.

MacEachern, S. (2000). 'Genes, tribes and African history'. *Current Anthropology* 41: 357–84.

McLain, D. K., Setters, D., Moulton, M. P. and Pratt, A. E. (2000). 'Ascription of newborns by parents and nonrelatives'. *Evolution and Human Behaviour* 21: 11–23.

MacLarnon, A. and Hewitt, G. (1999). 'The evolution of human speech: the role of enhanced breathing control'. *American Journal of Physical Anthropology* 109: 341–63.

MacNeilage, P. F. (1998). 'Evolution of the mechanisms of language output: comparative neurobiology of vocal and manual communication', in: J. R. Hurford, M. Studdert-Kennedy and C. Knight (eds) *Approaches to the Evolution of Language: Social and Cognitive Bases*, pp. 222–41. Cambridge: Cambridge University Press.

Mahdi, N. Q. (1986). 'Pukhtunwali: ostracism and honour among the Pathan hill tribes'. *Ethology and Sociobiology* 7: 295–304.

Malamuth, N. M. and Brown, M. (1994). 'Sexually aggressive men's perceptions of women's communications: testing three explanations'. *Journal of Personality and Social Psychology* 67: 699–712.

Malamuth, N. M., Heavey, C. L., and Linz, D. (1993). 'Predicting men's antisocial behaviour against women: the interaction model of sexual aggression', in: G. C. N. Hall, R. Hirschman, J. R. Graham and M. S. Zaragoza (eds) *Sexual Aggression: Issues in Etiology, Assessment and Treatment*, pp. 63–97. Washington: Taylor and Francis.

Malik, K. (2000). *Man, Beast and Zombie: What Science Can and Cannot Tell Us about Human Nature*. London: Weidenfeld and Nicolson.

Maljkovic, V. (1987). *Reasoning in Evolutionary Important Domains and Schizophrenia: Dissociation Between Content-Dependent and Content Independent Reasoning*. Unpublished undergraduate honours thesis, Department of Psychology, Harvard University, Cambridge (MA).

Mange, A. and Mange, E. (1980). *Genetic: Human Aspects*. New York: Saunders.

Mangel, M. and Clark, C. W. (1988). *Dynamic Modelling in Behavioural Ecology*. Princeton: Princeton University Press.

Manning, A. and Dawkins, M. S. (1992). *An Introduction to Animal Behaviour*. Cambridge: Cambridge University Press.

Manning, J. T. (1995). 'Fluctuating asymmetry and body weight in men and women: implications for sexual selection'. *Ethology and Sociobiology* 16: 143–53.

Manning, J. T. and Pickup L. J. (1998). 'Symmetry and performance in middle-distance runners'. *International Journal of Sports Medicine* 19: 205–09.

Manning, J. T. and Pickup, L. J. (2000). 'Symmetry and performance in middle-distance runners'. *International Journal of Sports Medicine* 19: 205–209.

Manning, J. T. and Taylor, R. P. (2001). 'Second to fourth digit ratio and male ability in sport: implications for sexual selection'. *Evolution and Human Behaviour* 22: 61–70.

Manning, J. T. and Wood, D. (1998). 'Fluctuating asymmetry and aggression in boys'. *Human Nature* 9: 53–66.

Margulis, S. W., Altmann, J. and Ober, C. (1993). 'Sex-biased lactation in a human population and its reproductive costs'. *Behavioral Ecology and Sociobiology* 32: 41–5.

Marlowe, F. (1999a). 'Male care and parental effort among Hadza foragers'. *Behavioural Ecology and Sociobiology* 32: 57–64.

Marlowe, F. (1999b). 'Showoffs or providers? The parenting effort of Hadza men'. *Evolution and Human Behaviour* 20: 391–404.

Marsden, P. V. (1987). 'Core discussion networks of Americans'. *American Sociological Review* 52: 122–31.

Marsh, P. and Paton, R. (1986). 'Gender, social class and conceptual schemes of aggression', in: A. Campbell and J. J. Gibbs (eds) Violent Transactions: the Limits of Personality. Oxford: Blackwell.

Martin, R. D. (1990). *Primate Origins and Evolution*. London: Chapman and Hall.

Maynard Smith, J. (1964). 'Group selection and kin selection'. *Nature*, London, 201: 1145–47.

Maynard Smith, J. (1976). 'Group selection'. *Quarterly Review of Biology* 51: 277–83.

Maynard Smith, J. (1982). *Evolution and the Theory of Games*. Cambridge: Cambridge University Press.

Maynard Smith, J. and Price, G. (1973). 'The logic of animal conflict'. *Nature*, London, 246: 15–18.

Mazur, A. (1986). 'US trends in feminine beauty and overadaptation'. *Journal of Sex Research* 22: 281–303.

McCannell, K. (1988). 'Social networks and the transition to motherhood', in: R. M. Milardo (ed.) *Families and Social Networks*, pp. 83–106. Newbury Park (CA): Sage.

McClelland, D. C. and Pilon, D. A. (1983). 'Sources of adult motives in patterns of parental behaviour in early childhood'. *Journal of Personality and Social Psychology* 44: 564–74.

McCullough, J. M. and York Barton, E. (1991). 'Relatedness and mortality risk during a crisis year: Plymouth colony, 1620–1621'. *Ethology and Sociobiology* 12: 195–209.

McGovern, T. H. (1981). 'The economics of extinction in Norse Greenland', in: T. M. L. Wrigley, M. J. Ingram and C. Farmer (eds) *Climate and History*, pp. 404–33. Cambridge: Cambridge University Press.

McGrew, W. C. (1992). *Chimpanzee Material Culture: Implications for Human Evolution*. Cambridge: Cambridge University Press.

McNamara, J. M. and Houston, A. I. (1996). 'State-dependent life histories'. *Nature, London*, 380: 215–21.

McNeill, D. (1992). *Hand and Mind: What Gestures Reveal About Thought*. Chicago: University of Chicago Press.

McVean, G. T. and Hurst, L. (1997). 'Molecular evolution of imprinted genes: no evidence for antagonistic coevolution'. *Proceedings of the Royal Society, London, B*, 264: 739–46.

Mealey, L. (1985). 'The relationship between social status and biological success: a case study of the Mormon religious hierarchy'. *Ethology and Sociobiology* 6: 249–57.

Mealey, L., Daood, C. and Krage, M. (1996). 'Enhanced memory for faces of cheaters'. *Ethology and Sociobiology* 17: 120–7.

Meggitt, M. J. (1965). *The Lineage System of the Mae-Enga of New Guinea*. Edinburgh: Oliver and Boyd.

Meggitt, M. J. (1977). *Blood is their Argument*. Mountain View (CA): Mayfield.

Miller, D. J. (1980). *Battered Women: Perceptions of their Problems and their Perception of Community Response*. MSc thesis: University of Windsor, Ontario.

Miller, G. F. (1999). 'Sexual selection for cultural displays', in: R. Dunbar, C. Knight and C. Power (eds) *The Evolution of Culture*, pp. 71–91. Edinburgh: Edinburgh University Press.

Miller, G. F. (2000). *The Mating Mind*. London: Heinemann.

Miller, G. F. (1997). 'Protean primates: the evolution of adaptive unpredictability in competition and courtship', in: A. Whiten and R. W. Byrne (eds) *Machiavellian Intelligence II*, pp. 312–40. Cambridge: Cambridge University Press.

Millgram, S. (1974). *Obedience to Authority*. London: Tavistock Publications.

Milner-Gulland, E. J. and Mace, R. (1998). *Conservation of Biological Resources*. Oxford: Blackwell.

Mitani, J. and Brandt, K. L. (1994). 'Social factors influence the acoustic variability in the long-distance calls of male chimpanzees'. *Ethology* 96: 233–52.

Mitchell, P. (1997). *Introduction to Theory of Mind*. London: Arnold.

Mithen, S. (1996). *The Prehistory of the Mind*. London: Thames and Hudson.

Moffitt, T., Caspi, A. and Belsky, J. (1992). 'Childhood experience and the onset of menarche: a test of sociobiological experience'. *Child Development* 63: 47–58.

Møller, A. P. (1990) 'Effects of a haematophagous mite on the barn swallow, *Hirundo rustica*: a test of the Hamilton and Zuk hypothesis'. *Evolution* 44: 771–84.

Monnot, M. (1999). 'The adaptive function of infant-directed speech'. *Human Nature* 10: 415–43.

Montross, L. (1975). 'Tactics'. *Encyclopedia Britannica*. 15th edition. London: Encyclopedia Britannica Ltd.

Moore, J. H. (1990). 'The reproductive success of Cheyenne war chiefs: a counter example to Chagnon'. *Current Anthropology* 31: 169–73.

Moore, M. M. (1985). 'Non-verbal courtship patterns in women: context and consequences'. *Ethology and Sociobiology* 6: 237–47.

Moore, T. and Haig, D. (1991). 'Genomic imprinting in mammalian development: a parental tug-of-war'. *Trends in Genetics* 7: 45–9.

Morbeck, M. E., Galloway, A. and Zihlman, A. I. (1996). *The Evolving Female*. Princeton: Princeton University Press.

Morgan, C. (1979). 'Eskimo hunting groups'. *Ethology and Sociobiology* 1: 83–6.

Morris, D. (1977). *Manwatching: A Field Guide to Human Behaviour*. London: Grafton Books.

Morris, P. H., Reddy, V. and Bunting, R. C. (1995). 'The survival of the cutest: who's responsible for the evolution of the teddy bear?' *Animal Behaviour* 50: 1697–1700.

Mottron, L. and Belleville, S. (1993). 'A study of perceptual analysis in a high-level autistic subject with exceptional graphical abilities'. *Brain and Cognition* 23: 279–309.

Mueller, U. and Mazur, A. (1998). 'Reproductive constraints on dominance competition in male *Homo sapiens*'. *Evolution and Human Behavior* 19: 387–96.

Mulcahy, N. J. (1999). *Altruism Towards Beggars as a Human Mating Strategy*. MSc thesis, University of Liverpool.

Mundinger, P. C. (1980). 'Animal culture and a general theory of cultural evolution'. *Ethology and Sociobiology* 1: 183–223.

Naroll, R. (1956). 'A preliminary index of social development'. *American Anthropologist* 58: 687–715.

Navon, D. (1977). 'Forest before trees: the precedence of global features in visual perception'. *Cognitive Psychology* 9: 353–83.

Nelson, M. (1997). 'Children by Choice: The Evolutionary Implications of Choosing Childlessness.' MSc thesis, University of Liverpool.

Nettle, D. (1996). *The Evolution of Linguistic Diversity*. PhD thesis, University of London.

Nettle, D. (1999a). Using Social Impact Theory to simulate language change. *Lingua* 108: 95–117.

Nettle, D. (1999b). *Linguistic Diversity*. Oxford: Oxford University Press.

Nettle, D. and Dunbar, R. I. M. (1997). 'Social markers and the evolution of reciprocal exchange'. *Current Anthropology* 38: 93–8.

Neville, H. J., Coffey, S. A., Holcomb, P. J., and Tallal, P. (1993). 'The Neurobiology of Sensory and Language Processing in Language Impaired Children'. *Journal of Cognitive Neuroscience*, 5, 235–53.

Niamir, M. (1990). *Herders' Decision-Making in Natural Resources Management in Arid and Semi-Arid Africa*. Rome: FAO.

Nisbet, I. C. T. (1977) 'Courtship feeding and clutch size in common terns, *Sterna hirundo*', in: B. Stonehouse and C. M. Perrins (eds) *Evolutionary Ecology*, pp. 101–9. London: Macmillan.

Nishida, T. (1987) 'Local traditions and cultural transmission', in: B. S. Smuts, D. L. Cheney, R. M. Seyfarth, R. W. Wrangham and T. T. Struhsaker (eds.) *Primate Societies*, pp. 462–74. Chicago: University of Chicago Press.

Noble, W. and Davidson, I. (1991). 'The evolutionary emergence of modern human behaviour. I. Language and its archaeology'. *Man* 26: 222–53.

Noble, W. and Davidson, I. (1996). *Human evolution, Language and Mind*. Cambridge: Cambridge University Press.

Noë, R. and Hammerstein, P. (1995). 'Biological markets'. *Trends in Ecology and Evolution* 10: 336–40.

Nowak, A., Szamrej, J., and Latané, B. (1990). 'From private to public opinion: a dynamical theory of social impact'. *Psychological Review* 97: 362–76.

Nowak, M. A. and Krakauer, D. C. (1999). 'The evolution of language'. *Proceedings of the National Academy of Sciences, USA*, 96: 8028–33.

Nowak, M. A., Krakauer, D. C. and Dress, A. (1999). 'An error limit for the evolution of language'. *Proceedings of the Royal Society, London, B*, 266: 2131–6.

Nowak, M. A., Plotkin, J. B. and Jansen, B. B. (2000). 'The evolution of syntactic communication'. *Nature, London*, 404: 495–8.

Nunn, C. L. (1999). 'The evolution of exaggerated sexual swellings in primates and the graded-signal hypothesis'. *Animal Behaviour* 58: 229–46.

Oates, J. (1977). 'Mesopotamian social organisation: archaeological and philogical evidence', in: J. Friedman and M. J. Rowlands (eds) *The Evolution of Social Systems*, pp. 457–85. London: Duckworth.

O'Connell, S. and Dunbar, R. I. M. (submitted). 'Comprehension of false belief: a comparative study of humans and chimpanzees'.

Olivieri, M. E. and Reiss, D. (1987). 'Social networks of family members: distinctive roles of mothers and fathers'. *Sex Roles* 17: 719–36.

OPCS (1993). *Mortality statistics: General*. London: HMSO.

OPCS (1996). *Births Statistics*. London: HMSO.

Orians, G. H. (1969). 'On the evolution of mating systems in birds and mammals'. *American Naturalist* 103: 589–603.

Orstrom, E., Gardner, R. and Walker, J. (1994). *Rules, Games and Common-Pool Resources*. Ann Arbor: University of Michigan Press.

Otta, E., da Silva Queiroz, R., de Sousa Campos, L., da Silva, M. W. D. and Silveira, M. T. (1999). 'Age differences between spouses in a Brazilian marriage sample'. *Evolution and Human Behaviour* 20: 99–103.

Otterbein, K. F. and Otterbein, C. S. (1965). 'An eye for an eye, a tooth for a tooth: a cross-cultural study of feuding'. *American Anthropologist* 67: 1470–82.

Ozonoff, S., Strayer, D. L., Mc Mahon, W. M. and Filloux, F. (1994). 'Executive function abilities in autism and Tourette's syndrome: an information processing approach'. *Journal of Child Psychology and Psychiatry* 35: 1015–32.

Packer, C. (1977) 'Reciprocal altruism in Papio anubis'. *Nature, London*, 265: 441–3.

Pagel, M. (1994). 'The evolution of conspicuous oestrous advertisement in Old World monkeys'. *Animal Behaviour* 47: 1333–41.

Pagel, M. (1997). 'Desperately concealing father: a theory of parent–infant resemblance'. *Animal Behaviour* 53: 973–81.

Paige, K. E. and Paige, J. M. (1981). *The Politics of Reproductive Ritual*. Berkeley: University of California Press.

Palmer, C. T. (1991a). 'Kin selection, reciprocal altruism and information sharing among marine lobstermen'. *Ethology and Sociobiology* 12: 221–35.

Palmer, C. T. (1991b). 'Human rape: adaptation or by-product?' *Journal of Sex Research* 28: 365–86.

Panter-Brick, C. (1989). 'Motherhood and subsistence work: the Tamang of rural Nepal'. *Human Ecology* 17: 205–28.

Papousek, H. and Papousek, M. (1987). 'Intuitive parenting: a dialectic counterpart to the infant's integrative competence', in: J. D. Osofsky (ed.) *Handbook of Infant Development*, 2nd edition, pp. 669–720. New York: Wiley.

Parker, G. A. (1970). 'Sperm competition and its evolutionary consequences in the insects'. *Biological Review* 45: 525–67.

Paul, L., Foss, M. A. and Baenninger, M. A. (1976). 'Double standards for sexual jealousy: manipulative morality or a reflection of evolved sex differences?' *Human Nature* 7: 291–321.

Pawlowski, B. and Dunbar, R. I. M. (1999a). 'Impact of market value on human mate choice decisions'. *Proceedings of the Royal Society, London, B*, 266 (1416): 281–5.

Pawlowski, B. and Dunbar, R. I. M. (1999b). 'Withholding age as putative deception in mate search tactics'. *Evolution and Human Behavior* 20: 52–9.

Pawlowski, B. and Dunbar, R. I. M. (2001). 'Human mate choice decisions', in: R. Noë, P. Hammerstein and J. A. R. A. M. van Hooff (eds) *Economic Models of Human and Animal Behaviour*, pp. 187–202. Cambridge: Cambridge University Press.

Pawlowski, B., Dunbar, R. I. M. and Lipowicz, A. (2000). 'Tall men have more reproductive success'. *Nature, London*, 403: 156.

Pawlowski, B., Lowen, C. L. and Dunbar, R. I. M. (1998). 'Neocortex size, social skills and mating success in primates'. *Behaviour* 135: 357–68.

Pearce, H. S. (1982). *Contemporary Matchmaking: a Sociological Study of Dating Agencies and 'Lonely Hearts' Columns*. PhD thesis: University of London.

Peccei, J. S. (1995) 'The origin and evolution of menopause: the altriciality-lifespan hypothesis'. *Ethology and Sociobiology* 16: 425–49.

Pennebaker, J. W. (1979) 'Truckin' with country-western psychology'. *Psychology Today*, November, pp. 18–19.

Pennington, R. (1991). 'Child fostering as a reproductive strategy among Southern African pastoralists'. *Ethology and Sociobiology* 12: 83–104.

Pennington, R. (1992). 'Did food increase fertility? An evaluation of !Kung and Herero history'. *Human Biology* 64: 497–521.

Pennington, R. and Harpending, H. (1993). *The Structure of an African Pastoralist Community: Demography, History and Ecology of the Ngamiland Herero*. Oxford: Oxford University Press.

Penton-Voak, I. Perrett, D. I., Castles, D. L., Kobayashi T., Burt D. M., Murray L. K. and Minamisawa, R. (1999). 'Menstrual cycle alters face preference'. *Nature, London*, 399: 741–2.

Perls, T. T., Alpert, L. and Fretts, R. C. (1997). 'Middle aged mothers live longer'. *Nature, London*, 389: 133.

Perner, J. (1991). *Understanding the Representational Mind*. Cambridge (MA): MIT Press.

Perner, J., Frith, U., Leslie, A. M. and Leekam, S. (1989). 'Exploration of the autistic child's theory of mind: knowledge, belief and communication'. *Child Development* 60: 689–700.

Perrett, D. I. and Emery, N. J. (1994). 'Understanding the intentions of others from visual signals: neurophysiological evidence'. *Current Psychology of Cognition* 13: 683–94.

Perrett, D. I., Lee, K. J., Penton-Voak, I., Rowland, D.,Yoshikawa, S., Burt, M., Henzi, S. P., Castles, D. and Akamatsu, S. (1998). 'Effects of sexual dimorphism on facial attractiveness'. *Nature, London,* 394: 884–7.

Pert, C. B. (1997). *Molecules of Emotion: Why You Feel the Way You Feel*. New York: Scribner.

Pérusse, D. (1993). 'Cultural and reproductive success in industrial societies: testing the relationship at the proximate and ultimate levels'. *Behavioral and Brain Sciences* 16: 267–322.

Pérusse, D. (1994). 'Mate choice in modern societies: testing evolutionary hypotheses with behavioral data'. *Human Nature* 5: 256–78.

Peyronnet, J. C. (1976). 'Les enfants abandonnés et leurs nourrices à Limoges au XVIII-siècle'. *Revue d'Histoire Mod. Contemp.* 23: 418–41.

Piazza, A., Rendine, S., Minch, E., Menozzi, P. Mountain, J. and Cavalli-Sforza, L. L. (1995). 'Genetics and the origin of European languages'. *Proceedings of the National Academy of Sciences, USA*, 92: 5836–40.

Pike, I. L. (2000). 'The nutritional consequences of pregnancy sickness: a critique of a hypothesis'. *Human Nature* 11: 207–32.

Pinker, S. (1994). *The Language Instinct*. London: Allen Lane.

Pinker, S. (1997). *How the Mind Works*. Harmondsworth: Penguin.

Pinker, S. and Bloom, P. (1990). 'Natural language and natural selection'. *Behavioral and Brain Sciences* 13: 707–84.

Plaisted, K. C. (2000). 'Aspects of autism that theory of mind cannot explain', in: S. Baron-Cohen, H. Tager-Flusberg and D. J. Cohen (eds) *Understanding Other Minds: Perspectives from Developmental Neuroscience*, pp. 222–50. Oxford: Oxford University Press.

Plotkin, H. (1998). *Evolution in Mind*. London: Penguin.

Poppen, P. J. (1995). 'Gender and patterns of sexual risk-taking in college students'. *Sex Roles* 32: 545–55.

Porter, L. W. and Lawlor, E. E. (1965). 'Properties of organisation structure in relation to job attributes and job behaviour'. *Psychological Bulletin* 64: 23–51.

Powell, M. and Ansic, D. (1997). 'Gender differences in risk behaviour in financial decision-making: an experimental analysis'. *Journal of Economic Psychology* 18: 605–28.

Power, C. (1998). 'Old wives' tales: the gossip hypothesis and the reliability of cheap signals', in: J. R. Hurford, M. Studdert-Kennedy and C. Knight (eds). *Approaches to the Evolution of Language: Social and Cognitive Bases*, pp. 111–29. Cambridge: Cambridge University Press.

Prebble, J. (1963). *The Highland Clearances*. London: Secker and Warburg.

Premack, D. and Woodruff, G. (1978). 'Does the chimpanzee have a theory of mind?' *Behavioral and Brain Sciences* 1: 515–26.

Preuschoft, S. (1992). '"Laughter" and "smile" in Barbary macaues (*Macaca sylvanus*)'. *Ethology* 91: 220–36.

Profet, M. (1988). 'The evolution of pregnancy sickness as protection to the embryo against Pleistocene teratogens'. *Evolutionary Theory* 8: 177–90.

Profet, M. (1992). 'Pregnancy sickness as adaptation: a deterrent to maternal ingestion of teratogens', in:. J. Barkow, L. Cosmides, and J. Tooby (eds) *The Adapted Mind*, pp. 327–66. Oxford: Oxford University Press.

Provine, R. (1997). *Laughter*. London: Faber and Faber.

Randall, S. (1995) 'Low fertility in a pastoral population: constraints or choice?', in: R. I. M. Dunbar (ed.) *Human Reproductive Decisions: Biological and Social Aspects*, pp. 279–96. London: Macmillan.

Rands, M. (1988). 'Changes in social networks following marital separation and divorce', in: R. M. Milardo (ed.) *Families and Social Networks*, pp. 127–46. Newbury Park (CA): Sage.

Rasa, A. E., Vogel, C. and Voland, E. (eds) (1989). *The Sociobiology of Sexual and Reproductive Strategies*. London: Chapman and Hall.

Rasmussen, D. R. (1981). 'Pairbond strength and stability and reproductive success'. *Psychological Reviews* 88: 274–90.

Regalski, J. M. and Gaulin, S. J. C. (1993). 'Whom are Mexican infants said to resemble? Monitoring and fostering paternal confidence in the Yucatan'. *Ethology and Sociobiology* 14: 97–113.

Relethford, J. H. (1995). 'Genetics and modern human origins'. *Evolutionary Anthropology* 4: 53–63.

Rendell, L. and Whitehead, H. (2001). 'Culture in whales and dolphins'. *Behavioral and Brain Sciences* 25.

Renfrew, C. (1987). *Archaeology and Language: the Puzzle of Indo-European Origins*. London: Jonathan Cape.

Richerson, P. J. and Boyd, R. (1998). 'The evolution of human ultra-sociality', in: I. Eibl-

Eibesfeldt and F. Salter (eds) *Indoctrinability, Ideology and Warfare: Evolutionary Perspectives*, pp. 71–95. New York: Berghahn.

Ridley, M. (1993). *The Red Queen: Sex and the Evolution of Human Nature*. London, Viking.

Ridley, M. (1997). *The Origins of Virtue*. Harmondsworth: Penguin.

Ridley, M. (1999). *Genome: The Autobiography of a Species in 23 Chapters*. London: Fourth Estate.

Rizzolatti, G. and Arbib, M. (1998) 'Language within our grasp'. *Trends in Neuroscience* 21: 188.

Roberts, G. (1998). 'Competitive altruism: from reciprocity to the handicap principle'. *Proceedings of the Royal Society, London, B*, 265, 427–31.

Rodseth, L., Wrangham, R. W., Harrigan, A. M. and Smuts, B. B. (1991). 'The human community as a primate society'. *Current Anthropology* 32: 221–55.

Rogers, A. R. (1989). 'Does biology constrain culture?' *American Anthropologist* 90: 819–31.

Rogers, A. R. (1990). 'Evolutionary economies of human reproduction'. *Ethology and Sociobiology* 11: 479–95.

Rogers, E. M. and Shoemaker, F. F. (1971). *The Communication of Innovations*. New York: Free Press.

Rose, H. and Rose, S. (eds) (2000). *Alas, Poor Darwin*. London: Jonathan Cape.

Ross, W. D. and Ward, R. (1982). 'Human proportionality and sexual dimorphism', in: R. L. Hall (ed.) *Sexual Dimorphism in* Homo Sapiens*: a Question of Size*, pp. 317–61. New York: Praeger.

Rounsaville, B. J. (1978). 'Theories in marital violence: evidence from a study of battered women'. *Victimology* 3: 11–31.

Rowanchilde, R. (1996). 'Male genital modification: a sexual selection interpretation'. *Human Nature* 7: 189–215.

Rowe, A. D., Bullock, P. R., Polkey, C. E. and Morris, R. G. (2001). '"Theory of mind" impairments and their relationship to executive functioning frontal lobe excisions'. *Brain*, 124: 1062–62.

Ruhlen, M. (1987). *A Guide to the World's Languages. I. Classification*. London: Edward Arnold.

Ruhlen, M. (1994). *The Origin of Language*. New York: Wiley.

Russell, M. J. and Wells, P. A. (1987). 'Estimating paternity confidence'. *Ethology and Sociobiology* 8: 215–20.

Rutter, M. and Madge, N. (1976). *Cycles of Disadvantage: a Review of Research*. London: Heinemann.

Sabean, D. (1976). 'Aspects of kinship behaviour and property in rural Western Europe before 1800', in: J. Goody, J. Thirsk and E. P. Thomson (eds) *Family and Inheritance: Rural Society in Western Europe 1200–1800*, pp. 96–111. Cambridge: Cambridge University Press.

Sacks, H., Schlegloff, E. A. and Jefferson, G. (1974). 'A simplest systematics for the organisation of turn-taking for conversation'. *Language* 50: 696–735.

Sahlins, M. (1976) *The Use and Abuse of Biology: an Anthropological Critique of Sociobiology*. Ann Arbor: University of Michigan Press.

Saidapur, S. K. and Girish, S. (2000). 'The ontogeny of kin recognition in tadpoles of the toad, *Bufo melanosticus* (Anura; Bufonidae)'. *Journal of Biosciences* 25: 267–73.

Salmon, C. A. (1998). 'The evocative nature of kin terminology in political rhetoric'. *Politics and the Life Sciences* 17: 51–7.

Salmon, C. A. and Daly, M. (1996). 'On the importance of kin relations to Canadian women'. *Ethology and Sociobiology* 5: 289–97.

Salmon, C. A. and Daly, M. (1998). 'Birth order and familial sentiment: middleborns are different'. *Evolution and Human Behavior* 19: 299–312.

Sanders, C. M. (1980). 'A comparison of adult bereavement in the death of a spouse, child and parent'. *Omega* 10: 303–22.

Scaife, M. and Bruner, J. S. (1975). 'The capacity for joint visual attention in the infant'. *Nature, London*, 253: 265.

Schiefenhovel, W. (1989). 'Reproduction and sex-ratio manipulation through preferential female infanticide among the Eipo, in the highlands of western New Guinea', in: A. E. Rasa, C. Vogel and E. Voland (eds) *The Sociobiology of Sexual and Reproductive Strategies*, pp. 170–93. New York: Chapman and Hall.

Schoener, T. W. (1971). 'Theory of feeding strategies'. *Annual Review of Ecology and Systematics* 2: 369–403.

Schult, C. A. and Wellman, H. M. (1997). 'Explaining human movements and actions: children's understanding of the limits of psychological explanation'. *Cognition* 62: 291–324.

Schumacher, A. (1982). 'On the significance of stature in human society'. *Journal of Human Evolution* 11: 697–701.

Schuster, I. M. G. (1983). 'Women's aggression: an African case study'. *Aggressive Behaviour* 9: 319–31.

Scott, S. and Duncan, C. J. (1999). 'Reproductive strategies and sex-biased investment: suggested roles of breast-feeding and wet-nursing'. *Human Nature* 10: 85–108.

Sear, R., Mace, R. and McGregor, I. A. (2000). 'Maternal grandmothers improve nutritional status and survival of children in rural Gambia'. *Proceedings of the Royal Society, London, B*, 267: 1641–7.

Seckler, D. (1980). 'Malnutrition: an intellectual odyssey'. *Western Journal of Agricultural Economics* 5: 219–27.

Segerstråle, U. (2000). *Defenders of the Truth: the Battle for Science in the Sociobiology Debate and Beyond*. Oxford: Oxford University Press.

Seielstad, M. T., Minch, E. and Cavalli-Sforza, L. L. (1998). 'Genetic evidence for a higher female migration rate in humans'. *Nature Genetics* 20: 278–80.

Service, E. R. (1962). *Primitive Social Organisation: an Evolutionary Perspective*. New York: Random House.

Shah, A. and Frith, U. (1983). 'An islet of ability in autistic children: a research note'. *Journal of Child Psychology and Psychiatry* 24: 613–20.

Shavit, Y., Fischer, C. S. and Koresh, Y. (1994). 'Kin and nonkin under collective threat: Israeli networks during the Gulf War'. *Social Forces* 72: 1197–1215.

Shaw, R. P. and Wong, Y. (1989). *Genetic Seeds of War: Evolution, Nationalism and Patriotism*. London: Unwin Hyman.

Shennan, S. (2000). 'Population, culture history and the dynamics of culture change'. *Current Anthropology* 41: 811–36.

Shepher, J. (1971). 'Mate selection among second generation kibbutz adolescents and adults: incest avoidance and negative imprinting'. *Archives of Sexual Behaviour* 1: 293–307.

Shepherd, J. A. and Strathman, A. J. (1989). 'Attractiveness and height: the role of stature in dating preference, frequency of dating and perceptions of attractiveness'. *Personality and Social Psychology Bulletin* 15: 617–27.

Sherman, P. W. (1977). 'Nepotism and the evolution of alarm calls'. *Science* 197: 1246–53.

Sherman, P. W. (1980) 'The limits of ground squirrel nepotism', in:. G. B. Barlow and J. Silverberg (eds) *Sociobiology: Beyond Nature–Nurture?* pp. 505–44. Boulder: Westview Press.

Sherman, P. W. and Reeve, H. K. (1997). 'Forward and backward: alternative approaches to studying human social evolution', in: L. Betzig (ed.) *Human Nature: a Critical Reader*, pp. 147–58. Oxford: Oxford University Press.

Sieff, D. (1990). 'Explaining biased sex ratios in human populations: a critique of recent studies'. *Current Anthropology* 31: 25–48.

Siegal, M., Beattie, K. (1991). 'Where to look first for children's knowledge of false beliefs'. *Cognition* 38: 1–12.

Siegel, L. S. (1994). 'The longterm prognosis of pre-term infants: conceptual, methodological and ethical issues'. *Human Nature* 5: 103–26.

Silk, J. B. (1980). 'Adoption and kinship in Oceania'. *American Anthropologist* 82: 799–820.

Silk, J. B. (1983). 'Local resource competition and facultative adjustment of sex ratios in relation to competitive abilities'. *American Naturalist* 121: 56–66.

Silk, J. B. (1987). 'Adoption among the Inuit'. *Ethos* 15: 320–30.

Silk, J. B. (1990). 'Which humans adopt adaptively and why does it matter?' *Ethology and Sociobiology* 11: 425–6.

Sillén-Tullberg, B. and Møller, A. P. (1993). 'The relationship between concealed ovulation and mating systems in anthropoid primates: a phylogenetic analysis'. *American Naturalist* 141: 1–25.

Silverman, D. (1970). *The Theory of Organisations*. London: Heinemann.

Silverstein, P., Perdue, L., Vogel, L. and Fantini, D. A. (1986). 'Possible causes of the thin standard of bodily attractiveness for women'. *The International Journal of Eating Disorders* 5: 907–16.

Singh, D. (1993). 'Adaptive significance of female physical attractiveness: role of waist-to-hip ratio'. *Journal of Personality and Social Psychology* 65: 293–307.

Singh, D. (1994). 'Is thin really beautiful and good? Relationship between waist-to-hip ratio (WHR) and female attractiveness'. *Personality and Individual Differences* 16: 123–132.

Singh, D. (1995). 'Female health, attractiveness and desirability for relationships: role of breast asymmetry and waist-to-hip ratio'. *Ethology and Sociobiology* 16: 465–81.

Singh, D. and Luis, S. (1995). 'Ethnic and gender consensus for the effect of waist-to-hip ratio on judgment of women's attractiveness'. *Human Nature* 6: 55–66.

Skuse, D. H. (1999). 'Genomic imprinting on the X-chromosome: a novel mechanism for the evolution of sexual dimorphism'. *Journal of Laboratory and Clinical Medicine* 133: 23–32.

Skuse, D. H., James, R. S., Bishop, D. V. M., Coppin, B., Dalton, P., Aamodt-Leeper, G., Bacarese-Hamilton, M., Creswell, C., McGurk, R. and Jacobs, P. A. (1997). 'Evidence from Turner's syndrome of an imprinted X-linked locus affecting cognitive function'. *Nature, London,* 387: 705–8.

Slumming, V. A. and Manning, J. T. (2000). 'Second to fourth digit ratio in elite musicians: evidence for musical ability as an honest signal of male fitness'. *Evolution and Human Behaviour* 21: 1–9.

Smith, E. (1991a). The influence of nutrition and postpartum mating on weaning and subsequent play behaviour of hooded rats. *Animal Behaviour* 41: 513–24.

Smith, E. A. (1991b). *Inujjuamiut Foraging Strategies.* New York: Aldine de Gruyter.

Smith, E. A. (1992a). Human Behavioural Ecology, I. *Evolutionary Anthropology* 1: 20–24.

Smith, E. A. (1992). Human Behavioural Ecology, II. *Evolutionary Anthropology* 1: 50–55.

Smith, E. A. (1993). 'Commentary'. *Current Anthropology* 34: 356.

Smith, E. A. (1998). 'Is Tibetan polyandry adaptive? Methodological and metatheoretical considerations'. *Human Nature* 9: 225–64.

Smith, E. A. (2000). 'Three styles in the evolutionary study of human behaviour', in: L. Cronk, W. Irons and N. Chagnon (eds) *Human Behaviour and Adaptation: an Anthropological Perspective,* pp 25–44. New York: Aldine de Gruyter.

Smith, E. A. and Smith, S. A. (1994). 'Inuit sex-ratio variation'. *Current Anthropology* 35: 595–624.

Smith, E. A. and Winterhalder, B. (1992). *Evolutionary Ecology and Human Behaviour.* New York: Aldine de Gruyter.

Smith, E. A., Borgerhoff Mulder, M. and Hill, K. (2000). 'Evolutionary analyses and human behaviour: a commentary on Daly and Wilson'. *Animal Behaviour* 60: F21–F26.

Smith, E. A., Borgerhoff Mulder, M. and Hill, K. (2001). 'Controversies in the evolutionary social sciences: a guide for the perplexed'. *Trends in Ecology and Evolution* 16: 128–34.

Smith, M., Kish, B. and Crawford, C. (1987). 'Inheritance of wealth as human kin investment'. *Ethology and Sociobiology* 8: 171–82.

Smuts, B. S. (1995). 'The evolutionary origins of patriarchy'. *Human Nature* 6: 1–32.

Sober, E. and Wilson, D. S. (1998). *Do Unto Others: the Evolution and Psychology of Unselfish Behaviour.* Cambridge (MA): Harvard University Press.

Sober, E. and Wilson, D. S. (2001). *Unto Others: the Evolution and Psychology of Unselfish Behaviour.* Cambridge (MA): Harvard University Press.

Soltis, J., Boyd, R. and Richerson, P. J. (1995). 'Can group-functional behaviors evolve by cultural group selection?' *Current Anthropology* 36: 473–94.

Spence, J. (1954). *One Thousand Families in Newcastle.* Oxford: Oxford University Press.

Sperber, D. (1994). 'The modularity of thought and the epidemiology of representations', in: L. Hirschfeld and R. Gelman (eds) *Mapping the Mind,* pp. 39–67. Cambridge: Cambridge University Press.

Sperber, D. (1996). *Explaining Culture.* Oxford: Blackwell.

Sperber, D., Cara, F. and Girotto, V. (1995). 'Relevance theory explains the selection task'. *Cognition* 57: 31–95.

Stack, C. B. (1974). *All Our Kin.* New York: Harper and Row.

Stack, C. B. (1975). *All Our Kin: Strategies for Survival in a Black Community.* New York: Harper Row.

Standen, V. and Foley, R. (eds) (1989). *Comparative Socioecology: the Behavioural Ecology of Humans and Other Mammals.* Oxford: Blackwell.

Staples, R. (1985). 'Changes in black family structure: the conflict between family ideology and structural conditions'. *Journal of Marriage and the Family* 47: 1005–13.

Starks, P. T. and Blackie, C. A. (2000). 'The relationship between serial monogamy and rape in the United States (1960–1995)'. *Proceedings of the Royal Society, London, B,* 267: 1259–63.

Stearns, S. C. (1992). *The Evolution of Life Histories.* Oxford: Oxford University Press.

Steward, J. H. (1972). *Theory of Culture Change.* Chicago: University of Chicago Press.

Stone, V. E. (2000). 'The role of the frontal lobes and amygdala in theory of mind', in:

S. Baron-Cohen, H. Tager-Flusberg and D. J. Cohen (eds) *Understanding Other Minds: Perspectives from Developmental Neuroscience*, pp. 254–73. Oxford: Oxford University Press.

Stone, V. E., Baron-Cohen, S. and Knight, R. T. (1998a). 'Frontal lobe contributions to theory of mind'. *Journal of Cognitive Neuroscience* 10: 640–56.

Stone, V. E., Baron-Cohen, S., Young, A. W., Calder, A. and Keane, J. (1998b). 'Impairments on social cognition following orbito-frontal or amygdala damage'. *Society for Neuroscience Abstracts* 24: 1176.

Stoneking, M. (1993). 'DNA and recent human evolution'. *Evolutionary Anthropology* 2: 60–73.

Stoneking, M. and Cann, R. L. (1989). 'African origin of human mitochondrial DNA', in: P. Mellars and C. Stringer (eds) *The Human Revolution*, pp. 17–30. Edinburgh: Edinburgh University Press.

Strakageiersbach, S. and Voland, E. (1990). 'On the influence of infant-mortality on marital fertility (Krummhörn, 18th-Century and 19th-Century). *Homo* 39: 171–85.

Strassman, B. I. (1992). 'The function of menstrual taboos among the Dogon: defense against cuckoldry?' *Human Nature* 2: 89–131.

Strassmann, B. I. (1996). 'Menstrual hut visits by Dogon women: a hormonal test distinguishes deceit from honest signalling'. *Behavioural Ecology* 7: 304–15.

Strassmann, B. I. and Clarke, A. L. (1998). 'Ecological constraints on marriage in rural Ireland'. *Evolution and Human Behaviour* 19: 33–55.

Strassmann, B. I. and Dunbar, R. I. M. (1999). 'Human evolution and disease: putting the stone age into perspective', in: S. C. Stearns (ed.) *Evolution in Health and Disease*, pp. 91–101. Oxford: Oxford University Press.

Strier, K. B. (2000). *Primate Behavioural Ecology*. Boston (MA): Allyn & Bacon.

Stringer, C. and McKie, R. (1996). *African Exodus: The Origins of Modern Humanity*. London: Jonathan Cape.

Stuss, D. T., Gallup, G. G. and Alexander, M. P. (2001). 'The frontal lobes are necessary for "theory of mind"'. *Brain* 124: 279–86.

Sugawara, K. (1984). 'Spatial proximity and bodily contact among the Central Kalahari San'. *African Study Monographs Supplement* 3: 1–43.

Sullivan, K. and Winner, E. (1993). 'Three-year-olds' understanding of mental states: the influence of trickery'. *Journal of Experimental Child Psychology* 56: 135–48.

Sulloway, F. (1996). *Born to rebel*. New York: Pantheon.

Suls, J. M. (1977). 'Gossip as social comparison'. *Journal of Communication* 27: 164–8.

Surbey, M. (1989). 'Family composition, stress and human menarche', in: F. Bercovitch and T. Ziegler (eds) *The Socioendocrinology of Primate Reproduction*, pp. 11–32. New York: Alan Liss.

Surian, L. and Leslie, A. M. (1999). 'Competence and performance in false belief understanding: a comparison of autistic and three-year-old children'. *British Journal of Developmental Psychology* 17: 141–55.

Swarbrick, R. (2000). *A Social Cognitive Model of Paranoid Delusions*. PhD thesis, University of Manchester.

Symons, D. (1987a). 'An evolutionary approach: can Darwin's view of life shed light on human sexuality?' in: G. H. Geer and W. T. O'Donohue (eds) *Theories of Human Sexuality*, pp. 91–125. New York: Plenum Press.

Symons, D. (1987). 'If we're all Darwinians, what's the fuss about?' in: C. B. Crawford, M. Smith and D. Krebs (eds) *Sociobiology and Psychology: Ideas, Issues and Applications*, pp. 121–46. Hillsdale (NJ): Erlbaum.

Symons, D. (1989). 'A critique of Darwinian Anthropology'. *Ethology and Sociobiology* 10: 131–44.

Symons, D. (1990). 'Adaptiveness and adaptation'. *Ethology and Sociobiology* 11: 427–44.

Tager-Flusberg, H. (2000). 'Language and understanding minds: connections in autism', in: S. Baron-Cohen, H. Tager-Flusberg and D. J. Cohen (eds) *Understanding Other Minds: Perspectives from Developmental Neuroscience*, pp. 124–49. Oxford: Oxford University Press.

Tager-Flusberg, H. and Sullivan, K. (2000). 'A componential view of theory of mind: evidence from Williams syndrome'. *Cognition* 76: 58–89.

Tager-Flusberg, H., Boshart, J. and Baron-Cohen, S. (1998). 'Reading the windows to the soul: evidence of domain-specific sparing in Williams syndrome'. *Journal of Cognitive Neuroscience* 10: 631–639.

Tajfel, H., Flament, C., Billig, M. and Bundy, R. P. (1971). 'Social categorisation and intergroup behaviour'. *European Journal of Social Psychology* 1: 149–78.

Tassinary, L. G. and Hansen, K. A. (1998). 'A critical test of the waist-to-hip ratio hypothesis of female physical attractiveness'. *Psychological Science* 9: 150–5.

Tattersall, I. (2000). 'Palaeoanthropology: the last half century'. *Evolutionary Anthropology* 9: 2–17.

Templeton, A. R. (1993). 'The "Eve" hypothesis: a genetic critique and reanalysis'. *American Anthropologist* 95: 51–72.

Temrin, H., Buchmayer, S. and Enquist, M. (2000). 'Step-parents and infanticide: new data contradict evolutionary predictions'. *Proceedings of the Royal Society, London, B*, 267: 943–5.

Terrien, F. W. and Mills, D. L. (1955). 'The effect of changing size upon the internal structure of organisations'. *American Sociological Review* 20: 11–13.

Thiessen, D. and Umezawa, Y. (1998). 'The sociobiology of everyday life: a new look at a very old novel'. *Human Nature* 9: 293–320.

Thiessen, D., Young, R. K., Burrough, R. (1993). 'Lonely hearts advertisements reflect sexually dimorphic mating strategies'. *Ethology and Sociobiology* 14: 209–29.

Thirsk, J. (1976). 'The European debate on customs of inheritance 1500–1700', in: J. Goody, J. Thirsk and E. P. Thompson (eds) *Family and Inheritance: Rural Society in Western Europe 1200–1800*, pp. 177–91. Cambridge: Cambridge University Press.

Thornhill, N. W. and Thornhill, R. (1990a). 'Evolutionary analysis of psychological pain of rape victims. I. The effects of victim's age and marital status'. *Ethology and Sociobiology* 11: 155–76.

Thornhill, N. W. and Thornhill, R. (1990b). 'Evolutionary analysis of psychological pain of rape victims. II. The effects of stranger, friend and family member offenders'. *Ethology and Sociobiology* 11: 177–93.

Thornhill, N. W. (1991). 'An evolutionary analysis of rules regulating human inbreeding and marriage'. *Behavioural and Brain Sciences* 14: 247–93.

Thornhill, R. A. (1976) 'Sexual selection and nuptial feeding behaviour in Bittacus apicalis (Insecta: Mecopetera)'. *American Naturalist* 110: 529–48.

Thornhill, R. A. (1981). '*Panorpa* (Mecoptera: Panorpidae) scorpionflies: systems for understanding resource-defence polygyny and alternative male reproductive efforts'. *Annual Review of Ecology and Systematics* 12: 355–86.

Thornhill, R. A. and Gangestad, S. W. (1993). 'Human facial beauty: averageness, symmetry and parasite resistance'. *Human Nature* 4: 237–69.

Thornhill, R. A. and Gangestad, S. W. (1996). 'The evolution of human sexuality'. *Trends in Ecology and Evolution* 11: 98–102.

Thornhill, R. A. and Gangestad, S. W. (1999). 'Facial attractiveness'. *Trends in Cognitive Science* 3: 452–60.

Thornhill, R. A. and Grammer, K. (1999). 'The body and face of woman: one ornament that signals quality?' *Evolution and Human Behavior* 20: 105–20.

Thornhill, R. A. and Thornhill, N. W. (1992). 'The evolutionary psychology of men's coercive sexuality'. *Behavioral and Brain Sciences* 15: 363–421.

Thornhill, R. A., Gangestad, S. W. and Comer, R. (1995). 'Human female orgasm and mate fluctuating asymmetry'. *Animal Behaviour* 50: 1601–15.

Thornton, A. (1991). 'Influence of marital history of parents on the marital and cohabitational experiences of children'. *American Journal of Sociology* 96: 868–94.

Tinbergen, N. (1963). 'On aims and methods of ethology'. *Zeitschrift fürTierpsychologie* 20: 410–33.

Tomasello, M. (2001). *The Cultural Origins of Human Cognition*. Cambridge (MA): Harvard University Press.

Tomasello, M. and Call, J. (1997). *Primate Social Cognition*. Oxford: Oxford University Press.

Tomasello, M., Kruger, A. and Ratner, H. (1993). 'Cultural learning'. *Behavioral and Brain Sciences* 16: 450–88.

Tooby, J. and Cosmides, L. (1990). 'The past explains the present'. *Ethology and Sociobiology* 11: 375–424.

Tooby, J. and Cosmides, L. (1992). 'The psychological foundations of culture', in: J. Barkow, L. Cosmides and J. Tooby (eds) *The Adapted Mind: Evolutionary Psychology and the Generation of Culture*, pp 19–136. New York: Oxford University Press.

Tooby, J. and Cosmides, L. (2000). 'Toward mapping the evolved functional organization of mind and brain', in: M. Gazzaniga (ed.) *The New Cognitive Neurosciences*, pp. 1167–78. Cambridge (MA): MIT Press.

Tooke, W. and Camire, L. (1991). 'Patterns of deception in intersexual and intrasexual mating strategies'. *Ethology and Sociobiology* 12: 345–64.

Tovée, M. J., Maisey, D. S., Emery, J. L. and Cornelissen, P. L. (1999). 'Visual cues to female physical attractiveness'. *Proceedings of the Royal Society, London, B*, 266: 211–18.

Trivers, R. L. (1971). 'The evolution of reciprocal altruism'. *Quarterly Review of Biology* 46: 35–57.

Trivers, R. L. (1972). 'Parental investment and sexual selection', in. B. Campbell (ed.) *Sexual Selection and the Descent of Man*, pp. 139–79. Chicago: Aldine.

Trivers, R. L. (1974). 'Parent-offspring conflict'. *American Zoologist* 14: 249–64.

Trivers, R. L. (1985). *Social Evolution*. Menlo Park, CA: Benjamin-Cummings.

Trivers, R. L. and Willard, D. (1973). 'Natural selection of parental ability to vary the sex ratio'. *Science* 179: 90–2.

Trudgill, P. (1974). *Sociolinguistics*. Harmondsworth: Penguin.

Trudgill, P. (1999). *The Dialects of England*. 2nd edition. Oxford: Blackwells.

Tullberg, B. S. and Lummaa, V. (2001). 'Induced abortion ratio in modern Sweden falls with age, but rises again before menopause'. *Evolution and Human Behaviour* 22: 1–10.

Turke, P. W. (1984). 'Effects of ovulatory concealment and synchrony on protohominid mating systems and parental roles'. *Ethology and Sociobiology* 5: 33–44.

Turke, P. W. (1988). 'Helpers at the nest: childcare networks on Ifaluk', in: L. Betzig, M. Borgerhoff-Mulder, and P. W. Turke (eds) *Human Reproductive Behaviour: a Darwinian Perspective*, pp. 173–88. Cambridge: Cambridge University Press.

Turke, P. W. (1989). 'Evolution and the demand for children'. *Population and Development Review* 15: 61–90.

Turke, P. W. (1990). 'Which humans behave adaptively and why does it matter?' *Ethology and Sociobiology* 11: 305–39.

Turke, P. W. and Betzig, L. L. (1985). 'Those who can do: wealth, status and reproductive success on Ifaluk'. *Ethology and Sociobiology* 6: 79–87.

Tversky, A. and Kahneman, D. (1981). 'The framing of decisions and the psychology of choice'. *Science* 211: 453–8.

Tversky, A. and Kahneman, D. (1983). 'Extensional versus intuitive reasoning: the conjuction fallacy in probability judgement'. *Psychological Review* 90: 293–315.

Tyler, J. and Lichtenstein, C. (1997). 'Risk, protective, AOD knowledge, attitude, and AOD behavior: factors associated with characteristics of high-risk youth'. *Evaluation and Program Planning* 20: 27–45.

Ujhelyi, M. (1998). 'Long-call structure in apes as a possible precursor for language', in: J. R. Hurford, M. Studdert-Kennedy and C. Knight (eds) *Approaches to the Evolution of Language: Social and Cognitive Bases*, pp. 177–89. Cambridge: Cambridge University Press.

Urban Church Project (1974). *Let My People Grow!* Workpaper No. 1. Unpublished report to the General Synod of the Church of England, London.

van den Berghe, P. L. (1979). *Human Family Systems: an Evolutionary View*. New York: Elsevier.

van den Berghe, P. L. (1980). 'Incest and exogamy: a sociobiological reconsideration'. *Ethology and Sociobiology* 1: 151–62.

van Hooff, J. A. R. A. M. (1972). 'A comparative approach to the phylogeny of laughter and smile', in: R. A. Hinde (ed.) *Nonverbal Communication*, pp. 209–41. Cambridge: Cambridge University Press.

van Schaik, C. P. (1983). 'Why are diurnal primates living in groups?' *Behaviour* 87: 120–44.

van Schaik, C. P. and Dunbar, R. I. M. (1990). 'The evolution of monogamy in large primates: a new hypothesis and some critical tests'. *Behaviour* 115: 30–62.

van Schaik, C. P. and Hrdy, S. B. (1991). 'Intensity of local resource competition shapes the relationship between maternal rank and sex ratios at birth in cercopithecine primates'. *American Naturalist* 138: 1555–62.

van Valen, L. (1973). 'A new evolutionary law'. *Evolutionary Theory* 1: 1–30.

Verner, J. and Willson, M. F. (1966). 'The influence of habitats on mating systems of North American passerine birds'. *Ecology* 47: 143–7.

Vines, G. (1997). 'Where did you get your brains?' *New Scientist* 154 (no. 2080): 34–9.

Vining, D. R. (1986). 'Social versus reproductive success: the central theoretical problem of human sociobiology'. *Behavioral and Brain Sciences* 9: 167–216.

Voland, E. (1988). 'Differential infant and child mortality in evolutionary perspective: data from 17th to19th century Ostfriesland (Germany)', in: L. Betzig, M. Borgerhoff Mulder and P. W.

Turke (eds) *Human Reproductive Behaviour: a Darwinian Perspective*, pp. 253–62. Cambridge: Cambridge University Press.

Voland, E. (1989). 'Differential parental investment: some ideas on the contact area of European social history and evolutionary biology', in: V. Standen and R. A. Foley (eds) *Comparative Socioecology: the Behavioural Ecology of Humans and Other Mammals*, pp. 391–403. Oxford: Blackwell.

Voland, E. (1990). 'Differential reproductive success in the Krummhorn population'. *Behavioural Ecology and Sociobiology* 26: 65–72.

Voland, E. (1997). 'Commentary on Levine and Silk'. *Current Anthropology* 38: 397.

Voland, E. and Dunbar, R. (1995). 'Resource competition and reproduction: the relationship between economic and parental strategies in the Krummhörn population'. *Human Nature* 6: 33–49.

Voland, E. and Engel, C. (1989). 'Female choice in humans: a conditional mate selection strategy of the Krummhörn women (Germany, 1720–1874)'. *Ethology* 84: 144–54.

Voland, E. and Voland, R. (1989). 'Evolutionary biology and psychiatry: the case of anorexia nervosa'. *Ethology and Sociobiology* 10: 223–40.

Voland, E., Siegelkow, E. and Engel, C. (1991). 'Cost/benefit oriented parental investment by high status families: the Krummhörn case'. *Ethology and Sociobiology* 12: 105–18.

Voland, E., Dunbar, R. I. M., Engel, C. and Stephan, P. (1997). 'Population increase and sex-biased parental investment in humans: evidence from 18th- and 19th-century Germany'. *Current Anthropology* 38: 129–35.

Wagstaff, G. F. (in press). *Making Sense of Justice: on the Psychology and Philosophy of Equity and Desert*. Lampeter: Edwin Mellen Press.

Wang, X. (1996). 'Evolutionary hypotheses of risk sensitive choice: age differences and perspective change'. *Ethology and Sociobiology* 17: 1–15.

Waser, P. M., Austad, S. and Keane, B. (1986). 'When should animals tolerate inbreeding?' *American Naturalist* 128: 529–37.

Washburn, S. L. (1981). 'Longevity in primates', in: J. March and J. McGaugh (eds) *Aging, Biology and Behavior*. New York: Academic Press.

Wason, P. C. (1966). 'Reasoning', in: B. M. Foss (ed.) *New Horizons in Psychology*, pp. 135–51. Harmondsworth: Penguin.

Wason, P. C. (1983). 'Realism and rationality in the selection task', in: J. St. B. T. Evans (ed.) *Thinking and Reasoning: Psychological Approaches*. London: Routledge and Kegan Paul.

Wasser, S. K. and Barash, D. P. (1983). 'Reproductive suppression among female mammals: implications for biomedicine and sexual selection theory'. *Quarterly Review of Biology* 58: 513–38.

Waynforth, D. and Dunbar, R. (1995). 'Conditional mate choice strategies in humans: evidence form lonely hearts advertisments'. *Behaviour* 132: 735–79.

Wedekind, C. and Füri, S. (1997). 'Body odour preference in men and women: aims for specific MHC-combinations or simply heterozygosity?' *Proceedings of the Royal Society, London, B*, 264: 1471–9.

Wedekind, C., Seebeck, T., Bettens, F. and Paepke, A. J. (1995). 'MHC-dependent mate preferences in humans'. *Proceedings of the Royal Society, London, B*, 260: 245–9.

Weiner, A. (1988). *The Trobrianders of Papua New Guinea*. New York: Holt Rinehart and Winston.

Weinrich, J. D. (1987). 'A new sociobiological theory of homosexuality applicable to societies with universal marriage'. *Ethology and Sociobiology* 8: 37–47.

Weisenberg, M., Tepper, I. and Schwarzwald, J. (1995). 'Humour as a cognitive technique for increasing pain tolerance'. *Pain* 63: 207–12.

Weiss, K. (1981). 'Evolutionary perspectives on human aging', in: P. Amoss and S. Harrell (eds) *Other Ways of Growing Old*. Stanford: Stanford University Press.

Wellman, H. M. (1990). *The Child's Theory of Mind*. Cambridge, (MA): MIT Press.

Wellman, H. M. (1991). 'From desires to beliefs: acquisition of a theory of mind', in: A. Whiten (ed.) *Natural Theories of Mind*, pp. 19–38. Oxford: Blackwell.

Wellman, H. M. and Woolley, J. D. (1990). 'From simple desires to ordinary beliefs: the early development of everyday psychology'. *Cognition* 35: 245–75.

Werner, E. (1989). 'High risk children in young adulthood: a longitudinal study from birth to 32 years'. *American Journal of Orthopsychiatry* 59: 72–81.

Werren, J. H. and Pulliam, H. R. (1981). 'An intergenerational transmission model for the cultural evolution of helping behaviour'. *Human Ecology* 9: 465–83.

Westendorp, R. G. J. and Kirkwood, T. B. L. (1998). 'Human longevity at the cost of repro-ductive success'. *Nature, London*, 396: 743–6.

Westermarck, E. A. (1891). *The History of Human Marriage*. New York: Macmillan.

Western, D. and Dunne, T. (1979). 'Environmental aspects of settlement site decisions among pastoral Maasai'. *Human Ecology* 7: 75–93.

Wetsman, A. and Marlowe, F. (1999). 'How universal are preferences for female waist-to-hip ratios? Evidence from the Hadza of Tanzania'. *Evolution and Human Behaviour* 20: 219–28.

Whissel, C. (1996). 'Mate selection in popular women's fiction'. *Human Nature* 7: 427–48.

White, D. R. and Burton, M. L. (1988). 'Causes of polygyny: ecology, economy, kinship and warfare'. *American Anthropologist* 90: 871–87.

White, R. (1982). 'Rethinking the Middle/Upper Palaeolithic transition'. *Current Anthropology* 23: 169–92.

Whitehurst, R. N. (1971). 'Violence potential in extramarital sexual responses'. *Journal of Marriage and the Family* 33: 683–91.

Whiten, A. (ed.) (1991). *Natural Theories of Mind*. Oxford: Blackwell.

Whiten, A., Goodall, J., McGrew, W. C., Nishida, T., Reynolds, V., Sugiyama, Y., Tutin, C. E. G., Wrangham, R. W., Boesch, C. (1999). 'Culture in chimpanzees'. *Nature, London*, 399: 682–5.

Wiederman, M. W. (1993). 'Evolved gender differences in mate preferences: evidence from personal advertisements'. *Ethology and Sociobiology* 14: 331–52.

Wiederman, M. W. and Kendall, E. (1999). 'Evolution, sex and jealousy: investigation with a sample from Sweden'. *Evolution and Human Behaviour* 20: 121–8.

Williams, G. C. (1966). *Adaptation and Natural Selection: a Critique of Some Current Evolutionary Thought*. Princeton: Princeton University Press.

Wilson, D. S. (1997). 'Incorporating group selection into the adaptationist program: a case study involving human decision making', in: J. Simpson and D. Kendrick (eds) *Evolutionary Social Psychology*, pp. 345–86. Mahwah (NJ): Erlbaum.

Wilson, D. S., Wilczynski, C., Well, A. and Weiser, L. (2000). 'Gossip and other aspects of language as group-level adaptations', in: C. Heyes and L. Huber (eds) *The Evolution of Cognition*, pp. 347–65. Cambridge (MA): MIT Press.

Wilson, E. O. (1975). *Sociobiology: the New Synthesis*. Cambridge (MA): Harvard University Press.

Wilson, M. and Daly, M. (1985). 'Competitiveness, risk-taking and violence: the young male syndrome'. *Ethology and Sociobiology* 6: 59–73.

Wilson, M. and Daly, M. (1997). 'Life expectancy, economic inequality, homicide and repro-ductive timing in Chicago neighbourhoods'. *British Medical Journal* 314: 1271–4.

Wilson, M., Daly, M., Gordon, S. and Pratt, A. (1996). 'Sex differences in valuations of the environment?' *Population and Environment* 18: 143–59.

Winterhalder, B. (1996). 'A marginal model of tolerated theft'. *Ethology and Sociobiology* 17: 37–53.

Wolf, A. P. (1995). *Sexual Attraction and Childhood Association: a Chinese Brief for Edward Westermarck*. Stanford: Stanford University Press.

Wolf, A. P. and Huang, C. (1980). *Marriage and Adoption in China*. Stanford: Stanford University Press.

Wood, B. and Collard, M. (1999). 'The human genus'. *Science* 284: 65–9.

Wood, B., White, T., Lovejoy, O., Latimer, B., Simpson, S. and Suwa, G. (1999). *Australopithecus garhi*: a new species of early hominid from Ethiopia'. *Science* 284: 629–35.

Wood, J. W. (1990). 'Fertility in anthropological populations'. *Annual Review of Anthropology* 19: 211–42.

Wood, J. W. (1994). *Dynamics of Human Reproduction*. Hawthorne: Aldine DeGruyter.

Worden, R. (1998). 'The evolution of language from social intelligence', in: J. R. Hurford, M. Studdart-Kennedy and C. Knight (eds) *Approaches to the Evolution of Language: Social and Cognitive Bases*, pp. 148–66. Cambridge: Cambridge University Press.

Wright, R. (1994). *The Moral Animal: Why We Are the Way We Are*. New York: Little Brown.

Wyatt, G. (1990). 'Changing influences on adolescent sexuality over the past forty years', in: J. Bancroft and J. Reinisch (eds) *Adolesence and Puberty*, pp. 182–206. Oxford: Oxford University Press.

Wynn, T. (1988). 'Tools and the evolution of human intelligence', in: R. W. Byrne and A. Whiten (eds) *Machiavellian Intelligence*, pp. 271–84. Oxford: Oxford University Press.

Yachanin, S. A. and Tweney, R. D. (1982). 'The effect of thematic content on cognitive strategies in the 4-card selection task'. *Bulletin of the Psychonomic Society* 19: 87–90.

Yamakazi, K., Beauchamp, G. K., Curran, M., Bard, J. and Boyse, E. A. (2000). 'Parent-progeny recognition as function on MHC-odour type identity'. *Proceedings of the National Academy of Sciences, USA*, 97: 10500–02.

Yoshimura, Y., Kighi, K., Matsumoto, Y. and Inoue, S. (1982). 'Quantitative effects of nitrogen and energy intakes on body weight and nitrogen retention in adult rats'. *Tokushima Journal of Experimental Medicine* 29: 163–72.

Young, A. W., Aggleton, J. P., Hellawell, D. J., Johnson, M., Broks, P. and Hanley, J. R. (1995). 'Face processing after amydalotomy'. *Brain* 118: 15–24.

Young, M. And Willmott, P. (1957). *Family and Kinship in East London*. London: Routledge and Kegan Paul.

Yu, D. W. and Shepherd, G. H. (1998). 'Is beauty in the eye of the beholder?' *Nature, London*, 396: 321–2.

Zahavi, A. (1975). 'Mate selection – a selection for a handicap'. *Journal of Theoretical Biology* 67: 603–5.

Zahavi, A. (1977). 'The cost of honesty (further remarks on the handicap principle)'. *Journal of Theoretical Biology* 67: 603–5.

Zahavi, A. and Zahavi, A. (1997). *The Handicap Principle: a Missing Part of Darwin's Puzzle*. Oxford: Oxford University Press.

Zaitchik, D. (1990). 'When representations conflict with reality: the preschoolers' problem with false belief and "false" photographs'. *Cognition* 35: 41–68.

Zillmann, D., Rockwell, S., Schweitzer, K. and Sundar, S. S. (1993). 'Does humour facilitate coping with physical discomfort?' *Motivation and Emotion* 17: 1–21.

Indexes

SUBJECT INDEX